In recent years, many widely dispersed climatic extremes around the globe, such as severe droughts and floods and the failure of the Indian monsoon, have been attributed to a common origin: the periodic warming of the sea surface in the central and eastern equatorial Pacific Ocean, termed the Southern Oscillation, or El Niño.

In this volume leading experts summarize information gained over the past decade which have led to marked improvements in our ability to forecast development of ENSO episodes months or seasons in advance. The book first examines the instrumental record of ENSO variability over the past 150 years, followed by topics related to large-scale signals associated with ENSO, impacts on hurricanes, hydrologic variability arising from changes in ENSO and newly emphasized aspects of climatic variability and human health in relation to the ENSO phenomenon. Historical and paleoclimatic aspects of ENSO are considered in the second section of this volume. The volume includes a theoretical examination of the causes of the ENSO cycle, the expression of the ENSO system on widely separated tree-ring and tropical ice core records, reconstruction of ENSO patterns using the available proxy climate record, and an interpretive synthesis of what the paleoclimatic data suggests about recent and long-term changes in ENSO. Possible linkages to other modes of regional and global-scale climatic variability, such as the Pacific Decadal Oscillation (PDO) and the North Atlantic Oscillation (NAO), are also discussed. This volume updates and expands the focus of Henry Diaz and Vera Markgraf's *El Niño: Historical and Paleoclimatic Aspects of the Southern Oscillation* (1992, Cambridge University Press).

This book will be of importance to a broad range of scientists working in several disciplines, including meteorology, oceanography, hydrology, geosciences, ecology, public health, emergency management response and mitigation, coastal zone management, and decision-making. It will also be used as a supplementary textbook and reference source in graduate courses in environmental studies.

Henry F. Diaz is a research climatologist at the Climate Diagnostics Center of the Environmental Research Laboratories of the National Oceanic and Atmospheric Administration (NOAA), U.S. Department of Commerce. He is a Fellow of the Cooperative Institute for Research in Environmental Sciences (CIRES), and Adjunct Associate Professor in the Department of Geography, at the University of Colorado (Boulder). He is a member of the editorial boards of the *International Journal of Climatology* and of *Arctic, Antarctic, and Alpine Research*. He has published extensively in the scientific literature on the behavior of the El Niño/Southern Oscillation (ENSO) phenomenon, and on other topics related to interannual to century-scale climatic variability on regional to global spatial scales.

Vera Markgraf is Research Professor at the Institute of Arctic and Alpine Research, University of Colorado, Boulder. From 1991 to 1998 she chaired the global change initiative Pole-Equator-Pole Paleoclimates of the Americas under the auspices of the International Geosphere-Biosphere Program-Past Global Change (IGBP-PAGES), funded by the U.S. National Science Foundation and PAGES. She is an editorial board member of the journals: *Palaeogeography, Palaeoclimatology, Palaeoecology; Review of Palaeobotany and Palynology*; and *Holocene*. She is the author of numerous articles and book chapters on late Quaternary climates of southern South America and circum-South Pacific land areas, and editor of *Inter-Hemispheric Climate Linkages* (to be published by Academic Press).

EL NIÑO AND THE SOUTHERN OSCILLATION

Multiscale Variability and Global and Regional Impacts

Edited by

HENRY F. DIAZ

Climate Diagnostics Center, Oceanic and Atmospheric Research, National Oceanic and Atmospheric Administration, Boulder, Colorado

and

VERA MARKGRAF

Institute of Arctic and Alpine Research, University of Colorado, Boulder, Colorado

CAMBRIDGE
UNIVERSITY PRESS

PUBLISHED BY THE PRESS SYNDICATE OF THE UNIVERSITY OF CAMBRIDGE
The Pitt Building, Trumpington Street, Cambridge, United Kingdom

CAMBRIDGE UNIVERSITY PRESS
The Edinburgh Building, Cambridge CB2 2RU, UK http://www.cup.cam.ac.uk
40 West 20th Street, New York, NY 10011-4211, USA http://www.cup.org
10 Stamford Road, Oakleigh, Melbourne 3166, Australia
Ruiz de Alarcón 13, 28014 Madrid, Spain

First published 2000

Printed in the United States of America

Typeface Times Roman 10/13 pt. *System* LaTeX 2_ε [TB]

A catalog record for this book is available from the British Library.

Library of Congress Cataloging in Publication Data

El Niño and the southern oscillation : multiscale variability and global and regional impacts / edited by Henry F. Diaz, Vera Markgraf.

p. cm.

ISBN 0-521-62138-0 (hb)

1. El Niño Current–Environmental aspects. 2. Southern Oscillation–Environmental aspects. I. Diaz, Henry F. II. Markgraf, Vera.

GC296.8.E4 E395 2000

551.6 – dc21 99-052939

ISBN 0 521 62138 0 hardback

Contents

Contributors

R. J. ALLAN
CSIRO-Atmospheric Research
Aspendale, Victoria
Australia

R. S. BRADLEY
Department of Geosciences
University of Massachusetts
Amherst, MA 01003-5820
U.S.A.

D. R. CAYAN
Climate Research Division, A-024
Scripps Institution of Oceanography
UCSD
La Jolla, CA 92093
U.S.A.

J. E. COLE
Geoscience Department
University of Arizona
Tucson, AZ 85721
U.S.A.

E. R. COOK
Lamont-Doherty Earth Observatory
Palisades, NY 10964
U.S.A.

R. D. D'ARRIGO
Lamont-Doherty Earth Observatory
Palisades, NY 10964
U.S.A.

M. D. DETTINGER
U.S. Geological Survey
San Diego, CA 92093
U.S.A.

H. F. DIAZ
NOAA/OAR/CDC
325 Broadway
Boulder, CO 80303
U.S.A.

D. B. ENFIELD
NOAA/AOML
Miami, FL 33149
U.S.A.

P. R. EPSTEIN
Center for Health and the
Global Environment
Harvard Medical School
Boston, MA 02115
U.S.A.

N. E. GRAHAM
Scripps Institution of Oceanography
UCSD
La Jolla, CA 92093
U.S.A.

K. A. HENDERSON
Byrd Polar Research Center
Ohio State University
Columbus, OH 43210-1002
U.S.A.

M. P. HOERLING
NOAA/OAR/CDC
325 Broadway
Boulder, CO 80303
U.S.A.

M. K. HUGHES
Laboratory of Tree-Ring Research
University of Arizona
Tuscon, AZ 85721
U.S.A.

R. KLEEMAN
Courant Institute for Mathematical Sciences
New York University, New York
NY 10012, U.S.A.

A. KUMAR
Environmental Modelling Center
NOAA/NCEP
Camp Springs, MD 20746
U.S.A.

C. W. LANDSEA
NOAA/AOML/HRD
Miami, FL 33149
U.S.A.

P.-N. LIN
Byrd Polar Research Center
Ohio State University
Columbus, OH 43210-1002
U.S.A.

M. E. MANN
Department of Geosciences
University of Massachusetts
Amherst, MA 01003-5820
U.S.A.

J. A. MARENGO
INPE
CEP 12630-000 Cachoeira Paulista
Sao Paulo
Brazil

V. MARKGRAF
INSTAAR
Campus Box 450
University of Colorado
Boulder, CO 80309
U.S.A.

W. J. M. MARTENS
International Centre for Integrative Studies
Maastricht University
P.O. Box 616
6200 MD, Maastricht
The Netherlands

G. J. McCABE
U.S. Geological Survey
Denver Federal Center
Denver, CO 80225
U.S.A.

A. MESTAS-NUÑEZ
CIMAS
University of Miami
Miami, FL 33149
U.S.A.

E. MOSLEY-THOMPSON
Byrd Polar Research Center
Ohio State University
Columbus, OH 43210-1002
U.S.A.

L. ORTLIEB
ORSTOM, IFRS
32 Avenue Henri-Varagnat
F-93143 Bondy-Cedex
France

G. POVEDA
Facultad de Minas
Universidad Nacional de Colombia
Medellín, Colombia

S. B. POWER
Bureau of Meteorology Research Centre
GPO Box 1289K
Melbourne 3000
Australia

M. L. QUIÑONEZ
PECET
Universidad de Antioquia
Calle 62 No. 52-19
Medellín, Colombia

W. ROJAS
Corporación para Investigaciones
Biologicas, CIB
Medellín, Colombia

D. W. STAHLE
Department of Geography
University of Arkansas
Fayetteville, AR 72701
U.S.A.

D.-Z. SUN
CIRES/CDC
Campus Box 216
University of Colorado
Boulder, CO 80309
U.S.A.

L. G. THOMPSON
Byrd Polar Research Center
Ohio State University
Columbus, OH 43210-1002
U.S.A.

I. D. VELÉZ
PECET
Universidad de Antioquia
Calle 62 No. 52-19
Medellín, Colombia

R. VILLALBA
Laboratorio de Dendrocronología
CRICYT-Mendoza
5500 Mendoza
Argentina

Acknowledgments

The editors wish to thank all of the contributing authors for their dedicated effort to assembling this book volume and for their patience in seeing this project to completion. We are very pleased with the results, and we hope that they are as proud of the outcome as we are.

We wish to thank the reviewers for giving generously of their time and expertise, and for their constructive suggestions throughout the evolution of this volume. Chapters were generally reviewed by two specialists in the field related to each chapter's main topic. We would like to single out, especially, the technical assistance of Craig Anderson, who helped us prepare a camera-ready manuscript, and the assistance of Diana Miller, our technical editor, for her meticulous review of each chapter manuscript.

The picture on the book cover illustrates the anomalously warm sea surface temperatures prevailing in the Pacific Ocean at the peak of the 1997–98 El Niño event. The sea surface temperature data are from the National Oceanic and Atmospheric Administration.

Preface

More than seven years have passed since the publication of our earlier monograph on historical and paleoclimatic aspects of the El Niño/Southern Oscillation (ENSO) phenomenon.[1] El Niño is now not only a household word but is used to sell everything from snowblowers to ski vacation packages! This entry of the term "El Niño" into the vernacular is partly the result of our "media culture" taking advantage of a catchy term to get myriad messages across, but it also derives from significant advances in our understanding and ability to predict aspects of this natural phenomenon months, or even seasons, in advance. At the same time, interpretations of the paleorecord from a variety of indices lead us to conclude that the ENSO phenomenon contributes the lion's share of the higher frequency variability in these records.

Events since 1992 – the year our previous book was published – have shown that improvements in our knowledge of various aspects of the long-term variability of ENSO are still needed. In the 1990s, events such as the persistent El Niño pattern from 1991 to 1994, and the recent development of another major El Niño episode in 1997–98 (only fifteen years after the great El Niño of 1982–83 first catapulted this phenomenon into the consciousness of the general public), underscore the need to understand long-term aspects of ENSO variability. A major theme of this volume is concerned with the relationship between global-scale climatic patterns, operating on different timescales, and the space–time behavior of the higher frequency ENSO system itself. The question of the long-term stability of the characteristics of the modern ENSO system is as relevant today as it was nearly a decade ago, when leading scientists in atmospheric and ocean sciences, together with a broad spectrum of paleoclimatologists,

[1] Diaz, H. F. and Markgraf, V. (eds.), 1992: *El Niño, Historical and Paleoclimatic Aspects of the Southern Oscillation.* Cambridge University Press, Cambridge, 476 pp.

first met in Boulder, Colorado, to examine the longer term aspects of the ENSO phenomenon.

This new edition of the book is aimed largely toward the same audience as the first volume, but in recognition of the increasing awareness of ENSO variability and its effects by a much greater number of people than was true seven years ago, this volume tries to examine important corollary effects of ENSO, such as changes in tropical cyclone frequency, worldwide hydrological signals, and impacts on public health.

The motivation for an updated volume on the topic of long-term variations in ENSO arose as new observational and modeling studies indicated that decadal, and possibly longer term, modes of variability in the Pacific Ocean likely interacted with the ENSO cycle to modulate the frequency and intensity of El Niño/La Niña events over time. New paleoclimate data sets have become available around the world that provide additional spatial and temporal constraints on our models of very long-term ENSO variability. This new volume, focusing on the longer frequency aspects of ENSO in the instrumental record, together with a judicious mix of paleoclimatic analyses to shed light on the workings of the ENSO phenomenon at different times in the past, should be of use to a broad spectrum of readers.

This book follows a format similar to that of our previous volume, being divided into a section dealing with analysis of different aspects of the modern ENSO record, followed by a section based on climatic reconstructions using a variety of climate proxy records, such as tree rings, ice cores, lake sediments, historical accounts, etc. One chapter examines the question of the initiation of the ENSO seesaw as a function of the basic oceanic state in a simplified, but consistent theoretical framework. A review of the paleoclimate literature by Markgraf and Diaz at the end of the book examines the historical and paleoclimatic record, with the aim of pulling together the various lines of evidence regarding the development of the ENSO phenomenon in late glacial times and during different intervals of the Holocene. The controversy regarding the early Holocene status of the ENSO system continues, although in our opinion, evidence for ENSO-induced climatic variability is stronger after ca. 6,000 years before present than it is prior to that time.

In our introduction to the first volume, we stated that "It is hoped that [the earlier book] would serve to stimulate further work both in the area of constructing additional reliable proxy chronologies from [various] ENSO-sensitive regions, as well as to stimulate the development of new, innovative ways to analyze existing proxy and instrumental records." We believe that our hope has been fulfilled by the great variety of studies, new observations, modeling, and other analytical studies that have been published since then on different aspects of ENSO variability and its climatic manifestations on a broad range of spatial and temporal scales. A synthesis of some of those recent studies can be found in several of the chapters in the current book – for instance, those by Rob Allan (on climatic variability and ENSO in the past 150 years), Enfield and Mestas-Nuñez (who examine global ENSO and non-ENSO modes of variability in sea surface temperature), Kleeman and Power (who consider the coupled response of the upper ocean and the atmosphere on different timescales), Dettinger et al. (who consider the nature of hydrologic responses to ENSO worldwide), and Mann et al. (who focus on multicentury reconstructions of ENSO based on the spatial pattern of a network of ENSO-sensitive proxy indices). A chapter by L. Ortlieb presents refinements to the El Niño chronology (and, occasionally, alternative interpretations for some El Niño years) developed by William (Bill) Quinn and colleagues from a variety of sources that had been published in a number of forums, including our earlier volume. On the paleoclimatology side, Cook et al. review advances in reconstructing ENSO and associated teleconnections using tree-ring records, and Thompson et al. provide an analogous perspective based on widely situated ice cores from tropical ice caps and glaciers.

New contributions consider the impacts of ENSO on problems relevant to society. One chapter (Poveda et al.) looks at the possible modulation of vector-borne diseases in the tropics (in particular, malaria incidence in Colombia). Another chapter by C. Landsea examines the impact of the ENSO phenomenon on the frequency and intensity of tropical cyclones worldwide.

Because the ENSO phenomenon affects global climatic patterns to such a large extent, improved understanding of the low-frequency aspects of the ENSO phenomenon is of great importance to societies in the affected regions. Furthermore, as the world economy has become highly interdependent over the past couple of decades, negative impacts in one region of the world can be felt in other regions where the direct effects of ENSO are not as pronounced. Changes in the frequency and intensity of ENSO events in the future, due to natural or anthropogenic forcing, may have severe impacts on societies throughout the world.

As was the case with our earlier effort, we hope that this book makes a contribution toward a broader understanding of the ENSO phenomenon, by providing an updated synthesis of some of the significant accomplishments made by many researchers in many parts of the world toward this goal, and by highlighting some of the areas where gaps in our knowledge still exist.

H. Diaz
V. Markgraf

1

ENSO and Climatic Variability in the Past 150 Years

ROBERT J. ALLAN

The Commonwealth Scientific and Industrial Research Organisation (CSIRO), Atmospheric Research, Aspendale, Victoria, Australia

Abstract

Efforts to improve our understanding of the various types of natural variability inherent in the global climate system have included a growing focus on interactions between the El Niño/Southern Oscillation (ENSO) phenomenon and lower frequency decadal- to secular-scale fluctuations in climate. New global historical instrumental data compilations are being analyzed by increasingly more sophisticated objective analysis techniques that are beginning to resolve important physical links and modulations involving dominant climatic signals. This chapter details the current historical observational evidence for interactions between ENSO and decadal- to secular-scale fluctuations in the climate system.

Spectral analyses of global historical sea surface temperature (SST) and mean sea level pressure (MSLP) anomalies reveal significant climatic signals at about 2–2.5, 2.5–7, 11–13, 15–20, 20–30, and 60–80 years and a long-term secular trend. The spatial and temporal characteristics of the SST and MSLP signals in these bands are resolved and examined by using joint empirical orthogonal function (EOF) and singular value decomposition (SVD) analyses. The ENSO signal is seen to consist of quasi-biennial (QB) (2- to 2.5-year) and lower frequency (LF) (2.5- to 7-year) components that interact to produce important modulations of the phenomenon. However, longer duration ENSO characteristics and climatic fluctuations are the result of decadal- to secular-scale influences. Protracted El Niño and La Niña phases are found to be a consequence of the "phasing" of quasi-decadal (11- to 13-year) and interdecadal (15- to 20-year) ENSO-like signals with higher frequency QB and LF ENSO components. Lower frequency ENSO-like patterns in global SST and MSLP anomalies extend through to multidecadal timescales and can be seen in the 60- to 80-year Sahelian rainfall/interhemispheric temperature contrast signal. The secular trend, reflecting the observed global warming signal, reveals SST and MSLP anomalies with neutral to slightly La Niña–like conditions in the Pacific sector.

Something of the impact of decadal- to secular-scale signals on the environment, relative to that caused by QB and LF ENSO components, can be seen in the patterns of correlations with global precipitation. Significant contributions to rainfall variability arising from these low-frequency components are evident not just in known ENSO-sensitive regions. However, it is the "phasing" of the various signals that dictates the bulk of the overall precipitation response. El Niño and La Niña episodes of the ENSO phenomenon can be both synchronous and asynchronous with El Niño–like and La Niña–like signals on various decadal to multidecadal timescales, resulting in permutations that lead to either enhanced or suppressed regional rainfall regimes.

Perhaps the major challenge in developing the above concepts further is the resolution of a more unified understanding of climatic variability and change involving features such as ENSO with the North Atlantic Oscillation (NAO), the Arctic Oscillation (AO), the North Pacific Oscillation (NPO), and the Antarctic Circumpolar Wave (ACW).

Introduction

The El Niño/Southern Oscillation (ENSO) phenomenon is a natural part of the global climate system and results from large-scale interactions between the oceans and the atmosphere that occur chiefly across its core region in the tropical–subtropical Pacific to Indian Ocean basins. As a consequence, the most direct climatic shifts and environmental impacts are found over, and in countries bordering, the Indo-Pacific sector of the planet (Ropelewski and Halpert 1987, 1989; Kiladis and Diaz 1989; Halpert and Ropelewski 1992; Allan et al. 1996; Glantz 1996). The importance of this phenomenon in the global climate system can be quantified to the extent that it is the next feature that explains a large amount of climatic variability after the seasonal cycle and the monsoon system.

El Niño/Southern Oscillation is an irregular/aperiodic phenomenon that tends to reoccur in the range of 2–7 years and is manifest by alternations between its two phases or extremes, often called El Niño and La Niña events. Once it begins, the "average" event tends to last for 18–24 months and shows characteristics of being locked to the seasonal cycle (Philander 1985, 1989, 1990; Yasunari 1985, 1987a,b; Allan et al. 1996; Glantz 1996). However, seasonal persistence is weakest in the boreal spring (austral autumn), the time of the so-called predictability barrier or spring frailty, when new events are likely to develop and existing conditions collapse (Torrence and Compo 1998; Torrence and Webster 1998, 1999).

Although composites of climatic patterns and impacts during El Niño and La Niña events tend to be of the opposite sign to one another, individual El Niño or La Niña events are never exactly the same and can vary in magnitude, spatial extent, onset, duration, cessation, etc. (Philander 1985, 1989, 1990; Ropelewski and Halpert 1987, 1989;

Kiladis and Diaz 1989; Allan et al. 1996). In reality, the phenomenon encompasses events that cover a wide range, or "family," of types and signatures (Allan et al. 1999).

During El Niño events, warming of tropical regions of the Pacific and Indian Oceans leads to the displacement of major rainfall-producing systems from the continents to over the previously mentioned oceanic areas, causing massive redistributions of climatic regimes. The opposite tendency with regard to continental and oceanic rainfall regimes occurs during La Niña episodes. As tropical regions are linked to mid- to high latitudes in both hemispheres via teleconnection patterns, any major variations in mass, energy, and momentum resulting from a redistribution of equatorial rainfall regimes are communicated into more temperate regions of the globe (Webster 1994). This effect has the potential to extend ENSO influence beyond the tropics and cause near-global modulations of climate (Glantz et al. 1991; Allan et al. 1996).

Within the tropical–subtropical regime, the ENSO phenomenon has shown a degree of interaction with the Indo-Asian monsoon system. This relationship is described as being "selectively interactive," in that during the boreal autumn to winter (austral spring to summer) ENSO is strong and the Indo-Asian monsoon is weak, with the opposite situation in the boreal spring to summer (austral autumn to winter) (Webster and Yang 1992; Webster 1995). Thus the nature of the Indo-Asian monsoon can influence ENSO and vice versa at various times of the year. In fact, it has been suggested that the "predictability barrier" may be a consequence of the nature of ENSO links with the Indo-Asian monsoon system (Webster and Yang 1992; Webster 1995; Torrence and Compo 1998; Torrence and Webster 1998, 1999).

Other features of the global climate system may also be coupled to ENSO at various space and time frames. Much remains to be unraveled concerning possible relationships between the phenomenon and climatic features such as the North Atlantic Oscillation (NAO) (Hurrell 1995, 1996; Hurrell and van Loon 1997; Huang et al. 1998), the Arctic Oscillation (AO) (Thompson and Wallace 1998), the North Pacific Oscillation (NPO) (Glantz et al. 1991), and the Antarctic Circumpolar Wave (ACW) (White and Peterson 1996; White et al. 1998). A better understanding of ENSO links to the above features may shed light on questions about the extension of the phenomenon into Europe (Fraedrich and Muller 1992; Fraedrich et al. 1992; Wilby 1993; Fraedrich 1994), the nature of the Pacific North American (PNA) teleconnection (Glantz et al. 1991), and the apparent propagational structure of ocean–atmosphere variables as ENSO phases evolve (Yasunari 1985, 1987a,b; Tourre and White 1995; Allan et al. 1996).

Further complicating this current understanding of ENSO, however, is growing evidence that the phenomenon is not spatially or temporally stable in the longer term and responds on a number of timescales (Gu and Philander 1995; Wang and Ropelewski 1995; Allan et al. 1996; Brassington 1997; Zhang et al. 1997; Torrence and Compo 1998; Torrence and Webster 1998, 1999; Xu et al. 1998). This response appears to result from the influence of a range of patterns of natural decadal- to secular-scale variability that modulate ENSO and, as a consequence, global climate. The current state of knowledge about the spatial and temporal nature and structure of ENSO and its modulation by natural decadal-scale features of the climate system, including possible anthropogenic influences, is examined in this chapter.

Background

Contemporary research has suggested that the ENSO phenomenon has undergone a change in mode and nature since the mid-1970s, with a predominance of El Niño over La Niña phases (Graham 1994; Kerr 1994; Latif and Barnett 1994a,b; Miller et al. 1994a,b; Trenberth and Hurrell 1994; Wang 1995), changes in ENSO and rainfall relationships (Kripalani and Kulkarni 1997; Nicholls et al. 1996, 1997; Suppiah 1996; Suppiah and Hennessy 1996), and a protracted El Niño or sequence of El Niños in the first half of the 1990s (Bigg 1995; McPhaden 1993; Jiang et al. 1995; Latif et al. 1995, 1997; Liu et al. 1995; Kleeman et al. 1996; Latif et al. 1996; Goddard and Graham 1997; Gu and Philander 1997; Webster and Palmer 1997). Many of these aspects of the phenomenon can be seen in a basic examination of the raw Southern Oscillation Index (SOI) since 1876 (Fig. 1.1). Debate continues as to whether these recent fluctuations in ENSO are unique in the instrumental, historical, and proxy records of the phenomenon and as to whether they may be a sign of the influence of the enhanced greenhouse effect (Gage et al. 1996; Trenberth and Hoar 1996, 1997; Harrison and Larkin 1997; Rajagopalan et al. 1997; Allan and D'Arrigo 1999; Dai et al. 1998).

Evidence for a distinct "climatic shift" across the Pacific Ocean sector of the globe since the mid-1970s has been reported by a number of authors (Ebbesmeyer et al. 1991; Graham 1994; Latif and Barnett 1994a,b; Miller et al. 1994a,b; Nitta and Kachi 1994; Trenberth and Hurrell 1994; Jiang et al. 1995). The theory that a climatic shift may be taking place has gained support from the findings of similar changes in climatic

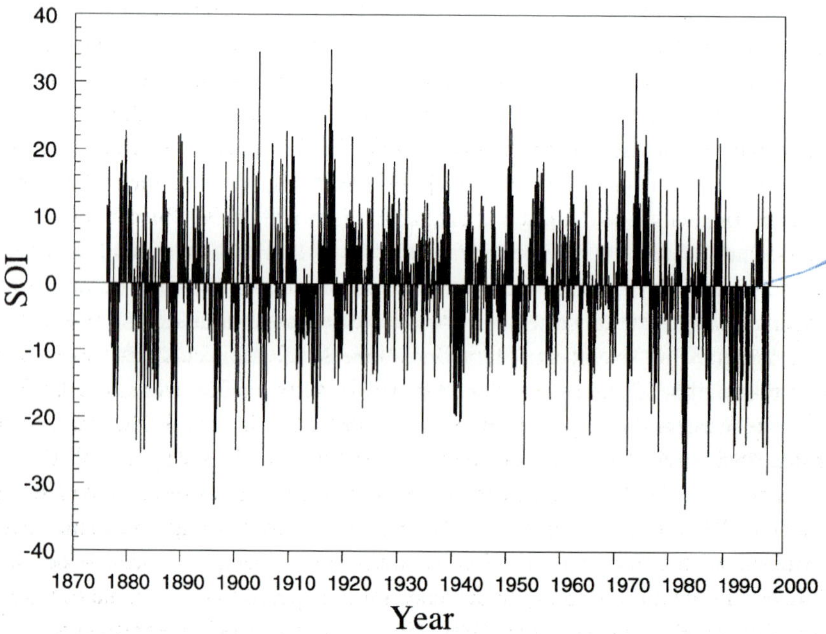

Fig. 1.1 Tahiti minus Darwin normalized monthly SOI from January 1876 to September 1998.

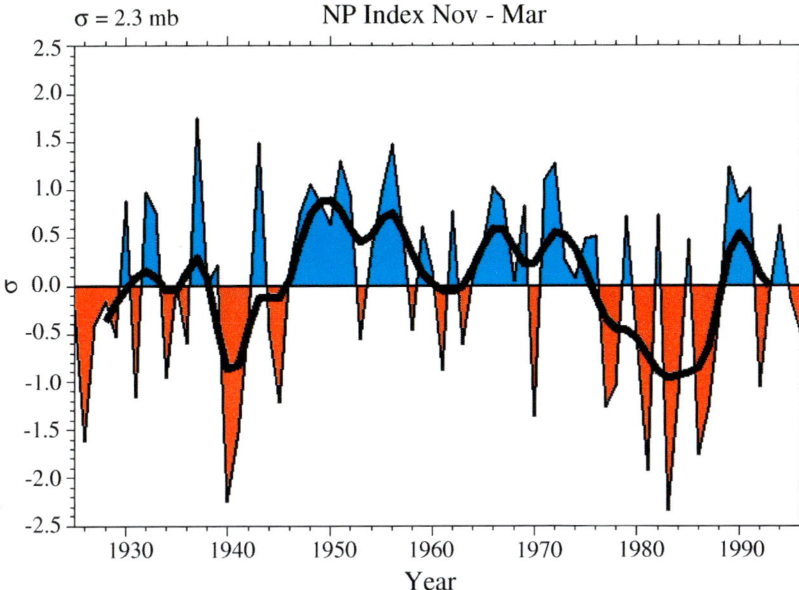

Fig. 1.2 The North Pacific (NP) index is the standardized area-weighted MSLP over the region 30°N–65°N, 160°E–140°W averaged for the months from November to March from 1899 to 1998. (Adapted from Trenberth and Hurrell 1994.)

patterns across the middle to higher latitudes of the Southern Hemisphere (van Loon et al. 1993; Hurrell and van Loon 1994; Karoly et al. 1996; Garreaud and Battisti 1999; Jones and Allan 1998), over the middle to higher latitudes of the North Pacific (Fig. 1.2) (Chen et al. 1992; Jacobs et al. 1994; Tanimoto et al. 1993; Dettinger and Cayan 1995; Lagerloef 1995; Mantua et al. 1997; Minobe 1997; Nakamura et al. 1997), in the North Atlantic and particularly with regard to the NAO (Fig. 1.3) (Deser and Blackmon 1993; Hurrell 1995, 1996; Hurrell and van Loon 1997), with respect to Northern Hemisphere circulation in general (Nitta and Yamada 1989; Trenberth 1990; Burnett 1993; Nitta 1993; Graham et al. 1994; Lejenas 1995; Moron et al. 1998), and globally (Parker et al. 1994; White and Cayan 1998). In addition, interest in the post-1970s "climatic shift" has been a catalyst for specific studies that have attempted to assess more about the nature, structure, and evolution of ENSO itself on decadal to secular timescales (Diaz and Pulwarty 1994; Gu and Philander 1995, 1997; Wang and Ropelewski 1995; Allan et al. 1996; Wang and Wang 1996; Brassington 1997; Kestin et al. 1998; Torrence and Webster 1998, 1999; Torrence and Compo 1998; Xu et al. 1998; Zhang et al. 1997).

It is also in the Pacific Ocean basin where the prime focus has been on the 1990–95 protracted El Niño or sequence of El Niño events, with ongoing discussions about the cause(s) of this long episode. Explanations have ranged from natural decadal–multidecadal climatic variability to evidence for the operation of the enhanced greenhouse effect on ENSO (Latif et al. 1995, 1997; Liu et al. 1995; Kleeman et al. 1996; Trenberth and Hoar 1996, 1997; Gu and Philander 1997; Harrison and Larkin 1997; Rajagopalan et al. 1997). More recently, Allan and D'Arrigo (1999) have shown that

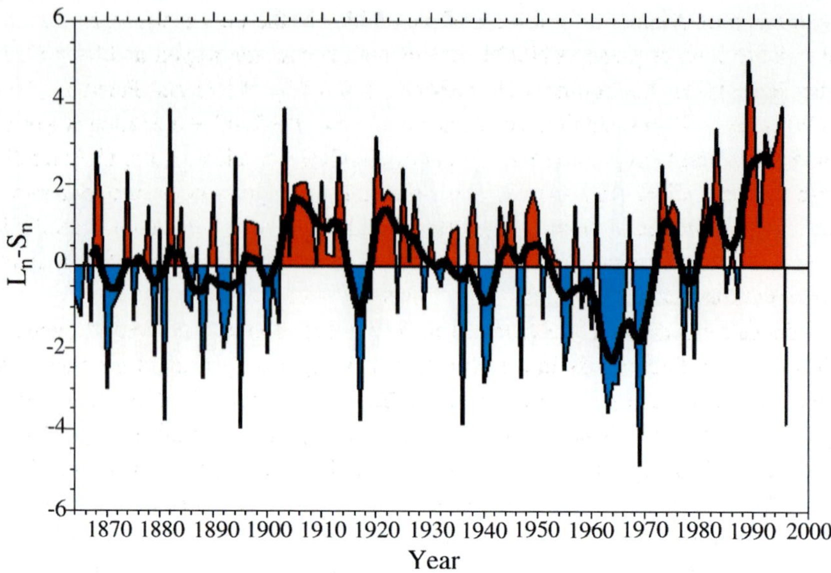

Fig. 1.3 The NAO index for December through March based on the difference in normalized MSLP between Lisbon, Portugal, and Stykkisholmur, Iceland, from 1864 to 1996. The MSLP anomalies at each station were normalized by division of each seasonal pressure by the long-term (1864–1983) standard deviation. (Adapted from Hurrell 1995.)

some four protracted El Niño and six protracted La Niña sequences have occurred over the past 100–120 years, and many others occurred in pre-instrumental times and in times for which proxy data are available, back to the 1700s. Of those in the proxy record, several surpassed or rivaled the 1990–95 El Niño in duration. Although an anthropogenic cause has been suggested for this event in the UCAR (University Corporation for Atmospheric Research) Quarterly (1997), and by Trenberth (1998), and Dai et al. (1998), there is a growing realization of the important role played by natural low-frequency fluctuations in the global climate system in modulating a wide range of ENSO characteristics.

A growing number of studies have now begun to investigate decadal to multidecadal variability using new long-term historical to contemporary data sets, more sophisticated analysis techniques, and the power of numerical model simulations to attempt to reveal not just the nature and structure of such fluctuations but also the physical mechanisms underlying them. Strong evidence for the influence of a decadal mode of the ENSO-like pattern observed during the 1990–95 period of the protracted El Niño or sequence of El Niño events, with a major sea surface temperature (SST) focus in the western-central Pacific and important influences on the global climate, has been detailed by Latif and Barnett (1994a,b), Latif et al. (1995, 1997), Kleeman et al. (1996), and Allan et al. (1999). The dominance of decadal to interdecadal climatic variability and ENSO-like patterns over the tropical and southern Atlantic Ocean has been addressed in papers by Mann and Park (1994), Venegas et al. (1996), and Chang et al. (1997). Allan et al. (1995) and Reason et al. (1996a,b, 1998a,b) have shown the presence over the Indian

Ocean basin of decadal to multidecadal variability in the ocean–atmosphere system and links to low-frequency ENSO behavior both across the region and beyond. For higher latitudes in the Southern Hemisphere, studies by White and Peterson (1996) and White et al. (1998) detail the existence of the so-called ACW operating at a period of 4–5 years and taking some 8–10 years to circulate completely around Antarctica. These studies indicate that the ACW is potentially a major source of interannual to decadal variability, but because only little more than 10 years' data on this oscillation are available, these researchers were thus unable to judge whether the ACW may also induce multidecadal fluctuations in the climate system.

Evidence for changes in the structure and general characteristics of ENSO, and possible links with fluctuations in the climate system operating at low frequencies, can be seen in the wavelet plots of the SOI and SST in the Niño−1+2, 3, and 4 regions together with a comparison of fractional variance among these indices (Figs. 1.4 and 1.5). The widely accepted explanation is that the ENSO phenomenon in the 2- to 8-year band has fluctuated between more robust periods in the 1870s–80s, 1910s, 1950s, and 1970s–80s and less energetic or quiescent epochs in the late 1890s–1900s, 1920s–30s, and 1960s. The phenomenon also appears to display variations in its periodicity, with a dominant period of 3–4 years from 1870 to 1910, a 5- to 7-year periodicity from 1910 to 1920 and again from 1930 to 1960, and a period of 4–5 years from 1970 to 1990. Such fluctuations in ENSO may be reflected in recently reported variations in its impacts. In the Australian, Indian, and Southeast Asian regions, studies have suggested that correlation patterns and statistics indicative of rainfall and SOI relationships have changed significantly over the past several decades (Parthasarathy et al. 1991; Vijayakumar and Kulkarni 1995; Allan et al. 1996; Nicholls et al. 1996, 1997; Suppiah 1996; Suppiah and Hennessy 1996; Kripalani and Kulkarni 1997). Broad indications of changing relationships with Indian summer monsoon rainfall in the first half of the instrumental record linked to major fluctuations in atmospheric circulation are suggested by the schematics in Parthasarathy et al. (1991). This evidence is supported by the patterns in wavelet analyses of all-India summer monsoon rainfall, taken together with similar analyses of the SOI and SST in the Niño−1+2, 3, and 4 regions, and by a comparison of fractional variance in the 2- to 8-year band. These data indicate a pronounced reduction in signal strength in all of these variables during the period 1920–40s.

An examination of Figures 1.4 and 1.5 shows that there is more to ENSO than the "classical" phenomenon that tends to occur every 2 to 7 years. Using a wide range of analysis techniques and data sets indicative of the phenomenon, recent work has confirmed the existence of various ENSO-like patterns in the climate system on decadal to multidecadal timescales (Diaz and Pulwarty 1994; Allan et al. 1996, 1999; Mann and Park 1996, 1999; Wang and Wang 1996; Brassington 1997; Kestin et al. 1998; Torrence and Webster 1998, 1999; Folland et al. 1998; Torrence and Compo 1998; Jones and Allan 1998; Xu et al. 1998). Such interest reflects the convergence of ENSO research with a long history of concern about broader historical climatic fluctuations and changes (Allan et al. 1996). This convergence has gained further momentum because regions of the globe that are impacted by "classical" ENSO events are also often those showing the most prominent longer term decadal to multidecadal modulations of their climate.

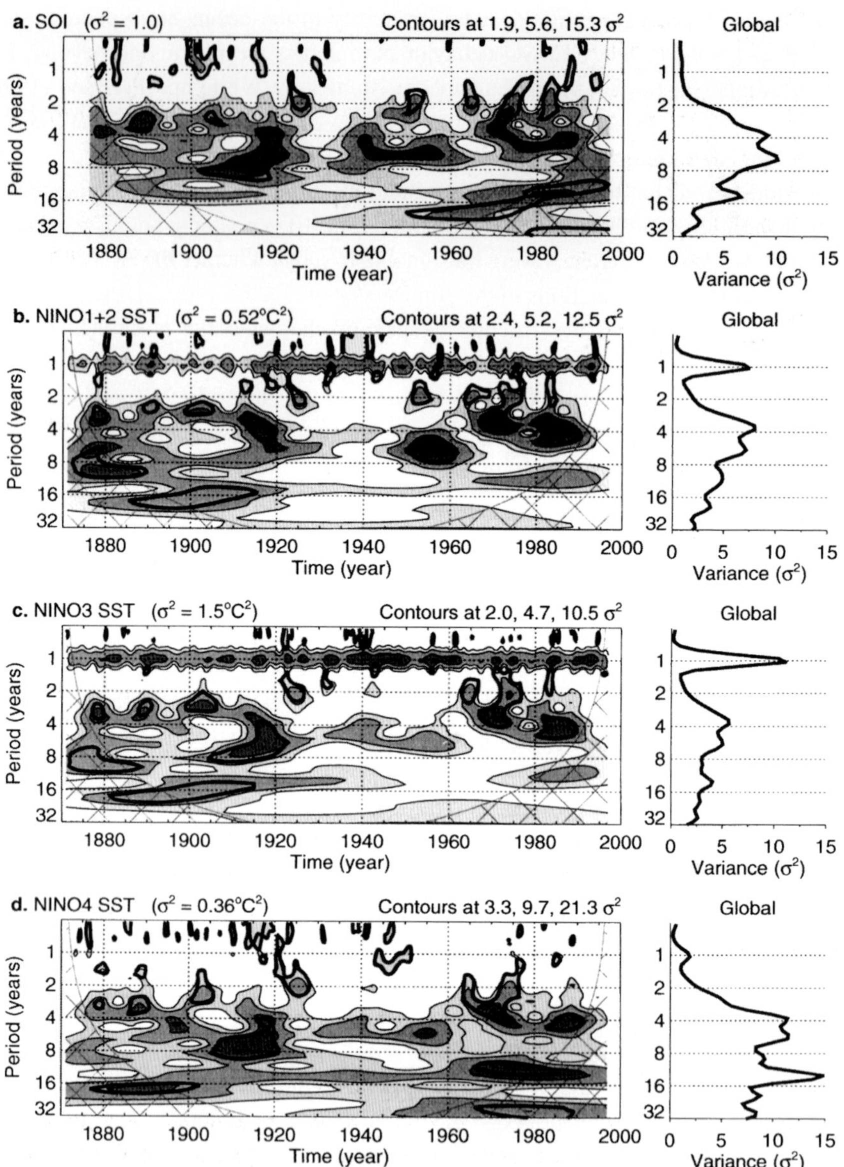

Fig. 1.4 (a) Wavelet power spectrum of SOI, 1876–1996, using the Morlet wavelet. The contour levels are in units of variance and are chosen so that 50%, 25%, and 5% of the wavelet power is above each level, respectively. The thick contour is the 10% significance level using the global wavelet spectrum as the background (significance determined from a Monte Carlo simulation of 100,000 white noise time series). Cross-hatched regions indicate the "cone-of-influence," where the padding with zeros has reduced the variance. The global wavelet spectrum is given in the plot on the right. (b) Same as (a) but for the Niño-1+2 SST. (c) Same as (a) but for the Niño-3 SST. (d) Same as (a) but for the Niño-4 SST. (From Torrence, pers. comm.)

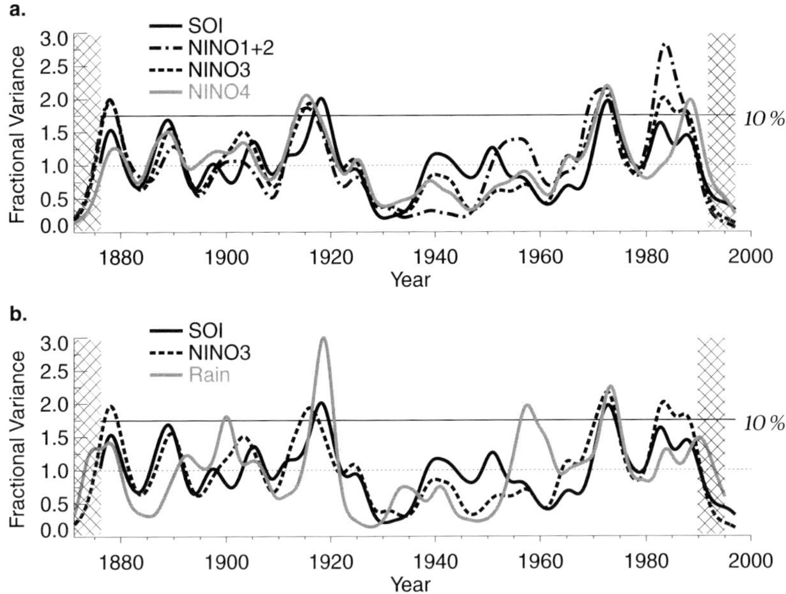

Fig. 1.5 Scale-average wavelet power in the 2- to 8-year band for the SOI, Niño-1+2 SST, Niño-3 SST, and Niño-4 SST. Each curve was normalized by dividing by its mean. The hatching indicates where zero padding has reduced the variance. The thin line is the 10% significance level for all three curves, using the global wavelet spectrum for each time series as the background. (From Torrence, pers. comm.)

Given the above, it is important to examine prominent climatic fluctuations observed and documented in the historical instrumental climate record and to assess their possible relationship with, and influence on, the "classical" nature and structure of the ENSO phenomenon.

Detailed Objective Analyses of Global Data

A wide range of papers by Mann and Park (1993, 1994, 1996, 1999), Mann et al. (1995), Folland et al. (1998), and Allan et al. (1999) have used high-quality near-global data sets of climatic variables such as precipitation, surface temperature (SST and land plus oceanic), and atmospheric pressure, together with more sophisticated objective analysis techniques such as singular spectrum analysis (SSA), empirical orthogonal function (EOF) analysis, and singular value decomposition (SVD), to investigate the nature and structure of climatic variability on decadal to multidecadal timescales. Mann and Park (1999) provide a comprehensive review of the various oscillatory spatiotemporal signal detection approaches that are being used to examine longer term fluctuations in various data sets that document aspects of the global climate system.

Recent papers examining oscillatory spatiotemporal signals in global climatic data by Mann and Park (1994, 1996, 1999) and Mann et al. (1995) have used multitaper method SVD (MTM-SVD) techniques to resolve distinct decadal to secular signals at

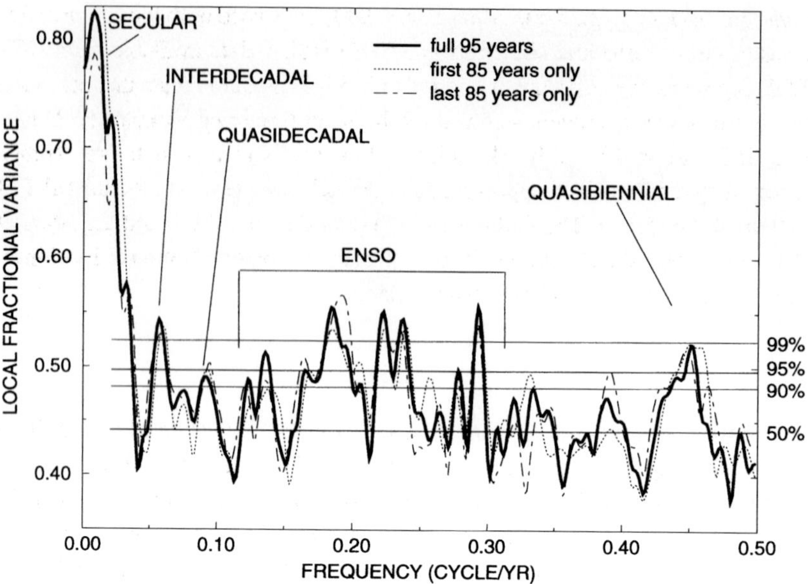

Fig. 1.6 Multitaper frequency-domain singular value decomposition (MTM-SVD) localized variance spectrum of joint historical Northern Hemisphere (17.5°N–72.5°N) air temperature and MSLP from 1899 to 1993 (relative variance is explained by the first eigenvalue of the SVD as a function of frequency). The 50, 90, 95, and 99% statistical confidence limits are shown as horizontal lines, and various significant climatic features in the spectrum are pointed out on the diagram. The full thick line is for the full 95 years, the light small dashed line is for the first 85 years of data only, and the light dash and dot line is for the past 85 years of data only. (From Mann and Park 1996.)

various frequencies in the climate system. The local fractional variance SVD spectra of historical joint Northern Hemisphere surface air temperature and sea level pressure (SLP) data in Figure 1.6 is one example of the application of this technique. The potential for a more complete understanding of various features in the climate system can be seen in the specific resolution of significant signals at quasi-biennial (QB), "classical" ENSO (with various peaks in the low-frequency [LF] band), quasi-decadal (~10–12 years), interdecadal (~15–18 years), and multidecadal (~70–90 years) frequencies, and on the secular timescale (the global temperature increase) (Mann and Park 1999). Such analyses provide a powerful advanced signal detection procedure.

Another approach to the resolution of decadal to multidecadal climate signals has been to use EOFs of regional to global historical data (see Mann and Park 1999). Much of this work has tended to focus on post-1950s records of climatic variables, where the best quality and quantity of observations are generally available (e.g., Kawamura 1994; Zhang et al. 1997). However, as Zhang et al. (1997) also show, there is value in examining longer data sets back to the beginning of this century – at least when one looks at the new versions of historical reconstructions now becoming available (Parker et al. 1994; Allan et al. 1996; Basnett and Parker 1997). A recent analysis of SST data by Folland et al. (1998) has employed EOFs to low-pass-filtered global historical observations back to 1860, while Allan et al. (1999) have used both joint EOF and SVD analyses of global historical SST and mean SLP (MSLP) data available from 1871. In the EOF approach, the technique is applied to band-pass-filtered data that contain each of the significant signal bands resolved by the local fractional variance SVD spectra of joint SST and MSLP fields. The joint SVD analysis gives results very similar to those found by the joint EOF approach, and the reader is directed to Allan et al. (1999) for details of the former. The findings of the Folland et al. (1998) and the Allan et al. (1999) studies form the core of the following examination of ENSO relationships with decadal to secular features of the climate system.

ENSO

Spectral analyses of atmospheric and oceanic variables indicative of the ENSO phenomenon show significant signals at about 2 years (QB) and in a band from 3–7 years (LF component) (Lau and Sheu 1988; Rasmusson et al. 1990; Ropelewski et al. 1992; Allan et al. 1996). The nature and structure of signals in the above bands can be seen in the 18- to 35-month QB and 32- to 88-month LF band-pass-filtered SOI, Niño-3, and Niño-4 series in Figure 1.7. The broad coherency of each of these components of ENSO across the three series is clearly evident. However, the superposition of QB and LF bands leads to instances of both strong positive and negative phasing (QB and LF being in phase and out of phase) of the signals. This very nature alone provides a degree of modulation of the overall ENSO signal. Not surprisingly, the LF band carries something of a propensity for longer sequences of both El Niño and La Niña extremes. In fact, there is some evidence for the range of protracted historical instrumental ENSO phases detailed in Allan and D'Arrigo (1999) in Figure 1.7.

Enveloping the interplay between QB and LF signals in Figure 1.7 is a longer term multidecadal fluctuation in variance, which is highlighted for the Niño-3 region SSTs on the following World Wide Web (WWW) page: *http://paos.colorado.edu/research/wavelets/wavelet1.html*. This feature displays some of the characteristics of broader ENSO nature discussed previously in this chapter, in that earlier and later portions of the time series show robust signals while the middle section experiences a period of quiescence. Such patterning provides a degree of lower frequency modulation of the 2- to 7-year band and appears to reflect the two teleconnection states discussed in Ward et al. (1994), Navarra et al. (1999), and Miyakoda et al. (1999). These fluctuations in variance need further physical explanation and understanding.

Expanded spatiotemporal resolution of the QB and LF components of the ENSO signal can be achieved through a joint EOF analysis of global historical SST and MSLP data band-pass filtered in the QB and LF year bands resolved by the MTM-SVD spectra in Figure 1.8. An analysis of these bands in Allan et al. (1999) reveals annual MSLP and SST anomaly patterns and joint time series indicative of QB and LF (or what might be termed "classical" ENSO) signals respectively (Figs. 1.9 and 1.10). Across the Pacific basin, both bands provide equally strong ENSO signals, while in the Indian and Atlantic Oceans more definitive MSLP and SST anomaly patterns are found in the LF year band. The superposition of the two joint annual EOF time

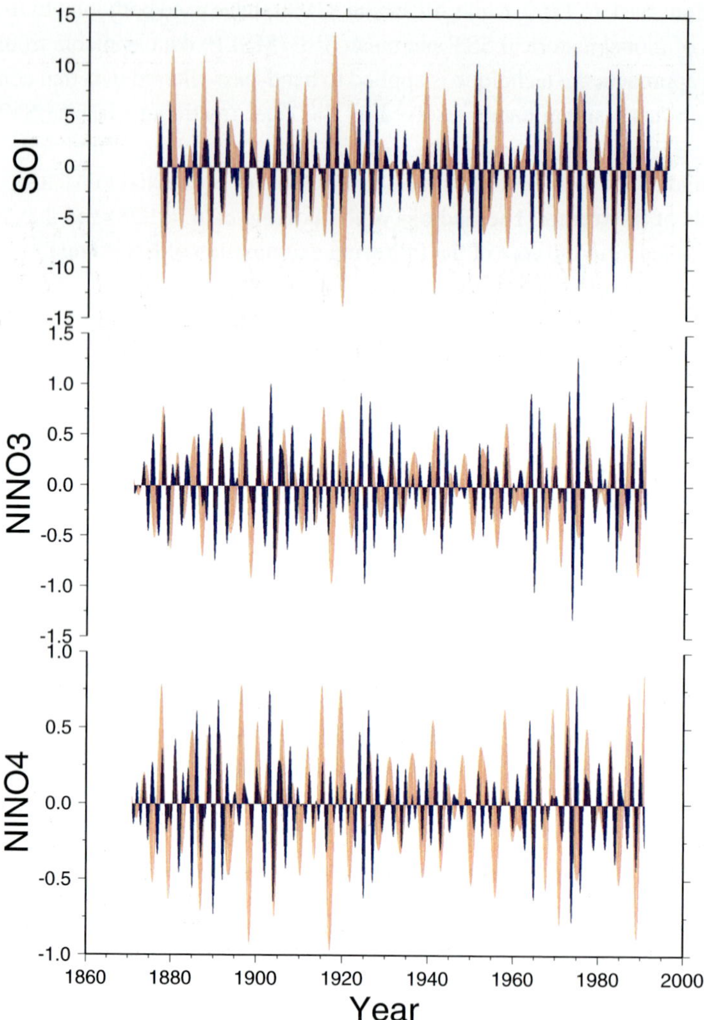

Fig. 1.7 18- to 35- (QB) and 32- to 88- (LF) month band-pass-filtered series of (a) the SOI and (b) GISST-based Niño-3 and Niño-4 region SSTs in the central to eastern equatorial Pacific Ocean (CEP-EEP) since 1871. All anomalies and normalizations are with respect to the 1961–90 period. 18- to 35- (32- to 88-) month filtered series are shown by blue (red) traces. (From Allan et al. 1996.)

series produces signal modulations that lead to ENSO phases (El Niños and La Niñas) displaying a wide range of types encompassing variations in frequency, magnitude, and duration. When the above is extended to include modulations of the spatial EOFs of MSLP and SST, this finding has considerable ramifications for climatic impacts. Although not shown, the joint seasonal EOFs reveal that these interactions between dominant signal bands also lead to modulations of the evolutionary characteristics of ENSO phases.

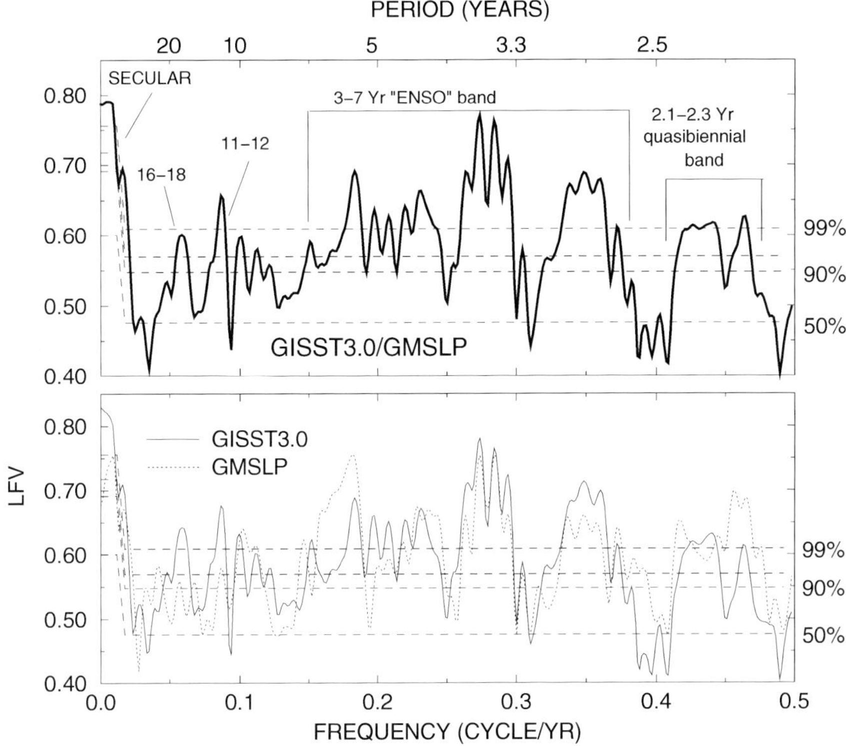

Fig. 1.8 Multitaper frequency-domain singular value decomposition (MTM-SVD) localized variance spectrum of (a) joint analysis of historical GISST3.0 and GMSLP2.1f and (b) separate analyses of historical GISST3.0 and GMSLP2.1f from 1871 to 1994 (relative variance is explained by the first eigenvalue of the SVD as a function of frequency). The 50, 90, and 99% statistical confidence limits are shown as horizontal lines, and various significant climatic features in the spectrum are pointed out on the diagram. In (b), the GISST3.0 spectra is shown by the solid line and the GMSLP2.1f spectra are shown by the dashed line. (From Allan et al. 1999.)

The influence of interactions between the QB and LF year bands on the seasonal nature of ENSO can be seen in Table 1.1. This simultaneous correlation table provides a broad picture of the representativeness of each of the joint QB and LF ENSO seasonal time series with regard to the QB and LF components of the SOI and the Niño-3 and Niño-4 SST indices (Allan et al. 1996). Although the majority of seasons show correlations that are statistically significant at the 99% level, an examination of the amount of variance that they explain indicates that the most robust ENSO signal throughout the year is to be found on the LF year band. However, the well-documented boreal spring (austral autumn) "breakdown" or "predictability barrier" in ENSO relationships (Webster and Yang 1992; Webster 1995; Allan et al. 1996; Torrence and Webster 1998) is most strongly manifest on the QB band. This result suggests that it is important to look more closely at fluctuations around 2–2.5 years in efforts to understand and improve ENSO forecasts.

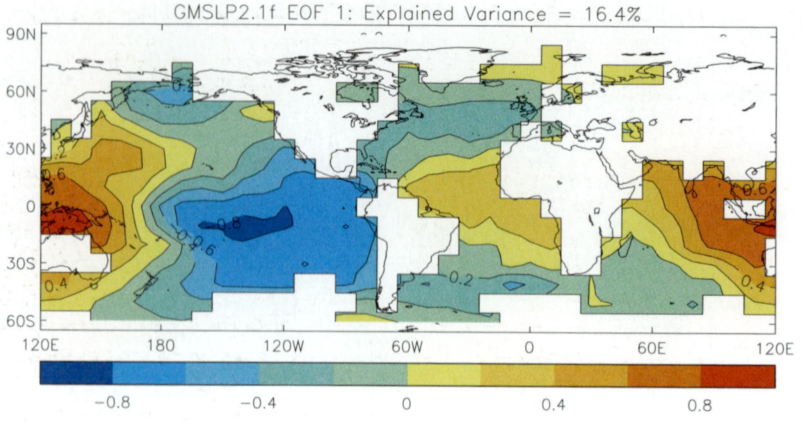

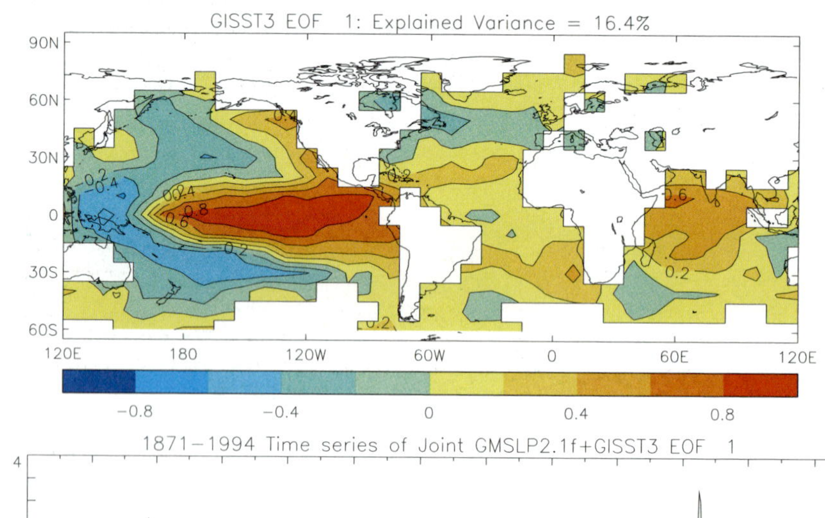

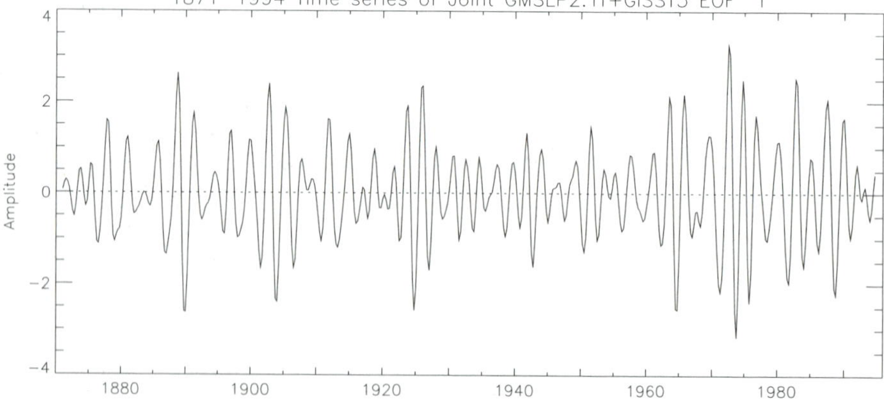

Fig. 1.9 Joint GMSLP2.1f and GISST3.0 EOF 1 spatial correlation fields for MSLP (top panel) and SST (middle panel) (relative to the base period 1901–90), and joint EOF 1 time series (bottom panel) in the 2- to 2.5-year (QB) band. The joint spatial EOFs explain 16.4% of the variance in the 2- to 2.5-year band. Regions where <40% of the observations occur are masked. (From Allan et al. 1999.)

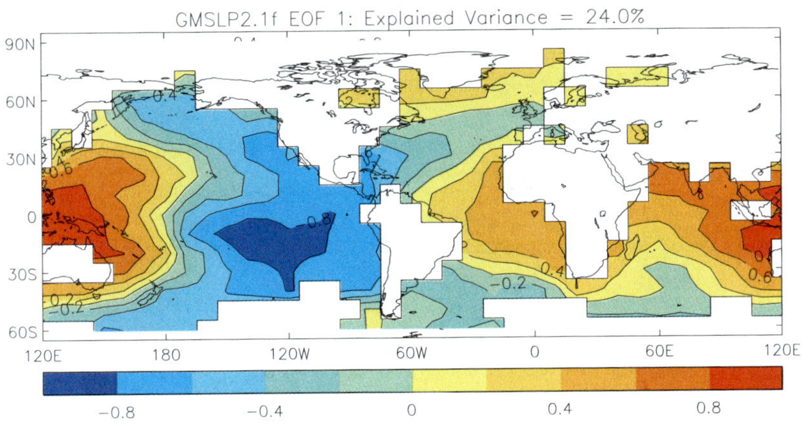

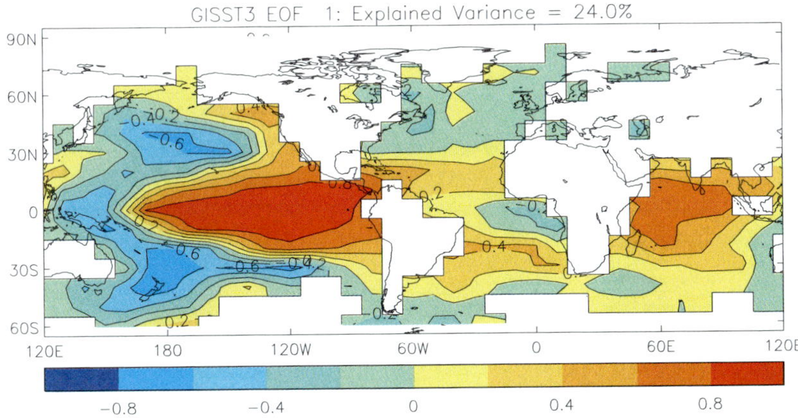

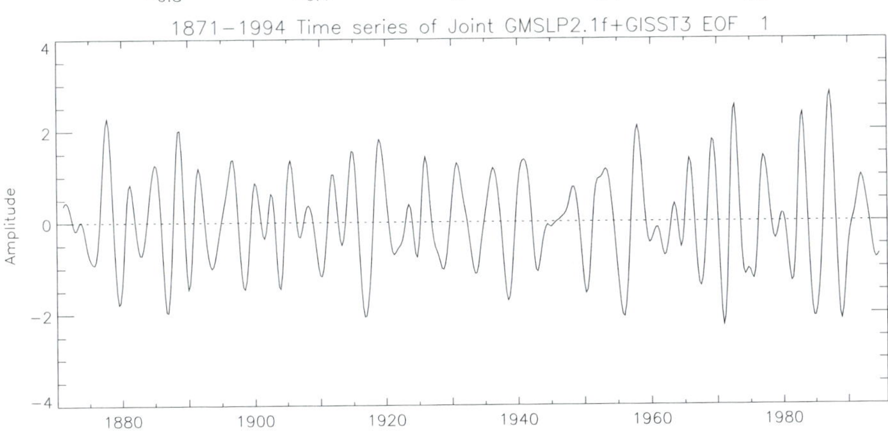

Fig. 1.10 Joint GMSLP2.1f and GISST3.0 EOF 1 spatial correlation fields for MSLP (top panel) and SST (middle panel) (relative to the base period 1901–90), and joint EOF 1 time series (bottom panel) in the 2.5- to 7- (LF) year band. The joint spatial EOFs explain 24% of the variance in the LF band. Regions where <40% of the observations occur are masked. (From Allan et al. 1999.)

Table 1.1 *Simultaneous seasonal correlations between joint GMSLP2.1f and GISST3.0 EOF 1 time series and observed SOI, Niño-3, and Niño-4 time series, both band-pass filtered in the QB band (top panel) and the LF band (bottom panel). Correlation values above the 95 and 99% statistical significance levels are shown in both panels by the * and ** symbols, respectively, with values that explain variance equal to or exceeding 50% highlighted in bold.*

2- to 2.5-year filtered quasi-biennial band

	JFM	AMJ	JAS	OND
SOI	**−0.89**(78%)**	−0.47**(22%)	**−0.91**(83%)**	**−0.87**(76%)**
Niño-3	**+0.82**(68%)**	+0.48**(24%)	**+0.83**(69%)**	**+0.95**(90%)**
Niño-4	**+0.86**(75%)**	+0.15(2%)	**+0.75**(56%)**	**+0.92**(84%)**

2.5- to 8-year filtered "classical" ENSO band

	JFM	AMJ	JAS	OND
SOI	**−0.89**(80%)**	**−0.71**(50%)**	**−0.89**(78%)**	**−0.83**(69%)**
Niño-3	**+0.92**(85%)**	**+0.80**(64%)**	**+0.87**(76%)**	**+0.93**(86%)**
Niño-4	**+0.92**(84%)**	**+0.87**(75%)**	**+0.86**(73%)**	**+0.91**(84%)**

Source: Allan et al. (1999).

A further examination of Figures 1.9 and 1.10 reveals that, as for Niño-3 SST, both the QB and LF ENSO bands are also enveloped by lower frequency modulations in variance. This nature is most clearly manifest on the higher frequency QB band. Such variance structure has often been seen in previous studies that have examined raw or mildly smoothed ENSO indices, and it has generally been attributed to lower frequency decadal–multidecadal fluctuations (Torrence and Compo 1998; Torrence and Webster 1998; Xu et al. 1998). Assessment of the implications of this for ENSO research is only just beginning, with suggestions that at least two "types" of ENSO teleconnection states are associated with more and less robust periods of variance in the ENSO phenomenon (Ward et al. 1994; Navarra et al. 1999; Miyakoda et al. 1999). Such teleconnection states have important implications for ENSO-monsoon modulations (Webster and Yang 1992; Webster 1995; Torrence and Compo 1998; Torrence and Webster 1998, 1999; Xu et al. 1998) and climatic impact patterns during El Niños and La Niñas (Allan et al. 1996).

A broader perspective of the above ENSO characteristics over time is provided by the evolutive MTM-SVD spectra shown in Figure 1.11. The MTM-SVD spectra supports the general thrust of numerous wavelet analyses of ENSO indices (Diaz and Pulwarty 1994; Wang and Wang 1996; Brassington 1997; Kestin et al. 1998; Torrence and Webster 1998, 1999; Torrence and Compo 1998; and Fig. 1.4), in showing more robust signals in the QB and LF ENSO bands during the first and last 40–50 years of the record and a period of very weak signal strength in the intervening epoch. The data

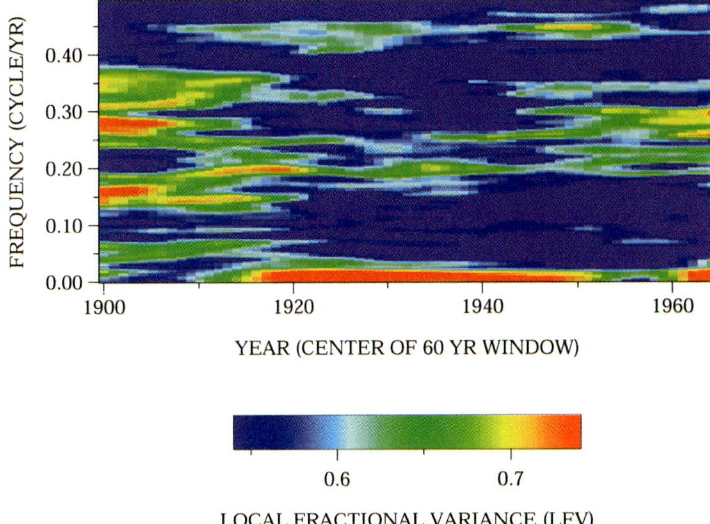

Fig. 1.11 Evolutive MTM-SVD spectral power based on performing the MTM-SVD analysis in a 60-year moving window through the joint GMSLP2.1f and GISST3.0 data. Frequency (cycles year^{-1}) on the vertical axis is plotted against time on the horizontal axis. The plots were truncated at 0.5 year^{-1}, or the biennial frequency. The range of local fractional variance is shown by the color bar, with values tending toward the yellow to red end of the spectrum being significant, and confidence levels being associated with the values given in Figure 1.6. (From Allan et al. 1999.)

show a collapse of the ENSO signal from a period of particularly strong fluctuations near 0.29 year^{-1} and 0.15 year^{-1} (3.4- and 6.7-year period) on the LF ENSO band in the specific period 1871–1931, to a more quiescent epoch apart from a power peak at 0.19 year^{-1} (5-year period) from 1906 to 1966, and then a renewal of power at 0.29 year^{-1} (3.4-year period) for the more recent 1936–96 epoch.

ENSO and Decadal–Multidecadal Signals

Recent work by Chang et al. (1997) on decadal climatic signals over the Atlantic Ocean sector has shown that the strongest ENSO-like delayed oscillator signals in SST for that region occur on about the quasi-decadal timescale. Using such findings as a guide to low-pass filtering of global SST data (MOHSST6C) to isolate interdecadal-secular signals from those associated with ENSO-like features, Folland et al. (1998) then applied EOF analysis to the data and resolved four major low-frequency patterns. Of these EOFs, the third and fourth describe fluctuations operating on multidecadal timescales. The third EOF is of a fluctuation on a 20- to 30-year time frame and has similarities to the patterns in Mantua et al. (1997) and Zhang et al. (1997), which display ENSO-like SST structures in the equatorial Pacific and Indian Oceans. Such SST anomaly patterning has potential ramifications for an enhancement of El Niño–like and La Niña–like conditions on long timescales. The fourth EOF is seen to be related to the 18- to 20-year rainfall signal in parts of southern Africa (Tyson 1986; Mason 1990, 1995).

In the Allan et al. (1999) analysis, joint EOFs of the 11- to 13-, 15- to 20-, and 20- to 30-year signal bands (Figs. 1.12, 1.13, and 1.14) resolved by the MTM-SVD spectra in Figure 1.8 have spatial patterns of SST and MSLP anomalies that all show ENSO–like characteristics. The spatial patterns and time series of these EOF bands, and their relationship to the SST EOFs in Folland et al. (1998), are discussed in the following subsections.

Quasi-Decadal and Interdecadal Bands

Fluctuations in the 11- to 13-year or quasi-decadal band (Fig. 1.12) display a pattern with the warmest SST anomalies displaced into the central-western equatorial Pacific in the current Niño-4 area. Mean SLP anomalies show a dipole structure across the central-eastern half of the Pacific basin, with distinct centers around 30°N and 40°S. This structure is very different from the QB and LF ENSO patterns in that region (Figs. 1.9 and 1.10), which show a single, distinct central-southeastern Pacific MSLP node of the Southern Oscillation. The 15- to 20-year or interdecadal signal (Fig. 1.13) has the warmest Pacific SSTs displaced slightly into the Southern Hemisphere and toward the South American coast in something like the contemporary Niño-1+2 and Niño-3 region configuration. As with the quasi-decadal band, MSLP anomalies in the Pacific region have nodes in both hemispheres in similar geographical locations.

Fluctuations in MSLP and SST anomaly patterns extend into the Indian and Atlantic Ocean basins on both quasi-decadal and interdecadal timescales. Over the former, MSLP and SST distributions vary between the bands and, in conjunction with the QB and LF signals, may go a long way toward explaining Indian Ocean SST dipole patterns (Nicholls 1989; Drosdowsky 1993a,b; Allan et al. 1996). In the Atlantic sector, both bands show a marked enhancement of spatial patterns indicative of signals described previously on decadal timescales by Chang et al. (1997).

Joint time series on these bands are suggestive of signals carrying an ENSO-like temporal structure responsible for protracted El Niño and La Niña phases. In a number of recent studies of the 1990–95 protracted El Niño sequence (Latif et al. 1997; Kleeman et al. 1996), the SST pattern across the Pacific during much of that period has a distribution matching that seen in the quasi-decadal spatial SST EOF (Fig. 1.12). In addition, the longer term fluctuations shown in Figures 1.12 and 1.13 are generally well aligned with protracted El Niño and La Niña events documented in the literature (Allan and D'Arrigo 1999). An even better matching is found when both the QB and LF joint time series are superimposed with those on the quasi-decadal and interdecadal bands, and the composite signal is constructed.

Multitaper method SVD evolutive spectral power (Fig. 1.11) carries something of the quasi-decadal and interdecadal signals, but not as clearly as for the higher frequency ENSO bands discussed previously. As with wavelet studies (Brassington 1997; Torrence and Compo 1998; Torrence and Webster 1998, 1999; Wang and Wang 1996), spectral power resolved at about the 0.09 year^{-1} (11.1-year period) quasi-decadal and the

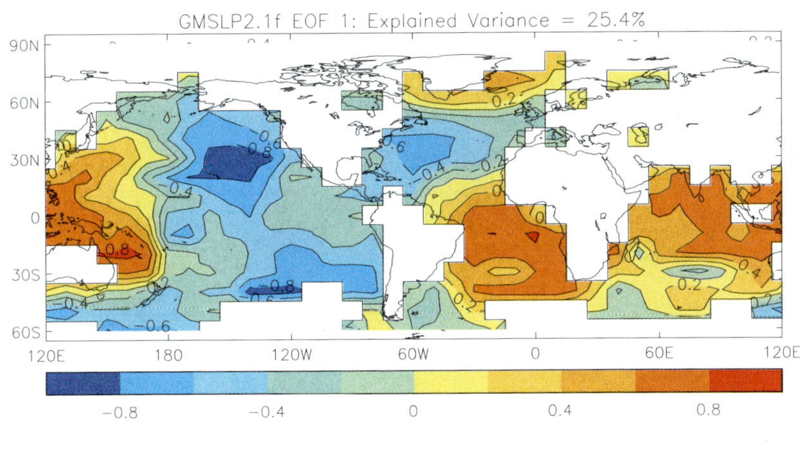

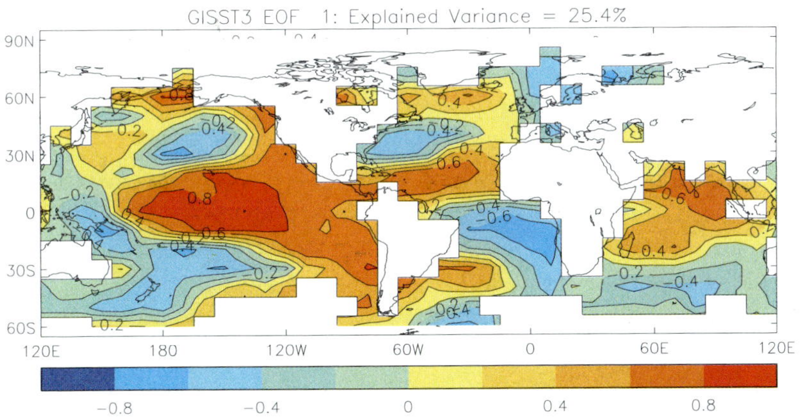

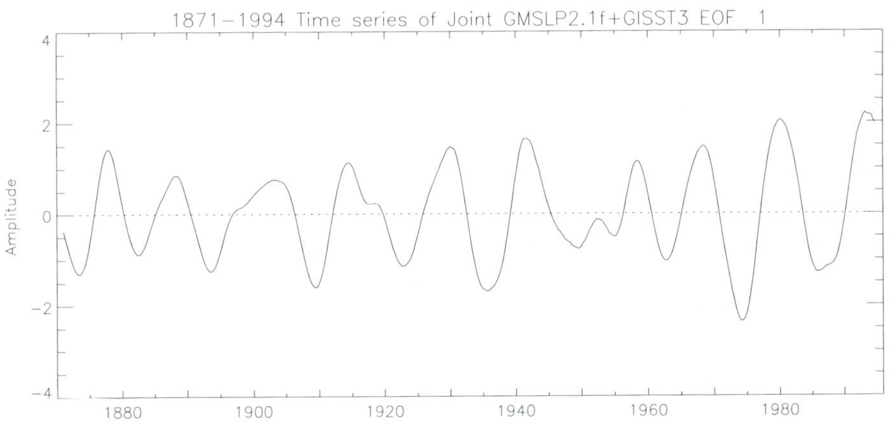

Fig. 1.12 Joint GMSLP2.1f and GISST3.0 EOF 1 spatial correlation fields for MSLP (top panel) and SST (middle panel) (relative to the base period 1901–90), and joint EOF 1 time series (bottom panel) in the 11- to 13-year band. The joint spatial EOFs explain 25.4% of the variance in the 11- to 13-year band. Regions where <40% of the observations occur are masked. (From Allan et al. 1999.)

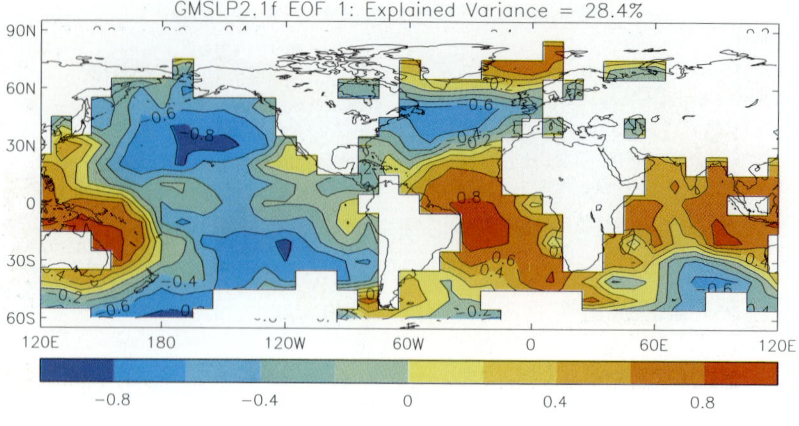

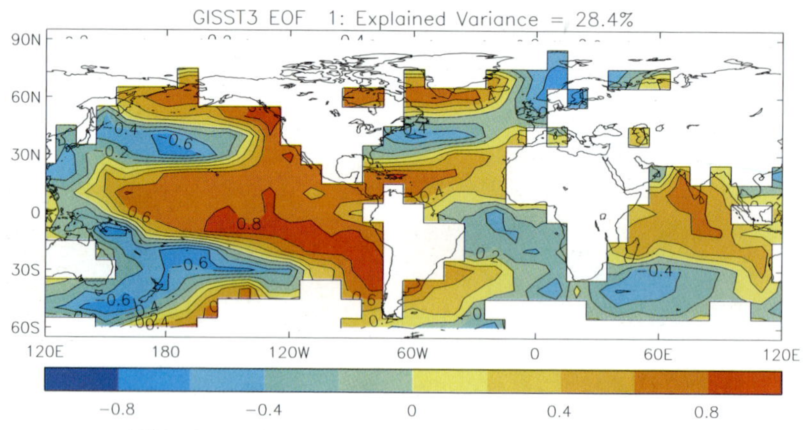

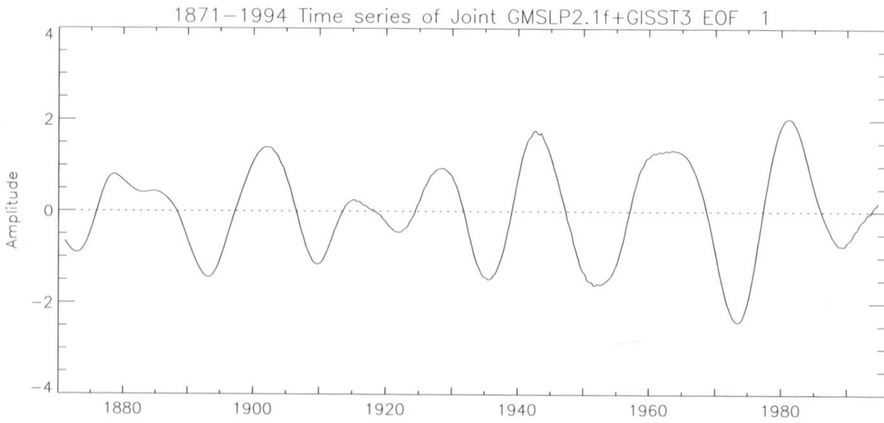

Fig. 1.13 Joint GMSLP2.1f and GISST3.0 EOF 1 spatial correlation fields for MSLP (top panel) and SST (middle panel) (relative to the base period 1901–90), and joint EOF 1 time series (bottom panel) in the 15- to 20-year band. The joint spatial EOFs explain 28.4% of the variance in the 15- to 20-year band. Regions where <40% of the observations occur are masked. (From Allan et al. 1999.)

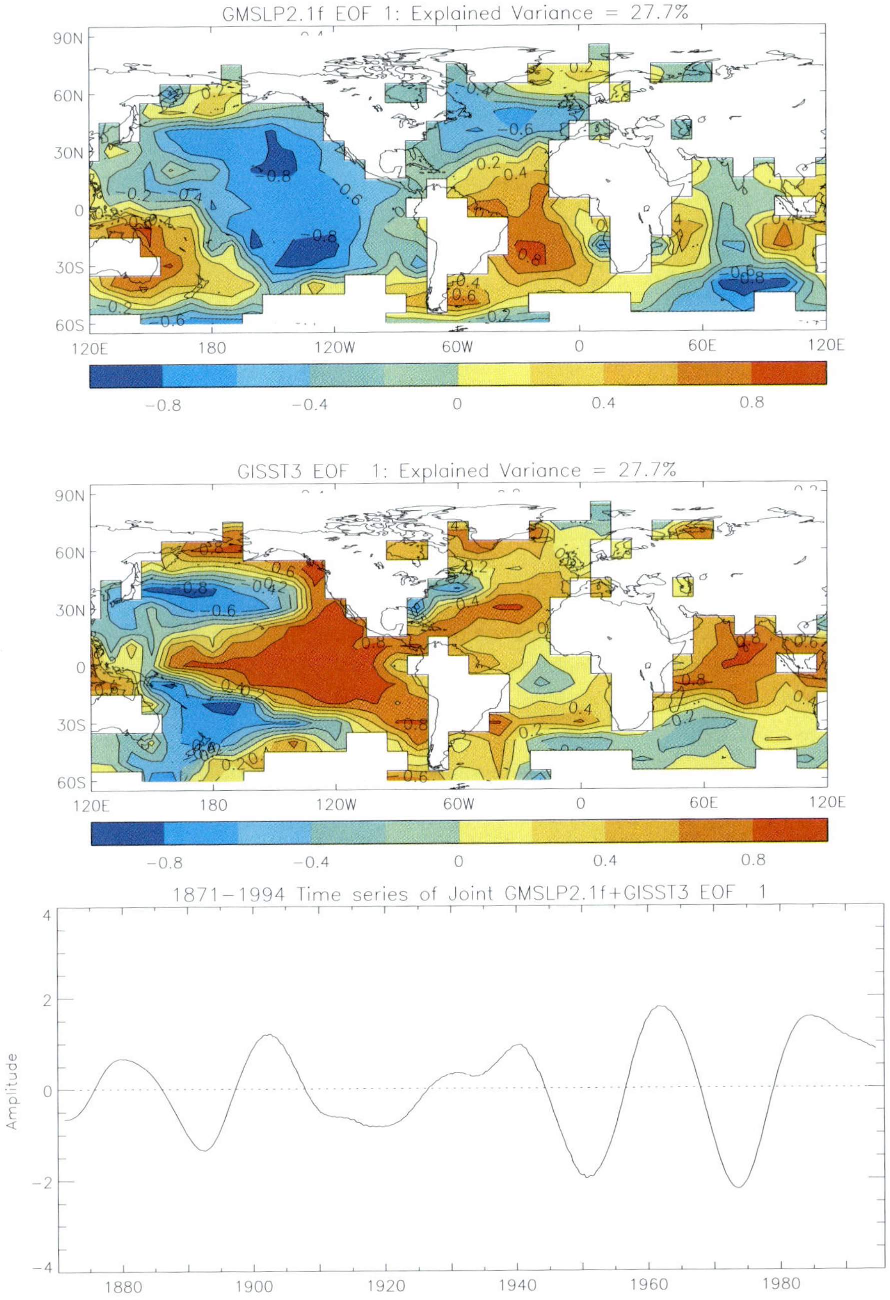

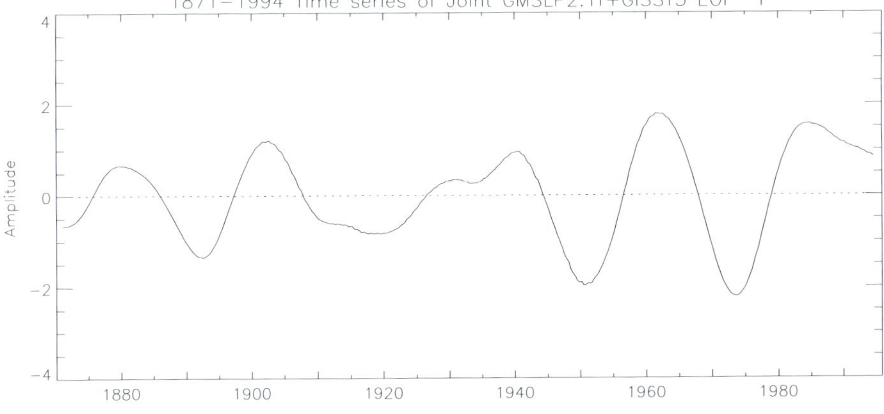

Fig. 1.14 Joint GMSLP2.1f and GISST3.0 EOF 1 spatial correlation fields for MSLP (top panel) and SST (middle panel) (relative to the base period 1901–90), and joint EOF 1 time series (bottom panel) in the 20- to 30-year band. The joint spatial EOFs explain 27.7% of the variance in the 20- to 30-year band. Regions where <40% of the observations occur are masked. (From Allan et al. 1999.)

0.06 year^{-1} (16.7-year period) interdecadal timescales shows fluctuations over time. In general, enhanced and quiescent signal strength tends to be aligned with the epochs of robust and weak QB and LF ENSO spectral power examined in the earlier section. This again reinforces the notion that ENSO modulations must be viewed in conjunction with fluctuations in the climate system on decadal–multidecadal time frames.

20- to 30-Year Band

The 20- to 30-year fluctuation in the joint EOF analysis of Allan et al. (1999) has limited realizations over the 124 years of data, and it is weak in the MTM-SVD spectra of Figure 1.11, so its statistical stability could thus be questioned. However, it is resolved as significant in the MTM-SVD fractional variance spectra for joint MSLP and surface air temperatures over the Northern Hemisphere in Mann and Park (1996) (Fig. 1.6), and a similar spatiotemporal pattern is evident as the third low-frequency EOF in global SST data in Folland et al. (1998). As a consequence, we have examined this signal in this chapter.

In Folland et al. (1998), the third low-frequency EOF shows strong signals at about 20–40 years in its time series, with peaks at about the turn of the century, in the 1920s–30s epoch, and in the period since the mid-1970s. All of these time frames exhibit patterns of ENSO-like SST anomaly response in the Indo-Pacific domain, while the second low-frequency (60- to 80-year period) EOF of Folland et al. (1998) certainly provides additional modulation for at least climatic signals in the Sahelian region. Interestingly, such findings cast doubt about the Chang et al. (1997) decadal signal in the Atlantic Ocean being the obvious "cutoff" for ENSO-like signals in the climate system. There is thus a need to reexamine new global historical data sets with broader perspectives of ENSO and low-frequency climatic variability in mind.

Basically, the 20- to 30-year band signal (Fig. 1.14) shows an enhancement and meridional broadening of equatorial Pacific SST anomalies, and more of a coalescence of the MSLP feature over much of the Pacific Ocean. In the Atlantic sector, signal strength and coherency are less than for either of the quasi-decadal and interdecadal bands. Across the Indian Ocean basin, MSLP and SST anomaly patterns are slightly more coherent than in the above bands, with a stronger MSLP feature at the middle to higher latitudes of the Southern Hemisphere. In general, the overall spatial pattern is dominated by a strong, Pacific-centered ENSO-like phenomenon. As is noted in Folland et al. (1998), similar findings of ENSO-like patterns in decadal–multidecadal SST data alone, and the maintenance of SST–rainfall relationships from interannual ENSO through to multidecadal ENSO-like timescales, suggest that teleconnection structures have a degree of very broad spatiotemporal coherency.

The joint EOF time series in the 20- to 30-year band is indicative of about five epochs of El Niño–like and La Niña–like patterns in the climate system over the period of record. Of particular interest is the tendency for El Niño–like conditions to be enhanced since the mid- to late 1970s, which coincides with the timing of the change in the ENSO regime across the Pacific Ocean basin (Graham 1994; Latif and Barnett 1994a,b; Miller et al. 1994a,b; Trenberth and Hurrell 1994; Wang 1995). Once again, the importance of interactions between climatic modes is revealed if the 20- to 30-year joint EOF time series (Fig. 1.14) is superimposed on those at 2–2.5, 2.5–7, 11–13, and 15–20 years (Figs. 1.9, 1.10, 1.12, and 1.13).

ENSO and Multidecadal–Secular Patterns

In Folland et al. (1998), the first EOF of low-pass-filtered global SST data is highly correlated with the global warming trend and shows a tendency for maximum warming in the middle to higher latitudes in both hemispheres, while displaying little evidence for coherent equatorial Pacific warming indicative of any modulation of the core region of ENSO SST fluctuations (Cane et al. 1997). Patterns of SST anomalies in the second EOF of Folland et al. (1998) closely resemble those associated with the long-term Sahelian rainfall series and the interhemispheric temperature contrast (Folland et al. 1986; Parker and Folland 1991; Folland et al. 1991) and appear to be operating on a timescale of about 60–80 years.

To resolve secular climatic signals in Allan et al. (1999), a low-pass filter (>29 years) was applied to the data prior to the joint EOF analysis. This approach reveals two dominant modes: a global warming trend pattern in EOF 1 and the Sahelian rainfall/interhemispheric temperature contrast signal (Folland et al. 1998) operating on a timescale of about 60–80 years in EOF 2. Together, the above signals account for some 72.4% of the variance on secular timescales.

The spatial pattern of SST anomalies in the joint EOF 1 diagram (Fig. 1.15) has low weights in the eastern equatorial Pacific and central Indian Ocean basins and in the North Atlantic south of Greenland. In general, the most coherent and strongest warming trend in SSTs occurs at middle to higher latitudes in both hemispheres. The spatial distribution of MSLP anomalies suggests an increase in values across most of the Indian Ocean basin, with mixed signals over the rest of the globe. The Pacific Ocean basin is the only region where coherent positive and negative MSLP anomaly features are found together. The joint EOF time series (Fig. 1.15) serves to illustrate that EOF 1 is capturing the observed global warming signal (Folland et al. 1998).

For EOF 2 (Fig. 1.16), the spatial patterns of MSLP and SST anomalies are indicative of the Sahelian rainfall/interhemispheric temperature contrast signal. These spatial patterns show a definitive ENSO-like fluctuation and a marked shift toward enhanced El Niño–like conditions since the 1970s. Whether this aspect of the recent climate is carried by this band alone, or the 20- to 30-year signal discussed earlier, or a combination of both, is still unclear.

Global Rainfall Correlations and Documentary Evidence

Resolution of the dominant signals and patterns in global historical MSLP and SST data at various interannual to secular timescales do not in themselves provide a picture of the degree of climatic impact they cause. In the literature, these data have generally been presented through maps and diagrams showing relationships and correlations between climatic phenomena and variables such as rainfall, temperature, streamflow, etc. (see

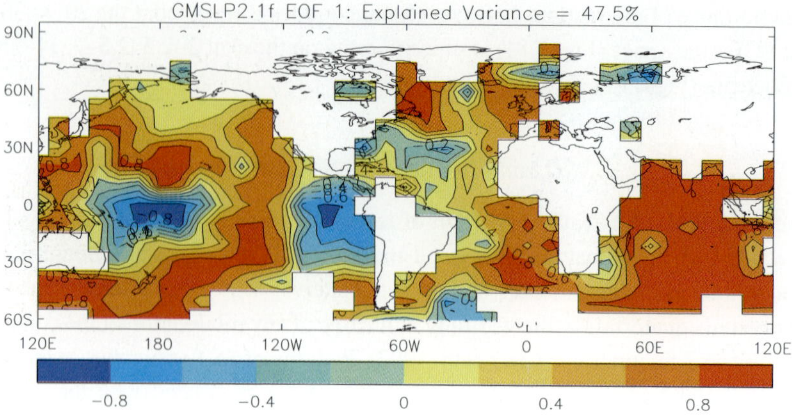

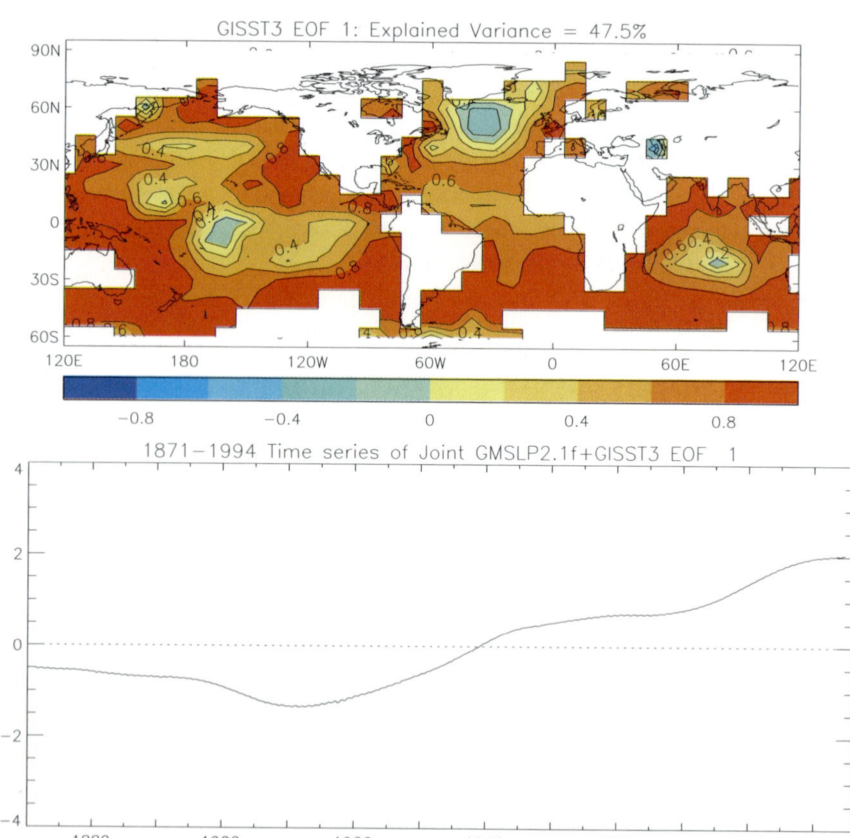

Fig. 1.15 Joint GMSLP2.1f and GISST3.0 EOF 1 (global warming trend signal) spatial correlation fields for MSLP (top panel) and SST (middle panel) (relative to the base period 1901–90), and joint EOF 1 time series (bottom panel) in the >29-year band. The joint spatial EOFs explain 47.5% of the variance in EOF 1 for the >29-year band. Regions where <40% of the observations occur are masked. (From Allan et al. 1999.)

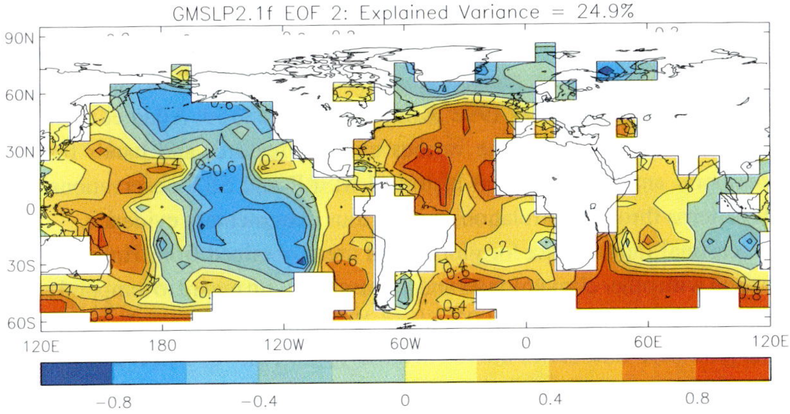

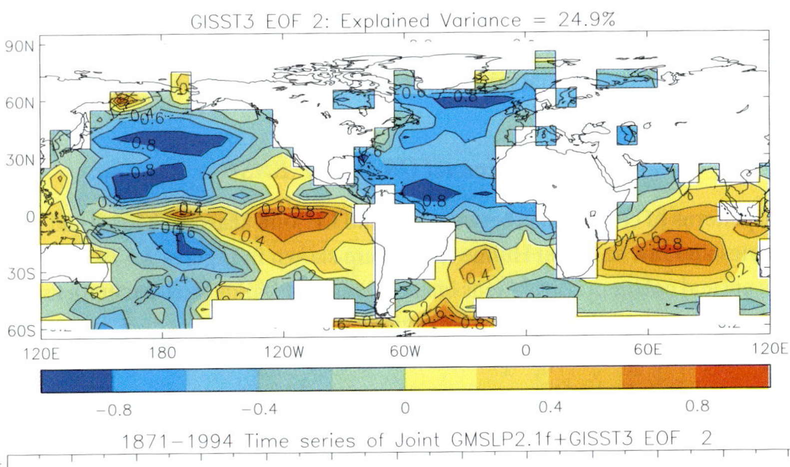

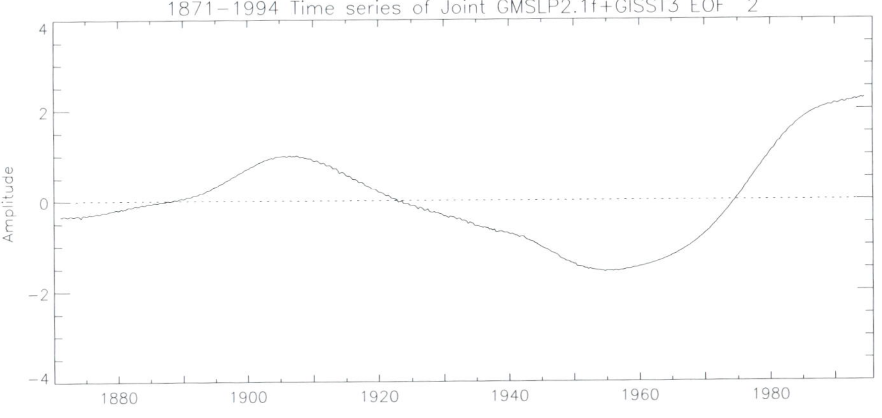

Fig. 1.16 Joint GMSLP2.1f and GISST3.0 EOF 2 (60- to 80-year Sahelian/interhemispheric temperature contrast signal) spatial correlation fields for MSLP (top panel) and SST (middle panel) (relative to the base period 1901–90), and joint EOF 2 time series (bottom panel) in the >29-year band. The joint spatial EOFs explain 24.9% of the variance in EOF 2 for the >29-year band. Regions where <40% of the observations occur are masked. (From Allan et al. 1999.)

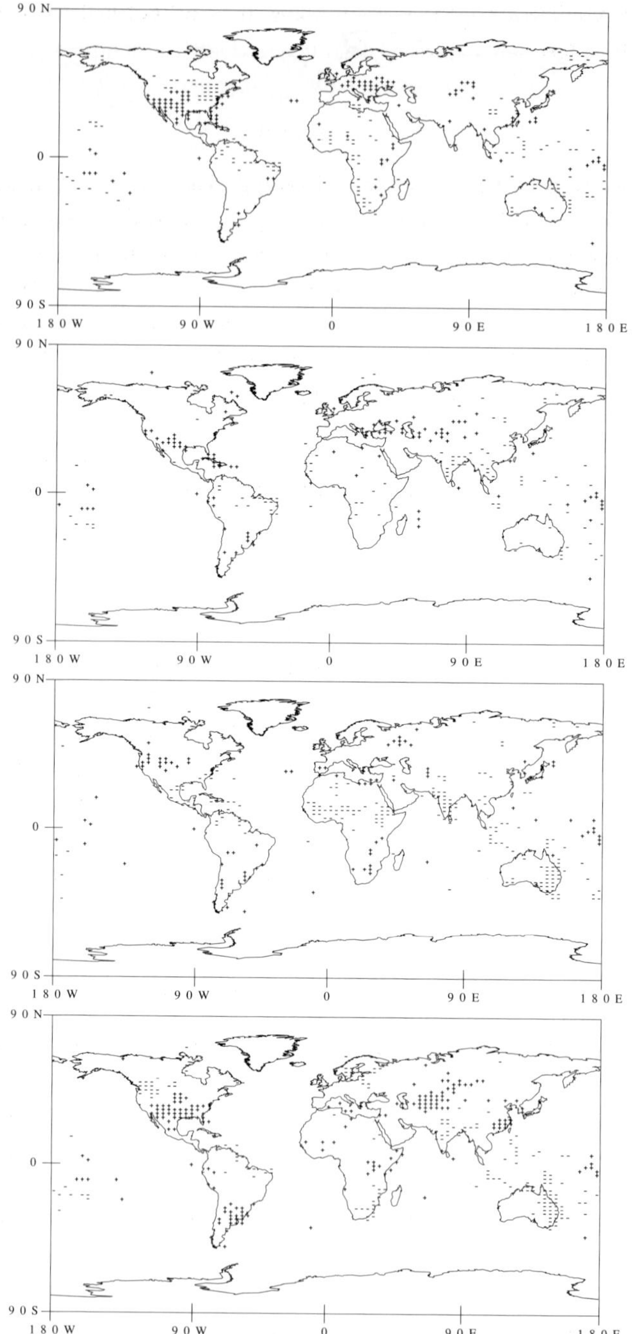

Fig. 1.17b Global maps of seasonal correlations between the joint EOF 1 time series (band-pass filtered in the LF band) and raw land-plus-island precipitation data from Hulme (1992) for the period 1900–94, in the 2.5–7 year band. Those correlations above $r = +0.2$ (marked with "+") or below $r = -0.2$ (marked with "−") are significant at the 95% level. Plots top to bottom are JFM, AMJ, JAS, OND.

frailty," or the "predictability barrier," has the least robust correlation patterns of any season globally in both the QB and LF bands.

Quasi-Decadal and Interdecadal Bands

Correlation patterns over the global land regions and islands on these bands are detailed in Figures 1.18a,b. As on the QB and LF bands (Figs. 1.17a,b), there is a strong tendency for the most coherent and significant correlations to be found in "classical" ENSO-sensitive areas. On the quasi-decadal band (Fig. 1.18a), parts of eastern Australia in all but the AMJ season, southern Africa in JFM, northern Argentina and Uruguay in AMJ and OND, East Africa in OND–JFM, and northeastern Brazil in AMJ display an ENSO-like rainfall response. Other strong correlations are found over the eastern United States in JFM. On the interdecadal timescale (Fig. 1.18b), strong and coherent ENSO-like responses are found over East Africa in OND, China and Mongolia in AMJ, and parts of eastern Australia in all seasons but JAS. Other well-organized responses are found over the eastern United States and western Canada in JFM.

The wider ramifications of the rainfall responses in Figures 1.18a,b can be seen when they are examined in conjunction with the ENSO patterns in Figures 1.17a,b. In regions such as East Africa in OND, southern Africa in JFM, northeastern Brazil in AMJ, and parts of eastern Australia throughout much of the year, there are strong modulations and reinforcements of ENSO-induced rainfall patterns by rainfall correlations on either or both bands. Regions that are not known as ENSO sensitive, but show strong and coherent QB and LF patterns that are reinforced by quasi-decadal and interdecadal correlations, are western parts of Canada and the eastern United States in OND–JFM. Thus the phasing of QB, LF, quasi-decadal, and interdecadal band signals is a major source of rainfall modulations and variations in overall climatic impacts from event to epoch timescales.

Multidecadal 20- to 30-Year Band

Global rainfall responses on the 20- to 30-year band (Fig. 1.19a) continue the tendency seen with the quasi-decadal and interdecadal bands in Figures 1.18a,b for ENSO-like rainfall responses to dominate the correlation patterns. This is evident over southern Africa and eastern Australia in JFM and over central regions of Australia, East Africa, and Argentina in OND. Apart from the above relationships, the only other consistent pattern on the 20- to 30-year band is a response over western Canada and the eastern United States in JFM.

Sahelian 60- to 80-Year Band and Global Warming Signals

The EOF 2–rainfall correlation pattern in Figure 1.19b is the first in any band examined to show a shift away from ENSO-like climatic impacts. The most coherent feature of this band is the strong rainfall correlation signal in JAS that extends in a band across Sahelian to East Africa. The African response continues to be concentrated over

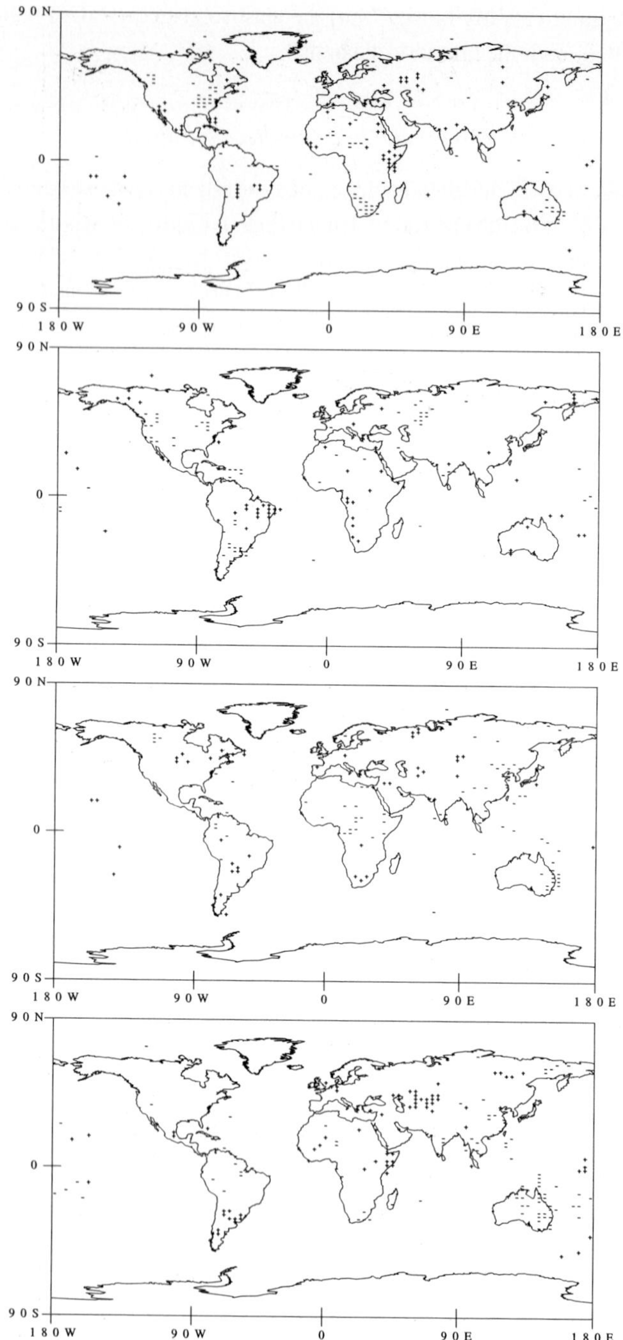

Fig. 1.18a Global maps of seasonal correlations between the joint EOF 1 time series (band-pass filtered in the 11- to 13-year band) and raw land-plus-island precipitation data from Hulme (1992) for the period 1900–94, in the 11–13 year band. Those correlations above $r = +0.2$ (marked with "+") or below $r = -0.2$ (marked with "−") are significant at the 95% level. Plots top to bottom are JFM, AMJ, JAS, OND.

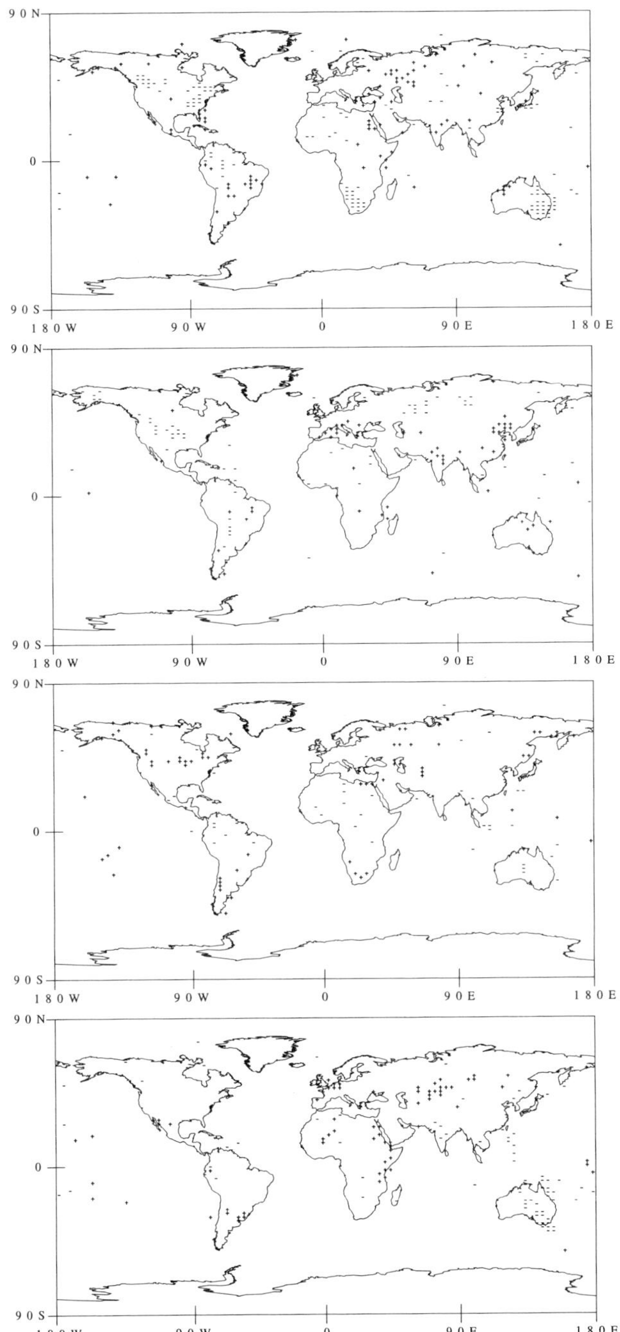

Fig. 1.18b Global maps of seasonal correlations between the joint EOF 1 time series (band-pass filtered in the 15- to 20-year band) and raw land-plus-island precipitation data from Hulme (1992) for the period 1900–94, in the 15–20 year band. Those correlations above $r = +0.2$ (marked with "+") or below $r = -0.2$ (marked with "–") are significant at the 95% level. Plots top to bottom are JFM, AMJ, JAS, OND.

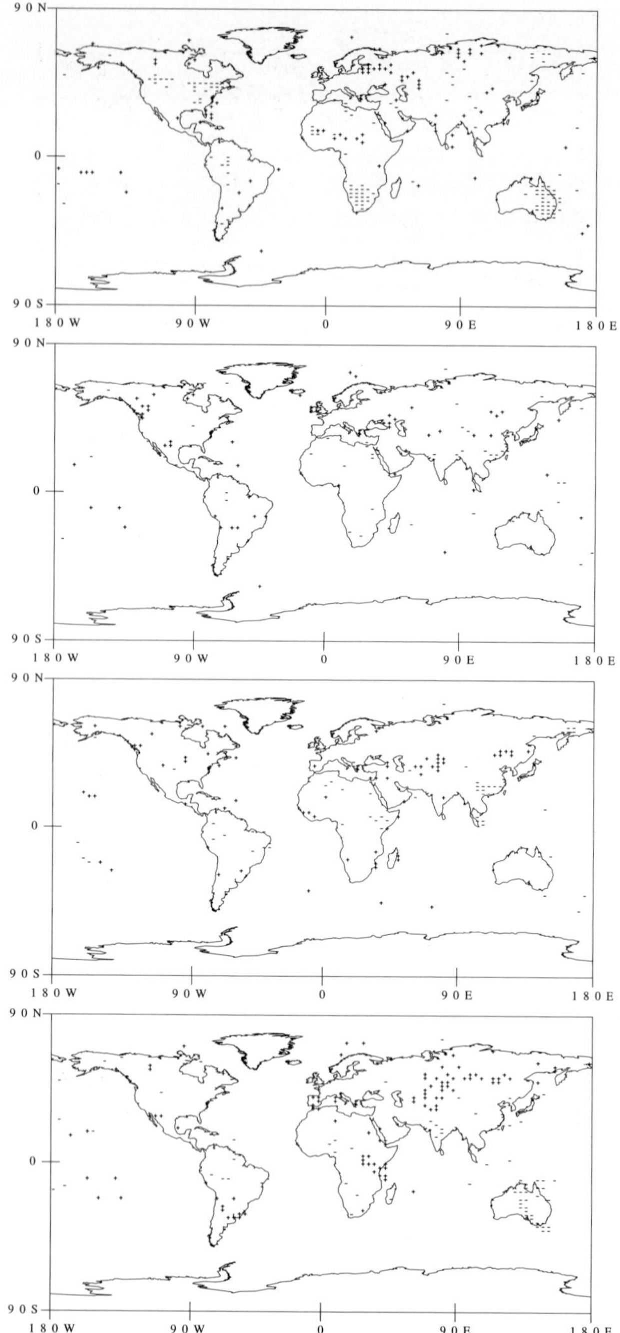

Fig. 1.19a Global maps of seasonal correlations between the joint EOF 1 time series (band-pass filtered in the 20- to 30-year band) and raw land-plus-island precipitation data from Hulme (1992) for the period 1900–94, in the 20–30 year band. Those correlations above $r = +0.2$ (marked with "+") or below $r = -0.2$ (marked with "–") are significant at the 95% level. Plots top to bottom are JFM, AMJ, JAS, OND.

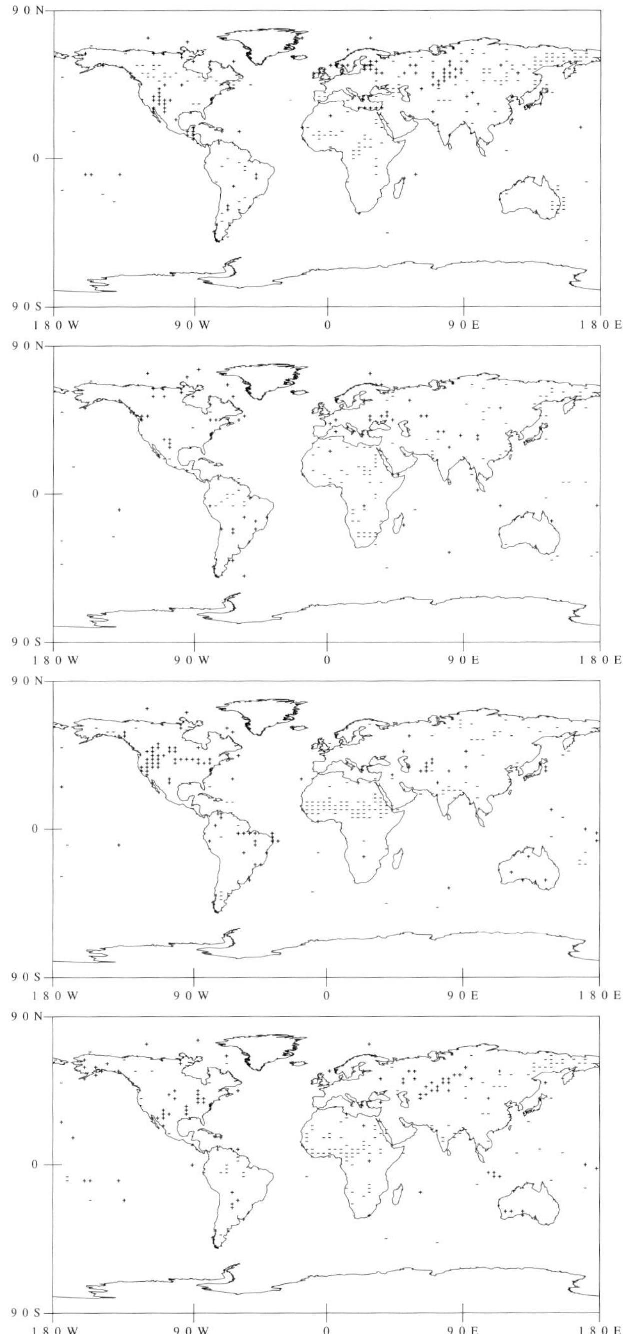

Fig. 1.19b Global maps of seasonal correlations between the joint EOF 2 time series (60- to 80-year signal in data low-pass-filtered for >29 years) and raw land-plus-island precipitation data from Hulme (1992) for the period 1900–94, in the 60–80 year band. Those correlations above $r = +0.2$ (marked with "+") or below $r = -0.2$ (marked with "−") are significant at the 95% level. Plots top to bottom are JFM, AMJ, JAS, OND.

parts of the Sahelian region in OND and confirms the dominance of low-frequency fluctuations over ENSO impacts in West Africa. Weaker correlation structures are found over Canada and the northwestern United States in JFM and JAS and over far eastern Russia in OND–JFM. Overall, the rainfall modulation provided by the 60- to 80-year signal appears to be a significant source for long-term variability in the West African monsoon system.

The distribution of rainfall correlations with the EOF 1 global warming trend time series is shown in Figure 1.20. Significant and coherent rainfall correlations are found over a number of global regions in various seasons. Well-documented droughts in Sahelian Africa (particularly far western West Africa) and the Ethiopian region appear to have been modulated by a combination of the 60- to 80-year and global warming signals (Figs. 1.19b and 1.20). Strong positive rainfall correlation patterns are suggested over northeastern Canada in AMJ–OND, the Argentinian region in OND–JFM, and parts of central to far eastern Europe and Asia for much of the year.

Important supportive evidence for the longer term climatic fluctuations seen in the rainfall correlation patterns over this century can be found in documentary sources and studies. Such material is detailed in the following subsection.

Documentary Evidence for Climatic Variability and Links to Rainfall Correlations

Unlike studies of the higher frequency ENSO phenomenon, which have included considerable objective analysis of regional to global impact patterns, longer term decadal to multidecadal fluctuations in the climate system have far fewer realizations. Although considerable instrumental data are available back to the early parts of this century, the most consistent type of evidence used to investigate long-term climatic patterns has tended to be documentary. Prior to concerns about changes in the nature of ENSO since the 1970s, three climatic fluctuations during this century caught the attention of researchers. These fluctuations are detailed here, and relationships with the various seasonal rainfall correlation bands are discussed.

The Turn of the Century

Perhaps the first concerted scientific effort to synthesize information on an apparent low-frequency fluctuation in the global climate system was the work by Sir John Eliot, the second British director of the Indian Meteorological Department (Eliot 1904, 1905). Having been involved in both meteorological observations and research in India since the 1870s and having presided over the department from 1889 to 1903, Eliot was well placed to see and influence the evolution of efforts to find reliable precursors of the Indian monsoon system (Allan et al. 1996). It was during Eliot's term that the Indian Meteorological Department expanded the range of parameters and regions examined in the formulation of predictors of the Indian monsoon to include conditions over the Indian Ocean, northeastern Africa, and southern Australia. However, by the turn of the century, the forecast structure had been found wanting with the failure of the forecasts from 1899 to 1901 and the 1902–05 efforts that were being provided confidentially

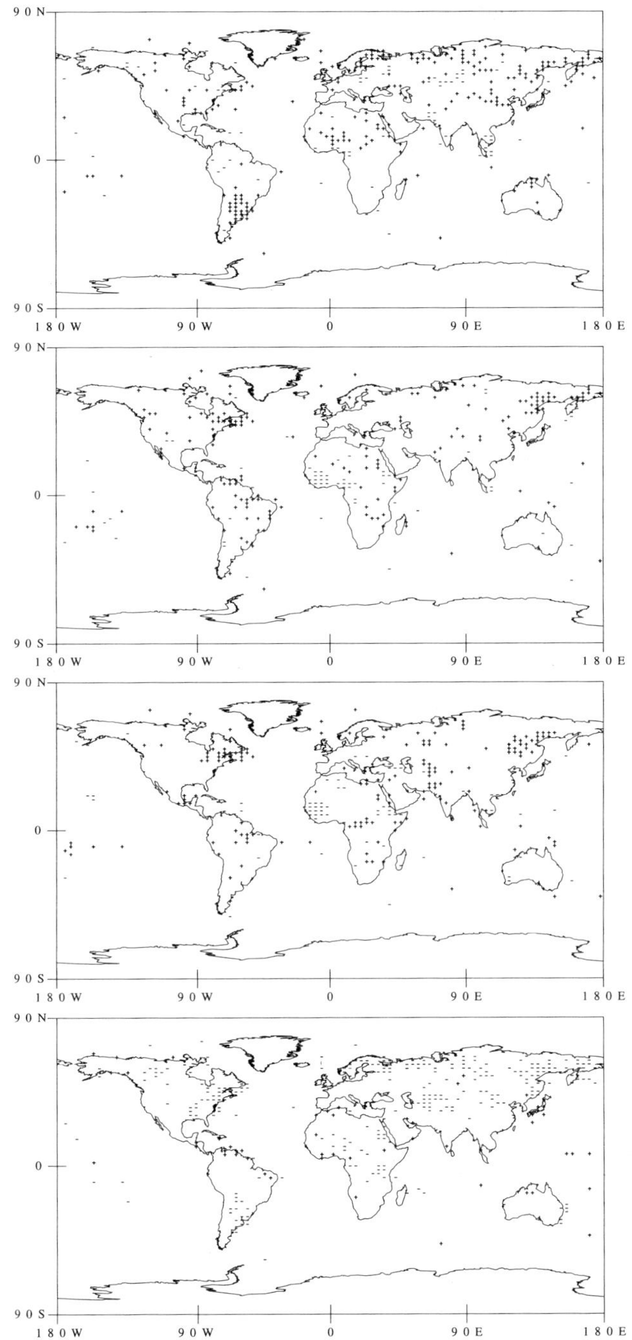

Fig. 1.20 Global maps of seasonal correlations between the joint EOF 1 time series (global warming trend in low-pass-filtered data for >29 years) and raw land-plus-island precipitation data from Hulme (1992) for the period 1900–94. Those correlations above $r = +0.2$ (marked with "+") or below $r = -0.2$ (marked with "−") are significant at the 95% level. Plots top to bottom are JFM, AMJ, JAS, OND.

to the Indian government. It was with such experiences, and about the time of his retirement from service in India in 1904, that Eliot put together papers indicating the "uniqueness" of the climate from about 1894 to 1902 throughout much of the then British Empire (Eliot 1904, 1905). By 1908, Sir Gilbert Walker, Eliot's successor in the Indian Meteorological Department, had been asked by the Indian government to explore the nature of possible changes in Indian climate and the likely role of human activities in such changes (Walker 1910). This work indicated the presence of some type of natural fluctuation in climatic patterns across the Indian subcontinent but concluded that no human-induced component was responsible.

The impacts and consequences of this turn-of-the-century fluctuation in the climate system can be seen in the extent of crop and animal losses in countries surrounding the Indian Ocean basin (southern, eastern, and central Africa; Australia; Mauritius; and India) (Eliot 1905). Its wider extension led to rainfall deficiencies and major displacements of farming populations across the Great Plains region of the United States in the mid- to late 1890s (Warrick 1980; Stockton and Meko 1983) and droughts in northern China in the 1890s (Whetton and Rutherfurd 1994), in Indonesia in the mid-1890s to 1903, as shown by tree-ring studies (Jacoby and D'Arrigo 1990; D'Arrigo et al. 1994), in northeastern Brazil for the first 5 years of the twentieth century (Diaz et al. 1989), in southeastern England during the 7 years prior to 1901 (Noble 1903), and across central to eastern Russia in 1901–03, resulting in famine in that region (Noble 1903).

In Australia, it became known as the "Federation Drought" and was most strongly manifest from coastal to inland regions of the states of Queensland, New South Wales, and South Australia (Foley 1957). The drought during 1895–1903 was responsible for an Australian wheat shortfall estimated at 13 million bushels and losses of some 50 million sheep in New South Wales alone (Eliot 1905). According to Foley (1957), in Australia as a whole, sheep numbers were halved and cattle losses were of the order of 30%. This period was also one of increased dust storms in Australia that led to reports of dust fallout over New Zealand from Australian continental sources (Noble 1904). Haze, smoke, and dust observed over northern Australian ports and by steamers operating to and from Hong Kong and the Philippines were discussed in terms of possible volcanic eruptions in Indonesia but could well have been products of large wildfires in the jungles of Borneo and Sumatra (Noble 1904). Although it was written in the late 1950s, the following comment from Foley (1957; p. 208) is still relevant today: "It is difficult for present day Australians to realise the magnitude of the effects of this drought on the economy of the country."

Over southern Africa, the period was generally punctuated by years of extreme drought and cold in which cattle and sheep losses were in the millions and maize and wheat crops failed (Eliot 1905; Lindesay and Vogel 1990). The severest drought conditions over southern Africa occurred during 1902–03 (Lindesay and Vogel 1990). In central Africa, drought conditions tended to persist from 1898 to 1903, while in eastern Africa the Nile discharge was very low for the year 1898 and for the whole period from 1900 to 1902 (Noble 1903; Eliot 1905).

During 1896–97, food crops in British India were some 33% below normal (about 18–19 million tons) and the country needed to import about 6 million tons of rice from

Burma (Eliot 1905). Some 62 million people were affected, with relief efforts focusing on the 34 million people most severely hit by famine conditions. Cattle losses were massive, with estimates in the Province of Hisar alone that 92% of the cattle used to plow fields and to provide the power to lift water from wells had been lost between April 1895 and October 1987 (Eliot 1905). For the period 1899 to 1900, the effects were less severe and more scattered, but still nearly 4 million cattle were lost in the Central Provinces and in and around Bombay.

The influence of this climatic anomaly extended to the high latitudes of Asia, leading to heavy cattle losses and some 15 million starving peasants in central-eastern and southeastern Russia (Noble 1903). The cost of cereal produce (rye and wheat) used by the Russian government to alleviate the famine was estimated at about 1.7 million pounds sterling alone.

In the Great Plains region of the United States, the 1890s drought peaked in the years 1894–95, with estimates that some 300,000 farmers were displaced and that wheat yields fell to as low as 9 bushels/acre (Warrick 1980). Population decline due to drought and malnutrition during the 1890s was particularly severe over a large portion of the Great Plains region. These impacts were compounded by a lack of government or public assistance, and a general vulnerability to natural disasters. According to the summary of Warrick (1980; p. 107): "Many areas experienced a dramatic loss of 50 to 75 percent of their population."

Research on this climatic fluctuation at the time is best summed up by Eliot (1905; p. 276): "Hence it is evident that during the dry period, from 1895 to 1901, there were disturbing actions of great magnitude which gave rise to large general variations of pressure probably over at least half the Eastern Hemisphere. The data at present available give no indication of the region in which the opposite and compensatory variations of air mass and pressure occurred."

Since that time, the most thorough investigations of the nature and extent of this climatic fluctuation have been undertaken mainly by Indian scientists (Pant et al. 1988; Parthasarathy et al. 1991, 1992; Thapliyal and Kulshrestha 1991; Vijayakumar and Kulkarni 1995; see also Fu and Fletcher 1988). The overall findings of this work are indicative of a change in the broad characteristics of the Indian monsoon system from meridional in the pre-1890s period to the development of more zonal atmospheric flow regimes in the 1890s–1900s epoch. This conclusion is supported by spatiotemporal fluctuations in the areas of statistically significant ENSO and rainfall correlations over India, Sri Lanka, and Southeast Asia (Parthasarathy et al. 1991; Vijayakumar and Kulkarni 1995; Suppiah 1996; Kripalani and Kulkarni 1997).

From the various decadal to multidecadal rainfall correlation patterns detailed in Figures 1.18a,b and 1.19a,b, and discussed earlier, the most likely candidates to explain the turn-of-the-century climatic fluctuation are the 60- to 80-year Sahelian/interhemispheric thermal contrast mode (Fig. 1.19b) in combination with the 20- to 30-year band signal (Fig. 1.19a). In addition, these modes were reinforced by ENSO-like patterns on the quasi-decadal and interdecadal timescales (Fig. 1.18a,b).

As was documented earlier in this chapter, large-scale drought conditions with crop and animal losses extended into the first three to five years of the twentieth century in Africa, India, China, Russia, northeastern Brazil, and eastern Australia. Correlation patterns for southern Africa suggest that the combined influence of low rainfall regimes from the quasi-decadal, interdecadal, and 20- to 30-year bands during JFM provided the strongest influence in that region. For Sahelian Africa the 60- to 80-year band carries the dominant rainfall suppression signal, while the global warming trend signal was generally the most coherent mode working against this pattern in far western West Africa. Across East Africa, enhanced precipitation on the interdecadal and 20- to 30-year bands in OND was countered by reduced rainfall as a result of the global warming signal in that season (see Fig. 1.20). Farther north in Ethiopia and the Horn of Africa, mixed signal responses appear to be operating, although the 20- to 30- and 60- to 80-year bands in JAS would favor dry conditions.

In northern India, the strongest signal reflecting a fluctuation in the Indian summer and reduced rainfall is evident on the 60- to 80-year band in JAS. Some evidence for premonsoon suppression of precipitation is found on the interdecadal band in AMJ. Farther east in China, the major reduction in rainfall would appear to be a result of the combined influence of the quasi-decadal and 20- to 30-year bands in JAS and the interdecadal band in AMJ. The situation in central to eastern Russia appears to be the result of rainfall suppression brought about by the influence of the global warming trend time series (inverse of Fig. 1.20) in OND.

In northeastern Brazil, the influence on rainfall deficits appears to be mostly derived from ENSO-like patterns on the quasi-decadal band in AMJ. Some reinforcement of this tendency may come from the EOF 1, greater than 29 years low-pass-filtered global warming trend signal. However, the rainfall modulation in this region has far less discernible low-frequency influences than for others discussed above.

Across eastern Australia, there is strong modulation of rainfall patterns toward dry conditions on the quasi-decadal band in JAS–JFM, on the interdecadal band in OND–AMJ, on the 20- to 30-year band in OND–JFM, and on the 60- to 80-year band in JFM. Thus a conjunction of such signals, as in the turn-of-the-century epoch, would work to reinforce drought conditions in eastern margins of the continent.

The 1920s–30s Period

The next major climatic fluctuation with wide-ranging effects over many areas of the globe occurred during the 1920s–30s period. This epoch is most well known for the "Great Plains Drought" or "Dust Bowl" in the grain belt of the central United States (Worster 1979; Heathcote 1980). Nevertheless, wider drought impacts were sustained in southern Africa, Australia, parts of Amazonia, and eastern China.

Although we speak of the 1930s period of United States drought conditions (Warrick 1980), the Great Plains Drought was most intense in the years 1934, 1936, and 1939 (Stockton and Meko 1983). Impacts of the "Dust Bowl" are well documented in the literature, and the subject has been the source for numerous geographical, socioeconomic, and historical studies, mainly because of the inextricable influences of the Great Depression and bad farming practices (Skaggs 1975, 1988; Warrick 1980). Climatic manifestations at this time included not only drought and high temperatures but also the vast dust storms of windblown topsoil (Mattice 1935; Choun 1936; Parkinson 1936). Much of this situation was compounded by poor land practices (leading to higher potential for erosion) and heavy land use (overcropping and overproduction) following increased mechanization and high wheat prices (Russell 1988; Goudie and Middleton 1992). In one dust storm in May 1934, an estimated 300 million tons of topsoil was blown away (Hughes 1995), with the material being deposited across Boston and New York and even on ships as far as 500 km out to sea in the Atlantic Ocean (Goudie and Middleton 1992). As a consequence, the Great Plains Drought severely impacted crop production, with a 29% decline in wheat yields from the region during the 1930s epoch. Unlike earlier droughts in the 1890s and 1910s, there was a high proportion of farm transfers (both voluntary and forced), with up to one in ten farms changing hands at the height of the drought (Warrick 1980). By its conclusion, the Dust Bowl had resulted in the migration of some 350,000 farmers to California (Hughes 1995).

The extent of the impact that the Great Plains Drought had on the American psyche can be gauged by the fact that it inspired significant literary works such as Steinbeck's *The Grapes of Wrath* and the musical ballads of Woody Guthrie, while being a catalyst for President Roosevelt's works program and government assistance under the New Deal (Russell 1988). Hughes (1995; p. 33) notes that a significant impetus in climatic research was initiated by this disaster: "The drought and terrible dust storms also led the Department of Agriculture to fund Rossby's assault on the secrets of atmospheric circulation in the hope that such calamitous natural disasters could be forecast to help farmers."

However, concern about the cause of the Dust Bowl drought also led to many popular myths about it, as Russell (1988; p. 61) notes: "The disruptive effects of radio waves proved the most popular."

In Australia, various states experienced more regional, less severe droughts from 1935 to 1938 (Foley 1957). However, the most severe and prolonged drought in Australia at this time occurred from 1939 to 1945 and was obviously linked to the protracted 1939–42 El Niño sequence (Bigg and Inoue 1992; Jones and Allan 1998; Allan and D'Arrigo 1999). The most significant impacts occurred in 1940 and from 1943 to 1945 in the growing seasons across the cereal-producing regions of the states of Victoria, New South Wales, South Australia, and Western Australia (Foley 1957). Overall, the Australian wheat yield was reduced from 13.9 to 6.25 bushels per acre and sheep numbers fell by 29 million (from 125 million to 96 million) in the period from 1942 to 1945. There was also evidence that large amounts of dust from eroded areas of southern Australia were blown as far away as New Zealand and colored sunsets over eastern Australia in the 1940s (Heathcote 1983).

Other regions of the globe also experienced major climatic impacts at various times during the 1920s–30s epoch. For southern Africa, Tyson (1986) reports on a marked dry spell from 1925 to 1933 that is resolved as a response to the 18- to 20-year oscillation in austral summer precipitation in that region. Periods of less severe, or more regional, droughts are reported as occurring in most Australian states at various stages during the 1926 to 1929 epoch by Foley (1957). Much recent Chinese and Japanese research has focused on "climatic jumps" that occurred in many places around the world in the

1920s and 1960s (Ye and Yan 1993; Nitta and Yoshimura 1993; Qiang and Demaree 1993). The former time indicates the beginning of the 1920s–30s climatic fluctuation, which appears to have been synchronous with atmospheric warming in the Northern Hemisphere and a period of low global ocean cloudiness prior to a long-term increasing trend (Mingli and Fletcher 1993). The early 1920s were also part of a period of increasing Indian monsoon activity from about 1900 to 1940 (although there was a tendency for marginally drier conditions during the Indian summer monsoon) (Fu and Fletcher 1988; Fu and Qiang 1991; Parthasarathy et al. 1991, 1992; Vijayakumar and Kulkarni 1995; Suppiah 1996). It also marks a period of low rainfall at Tornoto that lasted into the 1930s (Sarker and Thapliyal 1988), the start of a wetter period in Sahelian Africa (Nicholson 1989); a period of pronounced dry conditions in north-central Amazonia from the 1920s to the mid-1940s (Marengo 1995); and a tendency for drier conditions in eastern China (Mingli 1993). Whetton et al. (1990) indicate particularly low (high) river flow in the 1920s–30s period for the Krishna River in India (the Sénégal River in West Africa). For South America, Marengo (1995) showed that the most prominent river flow changes were increased streamflow for the Paraná River in Argentina and the Río Negro in northeast Brazil from 1920 to 1925 and decreased flow in the Orinoco River in north Amazonia from 1925 to 1939 and in the Río Negro in northeast Brazil from 1928 to 1932.

Evidence from the rainfall correlation maps in Figures 1.18a,b and 1.19a,b for a climatic fluctuation during the 1920s–30s epoch leading to drought in the Great Plains of the United States, and in southern Africa, Australia, India, parts of South America, and China, is varied. As with the turn-of-the-century fluctuation in climate, it appears that the combined effects of decadal to multidecadal signals are responsible. Over the central United States, the most discernible rainfall correlations indicative of suppressed precipitation are on the 60- to 80-year band in AMJ–OND (inverse of Fig. 1.19b given the time series in Fig. 1.16 at this time) and on the 20- to 30-year band in AMJ. This overlapping amalgam of influences is probably the likely source of variations in the intensity of this drought (Stockton and Meko 1983).

In southern Africa, the drought signal appears to be centered on the 20- to 30-year band in JFM, with some influence from the quasi-decadal and interdecadal bands in that season. The role played in rainfall patterns by other fluctuations such as the 18- to 20-year oscillation (Tyson 1986; Mason 1990, 1995), resolved as EOF 4 in the low-frequency part of the analysis in Folland et al. (1998), is not addressed. However, over far western and Sahelian Africa, this epoch shows a tendency toward wetter conditions as seen in the rainfall correlations on the 60- to 80-year band in JAS–OND (inverse of Fig. 1.19b given the time series in Fig. 1.16 at this time). Interestingly, in central to eastern Africa, the combined influence of the rainfall correlations from the global warming trend to the 20- to 30-year and interdecadal bands in OND is suggestive of a period of wetter conditions in the late 1920s.

Over eastern Australia, most of the drought response in the second half of the 1920s seems to have been a result of the operation of the 20- to 30-year band in OND–JFM. The propensity for long periods of dry conditions during much of the 1930s was apparently ameliorated considerably by the inverse of the rainfall correlations on the quasi-decadal and interdecadal timescales (Figs. 1.18a,b with the time series in Figs. 1.12 and 1.13). The influence of the quasi-decadal and interdecadal bands during much of the year, in conjunction with the 20- to 30-year band in OND–JFM, appears to have been the major low-frequency source of the prolonged 1939–45 drought.

India during the 1920s–30s epoch experienced mixed rainfall conditions, with generally drier conditions in the 1920s perhaps being linked to the rainfall correlations on the interdecadal band in AMJ over the second half of the decade and on the 60- to 80-year band in JAS during the first half of the 1920s. For the 1930s, the opposite pattern on the above bands would appear to have contributed to slightly wetter conditions.

The situation in South America can best be deduced for Argentina and northeastern Brazil. Over northern Argentina during the first half of the 1920s, increased rainfall and river flow appear to be linked to the inverse of significant correlations on the quasi-decadal band in JAS. The reverse appears to be occurring over central Argentina at this time. In northeastern Brazil, the same epoch is one of positive rainfall correlations on the quasi-decadal band in AMJ. During the late 1920s to early 1930s, lower rainfall over northeastern Brazil is consistent with negative rainfall correlations on the quasi-decadal band in AMJ.

Over northern China, lower rainfall in the 1920s is reflected in the inverse rainfall correlations on the interdecadal band in AMJ. Across central to eastern China, a reduction in rainfall is related more to inverse correlation patterns on the above band in JFM–AMJ.

The Late 1960s to 1970s

Unlike the situation with previous climatic fluctuations this century, the intense scientific interest in the climate system and ENSO in recent decades has been able to draw on an increasingly vast amount of high-quality data from ground-based and satellite platforms that provide detailed information on the global environment. The two climatic variations during this time that have received particular attention in climatological and environmental research are the so-called Sahelian drought and the "climatic shift" in ENSO characteristics in the Pacific Ocean sector. The latter focus has also been paralleled by expanding interest in wider regional to global fluctuations in climate on decadal to multidecadal timescales, which have already been discussed extensively elsewhere in this chapter.

From the mid-1970s to 1980s, considerable scientific research was focused, and concern generated, by the prolonged drought conditions that developed over Sahelian Africa from about 1968 (Nicholson 1985, 1989; Lamb and Peppler 1991). Continuing into the early 1990s, this climatic event has been the major catalyst for debate about biogeophysical versus large-scale climatic fluctuations as the principal cause of extended drought conditions and climatic change (Nicholson 1993; Nicholson et al. 1996). Recent overviews, such as Druyan (1989), indicate that large-scale climate controls appear to play the dominant role in Sahelian drought, with more localized feedbacks having very much a lesser role in exacerbating the situation. The scope of work on this drought situation has extended from investigations of displacements in features of the West African monsoon system to investigations of influences of the broader pattern of Atlantic air–sea interactions and global SSTs, using both observational (contemporary and historical) data and modeling simulations (Druyan and Hastenrath 1991, 1992; Lamb and Peppler 1991, 1992; Ward 1992). As a result of such intensive research, efforts have been made to apply the knowledge gained to predict Sahelian rainfall fluctuations, with mixed results (Owen and Ward 1989; Parker et al. 1988; Ward et al. 1990; Rowell et al. 1991; Ward 1992; Folland 1992; Smith 1995). However, in the 1990s, the focus on longer term climatic fluctuations and change has moved to the Pacific Ocean basin.

Evidence for the Sahelian drought is found in the strong and coherent negative rainfall correlation patterns on the 60- to 80-year band in JAS–OND (Fig. 1.19b). This low-frequency response dominates most other climatic regimes over the region, including the ENSO phenomenon in the QB and LF bands (Figs. 1.17a,b).

Discussion and Conclusions

Numerous studies indicate that near-global climatic impacts are associated with the "classical" ENSO phenomenon, but they also stress the need to examine the influence and impact of decadal- to multidecadal-scale fluctuations. Objective techniques capable of resolving distinct spatiotemporal climatic modes, such as EOF and SVD analyses, are beginning to be used to improve current understanding of the nature and structure of the ENSO phenomenon and its relationship to decadal- to secular-scale modes in the climate system. The EOF approach detailed in this chapter is applied to significant frequency bands/signals resolved by an MTM-SVD spectral technique, and it reveals important interactions and modulations involving ENSO and lower frequency ENSO-like patterns. As with other studies in the literature, the superposition of the QB and LF ENSO signals is seen to be an integral component in the structure of the ENSO phenomenon. An important additional aspect of such interactions is the apparent role the QB signal plays in carrying the bulk of the boreal spring (austral autumn) "breakdown" or "predictability barrier." Nevertheless, further research is required to understand the nature of coherent modulations of interannual ENSO variance that result in epochs of robust and quiescent signal strength and appear to cause variations in teleconnection structure and magnitude.

However, the EOF study also reveals the presence of quasi-decadal and interdecadal ENSO-like signals in global historical marine surface data, such as the Global Sea-Ice and Sea Surface Temperature (GISST) and Global Mean Sea Level Pressure (GMSLP) data sets, with prominent SST fluctuations in the equatorial Pacific Ocean and, to a lesser degree, Indian Ocean regions. These quasi-decadal and interdecadal signals are seen to be responsible for modulating protracted El Niño and La Niña events (such as the 1990–95 episode) through interactions with the quasi-biennial and "classical" elements of the ENSO phenomenon. In addition, EOF analysis techniques reveal multidecadal- to secular-scale fluctuations that include the global warming trend, a 60- to 80-year signal linked to long-term Sahelian rainfall patterns, and a strong ENSO-like configuration across the Indo-Pacific domain operating on a 20- to 30-year timescale. The 20- to 30- and 60- to 80-year bands are seen to be modes that favor epochs of El Niño–like and La Niña–like patterns in the climate system. Joint EOF time series and spatial patterns of MSLP and SST anomalies for both the 20- to 30- and 60- to 80-year modes show a tendency toward a sharp transition to El Niño–like conditions during the 1970s. However, the global warming trend fails to show climatic patterns indicative of enhanced El Niño–like conditions in recent decades.

Overall, the nature of the climatic fluctuations detailed in both seasonal rainfall correlation patterns and documentary/observational evidence is generally consistent with the influence of the type of ENSO and ENSO-like patterns resolved in the EOF analysis. These findings support the need to foster the convergence of studies focusing on ENSO and climatic variability on all timescales. Future progress in our understanding of ENSO and its relationships to fluctuations in the climate system will require increased knowledge about the nature and spatiotemporal structure of such interactions, as well as the physical mechanisms underlying them.

In summary, it is very apparent that ENSO-like patterns are evident in the global climate system throughout much of its spectrum. This low-frequency natural variability provides an important modulation of climatic impacts, as has been illustrated by correlations with land and island precipitation data. Consequently, the influence of the "classical" 2- to 7-year ENSO phenomenon on climatic patterns must be seen as occurring within an envelope of concurrent decadal- to secular-scale fluctuations, and all of these features need to be addressed if a more complete picture of climatic variability and long-range forecasting is to be achieved. One of the most difficult challenges facing future work on ENSO and longer term variability is the question of if and how ENSO and the decadal- to secular-scale fluctuations relate to other low-frequency features of the climate system, such as the NAO, the AO, the NPO and the ACW.

Acknowledgments Dr. Allan is supported jointly through the CSIRO Climate Variability and Impacts Programme and Climate Change and Impacts Programme and is funded in part by the Australian Federal Government's National Greenhouse Research Programme of the Australian Greenhouse Office and the State Government of Queensland. Particular thanks for comments and suggestions on the manuscript go to Drs. Henry Diaz, George Kiladis, and Jim Hurrell, with the latter two also undertaking the task of examining the work for the internal CSIRO reviewing process. Dr. Ian Smith, CSIRO Atmospheric Research, Australia, provided valuable input and assistance on this chapter, for which the author is most grateful. Invaluable support with data processing was obtained from Ms. Tracy Basnett, Hadley Centre for Climate Prediction and Research, United Kingdom Meteorological Office, Bracknell, United Kingdom, and Ms. Tara Ansell, Department of Earth Sciences, The University of Melbourne, Australia.

References

ALLAN, R. J. and D'ARRIGO, R. D., 1999: Persistent ENSO sequences: How unusual was the 1990–1995 El Niño? *Holocene*, **9**, 101–118.

ALLAN, R. J., LINDESAY, J. A., and REASON, C. J. C., 1995: Multidecadal variability in the climate system over the Indian Ocean region during the austral summer. *Journal of Climate*, **8**, 1853–1873.

ALLAN, R. J., LINDESAY, J. A., and PARKER, D. E., 1996: *El Niño Southern Oscillation and Climatic Variability*. Melbourne: CSIRO Publishing, 405 pp.

ALLAN, R. J., MANN, M. E., FOLLAND, C. K., PARKER, D. E., SMITH, I. N., BASNETT, T. A., and RAYNER, N. A., 1999: Global warming, El Niño, and multidecadal climate variability. In review.

BASNETT, T. A. and PARKER, D. E., 1997: Development of the Global Mean Sea Level pressure data set GMSLP2. Climate Research Technical Note CRTN79, Hadley Centre, Meteorological Office, Bracknell, United Kingdom, 16 pp.

BIGG, G. R., 1995: The El Niño event of 1991–94. *Weather*, **50**, 117–124.

BIGG, G. R. and INOUE, M., 1992: Rossby waves and El Niño during 1935–46. *Quarterly Journal of the Royal Meteorological Society*, **118**, 125–152.

BRASSINGTON, G. B., 1997: The modal evolution of the Southern Oscillation. *Journal of Climate*, **10**, 1021–1034.

BRAZDIL, R. and ZOLOTOKRYLIN, A. N., 1995: The QBO signal in monthly precipitation fields over Europe. *Theoretical and Applied Climatology*, **51**, 3–12.

BURNETT, A. W., 1993: Size variations and long-wave circulation within the January Northern Hemisphere circumpolar vortex: 1946–89. *Journal of Climate*, **6**, 1914–1920.

CANE, M. A., CLEMENT, A. C., KAPLAN, A., KUSHNIR, Y., POZDNYAKOV, D., SEAGER, R., ZEBIAK, S. E., and MURTUGUDDE, R., 1997: Twentieth-century sea surface temperature trends. *Science*, **275**, 957–960.

CHANG, P., JI, L., and LI, H., 1997: A decadal climate variation in the tropical Atlantic Ocean from thermodynamic air–sea interactions. *Nature*, **385**, 516–518.

CHEN, T.-C., van LOON, H., WU, K.-D., and YEN, M.-C., 1992: Changes in the atmospheric circulation over the North Pacific–North America area since 1950. *Journal of the Meteorological Society of Japan*, **70**, 1137–1146.

CHOUN, H. F., 1936: Duststorms in the southwestern plains area. *Monthly Weather Review*, **64**, 195–199.

DAI, A., TRENBERTH, K. E., and KARL, T. R., 1998: Global variations in droughts and wet spells: 1900–1995. *Geophysical Research Letters*, **25**, 3367–3370.

D'ARRIGO, R. D., JACOBY, G. C., and KRUSIC, P. J., 1994: Progress in dendroclimatic studies in Indonesia. *Terrestrial, Atmospheric and Oceanic Sciences*, **5**, 349–363.

DESER, C. and BLACKMON, M. L., 1993: Surface climate variations over the North Atlantic Ocean during winter: 1900–1989. *Journal of Climate*, **6**, 1743–1753.

DETTINGER, M. D. and CAYAN, D. R., 1995: Large-scale atmospheric forcing of recent trends towards early snowmelt runoff in California. *Journal of Climate*, **8**, 606–623.

DIAZ, H. F., BRADLEY, R. S., and EISCHEID, J. K., 1989: Precipitation fluctuations over global land areas since the late 1800s. *Journal of Geophysical Research*, **94**, 1195–1210.

DIAZ, H. F. and PULWARTY, R. S., 1994: An analysis of the time scales of variability in centuries-long ENSO-sensitive records in the last 1000 years. *Climatic Change*, **26**, 317–342.

DROSDOWSKY, W., 1993a: An analysis of Australian seasonal rainfall anomalies: 1950–1987. II: Temporal variability and teleconnection patterns. *International Journal of Climatology*, **13**, 111–149.

DROSDOWSKY, W., 1993b: Potential predictability of winter rainfall over southern and eastern Australia using Indian Ocean sea surface temperature anomalies. *Australian Meteorological Magazine*, **42**, 1–6.

DRUYAN, L. M., 1989: Advances in the study of sub-Saharan drought. *International Journal of Climatology*, **9**, 77–90.

DRUYAN, L. M. and HASTENRATH, S., 1991: Modelling the differential impact of 1984 and 1950 sea surface temperatures on Sahel rainfall. *International Journal of Climatology*, **11**, 367–380.

DRUYAN, L. M. and HASTENRATH, S., 1992: GCM simulation of the Sahel 1984 drought with alternative specifications of observed SST. *International Journal of Climatology*, **12**, 521–526.

EBBESMEYER, C. C., CAYAN, D. R., McLAIN, D. R., NICHOLS, F. H., PETERSON, D. H., and REDMOND, K. T., 1991: 1976 step in the Pacific climate: Forty environmental changes between 1968–75 and 1977–84. *In* Betancourt, J. L. and Tharp, V. L. (eds.), *Proceedings of the Seventh Annual Climate (PACLIM) Workshop*. California Department of Water Resources, Interagency Ecological Studies Program Technical Report **26**, pp. 115–126.

ELIOT, J., 1904: The meteorology of the Empire during the unique period 1892–1902. *Broad Views*, **1**, 191–201.

ELIOT, J., 1905: A preliminary investigation of the more important features of the meteorology of southern Asia, the Indian Ocean and neighbouring countries during the period 1892–1902. *Memoirs India Meteorol. Dept.*, **XVI** (II), 185–307.

FOLEY, J. C., 1957: Droughts in Australia: Review of records from earliest years of settlement. Bureau of Meteorology, *Bulletin No. 43*, 281 pp.

FOLLAND, C. K., 1992: LEPS scores for assessing climate model simulations and long-range forecasts. Climate Research Technical Note CRTN33, Hadley Centre, Meteorological Office, Bracknell, United Kingdom, 17 pp.

FOLLAND, C. K., OWEN, J., WARD, M. N., and COLMAN, A. W., 1991: Prediction of seasonal rainfall in the Sahel region using empirical and dynamical methods. *Journal of Forecasting*, **10**, 21–56.

FOLLAND, C. K., PALMER, T. N., and PARKER, D. E., 1986: Sahel rainfall and worldwide sea temperatures. *Nature*, **320**, 602–607.

FOLLAND, C. K., PARKER, D. E., COLMAN, A. W., and WASHINGTON, R., 1998: Large scale modes of ocean surface temperature since the late nineteenth century. Climate Research Technical Note CRTN81, Hadley Centre, Meteorological Office, Bracknell, United Kingdom, 21 pp.

FRAEDRICH, K., 1994: An ENSO impact on Europe? *Tellus*, **46A**, 541–552.

FRAEDRICH, K., and MULLER, K., 1992: Climate anomalies in Europe associated with ENSO extremes. *International Journal of Climatology*, **12**, 25–31.

FRAEDRICH, K., MULLER, K., and KUGLIN, R., 1992: Northern Hemisphere circulation regimes during the extremes of the El Niño/Southern Oscillation. *Tellus*, **44A**, 33–40.

FU, C. and FLETCHER, J., 1988: Large signals of climatic variation over the ocean in the Asian monsoon region. *Advances in Atmospheric Science*, **5**, 389–404.

FU, C. and QIANG, W., 1991: The jump phenomenon of the South Asia summer monsoon and its synchronism with the abrupt warming in global scale. *Sciences in China*, **B6**, 666–672.

GAGE, K. S., McAFEE, J. R., and WILLIAMS, C. R., 1996: Recent changes in tropospheric circulation over the central equatorial Pacific. *Geophysical Research Letters*, **23**, 2149–2152.

GARREAUD, R. D. and BATTISTI, D. S., 1999: Interannual and interdecadal variability in the Southern Hemisphere. *Journal of Climate*, **12**, 2113–2123.

GLANTZ, M. H., 1996: *Currents of Change: El Niño's Impact on Climate and Society*. New York: Cambridge University Press, 208 pp.

GLANTZ, M. H., KATZ, R. W., and NICHOLLS, N., 1991: *Teleconnections Linking Worldwide Climate Anomalies*. New York: Cambridge University Press, 535 pp.

GODDARD, L. and GRAHAM, N. E., 1997: El Niño in the 1990s. *Journal of Geophysical Research*, **102**, 10423–10436.

GOUDIE, A. S. and MIDDLETON, N. J., 1992: The changing frequency of dust storms through time. *Climatic Change*, **20**, 197–225.

GRAHAM, N. E., 1994: Decadal-scale climate variability in the tropical and North Pacific during the 1970s and 1980s: Observations and model results. *Climate Dynamics*, **10**, 135–162.

GRAHAM, N. E., BARNETT, T. P., WILDE, R., PONATER, M., and SCHUBERT, S., 1994: On the roles of tropical and midlatitude SSTs in forcing interannual to interdecadal variability in the winter Northern Hemisphere circulation. *Journal of Climate*, **7**, 1416–1441.

GU, D. and PHILANDER, S. G. H., 1995: Secular changes of annual and interannual variability in the tropics during the past century. *Journal of Climate*, **8**, 864–876.

GU, D. and PHILANDER, S. G. H., 1997: Interdecadal climate fluctuations that depend on exchanges between the tropics and extratropics. *Science*, **275**, 805–807.

HALPERT, M. S. and ROPELEWSKI, C. F., 1992: Surface temperature patterns associated with the Southern Oscillation. *Journal of Climate*, **5**, 577–593.

HARRISON, D. E. and LARKIN, N. K., 1997: Darwin sea level pressure, 1876–1996: Evidence for climate change? *Geophysical Research Letters*, **24**, 1779–1782.

HEATHCOTE, R. L., 1980: Perception of desertification on the southern Great Plains: A preliminary enquiry. *In* Heathcote, R. L. (ed.), *Perception of Desertification*. United Nations University, Tokyo, 34–59.

HEATHCOTE, R. L., 1983: *The Arid Lands: Their Use and Abuse*. Longmans, London, 323 pp.

HUANG, J.-P., HIGUCHI, K., and SHABBAR, A., 1998: The relationship between the North Atlantic Oscillation and El Niño–Southern Oscillation. *Geophysical Research Letters*, **25**, 2707–2710.

HUGHES, P., 1995: Dust bowl days. *Weatherwise*, **48**, 32–33.

HULME, M., 1992: A 1951–80 global land precipitation climatology for the evaluation of general circulation models. *Climate Dynamics*, **7**, 57–72.

HULME, M. and NEW, M., 1997: Dependence of large-scale precipitation climatologies on temporal and spatial sampling. *Journal of Climate*, **10**, 1099–1113.

HURRELL, J. W., 1995: Decadal trends in the North Atlantic Oscillation: Regional temperatures and precipitation. *Science*, **269**, 676–679.

HURRELL, J. W., 1996: Influence of variations in extratropical wintertime teleconnections on Northern Hemisphere temperature. *Geophysical Research Letters*, **23**, 665–668.

HURRELL, J. W. and van LOON, H., 1994: A modulation of the atmospheric annual cycle in the Southern Hemisphere. *Tellus*, **46A**, 325–338.

HURRELL, J. W. and van LOON, H., 1997: Decadal variations in climate associated with the North Atlantic Oscillation. *Climatic Change* **36**, 301–326.

JACOBS, G. A., HURLBURT, H. E., KINDLE, J. C., METZGER, E. J., MITCHELL, J. L., TEAGUE, W. J., and WALLCRAFT, A. J., 1994: Decade-scale trans-Pacific propagation and warming effects of an El Niño anomaly. *Nature*, **370**, 360–363.

JACOBY, G. C. and D'ARRIGO, R. D., 1990: Teak (*Tectona grandis* L. F.): A tropical species of large-scale dendroclimatic potential. *Dendrocronologia*, **8**, 83–98.

JIANG, F.-F., NEELIN, J. D., and GHIL, M., 1995: Quasi-quadrennial and quasi-biennial variability in the equatorial Pacific. *Climate Dynamics*, **12**, 101–112.

JONES, P. D. and R. J. ALLAN, 1998: Climatic change and long-term climatic variability. Chapter 9, *In* Karoly, D. J. and Vincent, D. (eds.), *Meteorology of the Southern Hemisphere*, Meteorological Monograph Vol. 27, No. 49, American Meteorological Society, Boston, 337–363.

JURY, M. R., McQUEEN, C., and LEVY, K., 1994: SOI and QBO signals in the African region. *Theoretical and Applied Climatology*, **50**, 103–115.

KAROLY, D. J., HOPE, P., and JONES, P. D., 1996: Decadal variations of the Southern Hemisphere circulation. *International Journal of Climatology*, **16**, 723–738.

KAWAMURA, R., 1994: A rotated EOF analysis of global sea surface temperature variability with interannual and interdecadal scales. *Journal of Physical Oceanography*, **24**, 707–715.

KERR, R. A., 1994: Did the tropical Pacific drive the world's warming? *Science*, **266**, 544–545.

KESTIN, T. S., KAROLY, D. J., YANO, J.-I., and RAYNER, N. A., 1998: Time-frequency variability of ENSO and stochastic simulations. *Journal of Climate*, **11**, 2258–2272.

KILADIS, G. N. and DIAZ, H. F., 1989: Global climatic anomalies associated with extremes in the Southern Oscillation. *Journal of Climate*, **2**, 1069–1090.

KLEEMAN, R., COLMAN, R. A., SMITH, N. R., and POWER, S. B., 1996: A recent change in the mean state of the Pacific basin: Observational evidence and atmospheric and oceanic responses. *Journal of Geophysical Research*, **101**, 20483–20499.

KRIPALANI, R. H. and KULKARNI, A., 1997: Rainfall variability over south-east Asia – Connections with Indian monsoon and ENSO extremes: New perspectives. *International Journal of Climatology*, **17**, 1155–1168.

LAGERLOEF, G. S. E., 1995: Interdecadal variations in the Alaska Gyre. *Journal of Physical Oceanography*, **25**, 2242–2258.

LAMB, P. J. and PEPPLER, R. A., 1991: West Africa. *In* Glantz, M. H., Katz, R. W., and Nicholls, N. (eds.), *Teleconnections Linking Worldwide Climate Anomalies*. New York: Cambridge University Press, 121–189.

LAMB, P. J. and PEPPLER, R. A., 1992: Further case studies of tropical Atlantic surface atmospheric and oceanic patterns associated with sub-Saharan drought. *Journal of Climate*, **5**, 476–488.

LATIF, M. and BARNETT, T. P., 1994a: Causes of decadal climate variability over the North Pacific and North America. Max-Planck-Institut für Meteorologie, Report No. 187, 43 pp.

LATIF, M. and BARNETT, T. P., 1994b: Causes of decadal climate variability over the North Pacific and North America. *Science*, **266**, 634–637.

LATIF, M., GROETZNER, A., MUNNICH, M., MAIER-REIMER, E., VENZKE, S., and BARNETT, T. P., 1996: A mechanism of decadal climate variability. Max-Planck-Institut fur Meteorologie, Report No. 187, 43 pp.

LATIF, M., KLEEMAN, R., and ECKERT, C., 1995: Greenhouse warming, decadal variability, or El Niño? An attempt to understand the anomalous 1990s. Max-Planck-Institut fur Meteorologie, Report No. 175, 51 pp.

LATIF, M., KLEEMAN, R., and ECKERT, C., 1997: Greenhouse warming, decadal variability, or El Niño? An attempt to understand the anomalous 1990s. *Journal of Climate*, **10**, 2221–2239.

LAU, K.-M. and SHEU, P. J., 1988: Annual cycle, Quasi-Biennial Oscillation, and Southern Oscillation in global precipitation. *Journal of Geophysical Research*, **93**, 10975–10988.

LAU, K.-M. and SHEU, P. J., 1991: Teleconnections in global rainfall anomalies: Seasonal to inter-decadal time scales. *In* Glantz, M. H., Katz, R. W., and Nicholls, N. (eds.), *Teleconnections*

Linking Worldwide Climate Anomalies. New York: Cambridge University Press, 227–256.

LEJENAS, H., 1995: Long term variations of atmospheric blocking in the Northern Hemisphere. *Journal of the Meteorological Society of Japan*, **73**, 79–88.

LINDESAY, J. A. and VOGEL, C. H., 1990: Historical evidence for Southern Oscillation–southern African rainfall relationships. *International Journal of Climatology*, **10**, 679–689.

LIU, W. T., TANG, W., and FU, L.-L., 1995: Recent warming event in the Pacific may be an El Niño. *EOS, Transactions of the American Geophysical Union*, **76**, 429 and 437.

MANN, M. E. and PARK, J., 1993: Spatial correlations of interdecadal variation in global surface temperatures. *Geophysical Research Letters*, **20**, 1055–1058.

MANN, M. E. and PARK, J., 1994: Global-scale modes of surface temperature variability on interannual to century timescales. *Journal of Geophysical Research*, **99**, 25819–25833.

MANN, M. E. and PARK, J., 1996: Joint spatiotemporal modes of surface temperature and sea level pressure variability in the Northern Hemisphere during the last century. *Journal of Climate*, **9**, 2137–2162.

MANN, M. E. and PARK, J., 1999: Oscillatory spatiotemporal signal detection in climate studies. *Advances in Geophysics*, **41**, 1–131.

MANN, M. E., PARK, J., and BRADLEY, R. S., 1995: Global interdecadal and century-scale climate oscillations during the past five centuries. *Nature*, **378**, 266–270.

MANTUA, N. J., HARE, S. R., ZHANG, Y., WALLACE, J. M., and FRANCIS, R. C., 1997: A Pacific interdecadal climate oscillation with impacts on salmon production. *Bulletin of the American Meteorological Society*, **78**, 1069–1079.

MARENGO, J. A., 1995: Variations and change in South American streamflow. *Climatic Change*, **31**, 99–117.

MASON, S. J., 1990: Temporal variability of sea surface temperatures around southern Africa: A possible forcing mechanism for the 18-year rainfall oscillation? *South African Journal of Science*, **86**, 243–252.

MASON, S. J., 1995: Sea surface temperature–South African rainfall associations, 1910–1989. *International Journal of Climatology*, **15**, 119–135.

MASON, S. J. and LINDESAY, J. A., 1993: A note on the modulation of Southern Oscillation–southern African rainfall associations with the Quasi-Biennial Oscillation. *Journal of Geophysical Research*, **98**, 8847–8850.

MASON, S. J. and TYSON, P. D., 1992: The modulation of sea surface temperature and rainfall associations over southern Africa with solar activity and the Quasi-Biennial Oscillation. *Journal of Geophysical Research*, **97**, 5847–5856.

MATTICE, W. A., 1935: Dust storms, November 1933 to May 1934. *Monthly Weather Review*, **63**, 53–55.

McPHADEN, M. J., 1993: TOGA-TAO and the 1991–93 El Niño–Southern Oscillation event. *Oceanography*, **6**, 36–44.

MILLER, A. J., CAYAN, D. R., BARNETT, T. P., GRAHAM, N. E., and OBERHUBER, J. M., 1994a: Interdecadal variability of the Pacific Ocean: Model response to observed heat flux and wind stress anomalies. *Climate Dynamics*, **9**, 287–302.

MILLER, A. J., CAYAN, D. R., BARNETT, T. P., GRAHAM, N. E., and OBERHUBER, J. M., 1994b: The 1976–77 climate shift of the Pacific Ocean. *Oceanography*, **7**, 21–26.

MINGLI, Z., 1993: Variation of precipitation over eastern China during the last 100 years. *Chinese Journal of Atmospheric Science*, **17**, 247–258.

MINGLI, Z. and FLETCHER, J. O., 1993: Variation of global ocean cloudiness during 1900–1990 and its connection with sea surface temperature. *In* Ye, D., Zeng, Q., Zhang, R., Matsuno, T., and Huang, R. (eds.), *Climate Variability, Proceedings of International Workshop on Climate Variabilities* (IWCV). Beijing, China: Chinese Meteorological Press, 36–41.

MINOBE, S., 1997: A 50–70 year climatic oscillation over the North Pacific and North America. *Geophysical Research Letters*, **24**, 683–686.

MIYAKODA, K., NAVARRA, A., and WARD, M. N., 1999: Tropical-wide teleconnection and oscillation: Part II, the ENSO-monsoon system. *Quarterly Journal of the Royal Meteorological Society*, **125**, 2937–2964.

MOOLEY, D. A. and PARTHASARATHY, B., 1984: Fluctuations in all-India summer monsoon rainfall during 1871–1978. *Climatic Change*, **6**, 287–301.

MORON, V., VAUTARD, R., and GHIL, M., 1998: Trends, interdecadal and interannual oscillations in global sea surface temperatures. *Climate Dynamics*, **14**, 545–569.

NAKAMURA, H., LIN, G., and YAMAGATA, T., 1997: Decadal climate variability in the North Pacific during the recent decades. *Bulletin of the American Meteorological Society*, **78**, 2215–2225.

NAVARRA, A., WARD, M. N., and MIYAKODA, K., 1999: Tropical-wide teleconnection and oscillation: Part I, Teleconnection indices and Type I/Type II states. *Quarterly Journal of the Royal Meteorological Society*, **125**, 2909–2936.

NICHOLLS, N., 1989: Sea surface temperatures and Australian winter rainfall. *Journal of Climate*, **2**, 965–973.

NICHOLLS, N., DROSDOWSKY, W., and LAVERY, B., 1997: Australian rainfall variability and change. *Weather*, **52**, 66–72.

NICHOLLS, N., LAVERY, B., FREDERIKSEN, C., and DROSDOWSKY, W., 1996: Recent apparent changes in relationships between the El Niño–Southern Oscillation and Australian rainfall and temperature. *Geophysical Research Letters*, **23**, 3357–3360.

NICHOLSON, S. E., 1985: Sub-Saharan rainfall 1981–84. *Journal of Climate and Applied Meteorology*, **24**, 1388–1391.

NICHOLSON, S. E., 1989: Long-term changes in African rainfall. *Weather*, **44**, 46–56.

NICHOLSON, S. E., 1993: An overview of African rainfall fluctuations of the last decade. *Journal of Climate*, **6**, 1463–1466.

NICHOLSON, S. E., BA, M. B., and KIM, J. Y., 1996: Rainfall in the Sahel during 1994. *Journal of Climate*, **9**, 1673–1676.

NITTA, T., 1993: Interannual and decadal scale variations of atmospheric temperature and circulations. *In* Ye, D., Zeng, Q., Zhang, R., Matsuno, T., and Huang, R. (eds.), *Climate Variability, Proceedings of International Workshop on Climate Variabilities* (IWCV). Beijing, China: Chinese Meteorological Press, 15–22.

NITTA, T. and KACHI, M., 1994: Interdecadal variations in precipitation over the tropical Pacific and Indian Oceans. *Journal of the Meteorological Society of Japan*, **72**, 823–830.

NITTA, T. and YAMADA, S., 1989: Recent warming of tropical sea surface temperature and its relationship to the Northern Hemisphere circulation. *Journal of the Meteorological Society of Japan*, **67**, 375–383.

NITTA, T. and YOSHIMURA, J., 1993: Trends and interannual and interdecadal variations of global land surface air temperature. *Journal of the Meteorological Society of Japan*, **71**, 367–374.

THOMPSON, D. W. J. and WALLACE, J. M., 1998: The Arctic Oscillation signature in the wintertime geopotential height and temperature fields. *Geophysical Research Letters*, **25**, 1297–1300.

TORRENCE, C. and COMPO, G. P., 1998: A practical guide to wavelet analysis. *Bulletin of the American Meteorological Society*, **79**, 61–78.

TORRENCE, C. and WEBSTER, P. J., 1998: The annual cycle of persistence in the El Niño/Southern Oscillation. *Quarterly Journal of the Royal Meteorological Society*, **124**, 1985–2004.

TORRENCE, C. and WEBSTER, P. J., 1999: Interdecadal changes in the ENSO-monsoon system. *Journal of Climate*, **12**, 2679–2690.

TOURRE, Y. M. and WHITE, W. B., 1995: ENSO signals in global upper-ocean temperature. *Journal of Physical Oceanography*, **25**, 1317–1332.

TRENBERTH, K. E., 1990: Recent observed interdecadal climate changes in the Northern Hemisphere. *Bulletin of the American Meteorological Society*, **71**, 988–993.

TRENBERTH, K. E., 1998: Forecasts of the development of the 1997–98 El Niño event. *Seventh Session of the CLIVAR SSG,* Santiago, Chile, April 1998, 11 pp.

TRENBERTH, K. E. and HOAR, T. J., 1996: The 1990–1995 El Niño Southern Oscillation event. *Geophysical Research Letters*, **23**, 57–60.

TRENBERTH, K. E. and HOAR, T. J., 1997: El Niño and climate change. *Geophysical Research Letters*, **24**, 3057–3060.

TRENBERTH, K. E. and HURRELL, J. W., 1994: Decadal atmosphere–ocean variations in the Pacific. *Climate Dynamics*, **9**, 303–319.

TYSON, P. D., 1986: *Climatic Change and Variability in Southern Africa.* Cape Town: Oxford University Press, 220 pp.

UCAR Quarterly, 1997: El Niño and global warming: What's the connection? **24**(1), 8–9.

van LOON, H., KIDSON, J. W., and MULLAN, A. B., 1993: Decadal variation of the annual cycle in the Australian dataset. *Journal of Climate*, **6**, 1227–1231.

VENEGAS, S. A., MYSAK, L. A., and STRAUB, D. N., 1996: Evidence for interannual and interdecadal climate variability in the South Atlantic. *Geophysical Research Letters*, **23**, 2673–2676.

VIJAYAKUMAR, R. and KULKARNI, J. R., 1995: The variability of the interannual oscillations of the Indian summer monsoon rainfall. *Advances in Atmospheric Science*, **12**, 95–102.

WALKER, G. T., 1910: On the meteorological evidence for supposed changes of climate in India. *Memoirs of the India Meteorological Department*, **XXI**, Part I, 21 pp.

WANG, B., 1995: Interdecadal changes in El Niño onset in the last four decades. *Journal of Climate*, **8**, 267–285.

WANG, B. and WANG, Y., 1996: Temporal structure of the Southern Oscillation as revealed by waveform and wavelet analysis. *Journal of Climate*, **9**, 1586–1598.

WANG, X. L. and ROPELEWSKI, C. F., 1995: An assessment of ENSO-scale secular variability. *Journal of Climate*, **8**, 1584–1599.

WARD, M. N., 1992: Provisionally corrected surface wind data, worldwide ocean–atmosphere surface fields, and Sahelian rainfall variability. *Journal of Climate*, **5**, 454–475.

WARD, M. N., FOLLAND, C. K., MASKELL, K., OWEN, J. A., and ROWELL, D. P., 1990: Forecasting Sahel rainfall: An update. *Weather*, **45**, 122–125.

WARD, M. N, MASKELL, K., FOLLAND, C. K., ROWELL, D. P., and WASHINGTON, R., 1994: A tropic-wide oscillation of boreal summer rainfall and patterns of sea surface temperature. Climate Research Technical Note CRTN48, Hadley Centre, Meteorological Office, Bracknell, United Kingdom, 29 pp.

WARRICK, R. A., 1980: Drought in the Great Plains: A case study of research on climate and society in the USA. *In* Ausubel, J. and Biswas, A. K. (eds.), *Climatic Constraints on Human Activities*. Oxford: Pergamon Press, 93–123.

WEBSTER, P. J., 1994: The role of hydrological processes in ocean–atmosphere interactions. *Reviews of Geophysics*, **32**, 427–476.

WEBSTER, P. J., 1995: The annual cycle and the predictability of the tropical coupled ocean–atmosphere system. *Meteorology and Atmospheric Physics*, **56**, 33–55.

WEBSTER, P. J. and PALMER, T. N., 1997: The past and the future of El Niño. *Nature*, **390**, 562–564.

WEBSTER, P. J. and YANG, S., 1992: Monsoon and ENSO: Selectively interactive systems. *Quarterly Journal of the Royal Meteorological Society*, **118**, 877–926.

WHETTON, P. H., ADAMSON, D., and WILLIAMS, M. A. J., 1990: Rainfall and river flow variability in Africa, Australia and East Asia linked to El Niño–Southern Oscillation events. *In* Bishop, P., (ed.), Lessons for human survival: Nature's record from the Quaternary. *Geological Society of Australia Symposium Proceedings*, **1**, 71–82.

WHETTON, P. H. and RUTHERFURD, I., 1994: El Niño–Southern Oscillation teleconnections in the Eastern Hemisphere over the last 500 years. *Climatic Change*, **28**, 221–253.

WHITE, W. B. and CAYAN, D. R., 1998: Quasi-periodicity and global symmetries in inter-decadal upper ocean temperature variability. *Journal of Geophysical Research*, **103**, 21335–21354.

WHITE, W. B., CHEN, S.-C., and PETERSON, R., 1998: The Antarctic Circumpolar Wave: A beta-effect in ocean–atmosphere coupling over the Southern Ocean. *Journal of Physical Oceanography*, **28**, 2345–2361.

WHITE, W. B. and PETERSON, R., 1996: An Antarctic circumpolar wave in surface pressure, wind, temperature, and sea ice extent. *Nature*, **380**, 699–702.

WILBY, R., 1993: Evidence of ENSO in the synoptic climate of the British Isles since 1880. *Weather*, **48**, 234–239.

WORSTER, D., 1979: *Dust Bowl: The Southern Plains in the 1930's*. New York: Oxford University Press.

XU, J., ZHU, Q., and SUN, Z., 1998: Interrelation between East-Asian winter monsoon and Indian/Pacific SST with the interdecadal variation. *Acta Meteorologica Sinica*, **12**, 275–287.

YASUNARI, T., 1985: Zonally propagating modes of the global east-west circulation associated with the Southern Oscillation. *Journal of the Meteorological Society of Japan*, **63**, 1013–1029.

YASUNARI, T., 1987a: Global structure of the El Niño/Southern Oscillation. Part I. El Niño composites. *Journal of the Meteorological Society of Japan*, **65**, 67–80.

YASUNARI, T., 1987b: Global structure of the El Niño/Southern Oscillation. Part II. Time evolution. *Journal of the Meteorological Society of Japan*, **65**, 81–102.

YE, D. and YAN, Z., 1993: Climatic jumps in history. *In* Ye, D., Zeng, Q., Zhang, R., Matsuno, T., and Huang, R. (eds.), *Climate Variability, Proceedings of International Workshop on Climate Variabilities* (IWCV). Beijing China: Chinese Meteorological Press, 3–14.

ZHANG, Y., WALLACE, J. M., and BATTISTI, D. S., 1997: ENSO-like interdecadal variability: 1900–1993. *Journal of Climate*, **10**, 1004–1020.

2

Understanding and Predicting Extratropical Teleconnections Related to ENSO

MARTIN P. HOERLING

National Oceanic and Atmospheric Administration, Oceanic and Atmospheric Research, Climate Diagnostics Center, Boulder, Colorado 80303, U.S.A.

ARUN KUMAR

National Oceanic and Atmospheric Administration, National Centers for Environmental Prediction, Environmental Modelling Center, Camp Springs, Maryland 20746, U.S.A.

Abstract

It is now well established that the El Niño/Southern Oscillation (ENSO) impacts the climate of middle and high latitudes. Unanswered questions remain, however, and their resolution is central to assessing and fully harvesting the atmospheric predictability inherent in the ENSO phenomenon. Among these questions are the sensitivity of the extratropical response to the annual cycle, the nonlinearity of that response with respect to the sign and amplitude of tropical sea surface temperature (SST) anomalies, and the further sensitivity of the response to differences in the SST anomaly patterns that distinguish one El Niño event from another. Beyond these problems of seasonal to interannual variability, it is also important to understand multidecadal-scale variations of ENSO impacts and to assess their origins.

Our inquiry into these problems begins with an analysis of the last half-century of observed circulation data, using the upper tropospheric flow patterns to identify the teleconnections that link the tropics and the extratropics during ENSO. Several new aspects of the observed teleconnection behavior are highlighted; however, the data archive is undoubtedly too brief to offer a complete or even an accurate sample of the spectrum of atmospheric sensitivity to ENSO. Nor is it likely that the full spectrum of tropical SST variations themselves have been sampled by observations. We thus provide additional analyses based on the results of atmospheric general circulation model experiments that have been forced either with the recent 50-year record of observed global SST variations or with idealized SST anomalies. These analyses, when combined with the observational analyses, yield a richer image of teleconnections related to ENSO and also provide a means to quantify the potential for predicting such patterns.

Introduction

The atmosphere thousands of kilometers remote from the equatorial Pacific is sensitive to the warming of those waters during El Niño, a circumstance forming the primary basis for extratropical seasonal climate predictability. It is apparent, however, that not all extratropical regions are sensitive to El Niño, nor is its impact uniform throughout the year. Furthermore, different El Niño events have been accompanied by an assortment of climate anomalies in the extratropics. How do we understand this rich variety of behavior, and importantly, what component of it is predictable?

In the current chapter's treatment of these issues, it is helpful to keep in mind a simple picture of the origin for teleconnections associated with El Niño/Southern Oscillation (ENSO). A more detailed dynamical description can be found in the cited references, and a general survey appears in the book *Teleconnections Linking Worldwide Climate Anomalies*, edited by Glantz et al. (1991). The recently completed Tropical Ocean–Global Atmosphere (TOGA) program has significantly improved the understanding of teleconnections linked to ENSO, and the reader is also referred to Trenberth et al. (1998) for a review of that program's accomplishments.

For our purpose, imagine a chain of atmospheric processes, with each link in the chain carrying information from the local vicinity of the El Niño sea surface temperature (SST) anomalies throughout the global climate system. The first link is the tropical response of rain-producing cumulonimbus clouds – critical because these clouds are the principal agents for exchanging heat from the Earth's surface, and thereby communicating El Niño's presence to the free atmosphere. Anomalous warming of the east equatorial Pacific Ocean acts to increase the precipitation in that region, and this occurs at the expense of rainfall elsewhere, often resulting in drought over the western Pacific archipelago (e.g., Ropelewski and Halpert 1987, 1989; Kiladis and Diaz 1989).

The second link in the chain is the horizontal communication of El Niño's presence, and this occurs through the atmosphere's sensitivity to shifts in tropical rainfall. Observations reveal a wave train having alternating low and high pressure that follows a "great circle" route in the upper troposphere; its emanation from the region of anomalous equatorial Pacific precipitation can be viewed as the forced response of the atmosphere to the new tropical energy sources and sinks associated with El Niño (Bjerknes 1969; Gill 1980; Hoskins and Karoly 1981; Webster 1981; Simmons 1982). Its centers of action are located over both the Pacific–North American (PNA) (e.g., Horel and Wallace 1981) and Pacific–South American (PSA) (e.g., Karoly 1989) regions, and associated with them are shifts in the midlatitude jet stream.

This nominally direct response alters the course of storms (cyclones and anticyclones) that determine the daily weather fluctuations at higher latitudes. Interactions between storms and the stationary waves during the course of a season are important for determining the overall character of the teleconnection response (Kok and Opsteegh 1985; Held et al. 1989; Hoerling and Ting 1994). Indeed, changes in the storm tracks are the immediate cause for precipitation and temperature anomalies experienced over the extratropics during ENSO.

We will focus our attention on the upper tropospheric manifestations of ENSO-related teleconnections because they provide the clearest view of global linkages in climate anomalies. We begin with the annual cycle of observed teleconnections and demonstrate that, despite the typically slow evolution of the tropical SST anomalies, large seasonal changes in teleconnections exist. Furthermore, contrary to the conventional view of atmospheric teleconnections being equal and opposite for extreme phases of ENSO, the separate analysis of warm and cold events reveals departures from such a simple linear view. We explore the robustness of the teleconnection response itself, on the one hand assessing its reproducibility from one El Niño event to another, and on the other hand assessing its statistical stationarity on multidecadal timescales.

In many regards, brevity of the instrumental record hinders our ability to understand the teleconnection response to ENSO. This is particularly true with respect to the more subtle, but acute, problems such as month-to-month variations in teleconnections, their sensitivity to the SST variations that distinguish one ENSO event from another, and decadal- to century-scale changes in the teleconnection patterns. Thus, to supplement observations, we will discuss the results from atmospheric general circulation model (GCM) simulations. In these simulations, the observed post-1950 record of monthly evolving SSTs has been imposed as a boundary condition, and a 12-member ensemble of such forced experiments has been constructed. By providing a multiplicity of "perfect SST analogs" for many warm and cold events, we attempt to identify the extent to which the teleconnections themselves are sensitive to different patterns of tropical forcing. Also, the runs yield 600 years of data in total, making it feasible to address the problem of longer timescale variations. Finally, the experiments will also be used to measure the potential for predictability itself.

Data

Sea surface temperatures are analyzed for the period 1950–97 by using a combination of two data sets. A new monthly data set for the 1950–81 period uses in situ observations together with statistical eigenvector reconstruction to estimate SST values at all ocean points on a 2° latitude/longitude grid (Smith et al. 1996). After 1982, in situ measurements combined with satellite observations are adequate to resolve the global SST field, and monthly mean analyses for the period 1982–97 are available on a 1° latitude/longitude grid as described by Reynolds and Smith (1994).

The principal mode of interannual SST variability in the tropical Pacific Ocean is associated with ENSO, and the spatial pattern of that variability is succinctly described by the first empirical orthogonal function (EOF1) of the SSTs shown in the bottom panel of Figure 2.1. The EOF analysis is of the covariance matrix that consists of 576 monthly time samples of anomalous SSTs from January 1950 to December 1997, and the analysis domain is 30°N–30°S, 120°E–60°W. The EOF1 explains 45% of the total SST variance over this domain. The top panel of Figure 2.1 shows the principal component (PC) time series of EOF1, with large positive (negative) departures indicative of warm (cold) event conditions.

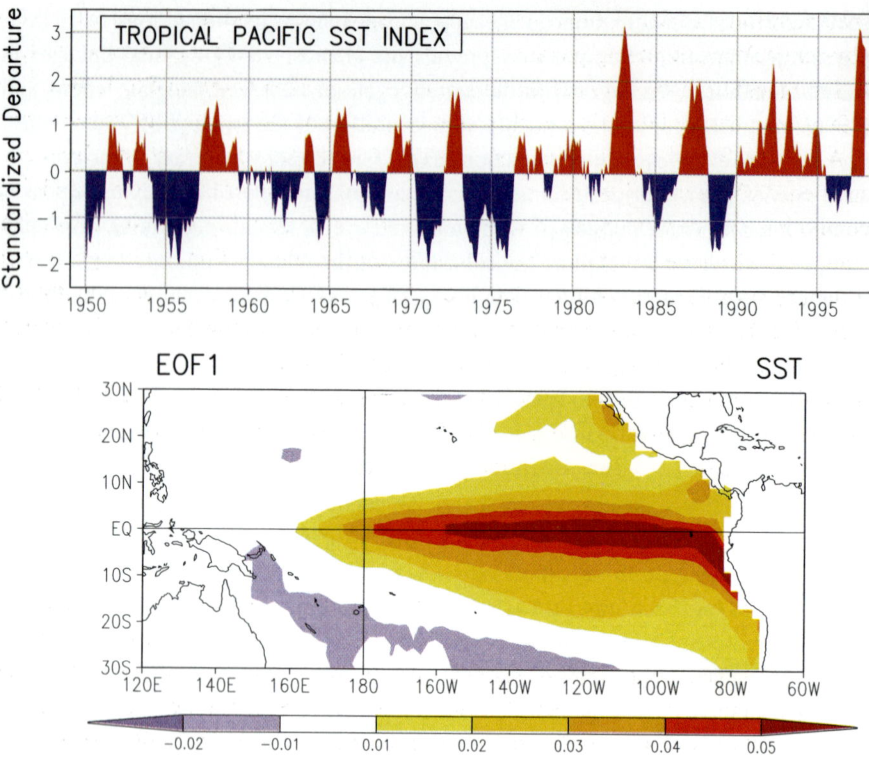

Fig. 2.1 The PC time series (top panel) and spatial structure function (bottom panel) of the first EOF of observed monthly averaged tropical Pacific SST variability. The analysis domain is 30°N–30°S, 120°E–60°W, and the analysis period is for all months during January 1950–December 1997. The time series displays standardized anomalies, with positive (negative) departures indicative of warm (cold) events. The contour labels used to display the structure function are dimensionless.

To examine teleconnection patterns associated with ENSO, we use the National Centers for Environmental Prediction/National Center for Atmospheric Research (NCEP/NCAR) reanalysis of 500 hPa heights available for 1958–97. An advantage of these data is their global extent, allowing for the first time an extensive historical analysis of the Southern Hemisphere (SH) teleconnections. A longer 500 hPa height data set from the NCEP covers the 1947–97 period. These data are available for areas northward of 20°N, and where appropriate they will be used to augment our analysis of the Northern Hemisphere (NH) teleconnections.

Estimates of tropical precipitation anomalies are derived from remote sensing data using polar orbiting satellite measurements of outgoing long-wave radiation (OLR). This data set is available for the period from March 1974 to the present as described in Liebmann and Smith (1996).

Atmospheric general circulation model simulations for 1950–95 are analyzed. The model has been forced at the lower boundary with the monthly evolving global SSTs as observed for this period by using the aforementioned SST data sets. The GCM has T40 spectral resolution with 18 unequally spaced sigma levels, and it employs a complete physics package, including prediction of clouds, shallow convection, deep convection, and turbulent exchange of heat and momentum. The reader is referred to Kumar et al. (1996) for a detailed description of the model, including its climatological behavior.

A 12-member ensemble of such simulations for the period 1950–95 has been constructed. These simulations differ only in their specification of the atmospheric initial conditions, and each realization experiences the same monthly evolving SSTs. The simulated teleconnection patterns will be compared to observations, although in some instances suitable observational data are simply unavailable. It is thus important to recognize the inevitable model dependency of some results shown here. It is expected that similar studies using a host of different GCMs will offer a more complete and definitive assessment of the sensitivity of teleconnection responses to ENSO. Additional experiments using idealized SST anomalies have also been performed with the same GCM; these are described in the results section.

Observed Teleconnections

The annual cycle of observed teleconnections is determined by regressing the monthly 500 hPa heights onto the equatorial Pacific SST index of Figure 2.1 for the 1958–97 period. This analysis procedure summarizes the linear relationship between the atmospheric circulation and SST. The results are shown in Figure 2.2 for both the PNA and PSA sectors, where such relationships are best defined. The patterns are displayed for two-month "seasons" beginning in July/August (top) and ending in May/June (bottom). Positive (negative) height anomalies are shown in solid (dashed) contours, and the phase of the circulation patterns are those associated with tropical Pacific warm events. The height anomaly amplitudes are those associated with a +1 standard deviation anomaly of the SST index to represent the warm-event phase. In this linear construct, the pattern associated with cold events would be the exact opposite. In subsequent sections, we will relax this linear constraint by analyzing the teleconnection patterns for warm and cold events separately using composite techniques.

Within the Northern Hemisphere (left side of Fig. 2.2), a distinct annual cycle of ENSO-related teleconnections occurs, with a weak response in boreal summer and a strong response in boreal winter. This cycle is consistent with the annual cycle of the climatological height variance itself, which also exhibits a summer minimum and a winter maximum. It is important to emphasize, however, that even at its wintertime peak, the extratropical response is considerably weaker than the overall interannual variability of 500 hPa heights. Hence, a significant fraction of the total variance of seasonal means in the extratropics is unaccounted for by the linear ENSO signal.

The phase of the teleconnection pattern is somewhat less variable from season to season than is its amplitude. In particular, a low-pressure anomaly (denoted by the symbol L) over the North Pacific is a characteristic feature of the response throughout the annual cycle. This center of action forms part of a wave train that arches from the tropical Pacific across North America. There is some indication of a modest eastward

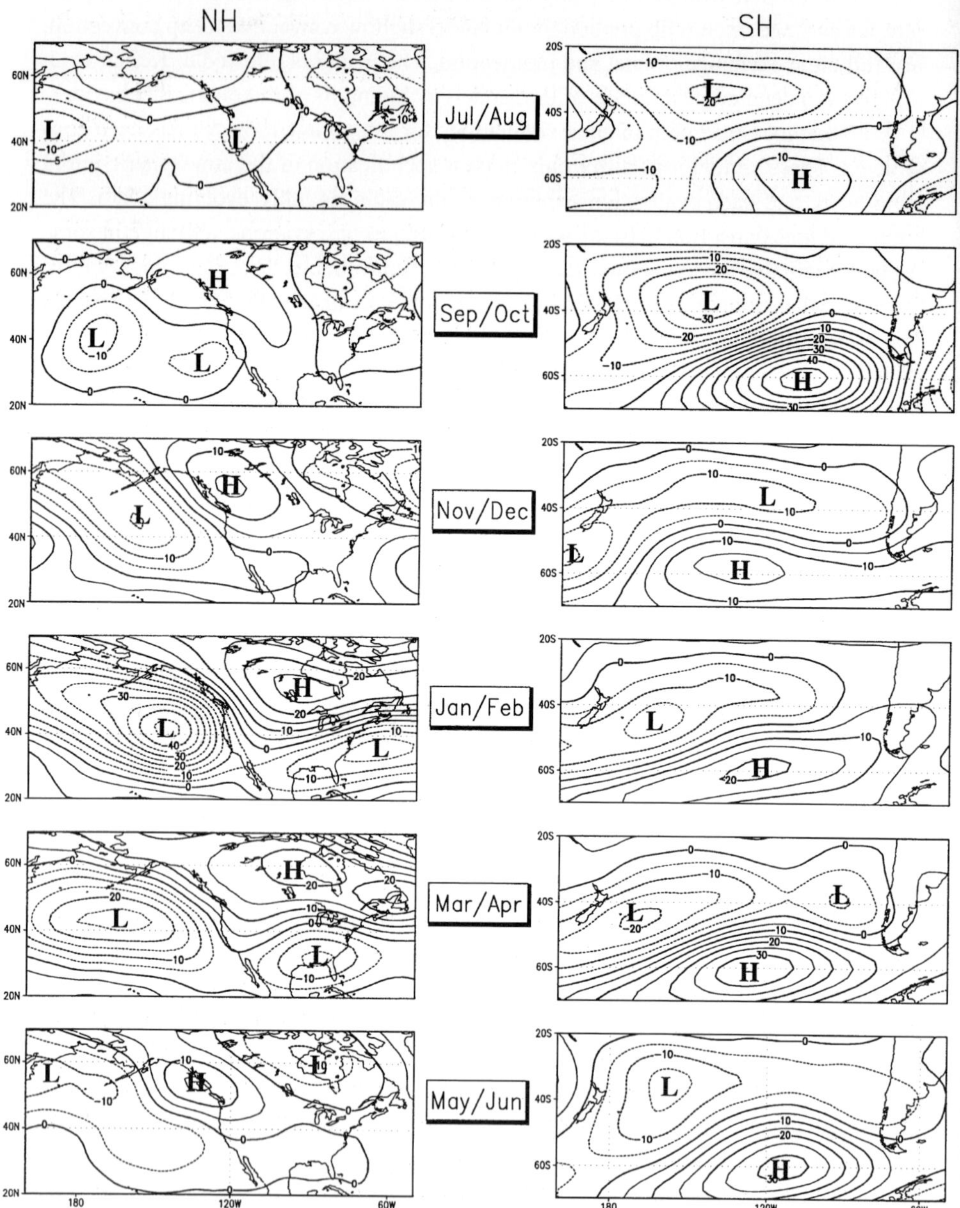

Fig. 2.2 Observed 500 hPa height patterns that are linearly related to EOF1 of tropical Pacific SST variations. The monthly averaged height anomalies have been regressed onto the PC time series of Figure 2.1 for each calendar month separately for the period January 1958–December 1997. Shown are consecutive two-month averages, with contours denoting the height anomalies associated with a +1 standard deviation of the EOF1 PC representative of warm-event conditions. Solid (dashed) contours denote high- (low-) pressure anomalies, and centers of high (low) pressure are denoted with the symbol H (L). The contour interval is 5 m.

shift of this wave train from fall to winter; note in particular that low-pressure anomalies over eastern Canada in fall are replaced by high-pressure anomalies (denoted by the symbol H) by winter/spring.

Within the Southern Hemisphere (right side of Fig. 2.2), the ENSO-related teleconnections exhibit much less seasonal variability. This subdued sensitivity to the annual cycle is entirely consistent with the small annual cycle of the Southern Hemisphere climatological circulation. Nonetheless, several noteworthy seasonal variations in amplitude occur. First, the weakest response throughout the year takes place during austral winter (July/August), in sharp contrast to the Northern Hemisphere behavior. Second, the largest amplitude response occurs in early austral spring (September/October). At that time, the height anomalies are double their wintertime counterparts. The El Niño composite analysis of Kiladis and Mo (1998) also shows this tendency for the largest amplitude 500 hPa height anomaly to occur in the austral spring season, and a similar characteristic of the seasonal variability of the SH response is found in sea level pressure (SLP) (van Loon and Shea 1987).

For the Pacific sector as a whole, there is considerable symmetry of the teleconnection response with respect to the equator, a feature previously emphasized in Karoly's (1989) case study analysis of three warm events. For example, low-pressure anomalies over the South Pacific during all months have mirrored counterparts over the North Pacific. Likewise, an apparent "turning" of wave trains occurs at subpolar latitudes in both hemispheres, and high pressure near 60° latitude marks their poleward extent. Thus, aside from the different seasonal cycles, this symmetry indicates that similar processes are responsible for the spatial structure of teleconnections in both hemispheres.

Observed Tropical Forcing of Teleconnections

The effective tropical forcing of the teleconnections is not the SST anomalies per se, but rather the influence of the SST on the tropical rainfall patterns. Figure 2.3 shows the annual cycle of observed tropical rainfall anomalies linearly related to the SST variations. This cycle is determined by regressing the monthly OLR onto the equatorial Pacific SST index for 1974–97. The empirical relation of Arkin and Meisner (1987) has been used to convert the satellite radiance anomalies into equivalent rainfall rates. Red (blue) shading indicates areas of enhanced (suppressed) rainfall. As in Figure 2.2, the rainfall anomalies are those associated with a +1 standard deviation anomaly of the SST index to represent the warm-event phase.

The rainfall response is dominated by a Pacific dipole pattern. During warm events, this pattern is characterized by increased precipitation in the central and eastern Pacific but reduced precipitation over the western Pacific and maritime continent. Little variation in this pattern is observed throughout the year – namely, the Pacific dipole structure is intact for all seasons. The large annual cycle of Northern Hemisphere teleconnections would thus appear to be controlled by the annual cycle in the climatological circulation itself, rather than by seasonal changes in the tropical forcing. Likewise, the comparatively small annual cycle of Southern Hemisphere teleconnections appears to be consistent with such modest seasonal changes in the tropical forcing, combined with

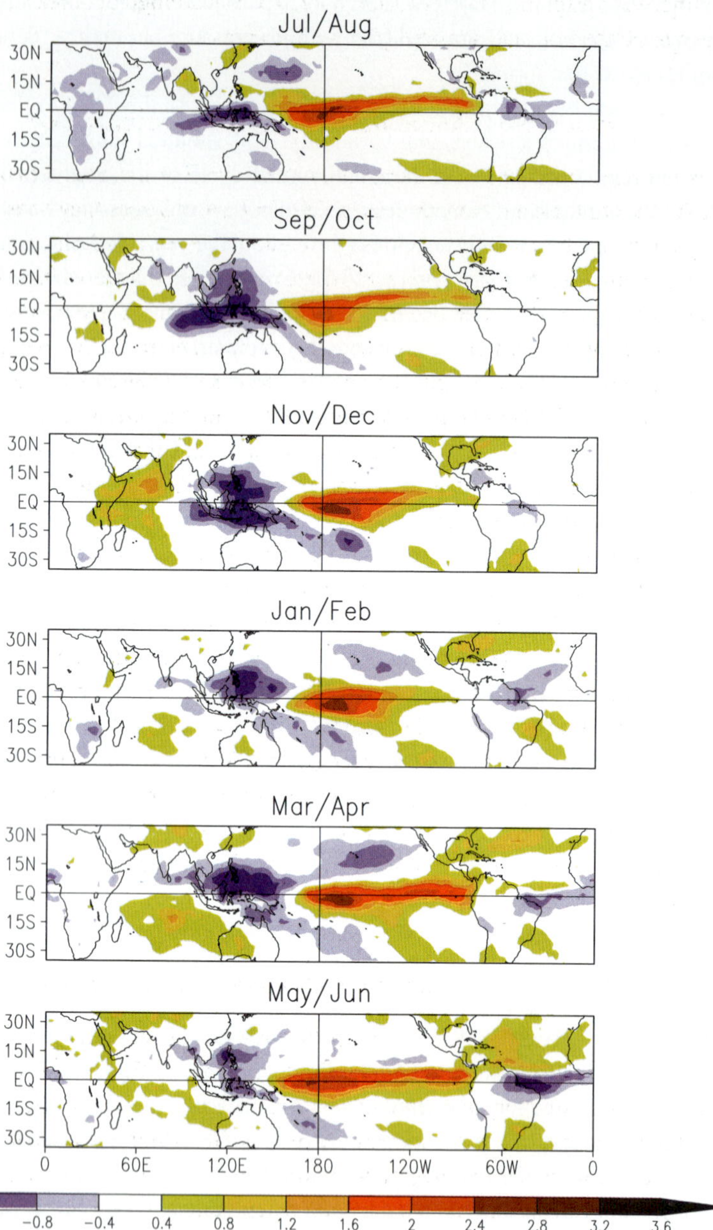

Fig. 2.3 Observed OLR-derived rainfall anomalies that are linearly related to EOF1 of tropical Pacific SST variations. The monthly averaged OLR anomalies have been regressed onto the principal component (PC) time series of Figure 2.1 for each calendar month separately for the period March 1974–December 1997. Equivalent rainfall anomalies are estimated from Arkin and Meisner's (1987) empirical formula that relates OLR to precipitation rates. Shown are consecutive two-month averages, with red (blue) shades denoting above (below) normal rainfall rates associated with a +1 standard deviation of the EOF1 PC representative of warm-event conditions. Shading interval is 0.4 mm/day.

the Southern Hemisphere's comparatively weaker climatological annual cycle, which may also mute the seasonal sensitivity of teleconnections.

Decadal-Scale Variations of Observed Teleconnections

An important question is whether the spatial patterns in Figure 2.2 are the characteristic features of the atmosphere's response to ENSO, and thus whether they constitute statistically stationary patterns. Is the recent 40-year record sufficient to define those features? What, if any, low-frequency variations of the teleconnection patterns have occurred in the instrumental record? We will address these matters in more detail in the future using the much longer record of teleconnections generated in the GCM. Here, we merely illustrate the reproducibility of the observed teleconnection patterns in the two 20-year periods 1958–77 and 1978–97. For each period, the monthly circulation anomalies are calculated with respect to their own 20-year means, and these results are regressed upon the EOF1 SST index of their respective periods. The results for the Northern Hemisphere are shown in Figure 2.4.

The essential features in the teleconnections recur during the two independent records. Note especially that low-pressure anomalies over the North Pacific occur during all seasons. The characteristic wave train structure is also robust throughout the record. However, some differences are apparent, in particular the stronger springtime response in the recent twenty years and various phase shifts in the centers of the teleconnection patterns between the two periods.

Whether these and other temporal changes in teleconnections across the last half-century are physically based owing to some change in external forcing, or are merely due to sampling errors, is unclear. Suffice it to say, however, that one's impression of ENSO's impact on extratropical interannual variability can be influenced by the time period used for analysis.

Inter-ENSO Variations of Observed Teleconnections

It is important to assess whether other spatial patterns of the atmosphere's response to tropical SST forcing exist besides the linear patterns in Figure 2.2. Perhaps most amenable to analysis from the limited instrumental record is whether asymmetries exist in the teleconnection patterns with respect to the extreme phases of ENSO. For this purpose, separate 500 hPa height composites for warm and cold events have been constructed based on the one standard deviation departures of the EOF1 SST time series for 1947–97 (see Fig. 2.1). We composited a total of nine warm and nine cold events for each season, and we constructed the composites based on four-month averages, rather than the two-month averages in the previous figures. The choice of four-month averages here is to achieve a smoother display of the seasonal evolution based on such small composite sizes. The results are presented in Figure 2.5, and the calendar labels refer to the two center months of the four-month averaging interval.

The teleconnection patterns are generally opposite in sign for warm and cold composites, and thus to zeroth order, the extratropical atmosphere responds linearly with

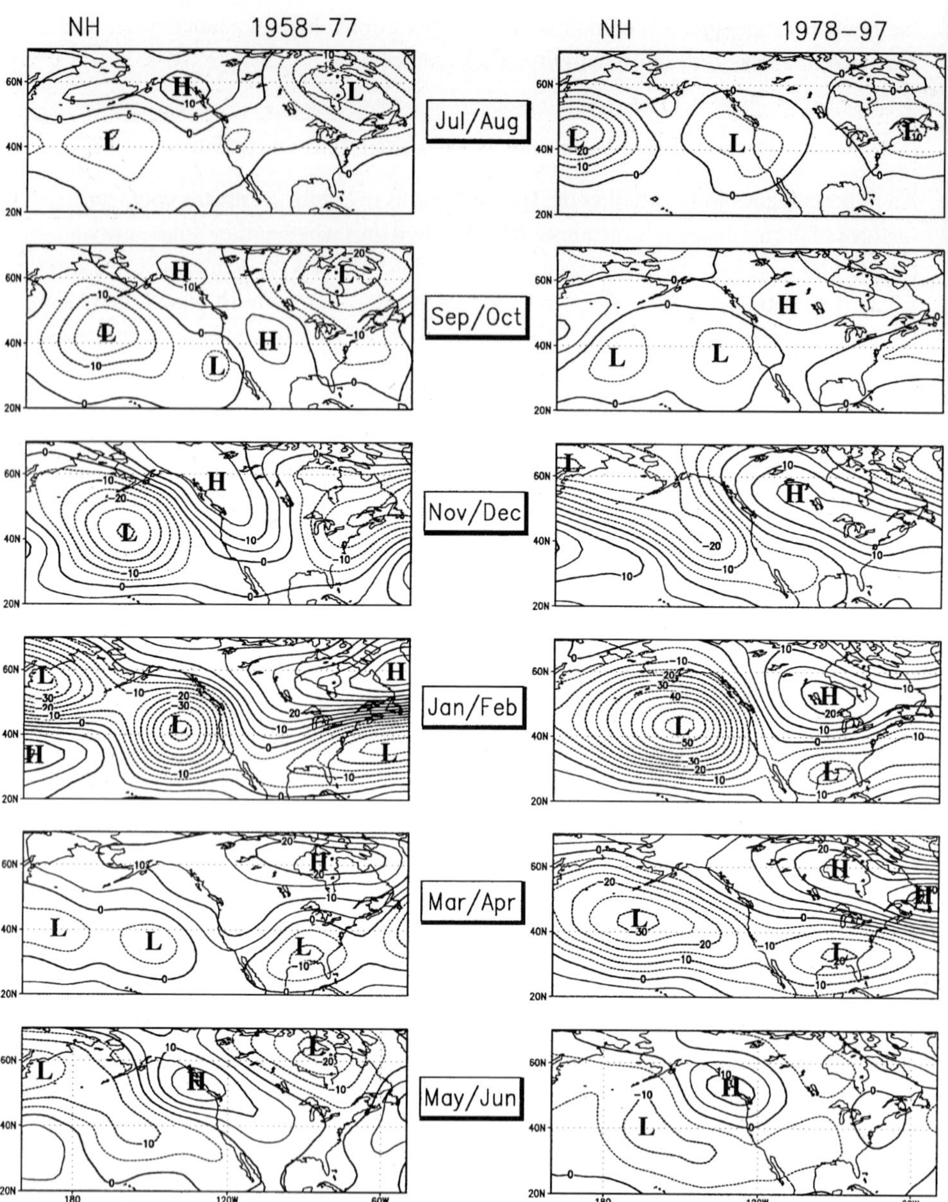

Fig. 2.4 Same as Figure 2.2, except that the observed monthly averaged height anomalies have been regressed on the PC EOF1 tropical Pacific SST variations for two separate 20-year periods: 1958–77 (left side), and 1978–97 (right side). Solid (dashed) contours denote high- (low-) pressure anomalies. The contour interval is 5 m.

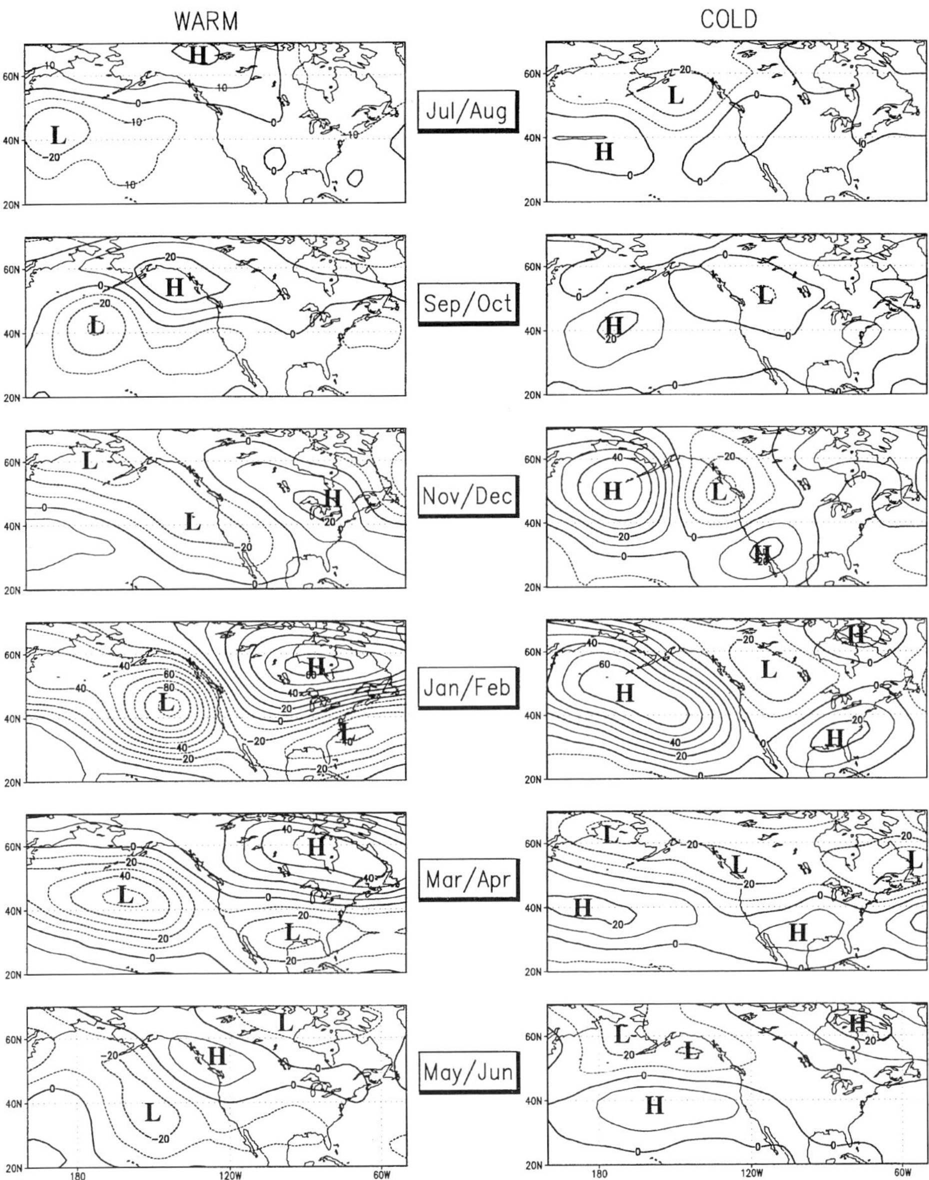

Fig. 2.5 Observed composite 500 hPa height anomalies for two-month averages for tropical Pacific warm (left side) and cold (right side) events. The composites are based on a selection of the nine strongest warm and nine strongest cold events for each two-month period during 1947–97. Solid (dashed) contours denote high- (low-) pressure anomalies. The contour interval is 10 m.

respect to ENSO's opposite phases. Note also that these composite patterns are very similar to the linear regression patterns of Figure 2.2, although no constraint of linearity exists for constructing the composites per se. The observations thus point to the possibility that the teleconnection response is largely describable by the linear association the basinwide pattern of tropical Pacific SST anomalies, and that nonlinear relations may be of secondary importance.

Some asymmetries between the warm- and cold-event composites are noteworthy, however. For example, throughout the northern cold seasons, the warm-event wave train is shifted east of its cold-event counterpart – the two patterns are nearly in quadrature with each other during the October–December period rather than being out of phase. During the core of boreal winter (December–March), the warm-event composites are also larger in amplitude. These asymmetries in the upper level circulation have surface counterparts over North America as shown by Hoerling et al. (1997) and Montroy et al. (1998).

Of further interest is whether, independent of ENSO's phase, the teleconnection patterns are also sensitive to the spatial pattern of tropical Pacific SST anomalies. For example, observations indicate large differences between the seasonal extratropical anomalies from one El Niño event to another, as demonstrated for the four different warm-event winters in Figure 2.6. Each event bears some resemblance to the El Niño composite, but the location of the primary centers of action vary greatly.

The occurrence of large inter–El Niño variations in circulation patterns raises the question of the robustness of the composite teleconnection signal itself. It is unclear whether the composite constitutes some blurred image of the atmosphere's sensitivity that results from averaging sharper patterns associated with the unique aspects of each El Niño event. Or, perhaps the composite is the one and only statistical signature of El Niño's impact, whereas variations from event to event are "climate noise" unrelated to the SSTs. The matter is once again not resolvable from observations; the brief instrumental record does not yield sufficient analogs for each El Niño that would make statistical analysis feasible.

Modeled Teleconnections

Ensemble atmospheric global circulation model (AGCM) simulations are indispensable for answering several of the aforementioned questions concerning the multi-timescale variations of ENSO teleconnections. Here we rely on one particular GCM as described earlier, and Figure 2.7 presents an appraisal of the model's overall sensitivity to tropical Pacific SST forcing. The analysis is identical to that shown in Figure 2.2, except that the model's 12-member ensemble averaged monthly height anomalies for 1950–95 have been regressed upon the EOF1 SST time series. Recall that the model is forced with the observed monthly mean global distribution of SSTs, a mode of simulation referred to as Global Ocean–Global Atmosphere (GOGA) experiments.

The GCM captures many of the observed teleconnection features. Note particularly the symmetry of responses with respect to the equator, and all seasons exhibit low-pressure anomalies over the midlatitude North and South Pacific during warm events. The model also confirms that the Northern Hemisphere response is phase locked with the annual cycle, acquiring the largest amplitude in the boreal winter/spring, whereas the Southern Hemisphere teleconnections are by comparison seasonally invariant. The

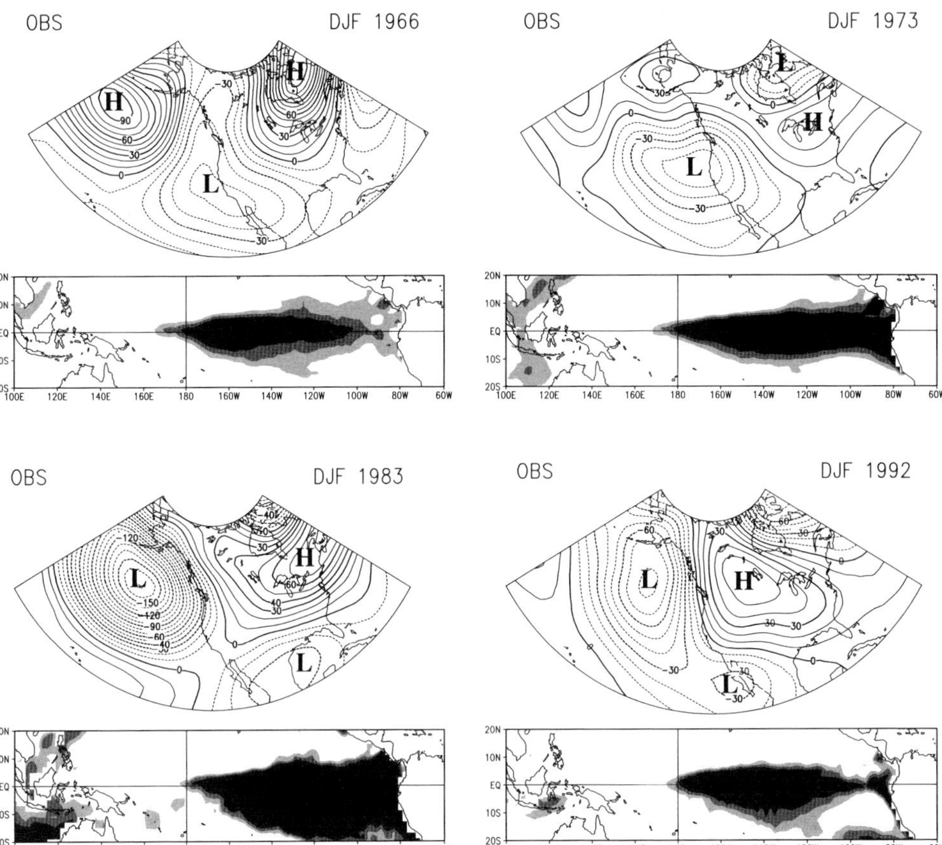

Fig. 2.6 Observed winter (December–February average) 500 hPa height (in meters) and tropical Pacific SST (in °C) anomalies during the indicated El Niño events of 1966, 1973, 1983, and 1992. (The year corresponds to January of the winter season.) Solid (dashed) contours denote high- (low-) pressure anomalies. The contour interval is 10 m. The darkest shading denotes SST anomalies of greater than 2°C.

modeled Southern Hemisphere climate response has received little attention in the literature, and it is thus encouraging to note the considerable agreement between GCM and observed teleconnections over the South Pacific and areas adjacent to South America. The GCM teleconnections, in particular the region of high pressure immediately west of the Drake Passage, provide dynamical confirmation for the observed preference for blocking activity in this region during warm events (Rutllant and Fuenzalida 1991).

Decadal-Scale Variations of Modeled Teleconnections

Such realism of the GCM's teleconnection responses motivates more detailed sensitivity analyses using the model data. First let us consider the statistical stationarity of the teleconnections themselves. We repeat the analysis of Figure 2.7, but using each separate model realization for the 1950–95 period. Since each such member

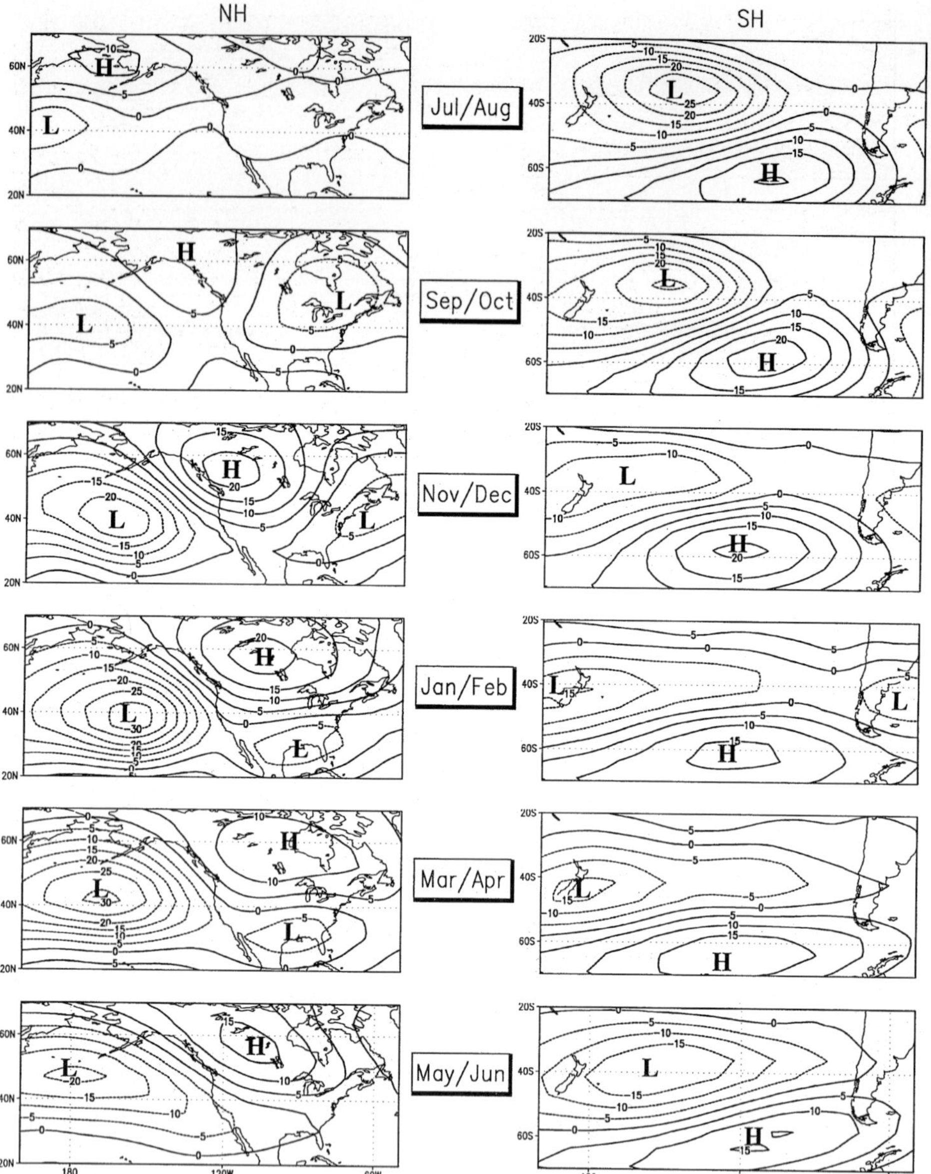

Fig. 2.7 Same as Figure 2.2, except for the GCM 500 hPa height patterns that are linearly related to EOF1 of tropical Pacific SST variations. The 12-member ensemble-averaged monthly height anomalies have been regressed on the PC time series of Figure 2.1 for each calendar month separately for the period February 1950–February 1995. Shown are consecutive two-month averages, with contours denoting the height anomalies associated with a +1 standard deviation of the EOF1 PC representative of warm-event conditions. Solid (dashed) contours denote high- (low-) pressure anomalies. The contour interval is 5 m.

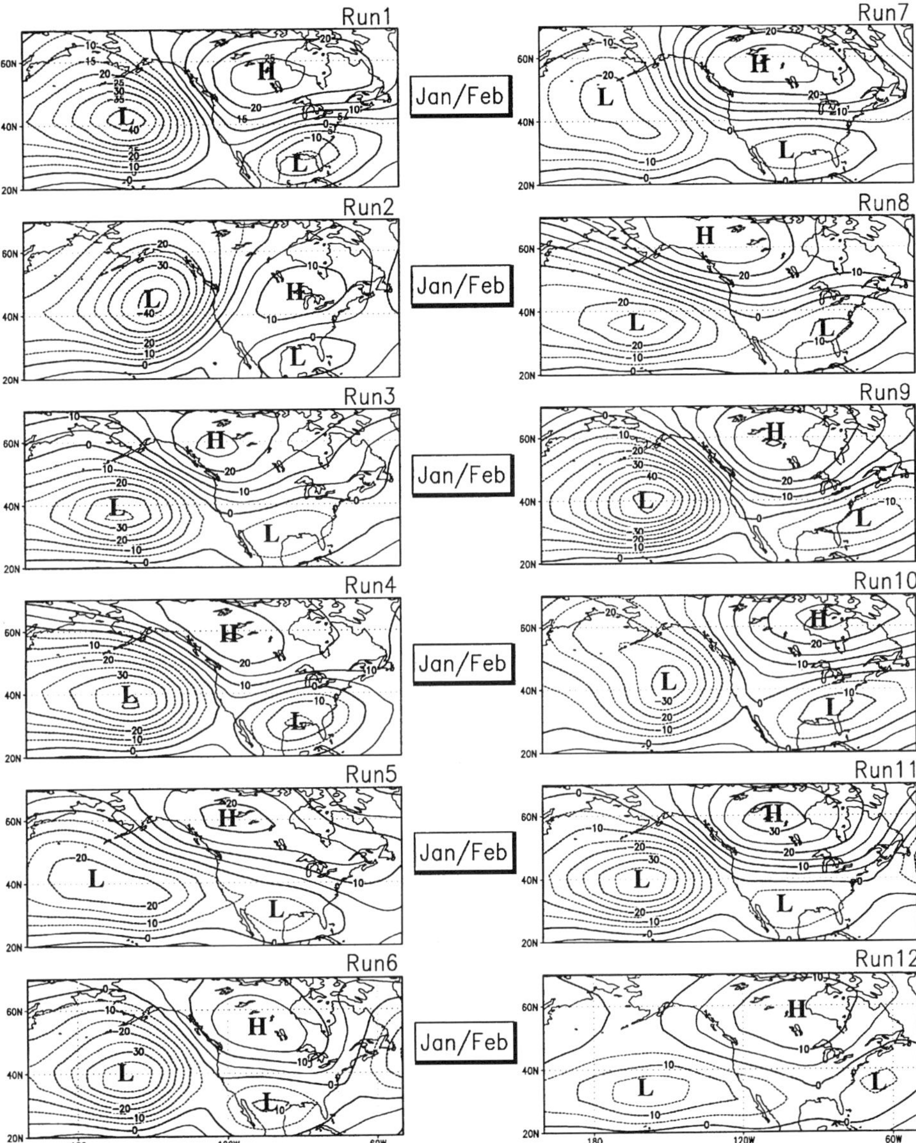

Fig. 2.8 As in Figure 2.7, but for the GCM 500 hPa height patterns that are linearly related to EOF1 of tropical Pacific SST variations calculated separately for each of the twelve model runs. The monthly averaged height anomalies have been regressed on the PC time series of Figure 2.1 for the period February 1950–February 1995. Shown are the two-month averages for January/February. Solid (dashed) contours denote high- (low-) pressure anomalies. The contour interval is 5 m.

experiences identical monthly varying SST forcing, the results illustrate the changing teleconnection patterns on half-century timescales that result purely from statistical sampling. Figure 2.8 shows results of the analysis for the Northern Hemisphere winter (January/February).

Although each realization possesses the same characteristic wave train structure, variability in the strength of teleconnections occurs from run to run. Note especially the weak response in model run 12, with amplitudes less than half those simulated in several other runs. Phase shifts in the centers of action also occur, with a high located over eastern North America in model run 2, but located over western North America in model runs 3 and 8. Such phase differences, due purely to sampling, can reverse the sign of regional climate impacts associated with ENSO. In run 2, for example, the proximity of the anomalous low to the Pacific Northwest implies increased storminess and rainfall during ENSO's warm phase. Yet, that same region in model run 8 would receive below-normal rainfall during warm events owing to the protective influence of high pressure.

What does one learn from these experiments that pertains to the observed multidecadal variations in ENSO-related teleconnections? The instrumental record effectively represents a single realization of 50-year duration, and the model simulations indicate that a variety of physical outcomes are plausible on half-century timescales, even when the climate system is subjected to identical external forcings. This finding alone may explain the differences in observed teleconnections between the 20-year periods 1958–77 and 1978–97 (see Fig. 2.4). The model results alert us to the possibility that ENSO's impact can reverse sign for multidecadal periods on regional scales due to purely random variations.

Of course, decadal changes in the external forcings themselves could also play a role in modifying the teleconnection patterns. Such forcings may include variations in solar output, volcanic activity, changes in the atmosphere's chemical composition, and low-frequency changes in tropical Pacific SSTs themselves. Concerning changes in tropical Pacific SSTs, the observed SST time series in Figure 2.1 indicates that ENSO undergoes multiscale variations, ranging from an irregular 2- to 7-year recurrence interval to multidecadal fluctuations. The latter are characterized by consecutive runs of a particular phase of ENSO, such as the prevalence of warm events in the post-1976 era.

Using the GCM, we analyze the extent to which the observed decadal variations in teleconnections during the post-1950 era that were described in Figure 2.4 may have been linked to such low-frequency variations in SST forcing. The model's ensemble-averaged heights are regressed separately on the interannual variations of EOF1 SSTs for 1950–75 and 1976–95, respectively. We once again display the height anomalies associated with a +1 standard deviation anomaly of the SST index to represent the warm-event phase. As is shown in Figure 2.9, the model teleconnections exhibit an increase in amplitude during the winter/spring seasons during the post-1976 period, which is qualitatively consistent with that occurring in nature. This increase is due to a doubling of the springtime interannual variability of equatorial Pacific SSTs that stems from the fact that warm events of the post-1976 era have been stronger and have achieved their peak SST amplitudes in late winter, rather than during early winter as occurred before 1976, and have thereby retained appreciable strength in the spring.

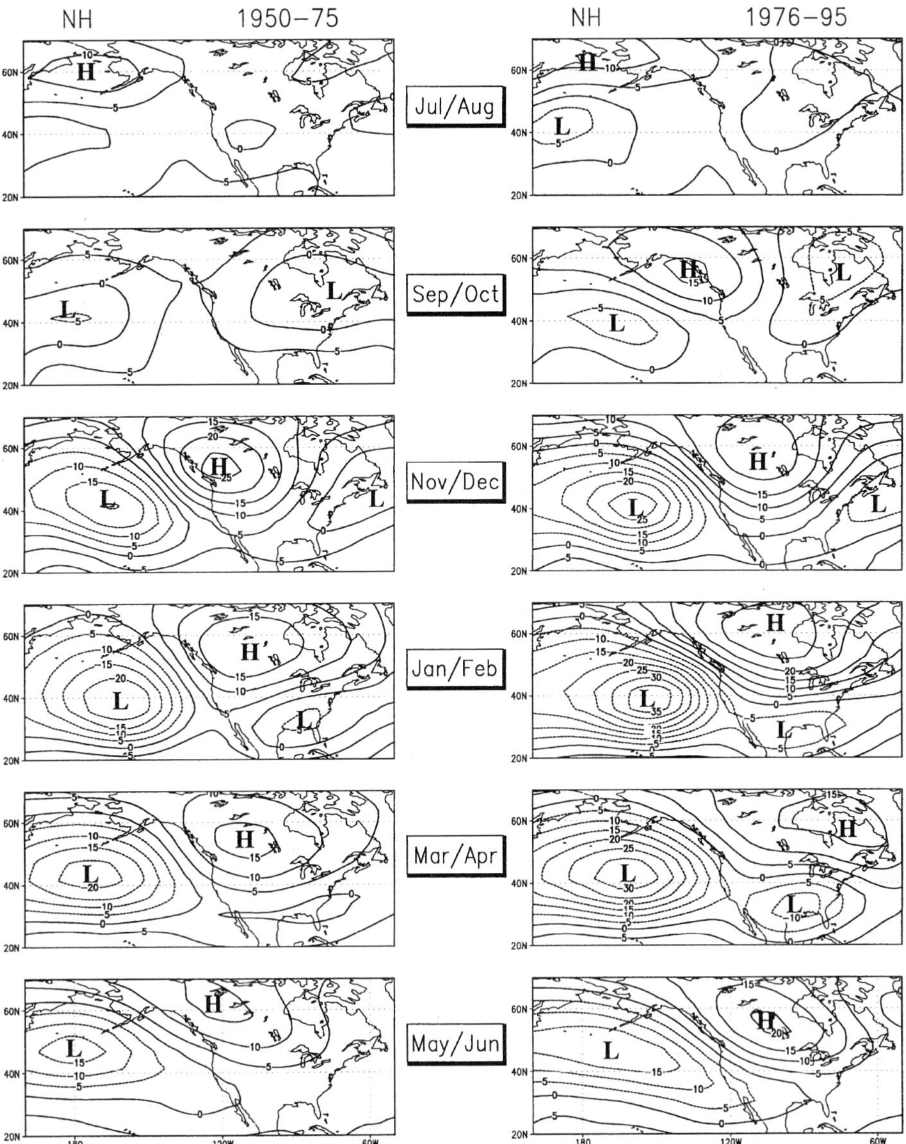

Fig. 2.9 Same as Figure 2.7, except that the ensemble GCM monthly averaged height anomalies have been regressed on the PC EOF1 tropical Pacific SST variations for two separate periods: 1950–75 (left side) and 1976–95 (right side). Solid (dashed) contours denote high- (low-) pressure anomalies. The contour interval is 5 m.

Inter-ENSO Variations of Modeled Teleconnections

Figure 2.10 presents the 12-member composite model responses for the same four El Niño events that were previously analyzed from the single realization of observations (see Fig. 2.6). The process of averaging a dozen "perfect analogs" for each event

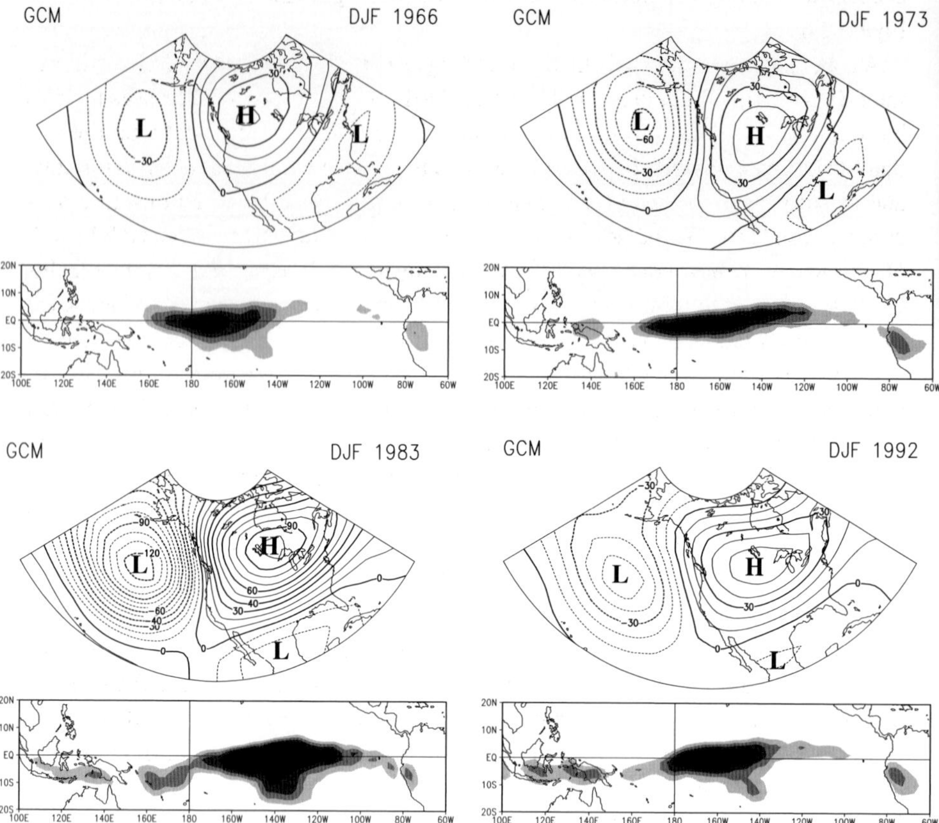

Fig. 2.10 Global circulation model simulated winter season height (in meters) and tropical rainfall (in mm/day) anomalies for the El Niño events of 1966, 1973, 1983, and 1992. (The year corresponds to January of the winter season.) The events are the same as those shown for observations in Figure 2.6. Solid (dashed) contours denote high- (low-) pressure anomalies. The contour interval is 10 m.

effectively eliminates the noise component of the atmosphere's behavior that is unrelated to SST forcing, thereby extracting the teleconnection response associated with each unique El Niño.

It is immediately evident from the model results that the *spatial structures* of those patterns are very similar for the various El Niño events, in contrast to the large inter–El Niño variations among nature's single realizations. It could be hypothesized that the tropical rainfall anomalies, as displayed in the lower panels for each year, are not sufficiently different from case to case to yield appreciably different teleconnection patterns. It is found, however, that the GCM realistically simulates the inter–El Niño variations of the tropical Pacific rainfall anomalies (e.g., Hoerling and Kumar 1997). This result raises the question as to how different the tropical forcing would need to be to yield different teleconnection patterns. Suffice it to say that inter–El Niño SST variations appear to be inadequate for that purpose, and the modest sensitivity in the GCM's spatial height responses in Figure 2.10 fails to account for the large variations in observed spatial height anomaly patterns shown in Figure 2.6.

A conclusion to be drawn from the preceding discussion is that much of this observed inter–El Niño variability in nature is unrelated to the event-to-event differences in SSTs but is primarily climate noise. This interpretation is entirely consistent with the fact that El Niño explains only a modest fraction of the extratropical interannual variability, and as such other sources of seasonal variability can mask the El Niño signal on a case-by-case basis. Of course, one may also question the reliability of the GCM itself. On this matter we have confirmed that the spatially fixed teleconnection response in this GCM appears not to stem from a spurious model insensitivity, since we find the same behavior in ensemble climate simulations using several different GCMs (not shown).

Large variations among the GCM's four teleconnections are evident in Figure 2.10 with respect to the *wave train amplitudes*, however. Thus, the response during 1983 is more than double the response during the other El Niños, which is perhaps not surprising given the unprecedented SST forcing during 1983. Further analysis of the model data for all the tropical Pacific warm events during 1950–95 reveals a quasi-linear relationship between the strength of the atmospheric response and the strength of the SST anomalies (Kumar and Hoerling 1997; Hoerling and Kumar 1997).

Also apparent in Figure 2.10 is a slight eastward displacement of the teleconnection patterns for the stronger El Niño events. This displacement occurs in unison with an eastward spread of enhanced rainfall across the equatorial Pacific during the stronger events, as is shown in the accompanying tropical strip analyses of the GCM's precipitation response. Regionally, such an outwardly modest phase shift can have large consequences. For example, along the Pacific West Coast of the United States, a 10° longitude shift of the wave train can alter the sign of the precipitation anomalies.

As an illustration of this regional sensitivity, additional idealized experiments have been performed using a perpetual January version of the GCM. One set employs an equatorial Pacific SST anomaly of 2°C peak amplitude, representative of a moderate El Niño event. A second set uses the same spatial pattern of the SST anomaly, but with double amplitude. Figure 2.11 shows the SST anomalies along the equator, together with the circulation and precipitation responses over the Pacific. These responses are based on an average of 48 months of January integration. Only in the strong El Niño experiment does enhanced rainfall cover the entire Pacific West Coast, whereas the more moderate El Niño yields enhanced rainfall over central and southern California only, while the Pacific Northwest is dry.

Also, for the large-amplitude SST events, there is nonlinearity in model responses with respect to ENSO's phase itself, and here we present examples of the GCM's teleconnections for several idealized scenarios of extreme warm and extreme cold events. Additional experiments have been performed using the annual cycle version of the GCM, but forcing it only with the tropical Pacific SST anomalies from 1950 to 1995 (Pacific Ocean–Global Atmosphere [POGA] runs). Furthermore, only EOF1 of those tropical Pacific SST variations is retained, such that the model is subjected only to a single SST pattern given by the structure function in Figure 2.1. The amplitude and phase variations of EOF1 are determined by the PC time series. A nine-member

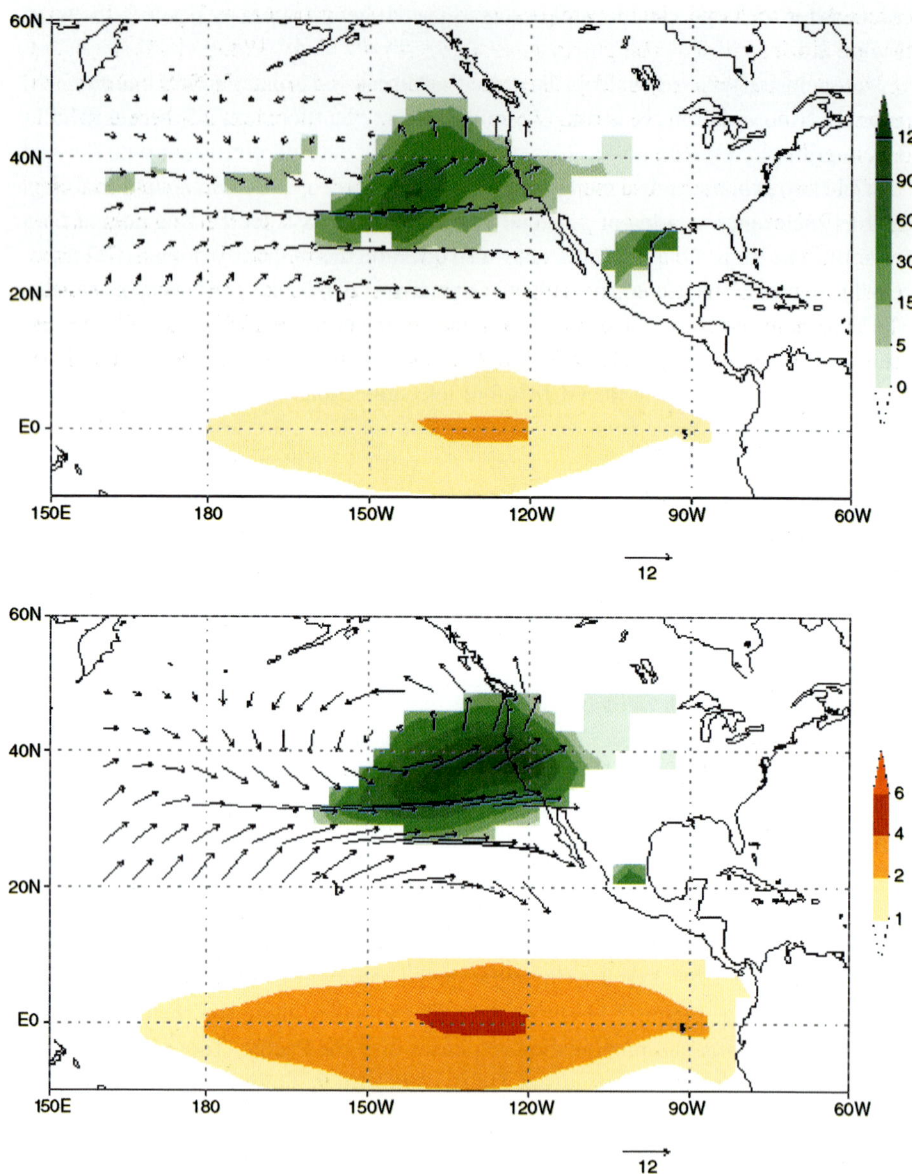

Fig. 2.11 Global circulation model simulated January response to idealized positive tropical Pacific SST anomalies. The tropical SST forcing, indicated in yellow shades, corresponds to a mature warm event having a peak +2°C amplitude in the top panel, and a +4°C amplitude in the lower panel. The GCM has been run in perpetual January mode, and results are for the ensemble average of 48 months of integration. Vectors of the 200 hPa wind response are shown with arrows poleward of 20°N. For the same region, green shading delineates positive rainfall anomalies (in mm/month).

ensemble of such experiments are performed, and the top panels of Figure 2.12 show the boreal winter 200 hPa height anomalies associated with the 1983 El Niño (left side) and the 1974 La Niña (right side). These are the two most extreme ENSOs in the recent record. Yet only during the strong warm event does the model atmosphere exhibit a large-amplitude response.

While suggestive, the results in the top panels of Figure 2.12 are inconclusive insofar as the 1983 event had a larger projection on EOF1 than its cold-event counterpart. A further nine-member ensemble of runs was performed with the sign of the PC time series reversed. Thus, for each event, an exact mirrored opposite phase was generated. The ensemble mean GCM responses for the reversed-phase 1983 and 1974 events are shown in the lower panels of Figure 2.12. Once again, a side-by-side comparison

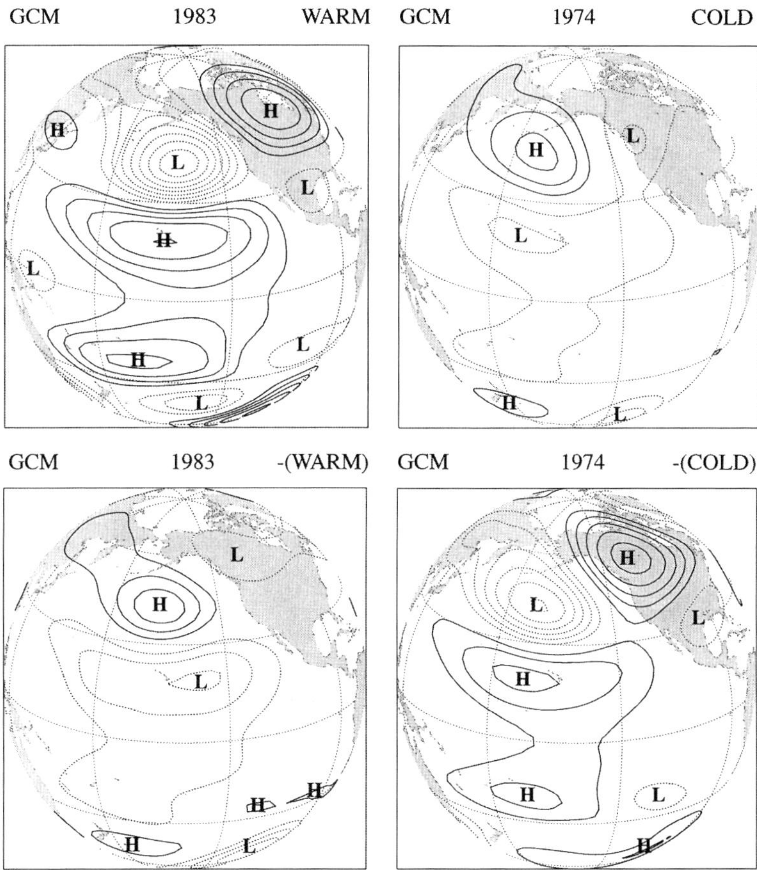

Fig. 2.12 Global circulation model 200 hPa winter season (December–February) height responses to idealized tropical Pacific SST anomalies. In the top panels, the SSTs are the projection of the 1983 extreme warm event and 1974 extreme cold event SST anomalies onto EOF1 of Figure 2.1. In the bottom panels, the opposite phases of the same EOF1 SST anomalies are used. Thus, the 1983 SST forcing denotes a cold event that is exactly equal and opposite to the actual 1983 warm event. Results are nine-member ensemble averages. The contour interval is 10 m; contours denote high- (low-) pressure anomalies. Satellite projection shows the PNA region, and the continents are shaded for reference.

reveals that the extreme warm-event phase generates an appreciably larger amplitude teleconnection response than does its cold-event counterpart. A top-to-bottom comparison reveals the sensitivity to the same amplitude SST forcing but of opposite sign, and this result confirms the nonlinear behavior in the GCM.

To put these model results into context, however, it is important to recognize that these simulated sensitivities are smaller than the observed variations of the extratropical climate anomalies from one El Niño event to another. Furthermore, although an appreciable nonlinearity of the model response was shown with respect to the extreme phases of ENSO, such events are by their very definition rare and thus do not dominate the overall statistics of ENSO teleconnection responses.

Seasonal Atmospheric Predictability Related to ENSO

The SST-forced teleconnections described previously are the cornerstones of seasonal predictions for the PNA and PSA sectors. For this reason, a significant effort continues to be placed on understanding those patterns and assessing the ability of GCMs to predict them accurately. However, the skill of seasonal climate predictions is inherently limited because most seasonal-interannual variability in the extratropics is unrelated to forcing from the ocean. These limitations exist regardless of GCM biases, which can further reduce skill. In other words, even "perfect" GCMs, when driven by perfect SST predictions, will not yield accurate seasonal forecasts for each and every ENSO event.

What then are the theoretical limits for seasonal climate forecast skill associated with ENSO? The answer will depend on the boundary forcing itself as discussed further below (see also Madden 1989; Kumar and Hoerling 1998a). For a general appraisal of the prospects of seasonal forecasts, however, it is customary to perform a ratio-of-variance analysis. Such a ratio compares the boundary-forced variability (e.g., the signal related to anomalous SSTs) to the total variability of the system (Madden 1976; Lau 1985; Shea and Madden 1990; Kumar and Hoerling 1995; Stern and Miyakoda 1995; Rowell 1998). The latter can be viewed as resulting from the linear combination of the internally generated variability (climate noise) and the boundary-forced climate signal. Small values of that ratio imply low predictability, whereas ratios near unity imply that seasonal predictions are virtually deterministic because SSTs determine the entire interannual variability.

To illustrate this point, we analyze the 200 hPa height variance from the ensemble of GCM simulations and assess the fraction of the model's total variance that is attributable to the interannual variability of global SSTs (see Kumar and Hoerling 1995). The total variance measures the year-to-year variability among all 552 (46 years times 12 runs) winters in the GCM record. The SST-forced component is simply the variance of the 46 ensemble-averaged winters during 1950–95. It should be noted that other sources of boundary-forced variability – for example, associated with anomalies in soil moisture, snow cover, or sea ice – may also be important, although these sources are not considered here.

The top panel of Figure 2.13 shows the GCM's total variance of 200 hPa heights averaged for December–February (DJF). The total variance is maximized in the extratropical latitudes, where it acquires amplitudes exceeding by an order of magnitude

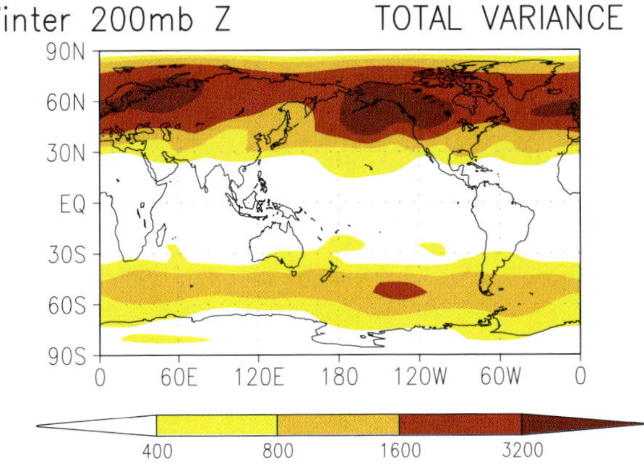

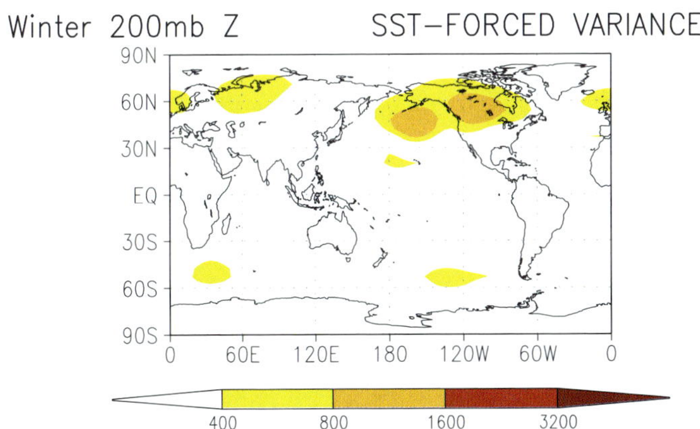

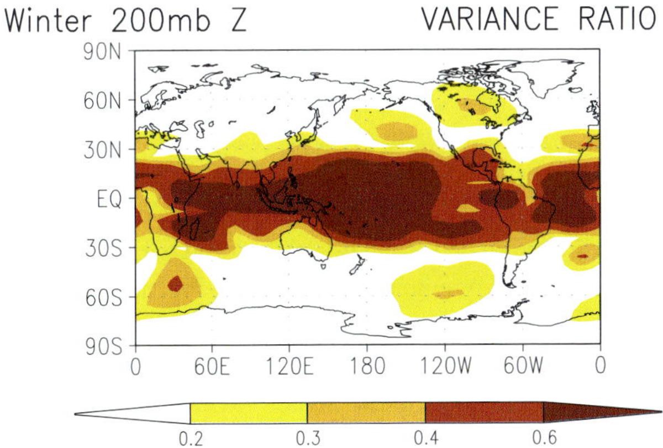

Fig. 2.13 Global circulation model simulated total variance (top panel) and SST-forced variance (middle panel) of the 200 hPa winter season (December–February) heights. The bottom panel shows the ratio of the variances that expresses the fraction of total variance related to the interannual SST variations. The results are based on the analysis of the 12-member GOGA simulations forced with the monthly SST variations from 1950 to 1995.

the tropical 200 hPa height variance. By comparison, the SST-forced variance (middle panel) is small in the extratropics and is confined to the PNA sector. Three distinct centers of maximum SST-forced variance trace out the teleconnection pattern discussed in previous sections.

The bottom panel of Figure 2.13 shows the fraction of total variance explained by the SST-forced variability. That ratio is large in the tropics, where greater than 50% of the total 200 hPa height variability is ENSO induced. (It should be noted that tropical rainfall does not exhibit such a widespread pattern of high predictability.) Even within the core of the extratropical teleconnection centers, however, ENSO explains less than 30% of the total height variance in the GCM. Similar estimates have been derived from observations (e.g., Horel and Wallace 1981), and the GCM confirms the fact that, on average, the potential for seasonal predictability in the extratropics is low.

Seasonal climate predictions in the extratropics are thus inherently probabilistic, rather than a deterministic outcome of tropical Pacific SST forcing, owing to this large and pervasive random component of seasonal variability superimposed upon a modest boundary-forced signal. The schematics of Figure 2.14 illustrate the philosophy of probabilistic climate predictions. Anomalies in the boundary forcing, such as occur during an El Niño, alter the frequency of occurrence of atmospheric states. Under "normal" conditions, for example, there is an equal likelihood of wet or dry seasonal means. Under El Niño conditions, the probability distribution shifts, and the frequency of one climate extreme is enhanced relative to its opposite extreme. Note in this example, however, that both extreme states are still plausible outcomes during El Niño despite the enhanced risk of dry conditions, and it is necessary to predict the probability of each outcome.

In Figure 2.14, the shift in the mean of the probability function (shown by the arrow) measures the strength of the SST-forced signal. It is evident that for small signals, the distribution of climate states during "normal" SST conditions is virtually indistinguishable from that occurring during El Niño conditions. Equivalently, the signal-to-noise ratio is low, and the potential for skillful predictions is low. However, for sufficiently large mean shifts, the two distributions possess no overlap. In practice, this situation is observed to occur in some tropical regions during El Niño, and the probability that a wetter than normal climate state will occur becomes a virtual certainty. In that case, the signal-to-noise ratio is large and predictability is high.

In fact, only a rudimentary understanding exists of how ENSO alters the probability distribution of climate states. It remains to be seen whether the shape of the distributions themselves changes or whether indeed the boundary forcing causes only a shift in the mean, as is illustrated for conceptual purposes in Figure 2.14. A related question is whether useful information exists in the higher moment statistics of ensemble GCM predictions.

An Evaluation of Seasonal Predictive Skill

The hindcast skill of seasonal mean predictions from 1950 to 1995 is assessed by using the ensemble of GCM simulations forced with the observed monthly mean global SSTs.

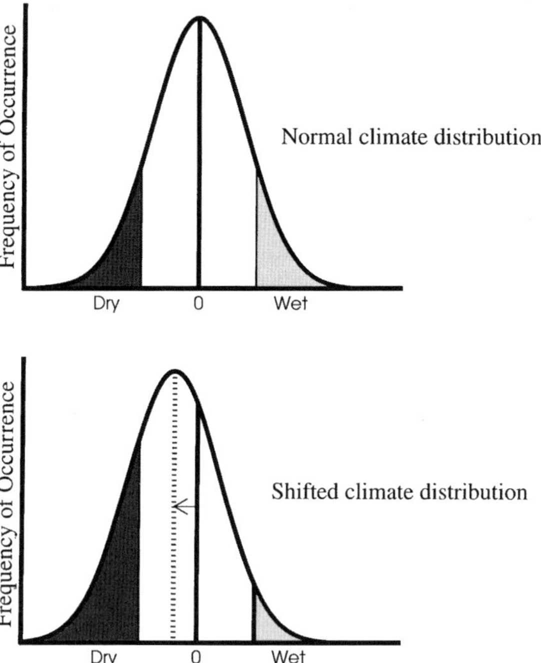

Fig. 2.14 Schematic diagrams of the impact of El Niño on the probability distribution function (PDF) of seasonal mean climate states. The top PDF shows a statistically normal distribution of climate states in the absence of anomalous boundary forcing. The frequencies of occurrence of extreme wet and of extreme dry seasonal means are thus assumed to have equal probability. The bottom PDF illustrates a shift in the mean, as indicated by the arrow, due to El Niño. The frequencies of occurrence of extreme wet and dry seasonal means have now changed, as determined by the mean shift (i.e., El Niño signal) itself. It should be understood that the relative magnitude of the mean shift and the spread of the distribution itself depend on the variable under consideration, time averaging interval, geographical location, and season. (Courtesy of R. Dole.)

Figure 2.15 presents a time series of the spatial correlation between model and observed 500 hPa height anomalies from 1950 through 1994. These time series are calculated over the PNA sector bounded by 20°N–70°N and 180°–60°W, corresponding exactly to the region shown in Figure 2.10. The correlations are for 90-day running means (dashed curve), and those results have been filtered with a three-point smoother (solid curve).

Considerable seasonal and interannual variability in the skill scores exists. The winter/early spring seasons exhibit highest skill, consistent with the large-amplitude teleconnection response to ENSO at that time. The year-to-year variations in skill result at least partly from the interannual variations in SST themselves. (Note the high skill scores during the warm event years of 1982–83 and 1986–87, in contrast to the low skill during the preceding years when tropical SSTs were near normal.) However, skill is not high during all ENSO winter seasons, as is evidenced by the model's poor performance during the 1965–66 and 1972–73 warm-event winters. Consistent with the aforementioned low signal-to-noise ratio, it is plausible that, due to chance alone, the observed

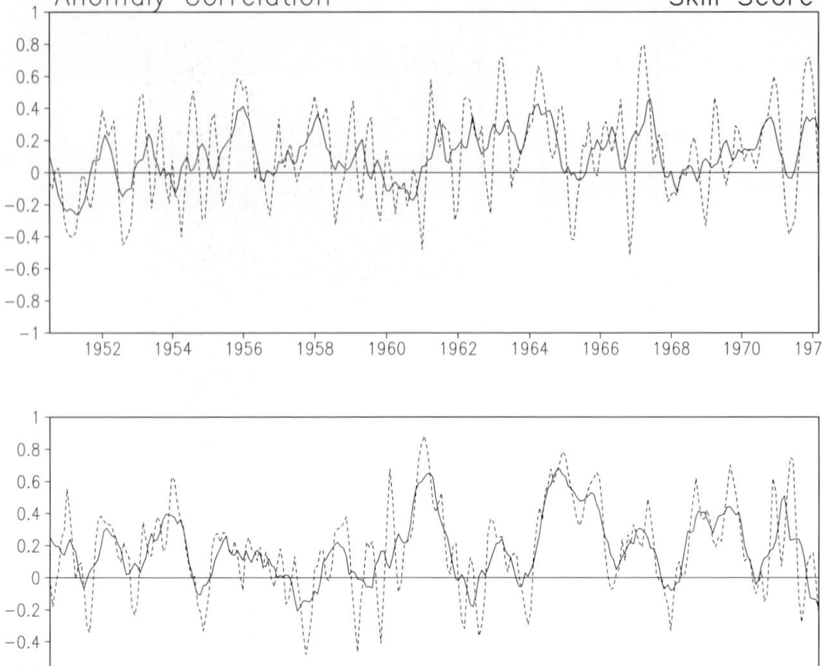

Fig. 2.15 Anomaly correlations between the spatial pattern of the observed and GCM ensemble mean 500 hPa height anomalies during 1950–94. Correlations are calculated over the PNA region 20°N–70°N, 180°–60°W. The GCM anomalies are the 12-member average of the simulations using the observed monthly global SSTs (GOGA runs). The dashed curve is for running three-month average skill scores, and the solid curve is the result of applying a three-point smoother.

500 hPa height anomalies during these cases were overwhelmed by the random internal variability. In other words, the single realizations of nature are not constrained to select the boundary-forced signal, and this is consistent with the probabilistic characteristic of seasonal climate predictions. For such occurrences, one can expect a priori that the observed anomalies and the ensemble mean GCM response will compare unfavorably, independent of model biases.

Owing to this appreciable random contribution to the individual seasonal mean skill scores, an appraisal of model performance requires averaging over many years. The graph in Figure 2.16 displays various estimates of seasonal predictive skill of the PNA-sector 500 hPa heights during the boreal winter season. These skill estimates represent the correlation skill scores averaged for the eighteen strongest warm and cold events during 1950–95.

Four bars are shown in Figure 2.16. The first is for the hindcast skill of the ensemble GOGA responses, based on averaging the skill scores for the eighteen ENSO winters

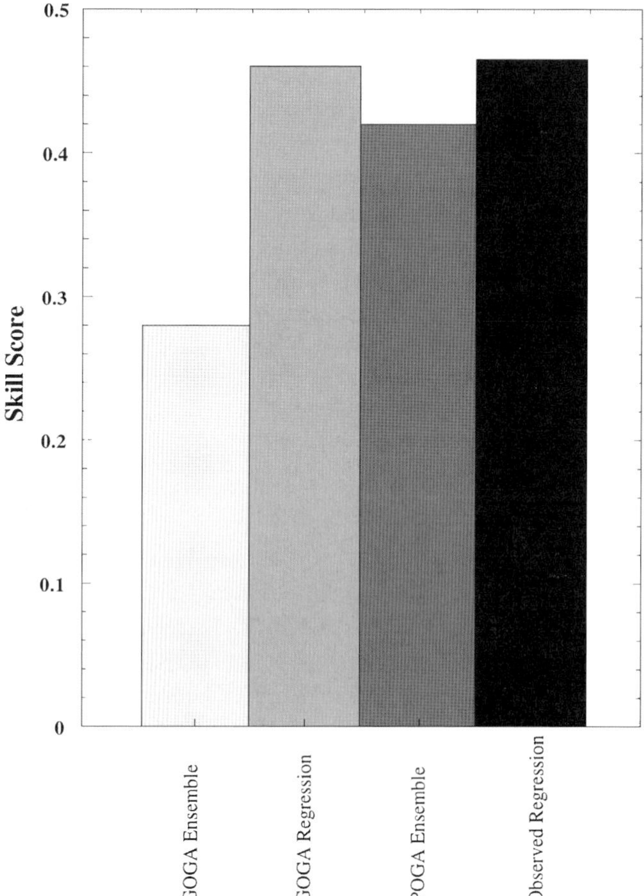

Fig. 2.16 Anomaly correlation skill scores for winter (December through February) season 500 hPa heights averaged for the eighteen strongest ENSO events during 1950–95. The spatial correlations are again calculated over the PNA region as in Figure 2.15. The results based on four different prediction methods are summarized by the individual bar graphs. The leftmost bar is derived from the GCM GOGA ensembles, the second bar is derived from specifying the GOGA 500 hPa regressions, the third bar is derived from the GCM POGA ensembles, and the rightmost bar is derived from specifying the observed 500 hPa regressions. See the text for further details.

of Figure 2.15. The second illustrates the skill of an alternative method in which the seasonal anomalies are predicted by merely specifying the GCM's 500 hPa linear regression patterns associated with EOF1 tropical Pacific SST forcing (see Fig. 2.7). In this approach, a single spatial pattern is correlated against the observed anomalies for each ENSO, and the only variation of the prediction is to reverse the polarity of that pattern for opposite phases of ENSO.

The skill scores based on the two methods are quite different, and it is evident that using the GOGA GCM ensemble mean as a prediction is actually inferior to using the model's statistical signal linearly related to EOF1. The skill score based on the ensembles is 0.28, as compared to 0.45 for the linear regressions. To understand this result, we have calculated the hindcast skill of ensemble POGA simulations, namely, model runs using only tropical Pacific SST forcing. (In these runs, the entire tropical Pacific SST variance is retained, as opposed to retaining only EOF1 as in the different set of POGA runs described earlier.) Those results are shown by the third bar. The skill score of the ensemble POGA hindcasts is 0.42, which is comparable to that based on the model's linear regressions and is appreciably higher than the ensemble GOGA hindcasts.

These results are counterintuitive at first glance: Greater skill is expected from GCM predictions based on globally complete knowledge of the boundary conditions, as opposed to runs in which a fraction of that information is withheld. Likewise, one expects skill based on using the GCM's ensemble mean anomalies to equal or exceed the skill of the linear statistics, at least in the absence of model biases. This will be true to the extent that nonlinear associations between tropical SSTs and the extratropical response are important, and furthermore to the extent that modes of SST variability independent of EOF1 also contribute to predictive skill over the PNA region.

The results reveal biases in the GOGA simulations, due either to the GOGA design itself or to systematic errors in the model's sensitivity to SST anomalies outside of the tropical Pacific. For example, Kumar and Hoerling (1998b) have found a spurious response of the atmospheric model to prescribed Indian Ocean SST anomalies. To the extent that some aspects of air–sea interactions are fundamentally coupled processes, an approach in which the SST anomalies are prescribed in an atmospheric GCM may yield spurious sources of energy, implying a design flaw in the so-called two-tiered system that is widely employed in climate forecast centers. The reasons for the low GOGA GCM skill scores are not completely known, however. A resolution to the problem is obviously important, both for fully harvesting the potential for seasonal predictability and for properly designing the forecast system itself.

Prospects for Extratropical Seasonal Climate Predictions

What are the prospects for ENSO-related seasonal predictions for the extratropics, and what role exists for dynamically based methods? Concerning the former, there is wide agreement that, owing to the atmosphere's large natural variability, the skill of seasonal predictions is inherently limited (e.g., Madden 1976; Chervin 1986; Kumar and Hoerling 1995; Palmer and Anderson 1994; Barnett 1995; Brankovic et al. 1994; Brankovic and Palmer 1997). These are not technological barriers imposed by data and model deficiencies, and thus they cannot be overcome by improvements in future observing or prediction systems.

Concerning the role of GCMs in seasonal predictions, it is widely assumed that dynamical methods are essential to harvest the modest theoretical skill levels. This notion is partly rooted in numerical weather prediction experience: The skill of daily weather forecasts has been elevated by advances in dynamical models, and the empirical-statistical tools that were so valuable in the past have been found to be inferior and thus have been largely supplanted. The notion is also rooted in the expectation that nonlinearity is an important factor in the seasonal mean extratropical response to boundary forcing, much as it is important in the evolution of daily weather patterns.

Thus, one of the outstanding problems in seasonal predictability research is determining how nonlinear the relationship between boundary forcing and the climate signal actually is. For example, do small changes in the SSTs induce large changes in the atmospheric response, or are the predictable signals essentially captured by the linear teleconnection patterns discussed in the previous sections? That the truth may lie closer to the latter is suggested by the considerable skill of linear statistical methods for seasonal predictions, highlighted by the fourth bar in Figure 2.16. This bar was derived by specifying the observed 500 hPa linear regression patterns associated with EOF1 tropical Pacific SST forcing (see Fig. 2.3) as the seasonal forecast. As with the other bars, the correlation represents the average for the eighteen ENSO winters during 1950–95, and the method of cross-validation has been used to ensure independence between predictor and predictand. This empirical-statistical method yields a skill score of 0.47, which is virtually identical to that based on the POGA GCM ensembles and considerably greater than that based on the GOGA ensembles.

It is immediately apparent that empirical methods are highly competitive with the dynamical ones on the seasonal timescale, and it remains to be determined whether improvements in GCMs will enhance skill above and beyond such benchmarks. Ultimately, the benefits of GCM methods for seasonal forecasting will hinge on the extent to which nonlinearities are important, because only then may one expect the dynamical methods to supplant the empirical tools.

If, however, the predictable signal is only a linear function of the large-scale tropical forcing, then the seasonal forecast problem is arguably one of merely forecasting the amplitude of the PC of EOF1 tropical Pacific SSTs, and then specifying the known teleconnection pattern. Analysis of other atmospheric GCMs will be needed to address this problem more thoroughly. However, the suite of model runs analyzed here suggest that the bulk of the extratropical predictability related to ENSO is embodied in such linear teleconnection signals.

It also remains to be seen whether additional predictability emerges from SST variations that are not linearly related to ENSO. For the tropics in particular, there is reason to believe that the seasonal prediction problem involves more than just ENSO. It is well known, for example, that seasonal rainfall anomalies in portions of Brazil, Africa, and Australia are associated with interannual SST variations in the adjacent tropical Atlantic and Indian Oceans (Folland et al. 1986; Nicholls 1989; Hastenrath 1990). Accurate predictions of global SSTs would therefore be essential for making seasonal climate forecasts for these regions.

Finally, only the seasonal mean extratropical response has been discussed here. Yet, one can imagine that even in regions where the seasonal mean impact of ENSO is weak, the subseasonal behavior may be appreciably modified. Here we wish to reemphasize the need to better understand the probability distributions in Figure 2.14 and to ascertain ENSO's impact on the higher statistical moments. Examples of these statistical parameters include the range of maximum and minimum temperature extremes during a season, first frost and freeze dates, and the frequency of daily rainfall events of varying intensities. These parameters, among others, are in many ways of greater societal interest than is the seasonal mean anomaly alone. It is more than just academic interest, therefore, that motivates the need to explore the full temporal spectrum of ENSO impacts.

Acknowledgments The support offered by NOAA's Climate Dynamics and Experimental Prediction (CDEP) Program and NOAA's Office of Global Programs Climate and Global Change (GOALS) program are gratefully acknowledged. We also wish to recognize the contributions by TaiYi Xu and Gary Bates of CIRES in the analysis and presentation of results.

References

ARKIN, P. A. and MEISNER, B. N., 1987: The relationship between large-scale convective rainfall and cloud cover over the Western Hemisphere during 1982–84. *Monthly Weather Review*, **115**, 51–74.

BARNETT, T. P., 1995: Monte Carlo climate forecasting. *Journal of Climate*, **8**, 1005–1022.

BJERKNES, J., 1969: Atmospheric teleconnections from the equatorial Pacific. *Monthly Weather Review*, **97**, 163–172.

BRANKOVIC, C. and PALMER, T. N., 1997: Atmospheric seasonal predictability and the estimates of ensemble size. *Monthly Weather Review*, **125**, 859–874.

BRANKOVIC, C., PALMER, T. N., and FERRANTI, L., 1994: Predictability of seasonal atmospheric variations. *Journal of Climate*, **7**, 217–237.

CHERVIN, R. M., 1986: Interannual variability and seasonal climate variability. *Journal of the Atmospheric Sciences*, **43**, 233–251.

FOLLAND, C. K., PALMER, T. N., and PARKER, D. E., 1986: Sahel rainfall and worldwide sea temperatures. *Nature*, **320**, 602–607.

GILL, A. E., 1980: Some simple solutions for heat-induced tropical circulations. *Quarterly Journal of the Royal Meteorological Society*, **106**, 447–462.

GLANTZ, M. H., KATZ, R. W., and NICHOLLS, N. (eds.), 1991: *Teleconnections Linking Worldwide Climate Anomalies*. Cambridge: Cambridge University Press, 535 pp.

HASTENRATH, S., 1990: Prediction of Northeast Brazil rainfall anomalies. *Journal of Climate*, **3**, 893–904.

HELD, I. M., LYONS, S. W., and NIGAM, S., 1989: Transients and the extratropical response to El Niño. *Journal of the Atmospheric Sciences*, **46**, 163–174.

HOERLING, M. P. and KUMAR, A., 1997: Why do North American climate anomalies differ from one El Niño event to another? *Geophysical Research Letters*, **24**, 1059–1062.

HOERLING, M. P., KUMAR, A., and ZHONG, M., 1997: El Niño, La Niña, and the nonlinearity of their teleconnections. *Journal of Climate*, **10**, 1769–1786.

HOERLING, M. P. and TING, M., 1994: Organization of extratropical transients during El Niño. *Journal of Climate*, **7**, 745–766.

HOREL, J. D. and WALLACE, J. M., 1981: Planetary-scale atmospheric phenomenon associated with the Southern Oscillation. *Monthly Weather Review*, **109**, 2080–2092.

HOSKINS, B. J. and KAROLY, D. J., 1981: The steady linear response of a spherical atmosphere to thermal and orographic forcing. *Journal of the Atmospheric Sciences*, **38**, 1179–1196.

KAROLY, D. J., 1989: Southern Hemisphere circulation features associated with El Niño–Southern Oscillation. *Journal of Climate*, **2**, 1239–1252.

KILADIS, G. N. and DIAZ, H. F., 1989: Global climate anomalies associated with extremes in the Southern Oscillation. *Journal of Climate*, **2**, 1069–1090.

KILADIS, G. N. and MO, K. C., 1998: Interannual and intraseasonal variability in the Southern Hemisphere. *In* Karoly, D. J. and Vincent, D. G. (eds.), *Meteorology of the Southern Hemisphere*. American Meteorological Society, Boston, MA, 307–336.

KOK, C. J. and OPSTEEGH, J. D., 1985: On the possible causes of anomalies in seasonal mean circulation pattern during the 1982–83 El Niño event. *Journal of the Atmospheric Sciences*, **42**, 677–694.

KUMAR, A. and HOERLING, M. P., 1995: Prospects and limitations of seasonal atmospheric GCM predictions. *Bulletin of the American Meteorological Society*, **76**, 335–345.

KUMAR, A. and HOERLING, M. P., 1997: Interpretation and implications of observed inter El Niño variability. *Journal of Climate*, **10**, 83–91.

KUMAR, A. and HOERLING, M. P., 1998a: Annual cycle of Pacific/North American seasonal predictability associated with different phases of ENSO. *Journal of Climate*, **11**, 3295–3308.

KUMAR, A. and HOERLING, M. P., 1998b: On the specification of regional SSTs in AGCM simulations. *Journal of Geophysical Research*, **103** (D8), 8901–8907.

KUMAR, A., HOERLING, M. P., JI, M., LEETMAA, A., and SARDESHMUKH, P. D., 1996: Assessing a GCM's suitability for making seasonal predictions. *Journal of Climate*, **9**, 115–129.

LAU, N.-C., 1985: Modeling the seasonal dependence of the atmospheric response to observed El Niños in 1962–76. *Monthly Weather Review*, **113**, 1970–1996.

LIEBMANN, B. and SMITH, C. A., 1996: Description of a complete (interpolated) outgoing longwave radiation dataset. *Bulletin of the American Meteorological Society*, **77**, 1275–1277.

MADDEN, R. A., 1976: Estimates of the natural variability of time-averaged sea-level pressure. *Monthly Weather Review*, **104**, 942–952.

MADDEN, R. A., 1989: On predicting probability distributions of time-averaged meteorological data. *Journal of Climate*, **2**, 922–925.

MONTROY, D. L., RICHMAN, M. B., and LAMB, P. J., 1998. Nonlinearities of observed monthly teleconnections between tropical Pacific sea surface temperature anomalies and central and eastern North American precipitation. *Journal of Climate*, **11**, 1812–1835.

NICHOLLS, N., 1989: Sea surface temperatures and Australian winter rainfall. *Journal of Climate*, **2**, 965–973.

PALMER, T. N. and ANDERSON, D. L. T., 1994: The prospects for seasonal forecasting – A review paper. *Quarterly Journal of the Royal Meteorological Society*, **120**, 755–793.

REYNOLDS, R. W. and SMITH, T. A., 1994: Improved global sea surface temperature analysis using optimum interpolation. *Journal of Climate*, **7**, 929–948.

ROPELEWSKI, C. F. and HALPERT, M. S., 1987: Global and regional scale precipitation patterns associated with El Niño/Southern Oscillation. *Monthly Weather Review*, **115**, 1606–1626.

ROPELEWSKI, C. F. and HALPERT, M. S., 1989: Precipitation patterns associated with the high-index phase of the Southern Oscillation. *Journal of Climate*, **2**, 268–284.

ROWELL, D. P., 1998: Assessing potential seasonal predictability with an ensemble of multidecadal GCM simulations. *Journal of Climate*, **11**, 109–120.

RUTLLANT, J. and FUENZALIDA, H., 1991: Synoptic aspects of the central Chili rainfall variability associated with the Southern Oscillation. *International Journal of Climatology*, **11**, 63–73.

SHEA, D. J. and MADDEN, R. A., 1990: Potential for long-range prediction of monthly mean surface temperatures over North America. *Journal of Climate*, **3**, 1444–1451.

SIMMONS, A. J., 1982: The forcing of stationary wave motions by tropical diabatic forcing. *Quarterly Journal of the Royal Meteorological Society*, **108**, 503–534.

SMITH, T. M., REYNOLDS, R. W., LIVEZEY, R. E., and STOKES, D., 1996: Reconstruction of historical sea surface temperatures using empirical orthogonal functions. *Journal of Climate*, **6**, 1403–1420.

STERN, W. and MIYAKODA, K., 1995: Feasibility of seasonal forecasts inferred from multiple GCM simulations. *Journal of Climate*, **8**, 1071–1085.

TRENBERTH, K. E., BRANSTATOR, G. W., KAROLY, D., KUMAR, A., LAU, N.-C., and ROPELEWSKI, C., 1998: Progress during TOGA in understanding and modeling global teleconnections associated with tropical sea surface temperatures. *Journal of Geophysical Research*, **107** (C7), 14291–14324.

van LOON, H. and SHEA, D. J., 1987: The Southern Oscillation. Part VI: Anomalies of sea level pressure in the Southern Hemisphere and of Pacific sea surface temperature during the development of a warm event. *Monthly Weather Review*, **115**, 370–379.

WEBSTER, P. J., 1981: Mechanisms determining the atmospheric response to sea surface temperature anomalies. *Journal of the Atmospheric Sciences*, **38**, 554–571.

3

Global Modes of ENSO and Non-ENSO Sea Surface Temperature Variability and Their Associations with Climate

DAVID B. ENFIELD

Atlantic Oceanographic and Meteorological Laboratory, NOAA, 4301 Rickenbacker Causeway, Miami, Florida 33149, U.S.A.

ALBERTO M. MESTAS-NUÑEZ

Cooperative Institute for Marine and Atmospheric Studies, Rosenstiel School of Marine and Atmospheric Sciences, University of Miami, Miami, Florida 33149, U.S.A.

Abstract

In this chapter we review much of the recent work by others regarding the nature of the global modes of sea surface temperature (SST) variability and the SST involvement in interannual to multidecadal climate variability. We also perform our own analysis of global SST so as to describe the SST variability associated with El Niño/Southern Oscillation (ENSO) and the low-frequency modes not associated with ENSO (non-ENSO). ENSO is a global phenomenon with significant phase propagation between basins, which we preserve and describe using complex multivariate analysis and subsequently remove from the global SST data. A similar analysis of the residuals reveals three non-ENSO modes of low-frequency variability related to signals described in the reviewed literature: (1) a secular trend representing the global warming signal with associated superimposed decadal variability; (2) an interdecadal mode with maximal realization in the extratropical North Pacific; and (3) a multidecadal mode with maximal realization in the extratropical North Atlantic. Relationships between SST and precipitation are analyzed with regression and multivariate analyses. These analyses show for the interannual-to-decadal timescales of the Western Hemisphere tropics that tropical Atlantic SST is comparable to the Pacific ENSO in its relevance to regional rainfall and is not redundant with respect to ENSO. Moreover, non-ENSO variability explains a significant fraction of the total cross-covariance between the two variables, and out-of-phase relationships between Pacific and tropical North Atlantic SST anomalies are associated with very strong rainfall departures over Central America and the Caribbean. We are led to conclude that present operational climate predictions can be significantly improved by extending numerical SST predictions from the Pacific to the world ocean and by enabling these models to emulate the observed non-ENSO modes of global variability.

Introduction

A great deal of research has been done on the Pacific El Niño/Southern Oscillation (ENSO) (see general reviews by Enfield 1989 and Philander 1990) and its climatic impacts (e.g., Ropelewski and Halpert 1987, 1989). ENSO is without a doubt the strongest, most globally coherent climate signal that exists in both the ocean and the atmosphere. Coupled ocean–atmosphere models developed in the past fifteen years have demonstrated the potential for prediction of ENSO-related Pacific sea surface temperature (SST) fluctuations (Barnston et al. 1994). However, the numerical prediction of anomalous SST (SSTA) variability in other oceans – and their impact upon precipitation, in general – is not yet within reach and may require at least another decade of research. To get beyond the mere prediction of equatorial Pacific "warm events," value-added forecasts of land precipitation based on global numerical SSTA predictions and the empirical relationships between rainfall and SSTA must become a future norm.

One of the most attractive ways to mine the untapped predictability in the ocean–atmosphere system is to predict SSTA in other ocean basins at useful lead times of one to several seasons. Most, but not all, of this additional SSTA variability is not related to ENSO, is less well understood, and is presently unpredictable with operational methods. In this chapter we briefly review what is known about this added dimension of SST variability and its associations with rainfall in the Western Hemisphere. In addition, we will present new analysis of more than a century of global SSTA, its ENSO and non-ENSO components, and their relationships to previous knowledge of climate variability. To do this, SSTA will be separated into a global, propagating ENSO component and the statistical modes of the non-ENSO residuals after removal of the ENSO component. We will describe the nature of both components and will show that the ways in which the Pacific and Atlantic SSTA combine are of fundamental importance to the nature of precipitation anomalies in the tropical Americas.

Non-ENSO Variability: A Review

It has been amply demonstrated that non-Pacific modes of SSTA are related to land climate, especially rainfall. Folland et al. (1986) showed that meridionally antisymmetric contrasts in global SSTA between the Northern and Southern Hemisphere have a significant influence on rainfall over the African Sahel. The many other studies of Atlantic SST and Northwest African rainfall include those by Citeau et al. (1989) and by Lamb and Peppler (1992), who show how the linkage works through the effects of meridional gradients in SSTA on the meridional position of the intertropical convergence zone (ITCZ, where the northern and southern trade wind regimes converge and convective activity and rainfall are large; see also Wagner 1996). More than a decade of research by Hastenrath and colleagues has documented the combined effects of eastern Pacific (mostly ENSO) and tropical Atlantic SSTA on inter-American rainfall (Hastenrath 1978, 1984; Hastenrath et al. 1987; Hastenrath and Greischar 1993). A number of studies have demonstrated the powerful effects of tropical Atlantic SSTA (especially its meridionally antisymmetric component) on the rainfall in Northeast Brazil (e.g., Moura and Shukla 1981; Nobre and Shukla 1996).

The above studies mainly document relationships between SSTA and rainfall in the near-equatorial tropics. As with ENSO, however, Atlantic SST variability can affect rainfall at higher latitudes as well. Enfield (1996) shows that tropical North Atlantic SSTA has an effect on rainfall comparable to that of the Pacific ENSO throughout the Intra-American Sea (IAS) and surrounding land regions from northern South America to the southern United States (Fig. 3.1).

While noting that equatorial Pacific and tropical Atlantic SSTA are intercorrelated through the extension of the Pacific ENSO signal into the Atlantic (Enfield and Mayer 1997), Enfield (1996) observes that the Atlantic SSTA–rainfall correlations over the IAS

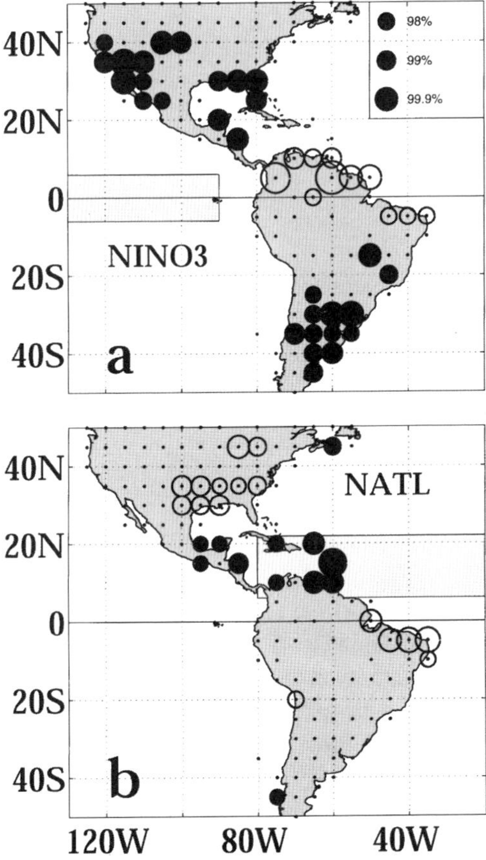

Fig. 3.1 Relative magnitudes (circle diameters) and sign (solid positive, unfilled negative) of maximum lagged (0–3 seasons) correlations between U.S. Department of Energy 5-degree by 5-degree gridded rainfall departures and the NINO3 (a) and NATL (b) indices of sea surface temperature anomaly variability, defined as simple averages over the regions $\pm 6^\circ$, 90–150°W and 6–22°N, 15–80°W, respectively. The inset legend shows the circle diameters corresponding to several significance levels (serial correlation accounted for). The very small dots show grid points where data exist.

region are mostly of the wrong sign to be explained by the intercorrelation of SSTA between basins. Apparently, the tropical Atlantic has non-ENSO seasonal-to-interannual SST variability related to climate, but the Atlantic extension of the ENSO SST signal through tropospheric forcing (Hameed et al. 1993; Enfield and Mayer 1997) tends to mask and complicate interpretation of the Atlantic-only processes that may exist. The correlation between equatorial Pacific SSTA and large-scale averages of tropical North Atlantic SSTA is about 0.5; hence about three fourths of the Atlantic variability is non-ENSO in nature. The global distribution of the Pacific ENSO SST signal is a problem for the detection and attribution of non-ENSO SST–climate relationships and preferably should be accounted for prior to further analysis.

One such Atlantic-only relationship is the tendency noted by some for the tropical North Atlantic (5°–25°N) and South Atlantic (0°–20°S) SSTA regions to correlate antisymmetrically across the ITCZ in what is sometimes referred to as the tropical Atlantic dipole (Weare 1977; Servain 1991) and for such covariability to be strongly related to rainfall anomalies in northwest Africa and northeast Brazil (e.g., Moura and Shukla 1981; Nobre and Shukla 1996). However, statistically significant dipole variability is difficult to detect at seasonal-to-interannual timescales (Houghton and Tourre 1992; Enfield and Mayer 1997) perhaps due to the masking influence of the extended ENSO signal, which is not antisymmetric in nature (Enfield and Mayer 1997).

The tropical Atlantic dipole is but one aspect of Atlantic variability from Antarctica to Greenland. Both observations and models indicate that the interannual variability associated with the tropical North Atlantic extends to the high northern latitudes in the form of broad, co-oscillating zonal bands of alternating phase, and with enhanced variance in the western portions off North America (e.g., Delworth 1996). North of the tropical North Atlantic lies an anticorrelated band, in the range of 30°–45°N, followed by an in-phase band south of Greenland (45°–65°N), while to the south of the tropical South Atlantic lies another in-phase band.

It appears that the primary cause of interannual SSTA variations at non-equatorial latitudes is through thermodynamical air–sea fluxes across the sea surface and not through ocean dynamics (Delworth 1996; Enfield and Mayer 1997). The anomalous fluxes are associated with wind fluctuations that are in turn a consequence of anomalous tropospheric circulation patterns such as the Pacific North American (PNA) and North Atlantic Oscillation (NAO). These are two of the more significant "teleconnection patterns" in tropospheric pressure fields, which have been documented by Barnston and Livezey (1987), Wallace et al. (1990), and others. The PNA pattern is the one associated with ENSO variability; it has its strongest node over the North Pacific, being associated with fluctuations in the Pacific subtropical jet stream and with ENSO-related climate anomalies over the continental United States. The NAO has a strong node over Greenland and an antinode over the subtropical North Atlantic and is associated with fluctuations in the North Atlantic westerly winds and with climate over central and northern Europe (Hurrell 1995). The PNA and NAO pressure patterns are both associated with Atlantic subtropical pressure fluctuations, which in turn affect the strength of the low-latitude North Atlantic trade winds. The ENSO-related PNA fluctuations are the most likely explanation for the extension of ENSO variability into the tropical North Atlantic SSTA, as noted by Enfield and Mayer (1997).

An important dimension to non-ENSO variability is its existence at decadal and longer time scales in both the Pacific and Atlantic. The tropical Atlantic dipole is possibly one example of this. While not ubiquitous at the ENSO seasonal-to-interannual timescale, the dipole emerges more clearly at longer, 10–20 year periodicities (e.g., Carton and Huang 1994; Mehta and Delworth 1995; Huang et al. 1995; Chang et al. 1997). Because these interdecadal variations affect the ITCZ and are associated with extended droughts in northeast Brazil and northwest Africa, there is a great need to improve our understanding of them.

Another example of interdecadal variability (from observations) is found in a Pacific SSTA mode, linearly independent of ENSO (Deser and Blackmon 1995; Latif and Barnett 1996). It has a strong nodal region centered along 40°N and west of 140°W and a co-oscillating companion region east of Australia. From 1949 through the early 1990s, the 40°N nodal region oscillated between multiyear warm periods and cool periods. Simultaneously, the tropical Pacific (±20) and the high latitudes off the west coasts of North and South America underwent shifts of opposite sign. A number of ocean–atmosphere interactive mechanisms have been proposed (using models) to account for this cyclicity (e.g., Latif and Barnett 1994; Gu and Philander 1997). The spatial pattern of the mode bears a certain resemblance to the ENSO SSTA pattern, but how they are related, or whether one merely provides a slowly varying background for the other, is not understood.

A longer, multidecadal cycle exists in the Atlantic SSTA north of about 30°N, which involves cool temperatures prior to about 1940, warm conditions during 1951–67, and cool conditions again during 1968–77 (Kushnir 1994; Hansen and Bezdek 1996). This oscillation coincides with a similar long-period fluctuation in the North Atlantic sea level pressure (SLP) field and midlatitude zonal winds (Kushnir 1994) and in the NAO index (McCartney et al. 1996; Hurrell 1995). It also agrees with the North Atlantic SSTA variability seen in the third rotated empirical orthogonal function (EOF) mode of Kawamura (1994). These features may be related through ocean–atmosphere interactions involving a waxing and waning of the mid–North Atlantic westerly winds, slow advection of SST anomalies around the subtropical and subpolar gyres, and subduction of SSTA to greater depths in the Labrador Sea (McCartney et al. 1996; Curry and McCartney 1996). The water mass pathways for this hypothesized mechanism are shown schematically in Figure 3.2. The primary climatic impact of the oscillation derives from its association with alternating, multiyear periods of mild and severe winters in central Europe and the Mediterranean Sea (McCartney et al. 1996; Kerr 1997).

The Global ENSO in SST

Before we can explore what is *not* ENSO in global SSTA, we must first define the ENSO SSTA signature, compute it, and describe it. Once a reasonable space-time

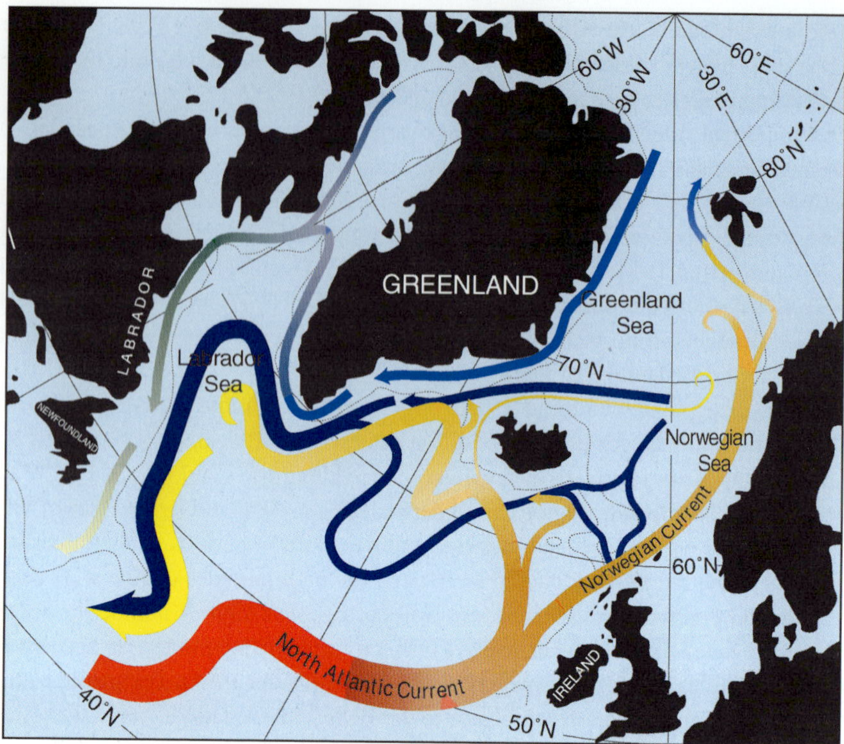

Fig. 3.2 Schematic illustration of the pathways associated with the advection and transformation of subtropical and subpolar surface water masses, wherein sea surface temperature anomalies eventually modify newly formed deep water through subduction, to complete a long-term climate cycle. The figure is a reproduction from McCartney et al. (1996), who describe this mechanism to explain the multidecadal climate variability.

representation is obtained, we will subtract it from the data and examine the residuals. We must use a global data set so as not to limit ENSO to the tropical Pacific, because tropospherically connected ENSO signals elsewhere in the world ocean (Covey and Hastenrath 1978; Hastenrath et al. 1987; Latif and Barnett 1995; Tourre and White 1995; Lanzante 1996; Enfield and Mayer 1997) also affect the global climate (Lau and Nath 1994). We also require a century-scale data set to detect secular trends and resolve decadal to multidecadal fluctuations.

The 1856–1991 reconstruction of historical ship-based data by Kaplan et al. (1998, henceforth K) is ideally suited for this purpose. The K analysis uses the covariance structure of the global SST anomaly field to fill missing data in a statistically optimal way, and it reconstructs the data based on the first 80 empirical modes of the modern portion of the data record in a manner analogous to that of Smith et al. (1996). Such reconstructions are not well suited to analysis on small regional scales, because they smooth out the variability at high frequencies and small space scales. However, for the basinwide scales typical of ENSO and interdecadal variability, they pose special advantages, especially for early periods that were more sparsely sampled than the modern era.

The K data are first smoothed with a low-pass filter that eliminates periodicities of 1.5 years or less. These smaller timescales include intraseasonal and month-to-month variations associated with the smaller regional space scales for which the reconstructed data are less appropriate. For the calculation of the global ENSO component we apply an additional, high-pass filter to temporarily remove periodicities longer than 8 years, yielding a 1.5- to 8-year band pass for the ENSO component. When the ENSO component is subsequently removed from the smoothed K data, the residuals contain non-ENSO variability at all periodicities longer than 1.5 years. Trends are not eliminated, because they are associated with secular variations that are probably real and correspond to observed global warming. They are removed from the ENSO analysis by the 8-year filter but reappear in the residual (non-ENSO) variability and are captured in one of the non-ENSO modes.

EOF analysis, also known as principal component analysis (PCA), consistently extracts from SSTA data a dominant ENSO-related mode as a first or second component. See, for example, the first rotated EOF of Kawamura (1994). Hence, one approach for defining and removing the global ENSO is to compute the EOFs of the band-passed global SSTA and subtract the first-mode data reconstruction from the original data. A similar approach was adopted by Cane et al. (1997) to isolate century-scale trends from the effects of low-frequency changes in ENSO behavior. However, whereas the dominant mode of Pacific variability associated with ENSO appears quasi-stationary (in or out of phase), meridional phase propagation of SSTA occurs over a significant region offshore of the eastern Pacific boundary, while other research indicates that tropospherically connected phase propagations to other ocean basins also exist (e.g., Latif and Barnett 1995). Ordinary EOF analysis misrepresents such propagations, and the quadrature component of variability associated with them will not be included in the dominant mode. Hence it will not be removed and will "contaminate" the residual data. Because we are especially interested in the non-ENSO variability found in other ocean basins, a modified analysis (without this drawback) must be applied.

To account for the phase propagations, we perform a standard eigenvector decomposition on a transformed, complex data set that comprises the band-passed K data (real part) and its Hilbert transform (imaginary part). The standard EOF procedure applied to the complex data then yields a complex EOF (CEOF) result in which spatial and temporal phase information is preserved in the ENSO-related first mode (Rasmusson et al. 1981). Lanzante (1996) used a very similar approach to analyze the amplitude and phase structure of the ENSO mode in the global tropics for the 1875–1979 time period, but without first eliminating decadal and longer timescales.

The spatial amplitude, spatial phase, and temporal realization of the first global CEOF mode are shown in Figures 3.3a, 3.4a, and 3.5a, respectively. The mode explains 34.4% of the global variance in the band-passed data and 17% of the low-passed K data. The local explained variance in the regions of high (low) amplitudes (Fig. 3.3a) is of course much larger (smaller) than the global amount. In the discussions of Figure 3.3 that follow, spatial amplitudes occurring in the upper half of the color palette denote regions of high, or primary, loading (importance to the mode), blue colors denote intermediate, or secondary, loading, and magenta denotes low, possibly insignificant, loading.

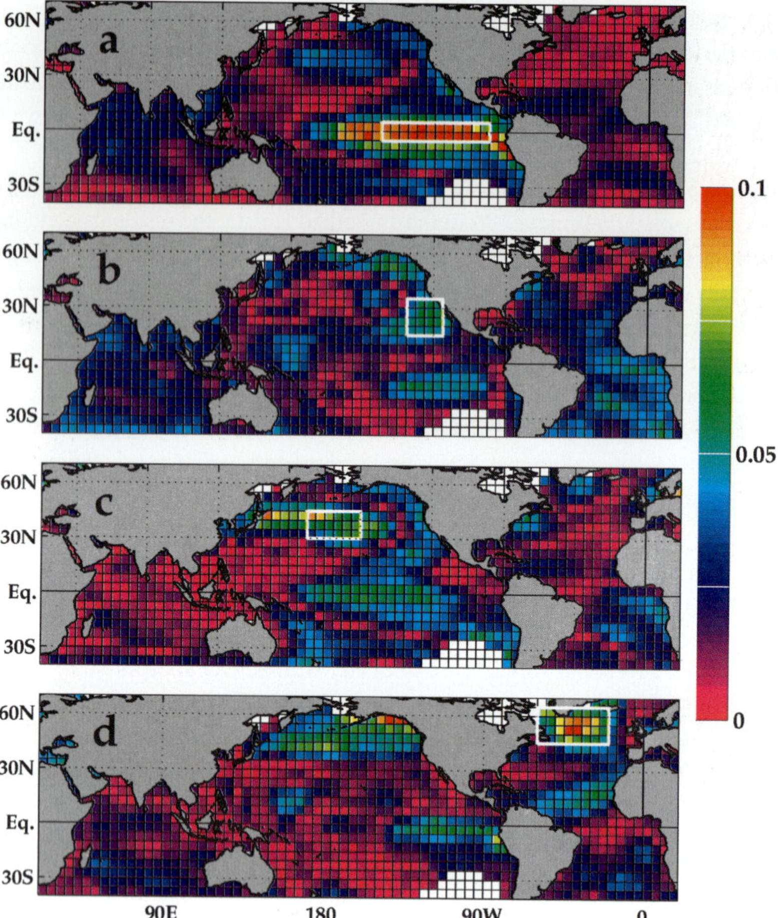

Fig. 3.3 Distributions of spatial amplitudes for the first complex empirical orthogonal function (CEOF) eigenvector describing global ENSO variability (a), and for the first three CEOF eigenvectors describing non-ENSO variability (b, c, d), as discussed in the text. The white rectangles on the maps outline areas of high amplitude discussed in the text and used to form the temporal realizations (Fig. 3.5).

Here (as also in the non-ENSO modes) we choose a rectangular index region (one not too large) where spatial amplitude is large and spatial phase is fairly uniform to serve as a reference for the temporal variability. The obvious reference rectangle for the ENSO mode is the well-known NINO3 region in the equatorial Pacific, bounded by 5°N–5°S, 90°W–150°W. In Figure 3.5a we show the temporal realization (temperature units) for phases and amplitudes combined over NINO3 rather than the more confusing temporal amplitude and phase functions from which they are derived (i.e., the complex expansion coefficients for the mode). The time variability elsewhere is merely the same series lagged by the appropriate spatial phase (Fig. 3.4a) and with amplitude scaled in proportion to the spatial amplitude (Fig. 3.3a). The color palette for the spatial phase

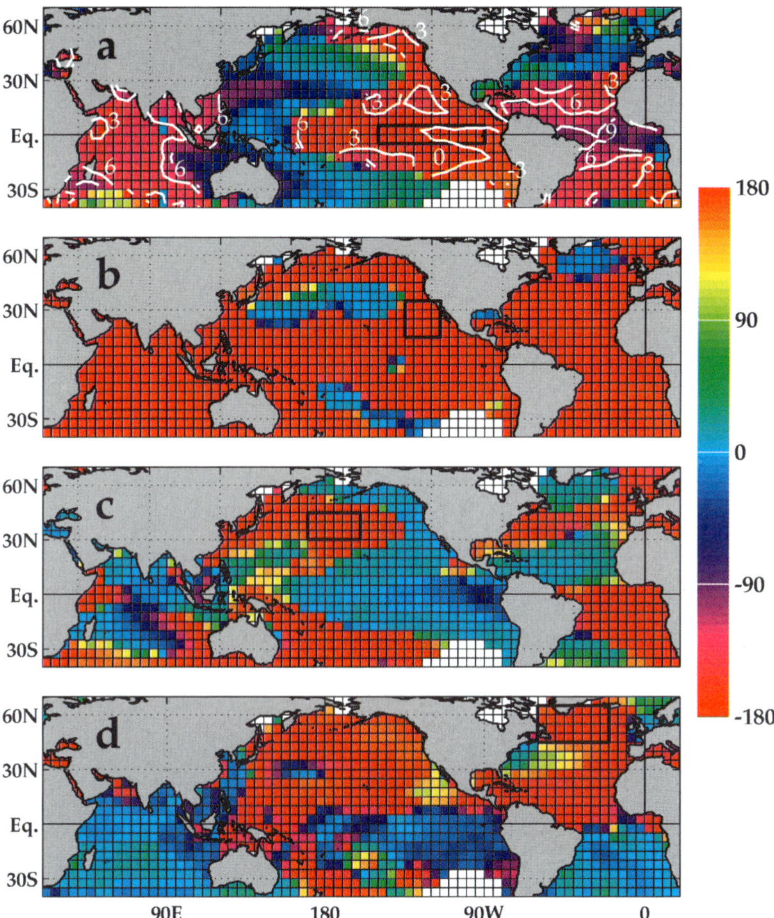

Fig. 3.4 Distributions of spatial phases (degrees) for the first complex empirical orthogonal function (CEOF) eigenvector describing global ENSO variability (a), and for the first three CEOF eigenvectors describing non-ENSO variability (b, c, d), as discussed in the text. The black rectangles on the maps outline areas for which the spatial phase is referenced to $\pm180°$. Phase advances in the direction of higher values. Lag contours are shown in the ENSO distribution (a) for lags of -3, 3, 6, and 9 months relative to the average phase for the NINO3 region.

(Fig. 3.4a) has been rotated so that the average spatial phase over the NINO3 region corresponds to ±180 degrees (red) and zero lag (contours in months). There are 37 zero upcrossings in the 135-year NINO3 series (Fig. 3.5a), yielding an average period of 43.7 months. This is about midway in the 3- to 4-year range usually attributed to ENSO variability and corresponds to 8 degrees of spatial phase per month of lag.

Our treatment of the time realization requires further explanation. It may be unfamiliar to specialists accustomed to seeing temporal amplitude and phase functions, but the latter would be difficult to interpret for the nonspecialist. One must understand that Figure 3.5a is not an average of the data over the NINO3 region. Rather, it is the average

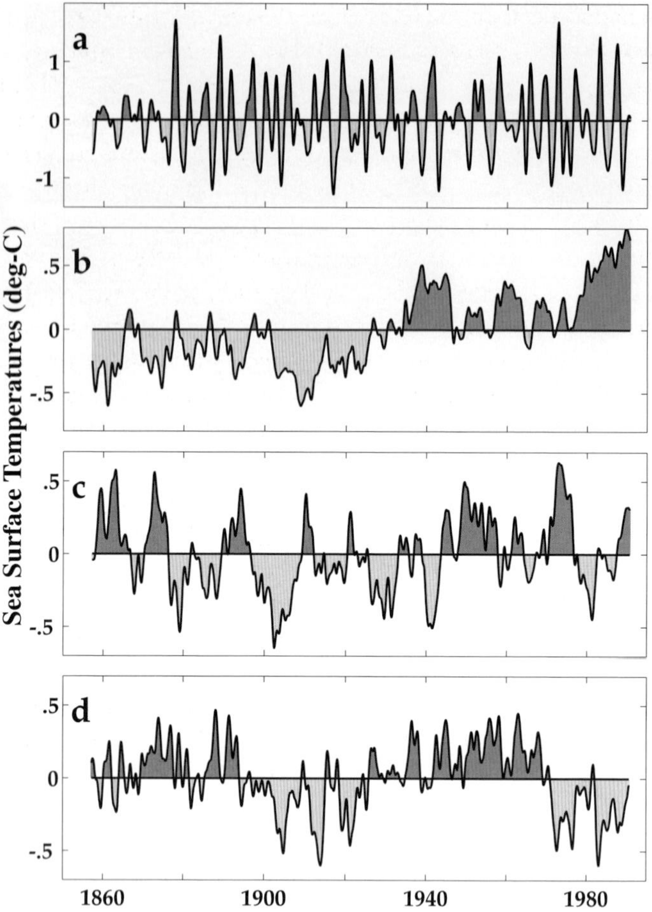

Fig. 3.5 Temporal realizations for each of the complex empirical orthogonal function (CEOF) modes in Figures 3.3 and 3.4, computed by averaging the modal reconstructions over the respective rectangular regions shown in the spatial distributions.

over the NINO3 region of the SSTA reconstructed from the ENSO mode. It is therefore the ENSO mode contribution to the data-based version of NINO3 and as such has some differences with respect to the latter. Notice, for example, that the 1982–83 warm event (El Niño) is not the strongest ENSO fluctuation in the record, as a similarly filtered average of the SSTA data would show. That is because part of that event's amplitude is actually in the slowly varying background state that was warm during the 1980s. That background condition is captured by the slower non-ENSO modes discussed in the next section.

The spatial functions (Figs. 3.3a, 3.4a) show the classical features normally associated with ENSO in the Pacific: (1) a region of intense amplitude within $\pm 5^{\circ}$–10° of the equator and east of the date line, decreasing to smaller amplitudes over a wide, wedge-shaped region that spreads poleward along the eastern boundary; (2) phase propagation northward along the coast of North America, with maximum lags of one to two seasons in the Gulf of Alaska and Bering Sea; and (3) regions of intermediate amplitude and opposite phase in the central North Pacific (30°–45°N), the western South Pacific (20°–40°S), and the tropical western Pacific. Less documented features include one-season lags in the central Pacific near ±20° and a precursor off central Chile with a one-season lead. Although a large region of the southeast Pacific has few data (unshaded area, offshore), the latter feature lies along the coastal ship route north of Cape Horn that has been relatively well sampled since the nineteenth century, and the precursor has been previously mentioned by Rasmusson and Carpenter (1982). A possible cause of the precursor is a weakening in the southeast trades (and associated surface heat fluxes) off Chile, prior to the main trade wind weakening at lower latitudes.

Both the tropical Atlantic and Indian Oceans show known coherence and lag structures relative to the equatorial Pacific. The tropical Atlantic is lagged by 3–9 months with a phase propagation from higher to lower latitudes. Regions of maximum amplitude occur near ±20° with a two-season lag, consistent with the results of Enfield and Mayer (1997). The Indian Ocean is similarly lagged with eastward phase propagation, consistent with Latif and Barnett (1995). Very little coherent variability is found in the North Atlantic, north of 20°–30°N. The lack of coherent variability off Angola (0°–20°S) and the somewhat larger lags in the western equatorial Atlantic are qualitatively consistent with the structures found by Enfield and Mayer (1997).

The above characteristics of the global ENSO are consistent with the similar but more detailed analysis of Lanzante (1996). Although Lanzante does not discuss this, his breakdown of the 1875–1979 data period into three shorter periods shows that the larger lags in the equatorial Atlantic existed only during 1950–79 and not during the previous 75 years. His tropical Atlantic lags are quite uniform during the earlier periods and are similar to the modern lags in the 5°–15°N band. It is not clear whether the meridional lag structure in the Atlantic (Fig. 3.4a) is a nonrobust feature of the system or one that is more ubiquitous but poorly captured under the sparser sampling of the earlier periods.

The non-ENSO SST Variability

To form a non-ENSO global data set, the SSTA reconstruction from the leading CEOF in the ENSO band (Figs. 3.3a, 3.4a, and 3.5a) was subtracted from the low-passed (1.5-year) K data, and the CEOF modes were recalculated. The first three CEOF modes account for 16%, 9%, and 7% of the variance in the non-ENSO residuals and 15%, 8%, and 5% of the low-passed K data (before removal of ENSO). The latter amounts can be compared with the 17% explained by the ENSO component. The spatial amplitudes, spatial phases, and temporal realizations for the first three non-ENSO modes (panels b, c, d) are shown in Figures 3.3, 3.4, and 3.5, respectively. The rectangular box shown in each spatial pattern is a region of large spatial amplitude (Fig. 3.3) and uniform spatial phase (Fig. 3.4) to which each temporal mode realization (expansion coefficient) was referenced. In the case of modes 2 and 3, the boxes overlap with key regions discussed by others and referred to in the review section.

All of the spatial patterns correspond to quasi-stationary oscillations wherein significantly loaded antipodal regions are 180° out of phase with respect to each other and propagating phases are infrequent. The areas of low loading frequently occur in regions of rapid phase change between spatial antipodes, as we would expect of a stationary oscillation. Also as expected of stationary oscillations, very similar patterns emerge from an ordinary EOF analysis of the low-passed data (not shown). As a check on the robustness of the modes to filtering, the CEOF procedure was also applied to both the raw (unfiltered) non-ENSO residuals and very low pass (8-year cutoff) K data. In both cases, all three of the residual modes shown emerged within the first four modes. Only in the case of the unfiltered data, a mode dominated by very high frequencies was promoted with respect to mode 3 of this analysis, while modes 1 and 2 always occurred as before.

Mode #1: Global Warming

It is clear from the temporal realization of mode 1 (Fig. 3.5b) that this mode contains the global warming signal. Comparison with Jones et al. (1986) and Houghton et al. (1996) confirms that this signal is the ocean counterpart to the global warming seen in surface air temperatures. Areas of significant loading (Fig. 3.3b) are as extensive as in the global ENSO mode. The dominance of red colors in the spatial phase (Fig. 3.4b) shows that warming occurs almost everywhere except for secondary cooling in the midlatitudes of the central and western North Pacific. More intense warming occurs in the Gulf of Alaska and in the core regions of the trade wind belts, especially in the east-central parts of the Pacific and South Atlantic. In his rotated EOF analysis of SSTA, Kawamura (1994) appears to have captured parts of our global warming mode in his second (Pacific and Indian Oceans) and fourth (South Atlantic) modes.

The least-squares-fitted linear trend to Figure 3.5b is 0.55°C per century, as compared with 0.22°C per century for the trend in the globally averaged SST (not shown). The difference reflects the greater degree of increase in the regions of primary warming (Fig. 3.5 is indexed to the box regions in the spatial distributions). After 1900, the trends are 0.93°C and 0.42°C per century, respectively, while prior to 1900 the warming is negligible. The residuals about the trend show significantly higher spectral energy at periodicities of 1–2 decades, also seen in the globally averaged data, and the oscillations are also larger in the regions of high loading than they are globally.

Mode #2: Pacific Interdecadal

Mode 2 is dominated by the central and eastern Pacific and has little significance in the Indian or western tropical Pacific Oceans. Warmings in the central North Pacific reference region are in phase with energetic fluctuations along an extensive midlatitude zonal band (30°–50°N) stretching from Japan to about 140°W. Another area of in-phase variability is found south of 10°S and east of Australia, near the date line. Antipodal variability, out of phase with the reference region, is strongest along the equator in the central Pacific and extends poleward and eastward over a large triangular region of the Tropics and along the extratropical Pacific coasts of North and South America. The Atlantic in-phase regions are the east-central portion of the tropical South Atlantic and the northwest North Atlantic region bounded by North America, the Gulf Stream extension, and the 55°W meridian. Atlantic antipodal variability occurs southeast of Greenland and to a lesser degree in the eastern Caribbean and east of southern Brazil near 20°W.

This pattern is coherent with the distribution of correlation between SST observations and a North Pacific SST index bounded by 25°–40°N, 170°E–160°W (Latif and Barnett 1996). The differences in the spatial distributions are very minor and can probably be explained by the much shorter time period (1949–92) used by Latif and Barnett. The extended time series for their index, calculated from the low-pass-filtered (1.5-year) K data, has a correlation of 0.65 with our mode 2 realization. Thus, mode 2 efficiently captures the Pacific interdecadal mode discussed in the review section.

Mode #3: Atlantic Multidecadal

The third mode is dominated by two realizations of a 70- to 80-year cycle with weaker, interannual variability superimposed (Fig. 3.5d). Its spatial amplitude is largest in the far North Atlantic southeast of Greenland and north of the Gulf Stream extension (Fig. 3.3d). The mode is also strong north of 40°N in the North Pacific and has secondary regions of moderately strong activity in the eastern equatorial Pacific and the tropical North Atlantic. Significant variability also occurs in the Indian Ocean and the South Atlantic. Especially interesting is the alternation of phase between the Northern and Southern Hemispheres (Fig. 3.4d).

In more respects than not, this mode corresponds to the second EOF of global SSTA (1949–92) computed by Nicholls et al. (1996), including its strength in the North Atlantic, the southern Indian and Atlantic Oceans, and the east-central equatorial Pacific, as well as in the alternation of phase between the Northern and Southern Hemispheres. The polarity shift (circa 1965–70) noted by Nicholls et al., from a cold to a warm (warm to a cold) background condition in the Southern (Northern) Hemisphere, is also reproduced in the temporal realization (Fig. 3.5d) but is now seen to be an alternating transition that occurs at roughly 35- to 40-year intervals. The temporal variation also corresponds to a similar cycle of warming and cooling noted by others for the western extratropical North Atlantic (Kushnir 1994; Hansen and Bezdek 1996; McCartney et al. 1996). These variations are known to have a strong association with tropospheric pressure fields (Kushnir 1994), wind strength in the midlatitude westerlies (Deser and Blackmon 1993), and the NAO (Hurrell 1995; McCartney et al. 1996). While the focus of ocean–atmosphere interaction appears to be in the Atlantic sector, its ramifications are clearly global in scope.

Relationships between SSTA and Rainfall

Difficulties with rainfall data have been an impediment to understanding the SST–rainfall relationships at the decadal to multidecadal timescales shown in the non-ENSO SST modes. Reliable rain gauge records of a century or more that can resolve these longer timescales are scarce, are difficult to obtain, or present special challenges due to changes associated with instrumentation, station location, and land use practices. However, the potential for the low-frequency changes in SST to influence climate can be seen in the many analyses of the joint variability of SSTA and tropospheric pressure and wind fields (e.g., Wallace et al. 1990; Deser and Blackmon 1993; Kushnir 1994; Trenberth and Hurrell 1994). It is possible that the ocean plays the role of a pacemaker, introducing persistence into the tropospheric variations, which in turn can condition longer term variations in rainfall. In this section we will recap other research regarding the relationships of rainfall at the long timescales to our non-ENSO modes of SST variability. In the following section we also present new analyses that illustrate certain principles as they apply to the interannual–decadal timescale.

Comprehension of global trends in rainfall and their attribution to the global warming mode (Figs. 3.3b, 3.4b, and 3.5b) are elusive at best and benefit from ancillary analysis, such as through modeling, to achieve added plausibility. Even where usable long rainfall records exist, the associations of secular changes in rainfall with possibly related SST trends, while common, have little or no statistical significance. Nevertheless, a few studies have documented precipitation effects more directly. Thus, when long rainfall records are averaged over land areas (post-1900), there is a small upward trend in rainfall globally of 1% and there are much larger areas of rainfall increase than of decrease (Houghton et al. 1996). According to Hurrell (1995), "the recent warming may be related to increasing tropical ocean temperatures that have led to an enhancement of the tropical hydrological cycle."

The Pacific interdecadal mode (Figs. 3.3c, 3.4c, and 3.5c) appears to be related to precipitation over North America, downstream from the Pacific focus of the principal ocean–atmosphere interactions. Latif and Barnett (1996) have described an association between this mode and interdecadal climate oscillations over North America. Ting and Wang (1997) note two modes of influence on summer rainfall in the central United States. One, related to the tropical Pacific, is interannual and clearly involves ENSO. The other mode involves the extratropical North Pacific and exhibits interdecadal variability with a spatial distribution of SSTA that strongly resembles our second non-ENSO mode in the Pacific. When SSTA is positive in our central North Pacific index region, their analysis shows a decreased SST gradient south of there, reduced intensity in the overlying mid-Pacific jet stream, and increased precipitation downstream over the east-central United States (30°–40°N). Neither the winter precipitation nor the effects of Atlantic SSTA were considered in their study.

In analogous fashion, climate variability in Europe is related to North Atlantic SSTA, which in turn corresponds to our third mode of non-ENSO SSTA (Figs. 3.3d and 3.5d). The relationship begins with the observation that the ocean temperatures in and near the Labrador Sea (Fig. 3.3d, box area) are inversely related to the NAO and the strength of the midlatitude westerly winds (McCartney et al. 1996; Kerr 1997). During winters in which SST is low and the NAO is significantly stronger than normal, the North Atlantic westerlies strengthen and tilt along an axis directed more to the northeast, bringing more moisture to Scandinavia and leaving central Europe and the Mediterranean relatively dry (Hurrell 1995). The covariations of rainfall and SSTA occur on the slow timescale seen in Figure 3.5d, possibly conditioned by North Atlantic Ocean circulation and the formation of anomalous deep water in the Labrador Sea (McCartney et al. 1996; see Fig. 3.2).

Interactions between Oceans

At present, operational climate predictions are based only on the prediction of ENSO-related SST in the Pacific. Neither the non-ENSO modes such as are discussed in this chapter nor the Atlantic and Indian Ocean extensions of ENSO are considered. In this section we argue that predictions should be global and that non-ENSO variability should be included.

The advantage of implementing predictions globally may seem obvious. But, since each of the various modes of variability (ENSO and non-ENSO) appears to be weighted more strongly in one of the three oceans, one might ask whether SSTA needs to be predicted globally, as the modal extension into the other basins might be considered redundant. The numerical model experiments of Lau and Nath (1994) illustrate why this is not the case. In their experiments, a model atmosphere was alternately forced by observed SST in a global ocean (Global Ocean–Global Atmosphere, GOGA; tropics and extratropics, all basins), a tropical ocean (Tropical Ocean–Global Atmosphere, TOGA; tropical Pacific), and a Midlatitude Ocean–Global Atmosphere (MOGA; extratropical Pacific). The outcomes were arbitrated by comparing the singular value decomposition (SVD) of SST and Northern Hemisphere 500 mb height for each model run against an SVD for global SST and Northern Hemisphere 500 mb data sets. Both GOGA and TOGA reproduced qualitatively the principal features in the data (both being dominated by the known SST and 500 mb patterns associated with ENSO) while MOGA came in a poor third. However, only GOGA quantitatively approached the observations, whereas the 500 mb pattern intensity for TOGA was significantly less intense. The implication is that ENSO-related SST variability outside the tropical Pacific, though redundant in the time domain, reinforces the Pacific and results in a more realistic atmospheric pattern. Although only the first SVD mode was examined, dominated by ENSO and interannual variability, a similar principle presumably applies to the non-ENSO modes discussed in this chapter.

We might also inquire as to what extent signals in other basins can compete with the ENSO signal in climate or add to ENSO-based predictability. This is not clear even if we choose alternative predictors in a non-Pacific basin, because the ENSO signal is global and therefore some degree of intercorrelation (involving ENSO) exists between basins, confounding the identification of other relationships. For example, both the NINO3 and tropical North Atlantic (NATL, see Fig. 3.1) have comparable rainfall correlation patterns over much of the Americas. But how does this change when the intercorrelations of SSTA between ocean basins are accounted for?

To check on this, we performed a multiple linear regression of gridded rainfall anomalies on three SSTA indices: NINO3 (±6°, 90°–150°W), NATL (6°–22°N, 15°–80°W), and SATL (22°S–2°N, 10°E–35°W). The 1979–95 merged (satellite-derived and land rain gauge) rainfall data set of Xie and Arkin (1996) was used for this purpose, and the data at each 2.5- by 2.5-degree square were separately regressed on the predictors. First,

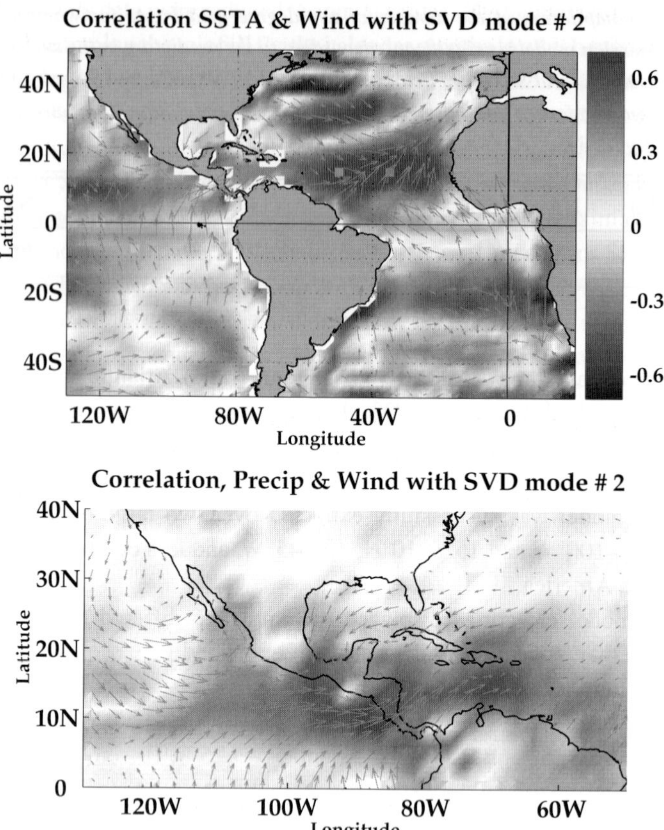

Fig. 3.7 Second mode of a singular vector decomposition (SVD) analysis of sea surface temperature anomalies (SSTA) and Xie and Arkin (1996) rainfall anomaly over the spatial domains shown in the upper and lower panels, respectively. The upper panel shows the homogeneous correlation map for gridded SSTA versus the temporal expansion coefficients for SSTA (colored shading). The lower panel shows the homogeneous correlation map for rainfall versus the temporal expansion coefficients for rainfall (colored shading). Arrows are the vector correlations for the gridded Comprehensive Ocean-Atmosphere Data Set (COADS) surface wind anomalies versus the respective expansion coefficients.

two variables are spatially (eigenvector maps) and temporally (expansion coefficients) associated with each other. However, for each mode there are two (not one) spatial maps and two sets of temporal expansion coefficients. In this case they correspond to SSTA (upper panel) and rainfall anomaly (lower panel).

The covariability of SSTA and rainfall anomaly associated with ENSO emerge in the first mode (not shown), which accounts for 72% of the squared cross-covariance. In terms of SSTA, this mode is virtually identical to what we think of as ENSO, with a correlation of 0.97 between the SSTA expansion coefficients and the NINO3 time series. In this mode, a strongly warm equatorial Pacific (El Niño) combines with a moderately warm tropical North Atlantic and cool tropical South Atlantic to produce a rainfall pattern that corresponds qualitatively to ENSO-related features already recognized from other studies (e.g., Ropelewski and Halpert 1987, 1989). The next three modes correspond to covariability between SSTA and rainfall that is unaccounted for by the first (ENSO) mode. They account for 16% of the squared cross-covariance but correlate significantly only with NATL and/or SATL. In the second mode (Fig. 3.7) the correlation of the SSTA expansion coefficients with NATL is 0.58, and the correlation with NATL-SATL (dipole) is 0.69.

Figure 3.7 shows that a strongly warm tropical North Atlantic combines with a moderately cool tropical Pacific and South Atlantic to produce positive departures of rainfall over Central America and the Caribbean as well as in the range of the eastern Pacific ITCZ. Under such circumstances, the vector correlations of the surface wind anomalies from the Comprehensive Ocean-Atmosphere Data Set (COADS; Woodruff et al. 1987) with the second mode expansion coefficients for SSTA (upper panel) show that the northeast trades are weakened over this region and over the tropical North Atlantic. This weakening is consistent with the analysis of Enfield and Mayer (1997), wherein reductions in the strength of the North Atlantic trade winds lead to warmings of the tropical North Atlantic.

The vector correlations of the wind anomalies with the temporal expansion coefficients for the rainfall side of the SVD mode are shown in the lower panel of Figure 3.7. They show that strong fluctuations in the easterly winds occur in the same area as the strong rainfall response, in the range of the ITCZ. In other words, a cool eastern Pacific, when combined with a strong positive dipole in the Atlantic, is associated with excess rainfall and reduced easterly wind flow over the Caribbean, Central America, and the eastern Pacific ITCZ. In that region, the rainfall response (i.e., the explained covariability between SSTA and rainfall) is much stronger than in the ENSO mode. The enhanced rainfall is consistent with greater tropospheric instability and convection and with greater low-level wind convergence within the ITCZ. The sense of the wind correlations suggests an accentuation of the southwest monsoonal flow south of the ITCZ that is normally associated with enhanced convection and rainfall during the boreal summer.

It is noteworthy that the sense of the Pacific SSTA in mode 2 is reversed with respect to that of the Atlantic dipole, as contrasted to the first (ENSO) mode, where the equatorial Pacific and the tropical North Atlantic SSTA have the same sign. In the third and fourth modes (not shown), the tropical Atlantic assumes a monopole configuration and combines with a cool or warm Pacific, respectively. We attribute no unique physical significance to any of the higher modes (2, 3, 4). However, these modes tell us that non-ENSO SST variability in the Atlantic combines in various ways with variability in the Pacific to explain a sizable fraction (0.16) of the squared cross-covariance between the SSTA and rainfall data sets. We can also see that when the east Pacific and tropical North Atlantic SSTAs are oppositely signed, the anomalous rainfall response over the Caribbean and most of Central America is enhanced. Further research will be required to find out why this is so.

There is an unresolved question of seasonality in the relationships described here, because the SVD analysis was performed for all months of the year. However, the first mode is overwhelmingly dominated by the ENSO-related SSTA variability in the

Pacific cold tongue (NINO3). Since every peak (positive or negative) in the NINO3 time series occurs in the November-December-January period, we can safely infer that the mode 1 rainfall relationships correspond to the boreal winter. The consistency of mode 1 with the results of Ropelewski and Halpert (1987) supports this inference. Because the large rainfall response in mode 2 occurs in a region where a sharp dichotomy exists between the very dry months of the boreal winter and the much wetter months of the summer, it is most likely a summer phenomenon (May through November).

Conclusions

In this chapter we have combined new analyses of SST and rainfall data, together with a review of previous work by others, to underline several important points:

(1) ENSO-related SST variability is strongest in the Pacific Ocean but extends to the tropical Atlantic and Indian Oceans as well, with systematic lags of one to three seasons consistent with tropospheric forcing between basins.
(2) After a global ENSO mode is properly removed from the SST data, the residual variability emerges in three additional modes that correspond to phenomena previously identified by others: (a) a mode dominated by global warming; (b) an interdecadal Pacific mode with strong loading in the extratropical North Pacific; and (c) a multidecadal mode with its greatest intensity in the extratropical North Atlantic.
(3) El Niño/Southern Oscillation variability in the other ocean basins is not redundant in regard to associated atmospheric responses.
(4) In the Western Hemisphere tropics, both the Atlantic and the Pacific covary with rainfall in significant and unique ways; although the global ENSO is the dominant process, non-ENSO variability provides a sizable increase in the explained cross-covariance between SSTA and rainfall.

While ENSO is clearly the dominant process, the non-ENSO variability that emerges from the CEOF analysis (Figs. 3.3, 3.4, and 3.5) is fairly robust, and the non-ENSO modes can be easily related in myriad ways to analyses previously done by others. Our conclusions (3) and (4) apply strictly to ENSO and non-ENSO fluctuations in the tropics, where the SSTA variability is dominated by the relatively short, interannual to decadal timescales. It is likely that the same principles can be extended to the non-ENSO modes of variability that dominate at higher latitudes and on longer timescales. Predictions of climate several months to several seasons ahead can be improved by also projecting the tendencies of the non-ENSO modes that form the slowly varying background to the interannual variability. Both the ENSO and the non-ENSO processes should be considered globally, and coupled models (with global model oceans) should be made to emulate them.

Acknowledgments We wish to thank T. Smith, A. Kaplan, and P. Xie for making their SST and rainfall data sets public and helping us obtain them. We also credit useful exchanges with H. Diaz, R. Houghton, M. Latif, D. Mayer, and Y. Tourre for contributing to our enlightenment. Our research has been supported by the National Oceanic and Atmospheric Administration (NOAA) through its Pan-American Climate Studies program, by the Inter-American Institute for Global Change Research (IAI), and by the NOAA Environmental Research Laboratories through their base funding of our laboratory.

References

BARNSTON, A. G. and LIVEZEY, R. E., 1987: Classification, seasonality and persistence of low-frequency atmospheric circulation patterns. *Monthly Weather Review*, **115**, 1083–1126.

BARNSTON, A., van den DOOL, H. M., ZEBIAK, S. E., BARNETT, T. P., JI, M., RODENHUIS, D. R,. CANE, M. A., LEETMAA, A., GRAHAM, N. E., ROPELEWSKI, C. R., KOUSKY, V. E., O'LENIC, E. A., and LIVEZEY, R. E., 1994: Long-lead seasonal forecasts – Where do we stand? *Bulletin of the American Meteorological Society*, **75**, 2097–2114.

CANE, M. A., CLEMENT, A. C., KAPLAN, A., KUSHNIR, Y., POZDNYAKOV, D., SEAGER, R., ZEBIAK, S. E., and MURTUGUDDE, R., 1997: Twentieth-century sea surface temperature trends. *Science*, **275**, 957–960.

CARTON, J. and HUANG, B., 1994: Warm events in the tropical Atlantic. *Journal of Physical Oceanography*, **24**, 888–903.

CHANG, P., JI, L., and LI, H., 1997: A decadal climate variation in the tropical Atlantic Ocean from thermodynamic air–sea interactions. *Nature*, **385**, 516–518.

CITEAU, J., FINAUD, L., CAMMAS, J. P., and DEMARCQ, H., 1989: Questions relative to ITCZ migrations over the tropical Atlantic Ocean, sea surface temperature and Senegal River runoff. *Meteorology & Atmospheric Physics*, **41**, 181–190.

COVEY, D. C. and HASTENRATH, S., 1978: The Pacific El Niño phenomenon and the Atlantic circulation. *Monthly Weather Review*, **106**, 1280–1287.

CURRY, R. G. and McCARTNEY, M. S., 1996: North Atlantic's transformation pipeline. *Oceanus*, **39**, 24–28.

DELWORTH, T. L., 1996: North Atlantic interannual variability in a coupled ocean–atmosphere model. *Journal of Climate*, **9**, 2356–2375.

DESER, C. and BLACKMON, M. L., 1993: Surface climate variations over the North Atlantic Ocean during winter: 1900–1989. *Journal of Climate*, **6**, 1743–1753.

DESER, C. and BLACKMON, M. L., 1995: On the relationship between tropical and North Pacific sea surface temperature variations. *Journal of Climate*, **8**, 1677–1680.

DRAPER, N. and SMITH, H., 1966: *Applied Regression Analysis*. New York: Wiley and Sons, 709 pp.

ENFIELD, D. B., 1989: El Niño, past and present. *Reviews of Geophysics*, **27**, 157–187.

ENFIELD, D. B., 1996: Relationships of inter-American rainfall to tropical Atlantic and Pacific SST variability. *Geophysical Research Letters*, **23**, 3505–3508.

ENFIELD, D. B. and MAYER, D. A., 1997: Tropical Atlantic SST variability and its relation to El Niño–Southern Oscillation. *Journal of Geophysical Research*, **102**, 929–945.

FOLLAND C. K., PALMER, T. N., and PARKER, D. E., 1986: Sahel rainfall and worldwide sea temperatures. *Nature*, **320**, 602–607.

GU, D. and PHILANDER, S. G. H., 1997: Interdecadal climate fluctuations that depend on exchanges between the tropics and extratropics. *Science*, **275**, 721–892.

HAMEED, S., SPERBER, K. R., and MEINSTER, A., 1993: Teleconnections of the Southern Oscillation in the tropical Atlantic sector in the OSU coupled upper ocean–atmosphere GCM. *Journal of Climate*, **6**, 487–498.

HANSEN, D. V. and BEZDEK, H. F., 1996: On the nature of decadal anomalies in North Atlantic sea surface temperature. *Journal of Geophysical Research*, **101**, 8749–8758.

HASTENRATH, S., 1978: On modes of tropical circulation and climate anomalies. *Journal of the Atmospheric Sciences*, **35**, 2222–2231.

HASTENRATH, S., 1984: Interannual variability and the annual cycle: Mechanisms of circulation and climate in the tropical Atlantic sector. *Monthly Weather Review*, **112**, 1097–1107.

HASTENRATH, S., de CASTRO, L. C., and ACEITUNO, P., 1987: The Southern Oscillation in the Atlantic sector. *Contributions in Atmospheric Physics*, **60**, 447–463.

HASTENRATH, S. and GREISCHAR, L., 1993: Further work on the prediction of northeast Brazil rainfall anomalies. *Journal of Climate*, **6**, 743–758.

HOUGHTON, J. T., MEIRA FILHO, L. G., CALLANDER, B. A., HARRIS, N., KATTENBERG, A., and MASKELL, K. (eds.), 1996: *Climate Change 1995, the Science of Climate Change*. Cambridge: Cambridge University Press, 572 pp.

HOUGHTON, R. W. and TOURRE, Y. M., 1992: Characteristics of low-frequency sea surface temperature fluctuations in the tropical Atlantic. *Journal of Climate*, **5**, 765–771.

HUANG, B., CARTON, J. A., and SHUKLA, J., 1995: A numerical simulation of the variability in the tropical Atlantic Ocean, 1980–88. *Journal of Physical Oceanography*, **25**, 836–854.

HURRELL, J. W., 1995: Decadal trends in the North Atlantic Oscillation: Regional temperatures and precipitation. *Science*, **269**, 676–679.

JONES, P. D., BRADLEY, R. S., DIAZ, H. F., KELLEY, P. M., and WIGLEY, T. M. L., 1986: Northern hemisphere surface air temperature variations: 1851–1984. *Journal of Climate and Applied Meteorology*, **25**, 161–179.

KAPLAN, A., CANE, M. A., KUSHNIR, Y., BLUMENTHAL, M. B., and RAJAGOPALAN, B., 1998: Analyses of global sea surface temperatures 1856–1991. *Journal of Geophysical Research*, **103**, 18567–18589.

KAWAMURA, R., 1994: A rotated analysis of global sea surface temperature variability with interannual and interdecadal scales. *Journal of Physical Oceanography*, **24**, 707–715.

KERR, R. A., 1997: A new driver for the Atlantic's moods and Europe's weather? *Science*, **275**, 754–755.

KUSHNIR, Y., 1994: Interdecadal variations in North Atlantic sea surface temperature and associated atmospheric conditions. *Journal of Climate*, **7**, 141–157.

LAMB, P. J. and PEPPLER, R. A., 1992: Further case studies of tropical Atlantic surface atmospheric and oceanic patterns associated with sub-Saharan drought. *Journal of Climate*, **5**, 476–488.

LANZANTE, J. L., 1996: Lag relationships involving tropical sea surface temperatures. *Journal of Physical Oceanography*, **9**, 2568–2578.

LATIF, M. and BARNETT, T. P., 1994: Causes of decadal climate variability over the North Pacific and North America. *Science*, **266**, 634–637.

LATIF, M. and BARNETT, T. P., 1995: Interactions of the tropical oceans. *Journal of Climate*, **8**, 952–964.

LATIF, M. and BARNETT, T. P., 1996: Decadal climate variability over the North Pacific and North America: Dynamics and predictability. *Journal of Climate*, **9**, 2407–2423.

LAU, N. C. and NATH, M. J., 1994: A modeling study of the relative roles of tropical and extratropical anomalies in the variability of the global atmosphere-ocean system. *Journal of Climate*, **7**, 1184–1207.

McCARTNEY, M. S., CURRY, R. G., and BEZDEK, H. F., 1996: North Atlantic's transformation pipeline chills and redistributes subtropical water. *Oceanus*, **39**, 19–23.

MEHTA, V. M. and DELWORTH, T., 1995: Decadal variability of the tropical Atlantic Ocean surface temperature in shipboard measurements and in a global ocean–atmosphere model. *Journal of Climate*, **8**, 172–190.

MOURA, A. D. and SHUKLA, J., 1981: On the dynamics of droughts in northeast Brazil: Observations, theory and numerical experiments with a general circulation model. *Journal of the Atmospheric Sciences*, **38**, 2653–2675.

NICHOLLS, N., LAVERY, B., FREDERIKSEN, C., and DROSDOWSKY, W., 1996: Recent apparent changes in relationships between the El Niño–Southern Oscillation and Australian rainfall and temperature. *Geophysical Research Letters*, **23**, 3357–3360.

NOBRE, P. and SHUKLA, J., 1996: Variations of sea surface temperature, wind stress, and rainfall over the tropical Atlantic and South America. *Journal of Climate*, **9**, 2464–2479.

PHILANDER, S. G. H., 1990: *El Niño, La Niña and the Southern Oscillation*. New York: Academic Press, 293 pp.

RASMUSSON, E. M., ARKIN, P. A., CHEN, W. Y., and JALLICKEE, J. B., 1981: Biennial variations in surface temperature over the United States as revealed by singular decomposition. *Monthly Weather Review*, **109**, 588–598.

RASMUSSON, E. M. and CARPENTER, T. C., 1982: Variations in tropical sea surface temperature and surface wind fields associated with the Southern Oscillation/El Niño. *Monthly Weather Review*, **110**, 354–384.

ROPELEWSKI, C. F. and HALPERT, M. S., 1987: Global and regional scale precipitation patterns associated with the El Niño–Southern Oscillation. *Monthly Weather Review*, **110**, 1606–1626.

ROPELEWSKI, C. F. and HALPERT, M. S., 1989: Precipitation patterns associated with the high index phase of the Southern Oscillation. *Journal of Climate*, **2**, 268–284.

SERVAIN, J., 1991: Simple climatic indices for the tropical Atlantic Ocean and some applications. *Journal of Geophysical Research*, **96**, 15137–15146.

SMITH, T. M., REYNOLDS, R. W., LIVEZEY, R. E., and STOKES, D. C., 1996: Reconstruction of historical sea surface temperatures using empirical orthogonal functions. *Journal of Climate*, **9**, 1403–1420.

TING, M. and WANG, H., 1997: Summertime United States precipitation variability and its relation to Pacific sea surface temperature. *Journal of Climate*, **10**, 1853–1873.

TOURRE, Y. M. and WHITE, W. B., 1995: ENSO signals in global upper-ocean temperature. *Journal of Physical Oceanography*, **25**, 1317–1332.

TRENBERTH, K. E. and HURRELL, J. W., 1994: Decadal atmosphere–ocean variations in the Pacific. *Climate Dynamics*, **9**, 303–319.

WAGNER, R. G., 1996: Mechanisms controlling variability of the interhemispheric sea surface temperature gradient in the tropical Atlantic. *Journal of Climate*, **9**, 2010–2019.

WALLACE, J. M., SMITH, C., and JIANG, Q., 1990: Spatial patterns of atmosphere–ocean interaction in northern winter. *Journal of Climate*, **3**, 990–998.

WEARE, B. C., 1977: Empirical orthogonal function analysis of Atlantic Ocean surface temperatures. *Quarterly Journal of the Royal Meteorological Society*, **103**, 467–478.

WOODRUFF, S. D., SLUTZ, R. J., JENNE, R. L., and STEURER, P. M., 1987: A comprehensive ocean–atmosphere data set. *Bulletin of the American Meteorological Society*, **68**, 1239–2278.

XIE, P. and ARKIN, P. A., 1996: Analyses of global monthly precipitation using gauge observations, satellite estimates, and numerical model predictions. *Journal of Climate*, **9**, 840–858.

4

Multiscale Streamflow Variability Associated with El Niño/Southern Oscillation

MICHAEL D. DETTINGER, DANIEL R. CAYAN, AND GREGORY J. McCABE

U.S. Geological Survey, San Diego, California 92093, and Denver, Colorado 80225, U.S.A.

JOSÉ A. MARENGO

Instituto Nacional de Pesquisas Espaciais,
CEP 12630-000 Cachoeira Paulista,
Sao Paulo, Brazil

Abstract

Streamflow responses to the El Niño/Southern Oscillation (ENSO) phenomenon in the tropical Pacific are detectable in many regions. During warm-tropical El Niño and cool-tropical La Niña episodes, streamflows are affected throughout the Americas and Australia, in northern Europe, and in parts of Africa and Asia. In North and South America, correlations between peak–flow season streamflows and seasonal Southern Oscillation Indices (SOIs) show considerable persistence. In South America, correlations between flows in other seasons with December–February SOIs also are notably persistent, whereas, in North America, correlations are smaller when other, non–peak season time periods are considered.

At least two modes of streamflow response to ENSO are present in the Western Hemisphere. When interannual North and South American streamflow variations are analyzed together in a single principal components analysis, two of the leading components are found to be associated with ENSO climate variability. The more powerful of these modes corresponds mostly to ENSO responses by the rivers of tropical South America east of the Andes, along with rivers in southern South America and the southwestern United States, with Brazil experiencing less runoff during El Niños and the other regions experiencing more runoff. This streamflow mode is correlated globally with ENSO-like sea surface temperature (SST) patterns on both interannual and interdecadal timescales; indeed, the tropical South American rivers east of the Andes are coherent with SOI on virtually all historical timescales. The second ENSO-related streamflow mode characterizes other parts of extratropical streamflow variation, emphasizing the north–south differences in streamflows in North America during ENSO extremes and (less robustly) streamflow variations along the central Andes. The relation of this extratropical streamflow mode to ENSO seems to be mostly from scattered interannual timescales and, overall, its decadal variations follow North Atlantic SSTs.

On decadal time scales, the most remarkable variation identified in the Western Hemisphere ENSO–streamflow correlations or teleconnections is a decades-long contrast between the teleconnections of recent decades and teleconnections from about the 1920s into the 1950s. Correlations between streamflows and SOI, Niño-3 SSTs, and even global SSTs nearly vanished in many regions of North and South America during the earlier period. The change appears to have been associated with weakening of ENSO and, possibly, a weakening of connections between the atmospheric and oceanic components of ENSO during the earlier period. The development of two ENSO-related principal components of North and South American streamflow, rather than one, may be an artifact of the differences in decadal scale responses of streamflows in the tropics and extratropics to multiscale ENSO forcings.

Introduction

Land–hydrologic systems are influenced by climatic changes and human activities on interannual and decadal timescales. A leading mode of global climate variation on interannual timescales is the El Niño/Southern Oscillation (ENSO) of the tropical ocean–air system. The ENSO affects seasonal climate throughout the tropics and in broad swaths of the extratropics, and it thus plays a major role in initiating hydrologic variations in many regions. Ropelewski and Halpert (1987, 1989, 1996) and Kiladis and Diaz (1989) have determined that tropical warm events (El Niños) and cool events (anti–El Niños or La Niñas) influence precipitation and temperatures in many parts of the globe. Depetris and Kempe (1990), Koch et al. (1991), Redmond and Koch (1991), Cayan and Webb (1992), Kahya and Dracup (1993, 1994), Chiew et al. (1994), Moss et al. (1994), Eltahir (1996), Guetter and Georgakakos (1996), Kazadi (1996), Piechota and Dracup (1996), Zorn and Waylen (1997), and Marengo et al. (1998) – among many others – have demonstrated that these ENSO-driven precipitation and temperature fluctuations translate into significant variations of streamflow in several regions of the extratropics. In this chapter, we investigate the relations of streamflow variability to El Niño conditions on global and hemispheric scales.

These interannual climate and hydrologic fluctuations are modulated by, or superimposed upon, lower frequency variations with decadal and longer timescales. The sources of these lower frequency climate variations are uncertain but may have roots in the tropics (Barnett et al. 1992; Trenberth and Hoar 1996), in the extratropical oceans (Pacific: Douglas et al. 1982, Trenberth 1990, Latif and Barnett 1994; Atlantic: Schlosser et al. 1991, Read and Gould 1992, Deser and Blackmon 1993, Tanimoto et al. 1993, Stocker and Broecker 1994, Chen and Ghil 1996, Houghton 1996), or in some interplay of the two (e.g., Graham 1994; Graham et al. 1994; Jacobs et al. 1994). Understanding of precipitation and streamflow variability on these extended timescales is developing (e.g., Probst and Tardy 1987, 1989; Lettenmaier et al. 1994; Marengo 1995; Cayan et al. 1998), motivated by the growing realization that extended droughts and periods of repeated flooding are of as much practical importance as ENSO timescale fluctuations and that not all ENSO episodes are alike. In this chapter, we investigate the relative importance of decadal climate variations on Western Hemisphere streamflow variations and show that the leading spatial modes of streamflow variation are related to ENSO processes in ways that vary dramatically on decadal timescales.

As streamflow is the surface hydrological variable most vital for understanding and predicting water supply and water hazards, this chapter focuses on the interannual and longer variations of streamflow associated with ENSO climate forcings, on global and hemispheric scales. After an introductory survey of ENSO influences on the global scale, the chapter focuses on streamflow in North and South America and delineates shared streamflow variations on interannual and decadal timescales that appear in response to ENSO, and ENSO-like (Zhang et al. 1997), climate variability. The dominant ENSO-driven streamflow variations are related to precipitation variations over the Americas, and spatial patterns of Western Hemisphere hydrologic variations associated with ENSO will be characterized by principal component analyses (PCAs) of both streamflow and precipitation. Streamflow variations in the dominant spatial modes are correlated with long-term sea surface temperature (SST) variations in order to identify some of their large-scale climatic underpinnings.

Data

The analyses presented here used instrumental records of streamflow, surface meteorological, atmospheric, and SST variables. The streamflow data either are part of a global collection of monthly streamflow records compiled by the senior author from public domain sources (to be documented elsewhere) or (for South America) are a mix of those public records with private sources provided by coauthor Marengo. In each case, the longest flow records available were used if, on visual inspection and comparisons with neighboring sites, they appeared to be free of large unnatural influences. From the global streamflow set, 732 stations that have more than 15 years of complete streamflow data were used in the next section.

Our North American streamflow analyses in subsequent sections focus mostly on a gridded subset of naturally varying streamflow series of the U.S. Geological Survey Hydroclimatic Data Network (HCDN; Slack and Landwehr 1992). Sites within the conterminous United States with 95% or better data availability for the 60-year period from 1925 to 1984 (to match availability of streamflow data from Canada and South America) were selected for gridding. Some HCDN sites from the central United States were added, despite their having somewhat less than 60 years of data, to provide data coverage in that subregion. Canadian streamflow series that met the 95% availability requirement and that appeared to be free from large human-induced changes for the same 60-year period were added to this North American set from public-domain World Meteorological Organization sources. Altogether, 143 sites are included in this North American set. A gridded version of these North American flows was developed by computing the simple (unnormalized) averages of the streamflow series within each $5^\circ \times 5^\circ$ box. The flows were gridded to prevent areas with many sites from overwhelming the pattern analysis, to reduce the influence of missing data points, and to more accurately reflect the degrees of freedom of streamflow variability in North America (see also Lins 1985a, b). Simple averages were gridded to represent total runoff as nearly as possible, rather than weighting small rivers as heavily as large.

The South American streamflow analyses in later sections focus on a 29-site dataset drawn from the global set and proprietary data provided by Marengo (from the Eletrobras and Eletronorte power agencies in Brazil); the series generally span the 62-year period from 1931 to 1992. Because coverage over much of South America was scarce and because several of the flow series represent very large basins (which would not necessarily reflect runoff at the nearest grid point), no gridding was performed for the South American PCA.

For analysis of annual and longer term streamflow variations, monthly streamflow time series like those used here must be summed into yearly totals. These totals need not correspond to calendar years but, rather, often are chosen to distinguish between complete wet seasons by beginning and ending "water years" during the driest months. In this chapter, water years are defined arbitrarily as totals from October though September. This definition is common in North America but is arbitrary when extrapolated to global analyses because, globally, the seasonality of streamflow hydrographs varies significantly from region to region and hemisphere to hemisphere. Other analyses in this chapter focus on streamflow variations during the time periods of maximum flow in each region. For these analyses, streamflow variations are measured in terms of peak–flow season totals, which vary from river to river, but which at each river contain the highest monthly flows in the river's long-term average annual hydrograph.

Global monthly precipitation anomalies for land areas from 1880 to 1992 (Eischeid et al. 1995) and for land *and* sea areas from 1979 to 1995 (Xie and Arkin 1996) were used to describe the hydrologic inputs that drive the streamflow variations considered and to verify those variations. Global SSTs reported by Parker et al. (1995) are a measure of the interannual state of the global climate system and are analyzed for their association with the streamflow variations considered here. The status of near-surface atmospheric circulations was inferred from monthly sea-level pressure (SLP) anomalies on a near-global, $5^\circ \times 5^\circ$ grid from 42.5°S to 72.5°N, during 1951–92, provided by Barnett et al. (1984; with updates since 1984 from the Scripps Institution of Oceanography).

The Southern Oscillation Index (SOI; Fig. 4.1) is a simple measure of the state of the atmosphere over the tropical Pacific and is a common index of ENSO conditions and effects. The SOI is the difference between deseasonalized, normalized SLP anomalies over Tahiti and Darwin, Australia, and measures the tendency for easterly winds to blow along the equatorial Pacific. When SOI is positive, easterly winds are strong in the tropics and the tropical Pacific is usually in its La Niña state, with unusually cool SSTs extending westward from the South American coast to the middle of the Pacific basin. When SOI is negative, easterly winds are weak and the tropical Pacific usually is in its El Niño state, with unusually warm SSTs along most of the equator. Seasonal averages of the SOI are considered here in various correlation and composite analyses to map ENSO-induced hydrologic variations. Composites of streamflows in El Niño and La Niña years are obtained by averaging over years in which the SOI is -1 and $+1$ standard deviations, respectively; correlations use the entire SOI series.

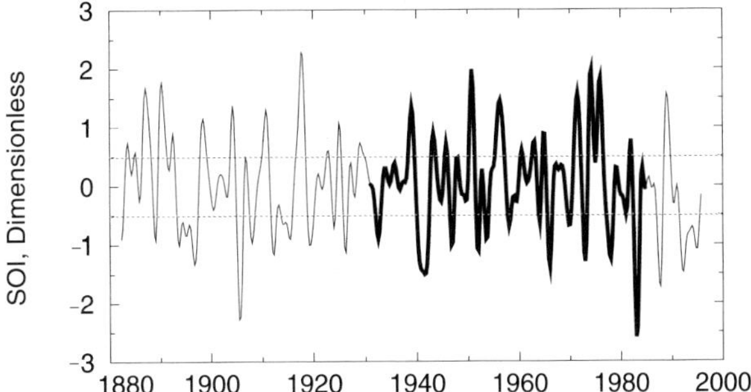

Fig. 4.1 Southern Oscillation Index (SOI), 18-month moving average; heavy during 1931–84 period studied in streamflow principal component analysis. Dashed horizontal lines show SOI = ±0.5.

Global ENSO–Streamflow Relations

In keeping with the global extent of ENSO influences on precipitation and surface temperatures, the effects of El Niños and La Niñas on streamflows also are global. Streamflow series that are correlated with SOI averages for December through February (DJF) at levels significantly different from zero (with $p < 0.05$ by a standard two-side t-test of the sampling errors in correlation estimates; Benjamin and Cornell 1970) are indicated in Figure 4.2 by colored circles; in the top panel, red-filled circles indicate positive correlations with SOI (so that, at red circles, El Niños correspond to less flow), blue circles indicate negative correlations, and open circles indicate sites at which the correlations are not significant at $p < 0.05$ levels. Underlying the gage symbols in the top panel is a map of the average deviations (from the climatological precipitation totals during non-ENSO DJFs) of DJF precipitation totals during four recent El Niño winters for which global coverage of precipitation series was available from Xie and Arkin (1996). The bottom panel shows deviations of precipitation during three La Niña DJF seasons and the same flow correlations as in the top panel, but with colors reversed so that red circles now indicate dry streamflow responses to La Niña and blue circles indicate wet La Niñas.

Streamflows and precipitation in western North America respond to ENSO with a pattern of dry El Niños in the Northwest and wet El Niños in the Southwest (Cayan and Webb 1992; Kahya and Dracup 1993, 1994). A generally opposite wet–dry pattern typically is associated with La Niñas, but the symmetry is only approximate. El Niños tend to be wetter than normal in more of the southwestern streams, and La Niñas are wetter than normal in more of the northwestern streams, than a strict mirror-image interpretation of the correlations would indicate. The precipitation difference patterns shown in Figure 4.2a suggest that the wet Southwest El Niño pattern is associated with a large band of higher than average precipitation that reaches from north of Hawaii to the southwestern United States. The tendency for northern and central Mexico also to be deluged by subtropical winter storms during El Niños is evident in the precipitation anomalies of Figure 4.2a (see also Rogers 1988 and Diaz and Kiladis 1992, Fig. 2.4). During La Niñas, positive precipitation anomalies over the eastern Pacific are diverted

(a) El Niño Precipitation with Streamflow Correlations

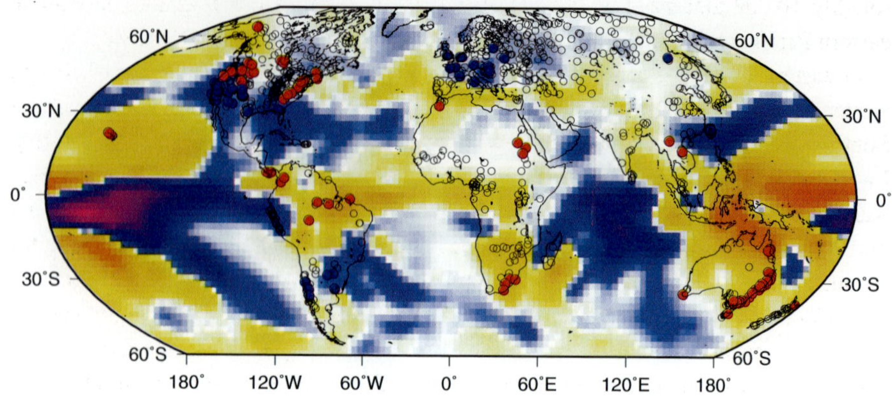

(b) La Niña Precipitation with Streamflow Correlations

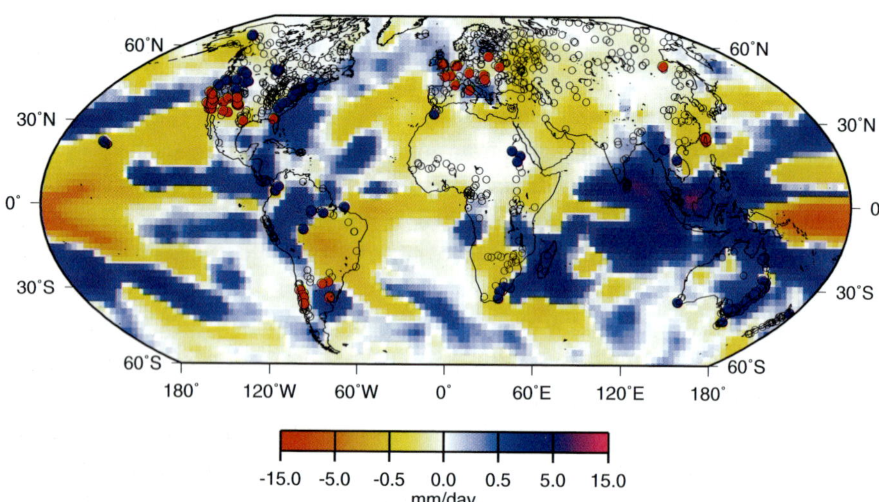

Fig. 4.2 Differences (shading) between average of October through September precipitation totals during selected El Niño/Southern Oscillation (ENSO) episodes and normal (1979–95) precipitation totals, and significance (circles) of correlations between streamflow totals for the same months and December through February Southern Oscillation Index (SOI) for streamflow periods of record. (a) Shading indicates average of deviations in El Niño years 1983, 1987, 1992, and 1995 from normal precipitation; red circles indicate streamflow series that are correlated positively with SOI ($p < 0.05$); blue circles are significantly correlated negatively; open circles are not significantly correlated. (b) Shading indicates average of deviations in La Niña years 1984, 1985, and 1989 from normal precipitation; color scheme for circles is reversed, with red circles indicating negative streamflow correlations and blue circles indicating positive correlations.

farther north and result in a wet Northwest. This influence is seen (Fig. 4.2b) in the roughly 10° latitude northward shift of precipitation anomalies over the midlatitude eastern Pacific and in higher than normal streamflows in the Pacific Northwest.

In eastern North America, significant positive correlations are observed between winter SOI and many streamflow series (indicating wet La Niñas), except in the extreme Southeast. These correlations reflect wet winters and summers in the Northeast during La Niñas and (less so) dry conditions during El Niños (Fig. 4.2; Diaz and Kiladis 1992; Kahya and Dracup 1993). The southeasternmost United States is wetter than normal during El Niños in response to a subtropical wet streak across Mexico into the states around the Gulf of Mexico (Piechota and Dracup 1996; Zorn and Waylen 1997).

In Central and South America, ENSO influences on streamflow are significant in several regions. Most notably, the Amazon basin (especially in the northern drainages) is drier in terms of both precipitation and flows during El Niño years than during La Niña years (Marengo et al. 1998), whereas Paraguay and Uruguay are wetter than normal in the same years (Rogers 1988; Depetris and Kempe 1990). Central America and northernmost South America also are drier than normal in El Niño years (Rogers 1988; Waylen et al. 1994). We find that the Amazonian and, to a lesser extent, Paraguayan streamflow influences of El Niño are replaced by approximate mirror images during La Niñas. In general, the South American streamflow correlations corroborate the precipitation patterns shown in Figure 4.2 as well as the less extensive patterns (based on longer precipitation histories) shown by Ropelewski and Halpert (1987, 1989), Kiladis and Diaz (1989), and Diaz and Kiladis (1992). The wet El Niño/dry La Niña precipitation relations of easternmost South America discussed by Rao and Hada (1990), Pisciottano et al. (1994), Enfield (1996), and Diaz et al. (1998) are represented only by small negative correlations in the regions of southeastern Brazil and Uruguay. Along the west coast of South America, precipitation and flows are related to El Niño in a seasonal manner. During both the El Niños and La Niñas composited in Figures 4.2a and 4.2b, Peru has been somewhat wet. During boreal summers while El Niños are gaining strength, southern Amazonian and equatorial Pacific precipitations appear to contribute to wet summers and thus to a negative correlation between streamflow and boreal summer SOI (not shown). During DJF, SOI is correlated negatively with Chilean flows (yielding wet El Niños), whereas during the preceding boreal summers, SOI is correlated negatively in southern Chile and positively in northern Chile.

African annual streamflows generally were not well correlated with SOI, except along the Nile and in South Africa (Fig. 4.2). The Nile River is near the western limit of the Old World limb of the seesaw described by the global SLP variations measured by SOI (Allan et al. 1996), so when SOI is negative (El Niño), pressure over eastern Africa is high and precipitation and streamflow are reduced. Indeed, the occurrence of low flows on the Nile has been used by Quinn (1992) to extend a chronology of El Niño episodes back into the seventh century. In eastern South Africa, low annual flows during El Niños (Fig. 4.2) are a response to warmer and drier conditions during much of the year leading up to a mature El Niño (Kiladis and Diaz 1989). Early in that year, however, wet conditions prevail (Kiladis and Diaz 1989). The tropical rivers of western

Africa experience a complicated regime of wet and dry periods during El Niños and do not yield significant correlations to contemporaneous, seasonal SOIs (Fig. 4.1; Kazadi 1996). Instead, flows in the rivers are correlated more (not shown) with climate in the equatorial Atlantic and even the North Atlantic.

European streamflows mostly are correlated negatively with winter SOI (indicating wet El Niños), except in Iberia. This streamflow response may correspond to increased winter precipitation, as is weakly suggested by Figure 4.2a, but Kiladis and Diaz (1989) found no statistically significant El Niño precipitation anomaly in an analysis of longer precipitation records. El Niño mean and La Niña mean streamflow deviations (not shown) indicate that the correlation of European streamflow with SOI comes from El Niño years that are wetter than non-ENSO years, especially in east central Europe.

In much of Asia, streamflow correlations with SOI are small or scattered, and no spatial pattern is found in Figure 4.2. The disturbance of Southeast Asian monsoons by El Niños is evidenced (modestly) by a few positive SOI–streamflow correlations in Southeast Asia, whereas higher flows than normal in Taiwan accompany El Niños in response to changes in the western Pacific jet. It is somewhat puzzling that ENSO-related fluctuations of the Indian monsoon (Dhar and Nandargi 1995) are not evidenced more in the streamflow correlations shown in Figure 4.1, but records for the subcontinent are short in the present data set, and the use of DJF SOIs may have reduced the significance levels of the relations shown in that figure.

Finally, eastern Australian streamflows respond with low flows to the El Niño droughts on that continent (Nicholls 1992; Chiew et al. 1994). The dry El Niño/wet La Niña patterns are evident in the precipitation anomalies of Figure 4.1; however, the correlations appear to derive more from wet La Niñas than from the dry El Niños, because flows are normally low in many of the rivers and El Niños cannot reduce them as much as wet conditions can enhance them.

Another view of these global streamflow–SOI relations is presented in Figure 4.3, which shows differences in runoff rates (streamflows/basin area) between boreal winters with low SOIs (El Niño) and those with high SOIs (La Niña), in mm/yr (Fig. 4.3a) and in standard deviations (Fig. 4.3b). The differences are significantly different from zero (by a standard two-sided t-test; Benjamin and Cornell 1970) in Hawaii, western North America, tropical and subtropical South America, eastern Europe, and eastern Australia. As expected, in all cases, the signs of these differences agree with the correlations of Figure 4.2. When differences between runoff in El Niño and normal years, and between runoff in La Niña and normal years, are mapped as in Figure 4.3, northwestern and eastern North American, Australian, and European flows responded with larger deviations in El Niño years than in La Niña years. Rivers in much of tropical South America respond with flow deviations from normal that are of nearly equal magnitude (but opposite signs) during La Niña and El Niño years, as do the flows of southwestern North America.

In proportion to their long-term averages, streamflow changes in Figure 4.3 are generally larger than the corresponding precipitation changes. However, such amplifications of precipitation variability in streamflow responses are not uncommon (e.g., Risbey

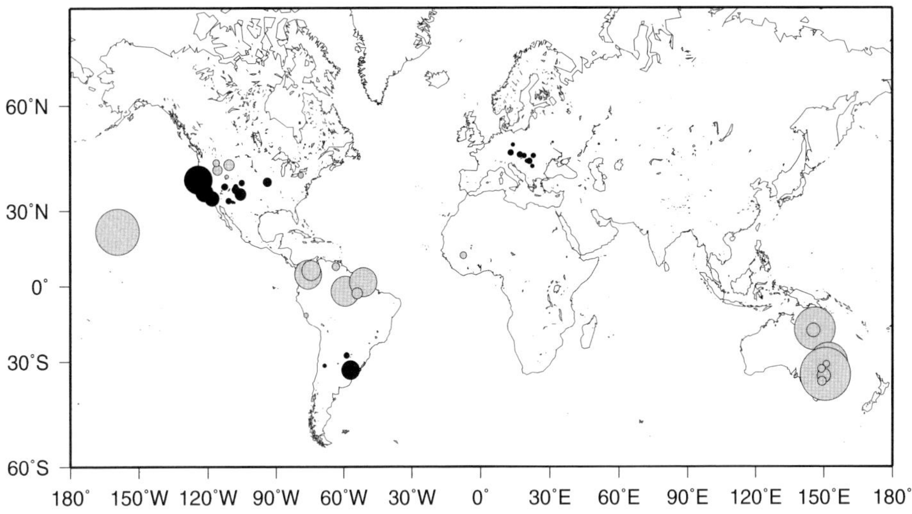

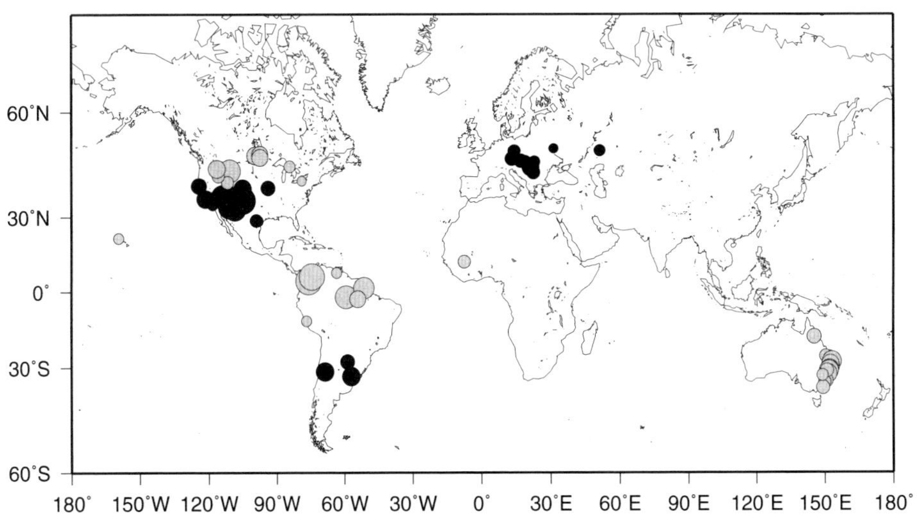

Fig. 4.3 Significant differences between annual runoff rates in years with positive and negative Southern Oscillation Indices ($p < 0.10$); gray circles indicate drier than normal El Niños; black circles indicate wetter than normal El Niños. (a) Radii of circles are scaled to differences in millimeters per year, with the Hawaiian circle representing 600 mm/yr. (b) Radii of circles are scaled to differences in local standard deviations of flow, with the Hawaiian circle representing 0.75 standard deviation.

Table 4.1 *Average fractional differences in precipitation and runoff between El Niño and La Niña episodes this century. Precipitation differences average only over grid cells in Eischeid et al. (1995) that are colocated with streamflow sites in the data set used here.*

	Significant differences only ($p < 0.05$)		All differences	
	Precipitation	Runoff	Precipitation	Runoff
Extratropics	+10%	+5%	+6%	0%
Tropics	−9%	−34%	−8%	−16%

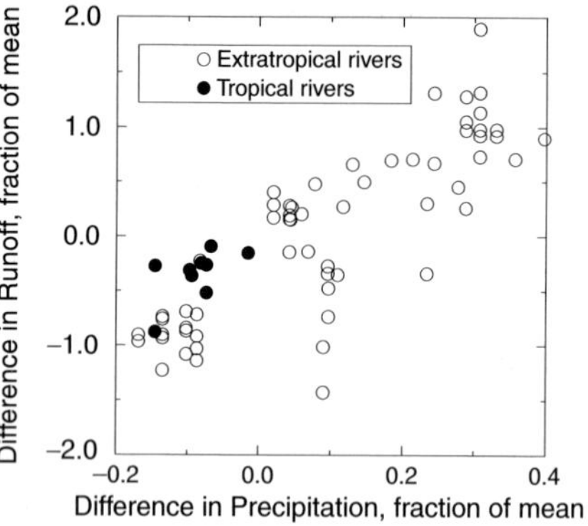

Fig. 4.4 Differences in runoff values from Figure 4.3b and differences in nearest neighbor precipitation values in Eischeid et al. (1995) grid during the same years as percentages of period-of-record mean annual flow and precipitation.

and Entekhabi 1996), and the ENSO responses are an important part of the overall streamflow variability. For each location with significant flow differences shown in Figure 4.3b, a corresponding precipitation difference between El Niños and La Niñas was calculated from the 5° × 5° gridded precipitation anomalies of Eischeid et al. (1995). These precipitation differences are compared to the corresponding runoff differences in Figure 4.4, as fractions of mean flows and precipitation. Overall, the runoff responses shown in Figure 4.4 are about 4 to 5 times larger than the (gridded) precipitation changes (Fig. 4.4, Table 4.1). The most extreme amplifications appear in the semi-arid rivers of Australia and the southwestern United States where flood years dominate the composite streamflow totals.

Broadly speaking, El Niños tend to yield drier than normal conditions in nearly all of the tropical rivers and a much more spatially varied response among the extratropical rivers. In our global streamflow set, all the tropical flow series that are correlated significantly with SOI are correlated positively with it (representing dry El Niños); when significance levels of the correlations are disregarded, 75% of all the tropical basins are correlated positively with SOI. A long-term estimate of the zonal averages of El Niño/La Niña tropical precipitation differences over land areas by Diaz and Kiladis (1992, Fig. 2.2) indicates that about 3% less tropical precipitation falls on land during El Niños than during La Niñas (including all tropical land areas). Average extratropical streamflow differences are −0.3% (Table 4.1), due mostly to widespread cancellation of positive and negative responses at the many extratropical basins.

Streamflow–ENSO Relations in the Western Hemisphere

To concentrate more on the temporal aspects of SOI–streamflow relations, we focus on interannual streamflow variations in the relatively data-rich Americas in the remainder of this chapter. The North American streamflows used are from the $5° \times 5°$ grid values described earlier, and the South American streamflows are measured rates from 29 sites. Seasonal correlations with SOI and a PCA of these flow series are presented in this section.

Seasonal Correlations with SOI

Because, when a significant ENSO–streamflow link exists, El Niños and La Niñas have qualitatively opposite influences on most of the streams, linear correlations of the North and South American flow series with SOI can be used as an approximate shorthand for the separate influences of El Niños and La Niñas on Western Hemisphere flows. Figure 4.5 shows the correlations between SOI during various seasons and streamflows during each streamflow series' climatological peak flow season. The correlation patterns of peak-season flows generally do not change much with the season of SOI being considered, although river-to-river differences in the timing of maximum correlations are evident. The long lead and lag correlations between SOI and streamflow arise because (i) ENSO processes typically develop slowly over months and seasons (e.g., the three-month autocorrelation of SOI is +0.67) and (ii) streamflow series also typically reflect precipitation inputs over several previous months or seasons. Consequently, by June through August (JJA) of the year preceding maturation of an El Niño (which is JJA of year 0 in the terminology of Rasmusson and Carpenter [1982]) the correlations between SOI and the following year's peak-season streamflows already have been established in many regions. As can be expected from the previous section and from studies listed there, flow in the Amazon basin, northernmost South America, and northwestern North America are correlated positively with DJF SOI; peak-season flows in southwestern and southeastern North America are correlated negatively with DJF SOI. (The strengths of the correlations mapped in Figure 4.5 differ from those indicated in Figure 4.2 because, in Figure 4.5, only peak-season flow variations are considered.)

Another way of viewing these correlations is in a tabular form, such as in Figure 4.6, intended to make the temporal evolution of SOI influences more obvious. Because of the relatively slow (interseasonal to interannual) pace of ENSO dynamics, most of the streamflows that correlate well with SOI maintain correlations with the same signs throughout the six seasons shown in each panel, and most of the significant correlations

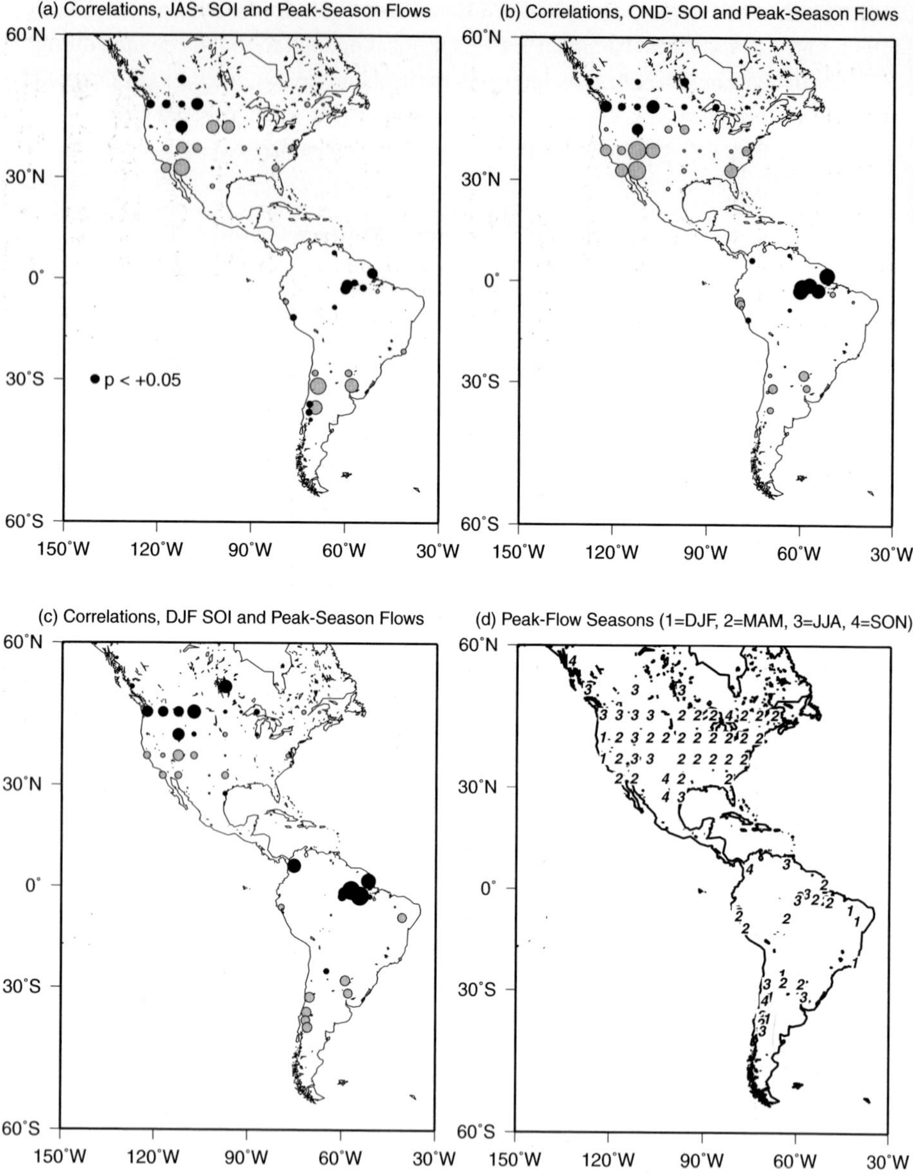

Fig. 4.5 (a)–(c) Correlations between seasonal Southern Oscillation Index (SOI) averages and peak-season streamflows, during 1931–84. Radii of circles are proportional to correlation [radius of minimally significant correlation shown in lower left in panel (a)]; gray circles indicate negative correlations and black circles indicate positive correlations. (d) Seasons of peak flows in average hydrographs. Seasons: DJF, December through February; MAM, March through May; JJA, June through August; SON, September through November; JAS-, July through September of the previous year; OND-, October through December of the previous year.

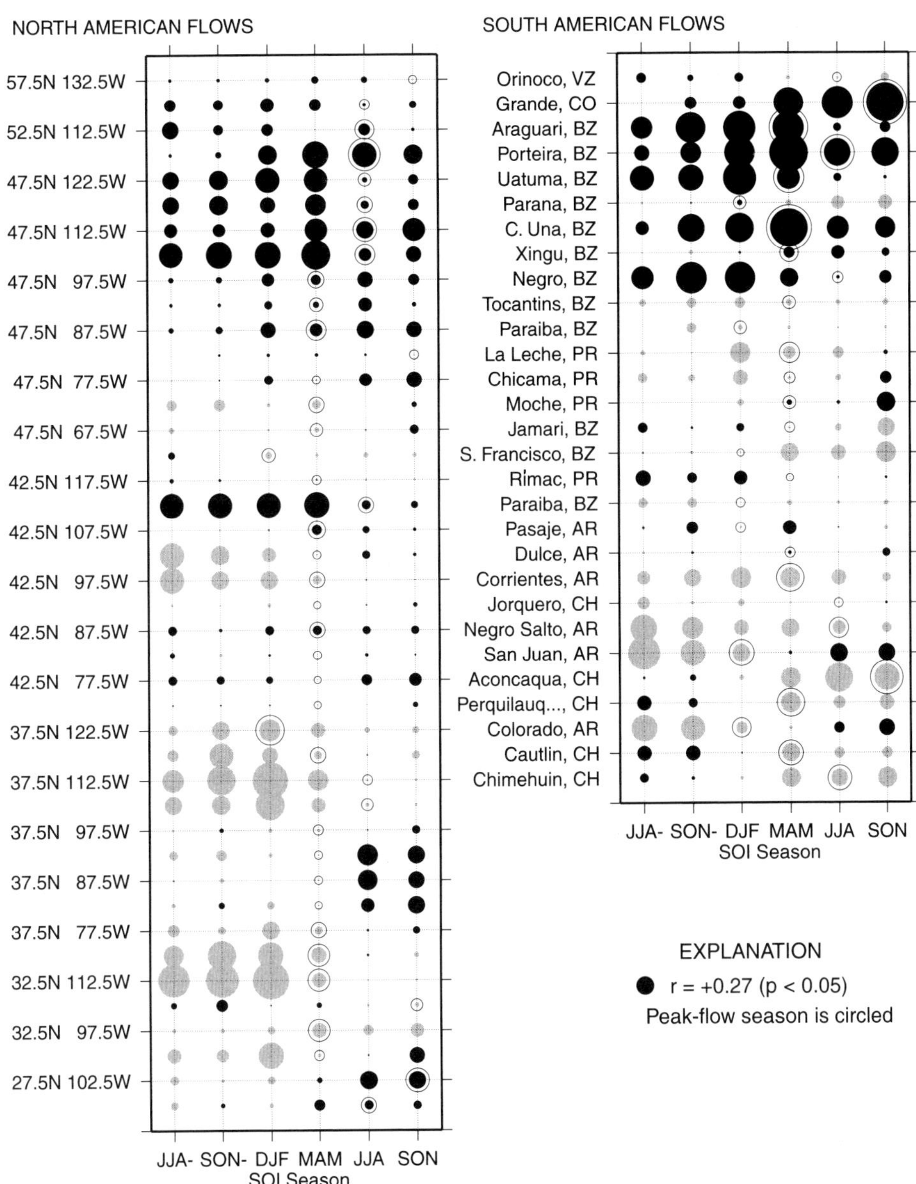

Fig. 4.6 Correlations between North and South American peak-season flows and seasonal Southern Oscillation Indices, during 1931–84. Radii of circles are proportional to correlation (radius of minimally significant correlation shown, in lower right); black circles indicate positive correlations; gray circles indicate negative correlations. Sites are sorted from north (top) to south (bottom) in each panel; North American sites are from the gridded series and intermediate sites are at intervening 5° × 5° grid points. Seasons: JJA-, June through August of preceding year; SON-, September through November of preceding year; DJF, December through February; MAM, March through May; JJA, June through August; SON, September through November.

remain significant from the preceding summer through at least the DJF of the mature El Niño or La Niña. In northern North America, peak-season streamflows (typically spring or summer) are correlated most with winter and spring SOI; farther south in North America, peak flows tend to correlate best with preceding autumn and winter SOIs. In South America, peak flows show maximum correlations with SOI in austral winter and spring in the northern tropical regions (top of right-hand panel of Figure 4.6) and in austral summer and fall just south of the equator (e.g., Rio Negro, Brazil). In the far south, peak flows show maximum correlations with SOI as it develops toward ENSO episodes in the preceding austral winter in Argentina on the east side of the Andes and in the subsequent winter in Chile on the west. In the extratropics, only the peak-season flows typically are well correlated with DJF SOI. In tropical South America east of the Andes, flows from most of the 18-month period shown are well (and positively) correlated to DJF SOI; along the Peruvian coast, correlations are less pronounced on these timescales (and for a shorter period of record, 1930s to 1960s) and of the opposite sign from the other rivers in tropical South America.

Principal Components of North and South American Streamflow

The large regions of persistently significant correlations of similar signs in both the North American and South American panels of Figure 4.6 indicate that flow on both continents varies together with SOI (and the ENSO dynamics) on interannual timescales and that those variations are organized on at least a hemispheric scale. The correlation patterns shown so far have represented long-term average streamflow responses to ENSO with no guarantees that those patterns of ENSO responses have occurred repeatedly. Determining the relative importance and reproducibility of these ENSO-driven streamflow patterns from episode to episode requires a more detailed analysis. One method for representing the most commonly occurring patterns within time-varying fields of scalar quantities (like streamflow) is spatial PCA (see, e.g., Manly 1986 and Barnston and Livezey 1987). This method computes a set of spatial patterns, or basis functions, that in linear combinations can be used to describe all variability in a set of concurrent time series. The patterns obtained from PCA are particularly useful because they are, by design, the patterns that can represent the maximum fractions of the variability using the fewest patterns. They are obtained by standard matrix manipulations to obtain the eigenvectors of the spatial cross-correlation matrix of the time series. These eigenvectors are the desired spatial basis functions (spatial patterns) and are called empirical orthogonal functions (EOFs). The strength of a particular EOF pattern in the original time series each year forms a time series called the principal component (PC) series, which is obtained by projecting the original time series onto that EOF. The initial EOF patterns are optimal for representing the collection of time series as a whole but typically do not distinguish between regions that vary together and those that have different time histories. For our purposes then, a useful "natural" division of the streamflow series into regions of shared variability was developed by "rotating" (Richman 1986) the EOFs and PCs to maximize the spatial variance of the EOF patterns. This rotation typically yields spatial basis functions (the rotated EOFs – REOFs) that delineate regions in which significant temporal variations are shared (or are in consistent contrast). Lins (1985a, b) has applied rotated PCA to streamflows in the conterminous United States and provides a discussion of the steps, limitations, and usefulness of this method for characterizing regional flow variability.

Because the discussion so far and the correlations in Figure 4.6 suggest that ENSO-correlated variations will be a mode of streamflow variability that is shared in broad areas of the Western Hemisphere, a spatial PCA was performed on the year-to-year streamflow variations from both continents. The PCA separated the streamflow variations into natural, uncorrelated modes, some of which were expected to be ENSO responses and others of which were likely to be interannual variations of (mostly) unknown origins. The PCA-identified modes of streamflow variation were analyzed by various correlations to identify those modes linked to ENSO and to determine the nature of those links.

Briefly, the PCA was done as follows: The common period of record for most of the streamflow series (1931–84; Fig. 4.1) was extracted from each series, and the annual flow cycle (which varied substantially from series to series, as in Fig. 4.5d) was removed by low-pass filtering each series with a linear filter (Kaylor 1977) with a half-power point at (18 months)$^{-1}$. Without this filtering, the differences in annual hydrographs from place to place dominated the analysis; also, Trenberth (1984) has shown that, in analyses related to the SOI, significant signal-to-noise improvements are provided by such filtering. Next, a standard EOF analysis of the cross-correlation matrix of the flow series was performed to obtain orthogonal EOFs and PCs. The eigenspectrum of the correlation matrix was inspected, and the number of PCs (6) that captured more variance than components of the noise background was estimated according to the ad hoc rules for sampling errors suggested by North et al. (1982). These six modes then were rotated by an orthogonal Varimax procedure (Richman 1986) to obtain rotated PCs (RPCs) that depict distinct but nonorthogonal indices of flow variability. The spatial patterns associated with these modes of flow variation are characterized here by the temporal correlation coefficients of the flow series with respect to the RPC time series.

The leading six components of (North and South) American streamflow variability constitute 47% of the interannual streamflow variance. In this chapter, however, only two RPCs (numbers 2 and 5 from among the six significant components) are of interest because together they appear to reflect regional ENSO responses by streamflow in the Americas; the other modes are discussed elsewhere (e.g., Dettinger and Cayan, 1998). The spatial pattern (REOF) and temporal progression (RPCs) of the first such component, RPC 2, which captures 9.0% of interannual streamflow variance, are shown in Figure 4.7. The spatial pattern includes the opposition of flow variations between the rivers of tropical and southern South America and, less prominently, between the southern United States and the northwestern and eastern United States. The RPC 2 dips to notably negative values in 1933, 1940–42, 1947, 1958, 1966, 1970, 1973, and 1983; all are associated with El Niño episodes in the tropical Pacific. Positive excursions of this mode are notable in 1936, 1939 (La Niña), 1945, 1950 (La Niña), 1963, 1972, 1974 (La Niña), and 1976 (La Niña), only half of which are La Niña episodes. Figure 4.8 shows how this ENSO-related RPC, and another, vary with SOI and Niño-3 SST (defined as

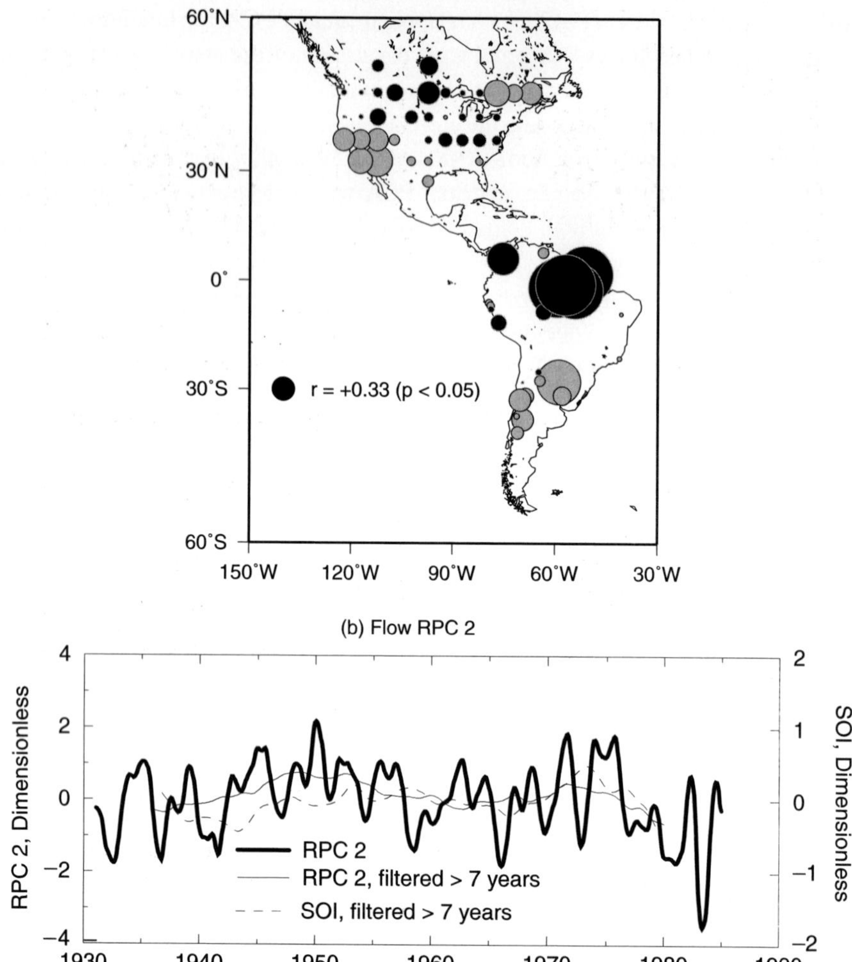

Fig. 4.7 Second rotated empirical orthogonal function (REOF) and principal component (RPC) of North and South American streamflows, during 1931–84. (a) REOF as correlations with the RPC series, black denotes positive correlations and gray negative; (b) RPC series (heavy curve), with RPC series (thin solid curve) and Southern Oscillations Index (thin dashed curve) low-pass filtered for variations slower than 7 years. Principal component analysis was performed on flows low-pass filtered with a half-power point at 18 months.

the mean SST anomaly in the region from 5°S to 5°N and from 150°W to 90°W). The four points in the lower left corner of Figure 4.8a are seasonal conditions and responses from the large El Niño of 1982–83. The correlation between the first ENSO streamflow mode (RPC 2) and SOI ($r = 0.67$; Fig. 4.8a) clearly is improved by the 1983 warm event but, even without that event, a linear relation between SOI and RPC 2 remains ($r = 0.61$ when the 1983 points are removed). The relation between RPC 2 and Niño-3 SSTs ($r = -0.46$; Fig. 4.8b) also is not so different when the 1983 episode is removed

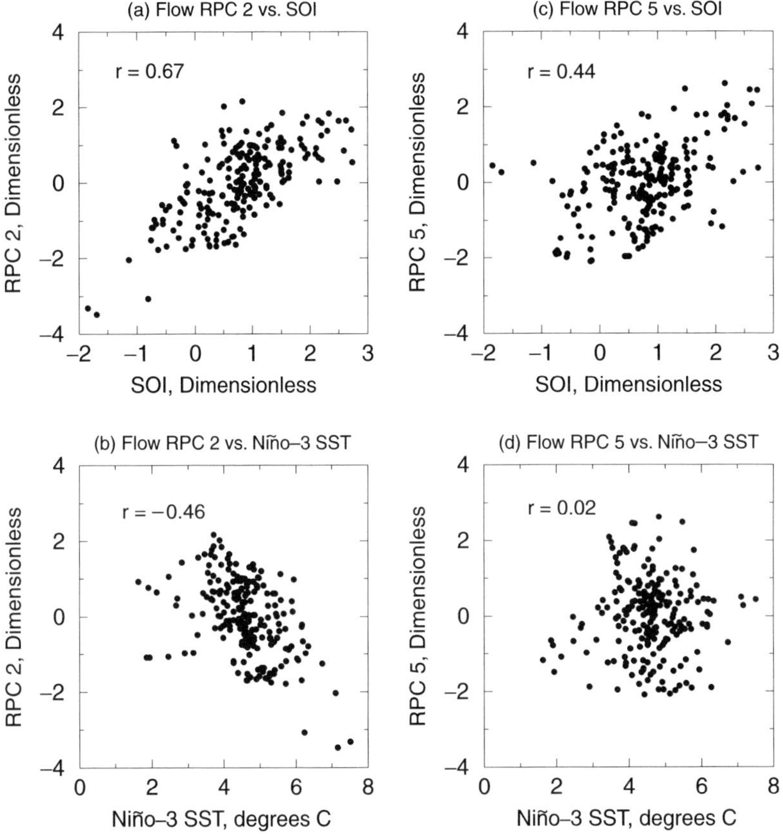

Fig. 4.8 Rotated principal components (RPCs) and seasonal Southern Oscillation Indices (SOIs) and Niño-3 sea surface temperatures (SSTs): (a) RPC 2 with SOI, (b) RPC 2 with Niño-3 SST, (c) RPC 5 with SOI, and (d) RPC 5 with Niño-3 SSTs.

($r = -0.36$). The robustness of this mode of streamflow variation also was indicated when the PCA was repeated with 1983 removed from the series. That PCA yielded the same mode with a similar spatial pattern and similar variance capture (8.3%).

Streamflow mode RPC 2 reflects precipitation variations over the same regions. Correlation of this ENSO-related streamflow PC with the gridded precipitation anomalies of Eischeid et al. (1995) indicates that the flow variations correspond to a significant fraction of precipitation variability in South America, with positive precipitation correlations (to this flow RPC) in Amazonia and negative correlations around Paraguay. Furthermore, a PCA of the gridded precipitation anomalies for North and South America yields corresponding precipitation variations as its leading mode (12% of interannual precipitation variance), with the spatial pattern shown in Figure 4.9. This precipitation pattern (REOF) represents dry El Niños in Central America and northernmost South America (through Amazonia) and wet El Niños in southeastern South America and the southern United States (see also the correlations in Fig. 4.2 of Enfield 1996). Precipitation in northern North America is modestly negatively correlated with SOI in this mode.

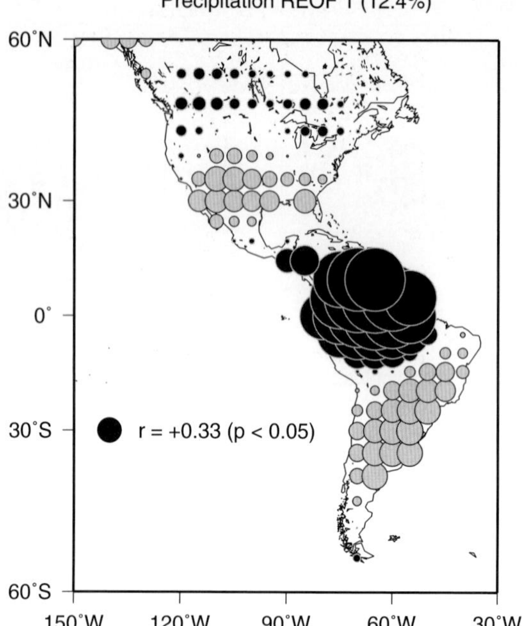

Fig. 4.9 Same as for Figures 4.7a and b, except for the first rotated empirical orthogonal function of North and South American precipitation, during 1900–92.

Another, less-powerful streamflow mode, RPC 5 (6.5% of interannual streamflow variance), also captures variability related to ENSO (correlation with SOI is $r = 0.44$; Fig. 4.8c). The associated spatial pattern of flow anomalies, REOF 5 (Fig. 4.10a), has contrasting streamflow variations in the northwestern and southwestern United States, with streamflow fluctuations along the Chilean coast (roughly from 28°S to 40°S latitude) that are in phase with those in the southwestern United States (Fig. 4.10a). The same pattern appears (capturing 7.4% of variance) when the PCA is repeated without 1983, except that the Chilean correlations are muted. Notable extremes of RPC 5 (Fig. 4.10b) are the strong negative case of 1973 (an El Niño) and strong positive peaks in 1974 and 1976 (two La Niñas). Figures 4.8c and 4.10d show that RPC 5 does not depend on the 1983 El Niño (i.e., the 1983 points, with SOI less than −10, correspond to near-zero RPCs in Fig. 4.8c). Overall, this RPC is associated more with the positive excursions of SOI (La Niñas) than with the negative (Fig. 4.8c) and is not related to Niño-3 SSTs (Fig. 4.8d). Correlation of this streamflow mode with the gridded precipitation anomalies of Eischeid et al. (1995) indicated that it corresponds more to precipitation variations in North America (especially in the Pacific Northwest) than elsewhere but that it captures only a small fraction of precipitation variability even in North America. No single, equivalent precipitation mode was found in the PCA of precipitation anomalies over both Americas.

Thus the broad spatial patterns of SOI–streamflow correlation shown in previous sections reflect two principal modes of interannual streamflow variation, one in which

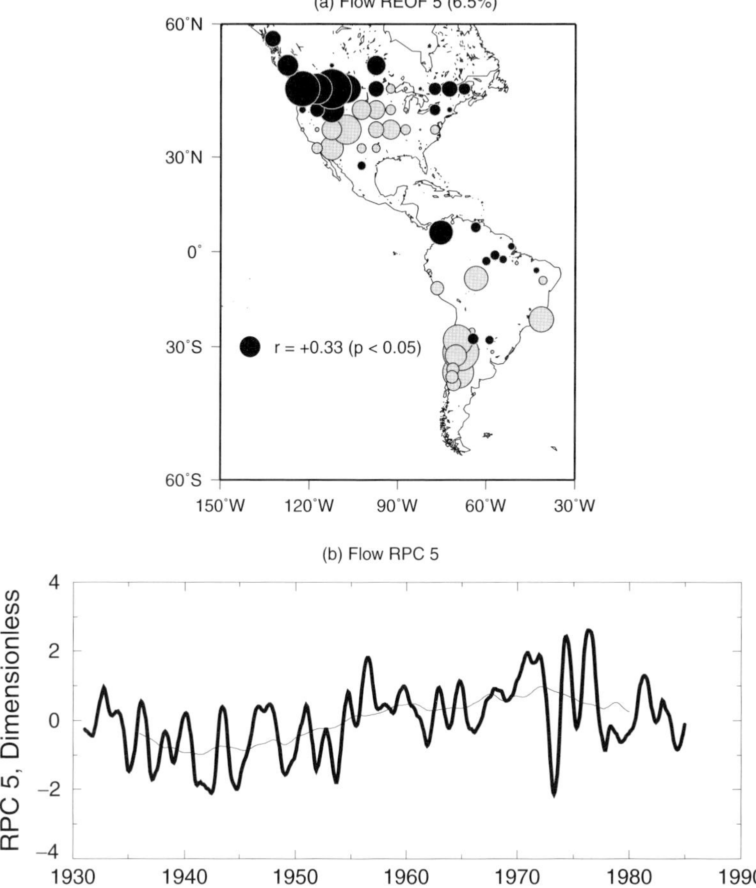

Fig. 4.10 Same as for Figure 4.7, except for the fifth rotated empirical orthogonal function and principal component of North and South American streamflows, during 1931–84.

the tropical South American streamflow responses to ENSO (excepting Peruvian and, presumably, Ecuadoran rivers) are consistently in the opposite sense from the subtropical streamflow responses, and another in which northern and southern North American streamflow responses are opposites of each other. These flow modes correspond directly to ENSO-driven precipitation variations and constitute about 15% of the overall interannual variance of the (normalized) American streamflows studied here. The two modes differ in their spatial emphases and in when they have been most prominent, but each is reflected in streamflow during several ENSO events. The global climatic conditions associated with the two modes are discussed further in the next two sections.

Multiple Timescale Aspects of ENSO–Streamflow Relations

Cayan et al. (1998) have shown that the variance of North American precipitation totals derives almost as much from decadal fluctuations as from interannual variations.

Marengo (1995) has found large decadal precipitation and streamflow fluctuations in South America, which can look like trends when records are short. Computation of the percentages of annual streamflow variance that are slower than (7 years)$^{-1}$ in the present data set also indicates that, in most of the Western Hemisphere, decadal and slower streamflow variations contribute almost as much to year-to-year variance as do streamflow variations in the frequency range from about (2 years)$^{-1}$ to (7 years)$^{-1}$.

Thus, the multiscale aspects of ENSO–streamflow relations must be considered to understand overall streamflow variability. Typically, the ENSO influences in climatic records appear mostly as interannual fluctuations, at least in part because records are short and decadal fluctuations in them are few. However, long-term variations in the character of ENSO are present. During this century, some decades have yielded frequent and vigorous ENSO episodes, whereas others have had relatively weak and infrequent episodes. Most notably, the time intervals from the 1930s through the 1960s had somewhat weaker ENSO variations relative to the first decades of the century and the decades beginning with the 1970s (Fig. 4.1; see also Rasmusson and Carpenter 1982 and Cole et al. 1993). Similarly, during some decades (e.g., 1950s and 1970s), cool tropical episodes (La Niñas, positive SOIs) were more common, whereas during others (e.g., 1980s and 1990s), warm episodes were common. Given these decadal differences in ENSO variability and because streamflow responds to positive and negative SOI conditions, it is natural to wonder if streamflow variations in response to ENSO are similar on all timescales.

Streamflow Coherency with SOI

The PCA of Western Hemisphere streamflow variations suggested that two distinct forms of ENSO response may exist, one emphasizing the tropics and the other emphasizing the extratropics. Some of the distinction may be due to the differing multiscale ENSO responses by rivers emphasized by the two RPCs. In particular, streamflows in the tropical rivers tend to follow SOI coherently over all historical timescales, whereas extratropical rivers appear to follow SOI in scattered, narrow frequency bands. The close interannual relations between ENSO and the tropical streamflows east of the Andes are suggested by the linear relations between ENSO indices and streamflow RPC 2 in Figures 4.8a and 4.8b. Some of the close connection on decadal timescales is evident in Figure 4.7b, where decadally filtered versions of RPC 2 and SOI can be compared as the two light curves. Examination of streamflow series from individual tropical rivers shows the decadal links to SOI even more clearly; for example, streamflows in the Araguari River of Brazil correlate with SOI at $r = 0.92$ on decadal timescales.

Extratropical rivers in the Americas are correlated to SOI within much narrower frequency bands. To show in detail how closely the eastern tropical streamflows follow SOI and how different the extratropical response is, coherency spectra of SOI with regional averages of streamflow variations in North and South America are presented in Figure 4.11. Coherency spectra are essentially correlations squared between two time series, calculated in each of many narrow frequency bands. In Figure 4.11, only those

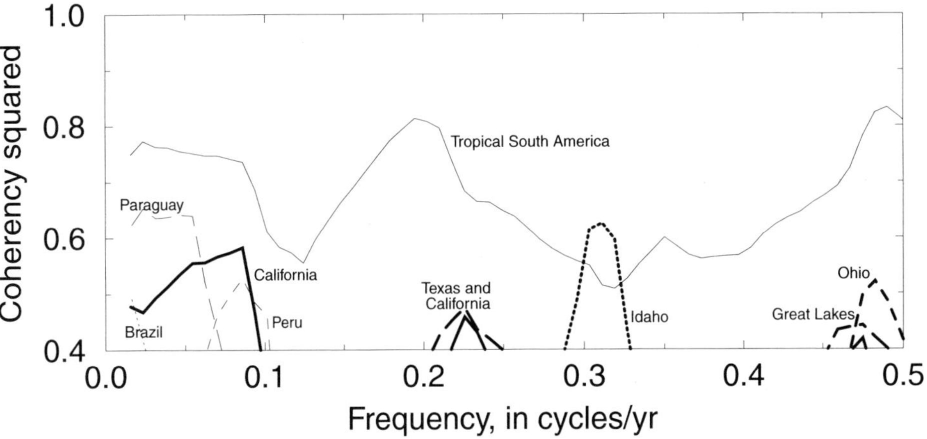

Fig. 4.11 Coherency squared of regional streamflow variations with respect to Southern Oscillation Index; lower bound marks significance level of roughly $p < 0.1$. "Tropical South America" refers to rivers east of the Andes.

coherencies that correspond to $p < 0.10$ or better significance levels are shown. Clearly, the thin solid curve for the tropical South American streamflow variations is the only one that is coherent with SOI at all frequencies shown. The other regional streamflow variations are weakly coherent with SOI and, then, only at isolated frequencies. Western and central North American flow modes are coherent with SOI at some interannual frequencies, notably around the quasi-quadrennial and quasi-biennial ENSO bands (Rasmusson et al. 1990; Jiang et al. 1995). Western North American (Californian) and all four extratropical South American regions are coherent with SOI at decadal and longer timescales.

Streamflow RPC Correlations with SSTs

Correlation of the streamflow RPCs identified in preceding sections with the long-term global SST anomalies provides another useful means for determining which temporal variations tend to follow ENSO processes on which timescales. Following the analyses of decadal precipitation by Cayan et al. (1998) and Dettinger et al. (1998), we next correlate filtered versions of global SSTs with similarly filtered versions of the two pan-American flow RPCs associated with ENSO.

The correlations of the streamflow RPC 2 (Fig. 4.7b) with global SST anomalies are shown in Figure 4.12 for variations (a) slower than 18 months, (b) between 18 months and 84 months (7 years), and (c) slower than 84 months. Notice that, taken together, the variations analyzed in Figures 4.12b and 4.12c make up the variations analyzed in Figure 4.12a. A dip in the coherencies of Figure 4.11 near 7 years ($f = 0.14$ cycle/yr) suggests that $(7 \text{ years})^{-1}$ is a reasonable divide between interannual and decadal frequencies. At all three timescales, the SST correlations are dominated by large negative correlations ($r < -0.6$) along the equatorial Pacific and equatorial Indian Ocean

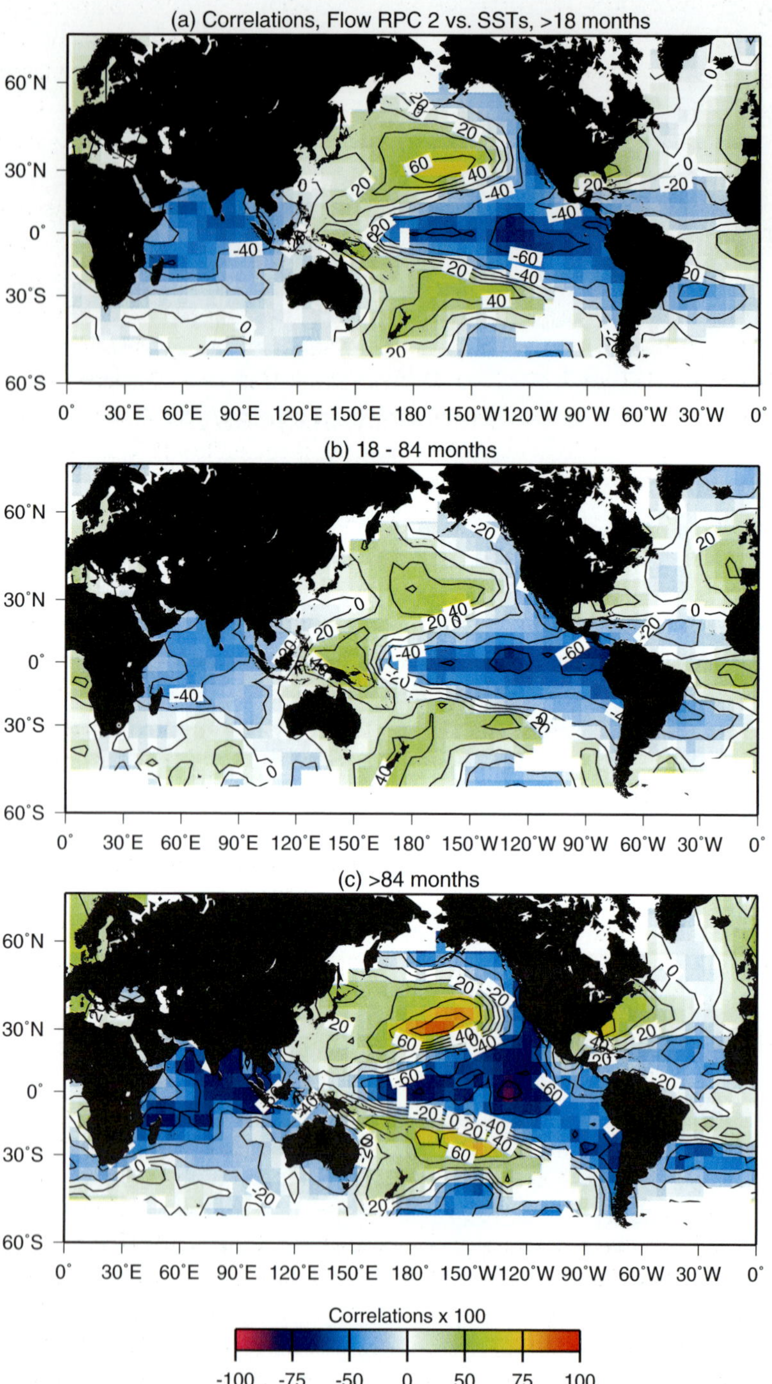

Fig. 4.12 Correlations (×100) between rotated principal component (RPC) 2 of North and South American streamflows with global sea surface temperature anomalies for variations (a) >18 months, (b) 18–84 months, and (c) >84 months (7 years).

and by positive correlations in the central North Pacific ($r > +0.6$) and central South Pacific. In Figures 4.12a and 4.12b, correlations greater than about 0.3 or less than about -0.3 are significantly different from zero at $p = 0.05$ levels. The larger correlations in Figure 4.12c are the result of far fewer realizations of these slow variations during the 1931–84 period. For those decadal variations, the largest correlations in Figure 4.12c ($|r| \sim 0.8$), however, do reach the level of significance with $p \sim 0.05$.

Interpreted as signs and strengths of the SST anomalies associated with positive excursions of the streamflow RPC, the pattern that is shared by all panels in Figure 4.12 is characteristic of well-developed ENSO episodes, with positive RPCs corresponding to cool tropical La Niña episodes and negative RPCs corresponding to El Niño episodes. The pattern, with its broad equatorial Pacific expression reaching essentially to Peru, is like the canonical (Type 1) ENSOs of Fu et al. (1986). Kahya and Dracup (1994) found El Niños of this form to be the most consistently expressed ENSO form in streamflows of the North American Southwest. Variations of this streamflow RPC are significantly associated with ENSO or ENSO-like SST and climate variations on both interannual and decadal timescales, which may indicate that the physical mechanisms that drive both timescales are similar (e.g., Zhang et al. 1997). Correlations (not shown here) between SSTs and the first precipitation RPC (corresponding to the REOF in Fig. 4.9) – which corresponds to this streamflow mode – also are ENSO-like on all three timescales, although they are less clearly so on the decadal scale. Correlations of this flow mode with similarly filtered SLP anomalies (not shown) yields, on all three timescales, a global Southern Oscillation pattern (a Walker cell with positive correlations radiating northwestward and southeastward from the eastern tropical Pacific region in opposition to negative correlations along the rest of the equator and over much of the rest of the globe; see Allan et al. 1996).

In contrast, streamflow RPC 5 (Fig. 4.10) is related to ENSO-like variations on interannual timescales but is associated with a different SST pattern on the decadal timescale (Fig. 4.12). The overall correlation pattern (Fig. 4.13a) is similar in the Pacific and Indian Oceans to the ENSO-like pattern of SST correlations to RPC 2 (Fig. 4.12a). Similarly, the interannual correlations (Fig. 4.13b) are dominated by negative correlations in the central equatorial Pacific (the Niño-3 region) with ENSO-like patterns evident in the tropics but weakly expressed in the extratropics (compared to Fig. 4.12b). The tropical part of this interannual pattern is localized in the central tropical Pacific, more like the Type 2 ENSO of Fu et al. (1986) than the pattern of Figure 4.12b. In contrast, the decadal SST correlations (Fig. 4.13c) are concentrated in the North Atlantic, which is known to exhibit considerable decadal variability (see Schlosser et al. 1991, Deser and Blackmon 1993, and Chen and Ghil 1996). The decadal variations of this RPC (thin line, Fig. 4.10b) are roughly the opposite of a gradual decline in the North Atlantic Oscillation Index (SLP differences between Iceland and Portugal) from the 1940s to the late 1960s, followed by a recent increase (see Hurrell 1995). Correlations of this flow mode with SLP anomalies (not shown) yield weak correlations in a Southern Oscillation pattern on interannual timescales and a correlation pattern dominated by SLP contrasts between the northern latitudes and the rest of the world, with emphases over North America.

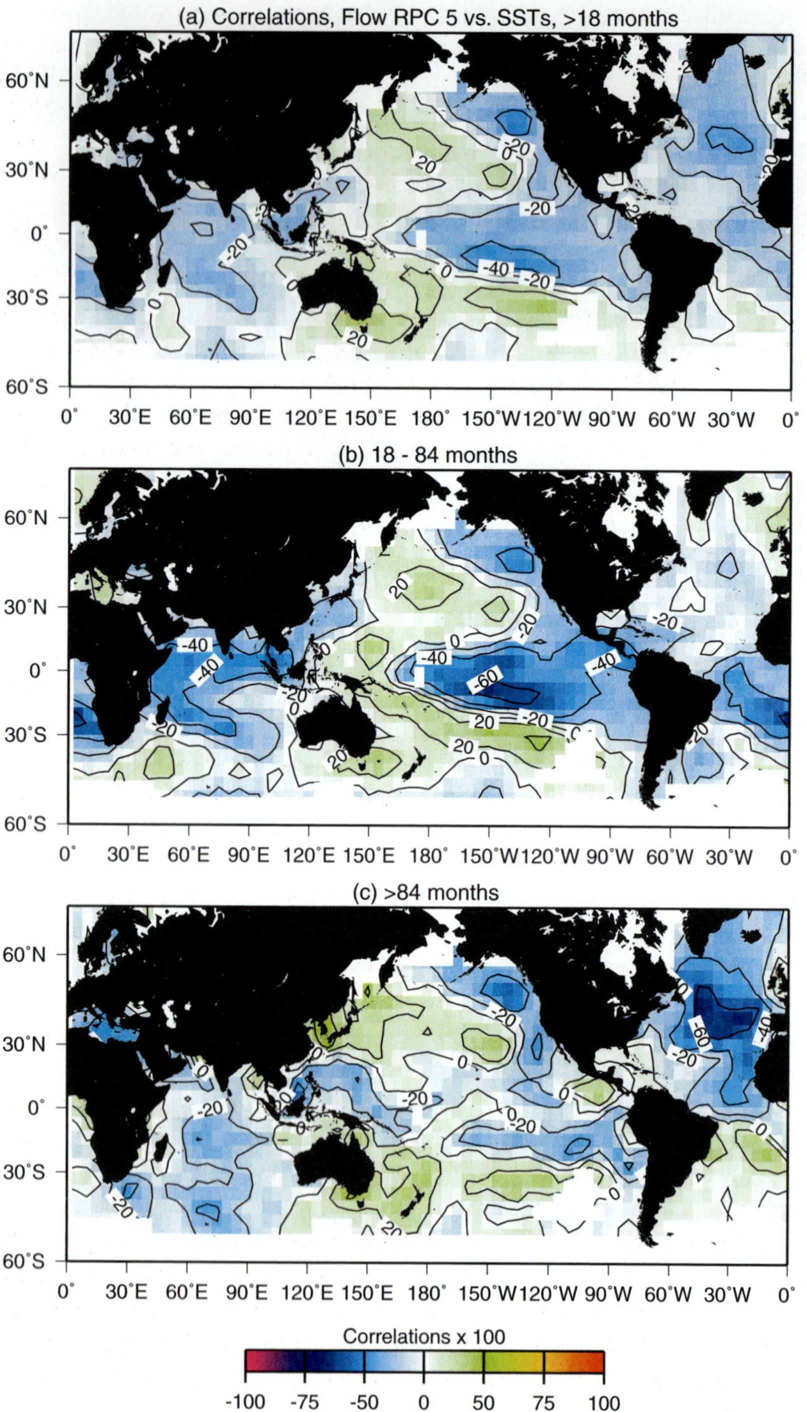

Fig. 4.13 Same as Figure 4.12, except for rotated streamflow principal component (RPC) 5.

Decadal Variations of ENSO Teleconnections

Within the time frame for which we have hemispheric streamflow data, the most important multiscale aspect of hydrologic variability associated with ENSO may be a marked change in the patterns of correlations between hydrologic variations and ENSO indices in the early to middle decades of this century. A decided weakening of ENSO processes from about the 1920s through (at least) the 1950s has been noted by several authors (e.g., Troup 1965; Rasmusson and Carpenter 1982; Trenberth 1984; Trenberth and Shea 1987; Elliot and Angell 1988; Lough 1991, 1993; Cole et al. 1993). This weakening has been described in terms of fewer and less-intense warmings, weaker SOI correlations, and changes in the pressure distributions and tropical precipitation.

The multidecadal weakening resulted in a marked change in streamflow relations, or teleconnections, to ENSO. To illustrate this change, the points plotted in Figures 4.8a and 4.8c were separated into 1931–55 and 1960–84 subsets and replotted in Figure 4.14. Most of the correlation between the streamflow RPC 2 and SOI in Figure 4.8a is shown in Figure 4.14 to be the result of a relatively strong linear relation between streamflow RPC 2 and SOI during the 1960–84 period with only a slight contribution from the 1931–55 period. Similarly, the more modest correlation between streamflow RPC 5 and SOI in Figure 4.8c was derived mostly from the recent decades (Fig. 4.14d) rather than the earlier epoch (Fig. 4.14c). Similar regional differences are evident when the RPC/Niño-3 SST scatter plots in Figures 4.8b and 4.8d are divided into the same epochs.

The long-term change in ENSO teleconnections early this century also interfered with the close multiscale correlation of tropical streamflows with SOI. Coherency spectra of averaged tropical streamflows (light solid curve, Fig. 4.11) during the time intervals 1931–55, 1960–84, and 1931–84 are shown in Figure 4.15a. Window widths were adjusted so that significance levels are similar from curve to curve. Clearly, coherencies at most frequencies (except around [4 year]$^{-1}$ to [6 year]$^{-1}$, in the lower interannual ENSO frequencies of Rasmusson et al. [1990] and Jiang et al. [1995]) have been much larger in recent decades (light solid curve, Fig. 4.15a) than in previous ones (dashed curve, Fig. 4.15a). In the Araguari River in Amazonia, this change in coherency was reflected in an abrupt change in 21-year correlations between annual flows and SOI from roughly +0.1 (and not significantly different from zero) prior to 1960 to about +0.75 in the 1970s and 1980s (Fig. 4.15b). For other regions, which were shown in Figure 4.11 to be coherent with SOI in narrow frequency bands, the coherency spectra (not shown) generally increase from the 1931–55 to 1960–84 time intervals, with particularly widespread coherency gains in the lower ENSO frequency band. Correlations between many streamflow and precipitation series outside the tropics and SOI (or other tropical ENSO indices) show decadal changes similar to those found in the tropical rivers; for example, 21-year correlations between Arizona precipitation and the preceding summer's Niño-3 SSTs shown in Figure 4.15b illustrate the extratropical hydrologic changes.

These changes may be associated partly with changes in the quality and representativeness of the SOI series during early parts of the century (Trenberth 1984; Tren-

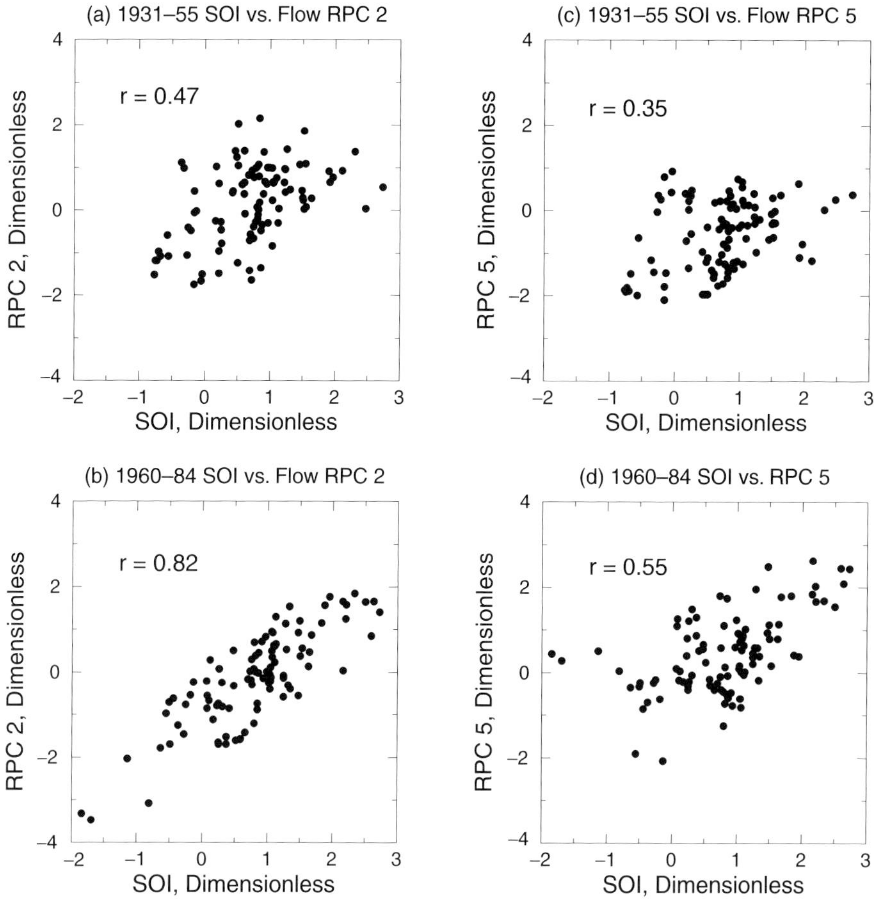

Fig. 4.14 Same as panels of Figure 4.8, except for (a) 1931–55 Southern Oscillation Index (SOI) and flow rotated principal component (RPC) 2, (b) 1960–84 SOI and flow RPC 2, (c) 1931–55 SOI and flow RPC 5, and (d) 1960–84 SOI and flow RPC 5.

berth and Shea 1987), but, in large part, they appear to reflect actual changes in the hemispheric influence of ENSO. As was mentioned previously, Niño-3 SSTs can be substituted for SOI in Figure 4.14 with similar disparity of correlations from earlier decades to recent decades.

Reasons for the reduced hydrologic/ENSO teleconnections during the 1930s to 1950s period are uncertain. The change in expressions of the ENSO phenomenon in the tropics has been ascribed to the weakening of the ENSO process during that earlier period, and comparison of the 21-year standard deviations of SOI with the corresponding correlations between SOI and Niño-3 SSTs, shown by solid circles in Figure 4.16, indicates that the oceanic (Niño-3 SST) and atmospheric (SOI) expressions of ENSO *did* become disconnected during decades when one or the other is weak (and usually when both are weak, because the 21-year standard deviations of SOI and Niño-3 are correlated with

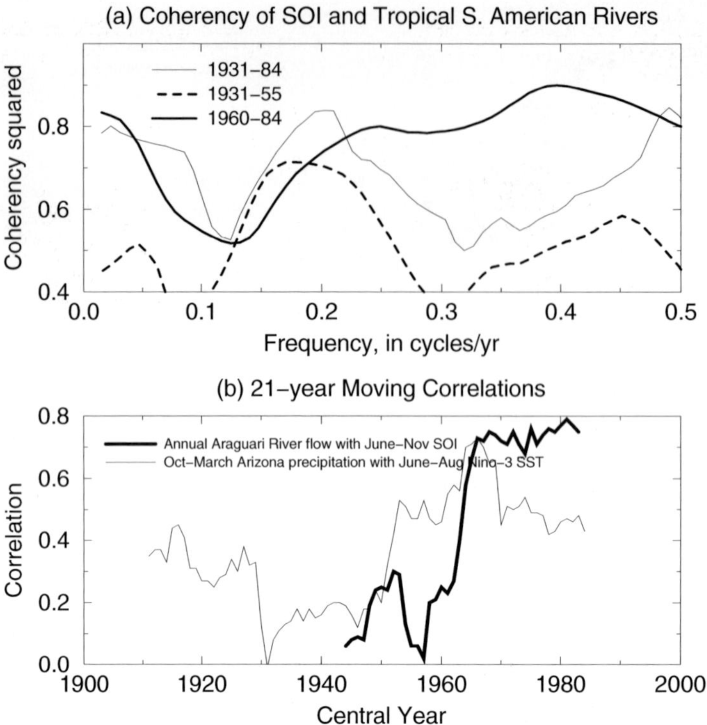

Fig. 4.15 Coherency squared of tropical South American streamflow variations (east of the Andes) with respect to Southern Oscillation Index (SOI) during several different epochs; lower bound marks significance level of roughly $p < 0.1$; (b) 21-year moving correlations of annual Araguari River flows with June–November SOI and of October–March Arizona precipitation with June–August Niño-3 sea surface temperatures; correlations greater than 0.4 are significantly different from zero at $p < 0.05$.

$r = +0.83$). A similar plot of SOI standard deviations versus correlations of SOI and Arizona precipitation (open circles in Fig. 4.16) shows that, when SOI variability is low (on decadal timescales), this (and essentially all other tropical and North American) hydrologic teleconnection weakens to the point of failing. Regionally, correlations of SOI with the streamflow RPCs decline along with the standard deviation of SOI, in 21-year windows like those used in Figure 4.16. On interannual timescales, Enfield and Luis Cid (1991) have suggested that one form of hydrologic response arises when SOI (and ENSO) is strongly positive or negative, and another when ENSO variation is weak. Kumar and Hoerling (1997) observed this kind of response in simulated responses to ENSO-like forcings.

Overall, streamflow RPC 2 was well correlated with ENSO on both interannual and decadal timescales, whereas RPC 5 was correlated to ENSO only on the interannual timescale. These different styles of ENSO response, together with evidence of decadally varying ENSO teleconnections, suggested that separate PCAs for the two epochs might uncover different streamflow-response patterns. Interestingly, however, the changes in hydrologic teleconnections as ENSO weakened on decadal timescales from about the

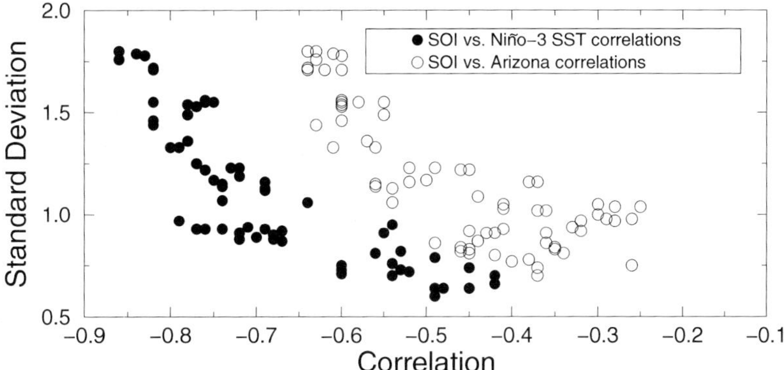

Fig. 4.16 Comparisons of the correlation coefficients (solid circles) of October through December (OND) Southern Oscillation Index (SOI) and Niño-3 sea surface temperatures (in 20-year windows) with OND SOI standard deviations in the same 20-year windows, during 1931–90 and the correlation coefficients (open circles) of OND SOI and Arizona-5 climate division October–March precipitation (in 20-year windows) with OND SOI standard deviations in the same 20-year windows, during 1931–90.

1920s into the 1950s had a relatively limited influence on the PCA patterns obtained from North and South American streamflows. When the PCA was repeated, but with only the time series from 1931–55, streamflow RPCs with spatial patterns quite similar to the original RPC 2 and RPC 5 modes were obtained and captured 9.9% and 9.3% of variance, respectively. Principal component analysis of streamflows from 1960 to 1984 reveals a recent ENSO flow mode with a spatial pattern that appears to be a sum of the original RPC 2 and RPC 5 (including cancellation of the contrasting streamflow correlations in eastern North America in Figs. 4.7 and 4.10).

The consolidation of ENSO-related flow PC patterns in the recent subset of American flows suggests that the two RPCs in the original analysis may have been a way for (inherently linear) PCA to capture a nonlinear or nonstationary ENSO relationship in which strong ENSOs, decades with strong ENSO variability, or decades with strong ENSO teleconnections elicit patterns of streamflow response that are different from those of weak ENSOs. Alternatively, the two streamflow RPCs, with their different spatial patterns, may be a PCA response to asymmetries in El Niño and La Niña flow patterns (Hoerling et al. 1997), a PCA response to ENSO-response patterns that are sometimes modified by subtle inter-ENSO differences or extratropical influences, or even statistical artifacts (e.g., Newman and Sardeshmukh 1995). In any case, the different styles of streamflow response to multiscale ENSO variability in the tropics and extratropics provides reason enough to believe that at least two modes of ENSO–streamflow response exist.

Conclusions

Anticipation of year-to-year streamflow variations has a vital role in long-term management of water supplies and hazards nearly everywhere, and an understanding of

Program Grant NA56GPO404 and the USGS Global Change Hydrology Program. MD benefited from his extended visit to the NOAA Climate Diagnostics Center, Boulder, Colorado, while preparing this chapter; that visit was made possible by Henry Diaz.

References

ALLAN, R., LINDESAY, J., and PARKER, D., 1996: *El Niño, Southern Oscillation and Climate Variability*. Collingwood, Australia: CSIRO Publishing, 405 pp.

BARNETT, T. P., BRENNECKE, K., LIMM, J., and TUBBS, A. M., 1984: Construction of a near-global sea-level pressure field. *Scripps Institution of Oceanography Reference Series* 84–7, La Jolla, CA.

BARNETT, T. P., DEL GENIO, A. D., and RUEDY, R. A., 1992: Unforced decadal fluctuations in a coupled model of the atmosphere and ocean mixed layer. *Journal of Geophysical Research*, **97**, 7341–7354.

BARNSTON, A. G. and LIVEZEY, R. E., 1987: Classification, seasonality, and persistence of low-frequency atmospheric circulation patterns. *Monthly Weather Review*, **115**, 1083–1126.

BENJAMIN, J. R. and CORNELL, C. A., 1970: *Probability, Statistics, and Decision for Civil Engineers*. New York: McGraw-Hill, 684 pp.

CAYAN, D. R., DETTINGER, M. D., DIAZ, H. F., and GRAHAM, N., 1998: Decadal variability of precipitation over western North America. *Journal of Climate*, **11**, 3148–3166.

CAYAN, D. R. and WEBB, R. H., 1992: El Niño/Southern Oscillation and streamflow in the western United States. *In* Diaz, H. F. and Markgraf, V. (eds.), *El Niño: Historical and Paleoclimatic Aspects of the Southern Oscillation*. Cambridge: Cambridge University Press, 29–68.

CHEN, F. and GHIL, M., 1996: Interdecadal variability in a hybrid coupled ocean–atmosphere model. *Journal of Physical Oceanography*, **26**, 1561–1578.

CHIEW, F. H. S., McMAHON, T. A., DRACUP, J., and PIECHOTA, T., 1994: El-Niño/Southern Oscillation and streamflow patterns in South-East Australia. *Transactions, Institution of Engineers, Australia*, **36**, 285–291.

COLE, J. E., FAIRBANKS, R. G., and SHEN, G. T., 1993: Recent variability in the Southern Oscillation: Isotopic results from a Tarawa atoll coral. *Science*, **260**, 1790–1793.

DEPETRIS, P. J. and KEMPE, S., 1990: The impact of the El Niño 1982 event on the Parana River, its discharge and carbon transport. *Global and Planetary Change*, **3**, 239–244.

DESER, C. and BLACKMON, M., 1993: Surface climate variations over the North Atlantic Ocean during winter: 1900–1989. *Journal of Climate*, **6**, 1743–1753.

DETTINGER, M. D. and CAYAN, D. R., 1998: Historical modes of interannual and decadal hydroclimatic-hydrologic variability in the Western Hemisphere (abstr.). *Pole-Equator-Pole Paleoclimate Workshop*, Merida, Venezuela, March 1998.

DETTINGER, M. D., CAYAN, D. R., DIAZ, H. F., and MEKO, D., 1998: North-south precipitation patterns in western North America on interannual-to-decadal time scales. *Journal of Climate*, **11**, 3095–3111.

DHAR, O. N. and NANDARGI, S., 1995: Monsoon activity and flows in Central Indian rivers. *Indian Journal of Power and River Valley Development*: **45**, 80–90.

DIAZ, A. F., STUDZINSKI, C. D., and MECHOSO, C. R., 1998: Relationships between precipitation anomalies in Uruguay and southern Brazil and sea surface temperatures in the Pacific and Atlantic Oceans. *Journal of Climate*, **11,** 251–271.

DIAZ, H. F. and KILADIS, G. N., 1992: Atmospheric teleconnections associated with the extreme phases of the Southern Oscillation. *In* H. F. Diaz and V. Markgraf (eds.), *El Niño: Historical and Paleoclimatic Aspects of the Southern Oscillation.* Cambridge: Cambridge University Press, 7–28.

DOUGLAS, A. V., CAYAN, D. R., and NAMIAS, J., 1982: Large-scale changes in North Pacific and North American weather patterns in recent decades. *Monthly Weather Review*, **110**, 1851–1862.

EISCHEID, J. K., BAKER, C. B., KARL, T. R., and DIAZ, H. F., 1995: The quality control of long-term climatological data using objective data analysis. *Journal of Applied Meteorology*, **34**, 2787–2795.

ELLIOT, W. P. and ANGELL, J. K., 1988: Evidence for changes in Southern Oscillation relationships during the last 100 years. *Journal of Climate*, **1**, 729–737.

ELTAHIR, E. A. B., 1996: El Niño and the natural variability in the flow of the Nile River. *Water Resources Research*, **32**, 131–137.

ENFIELD, D. B., 1996: Relationships of inter-American rainfall to tropical Atlantic and Pacific SST variability. *Geophysical Research Letters*, **23**, 3305–3308.

ENFIELD, D. B. and LUIS CID, S., 1991: Low frequency changes in El Niño–Southern Oscillation. *Journal of Climate*, **4**, 1137–1146.

FU, C., DIAZ, H. F., and FLETCHER, J. O., 1986: Characteristics of the response of sea surface temperature in the central Pacific associated with warm episodes of the Southern Oscillation. *Monthly Weather Review*, **114**, 1716–1738.

GRAHAM, N. E., 1994: Decadal-scale climate variability in the 1970s and 1980s: Observations and model results. *Climate Dynamics*, **10**, 135–162.

GRAHAM, N. E., BARNETT, T. P., WILDE, R., PONATER, M., and SCHUBERT, S., 1994: On the roles of tropical and midlatitude SSTs in forcing interannual to interdecadal variability in the winter Northern Hemisphere circulation. *Journal of Climate*, **7**, 1416–1441.

GUETTER, A. K. and GEORGAKAKOS, K. P., 1996: Are the El Niño and La Niña predictors of the Iowa River seasonal flow? *Journal of Applied Meteorology*, **35**, 690–705.

HOERLING, M. P., KUMAR, A., and ZHONG, M., 1997: El Niño, La Niña, and the nonlinearity of their teleconnections. *Journal of Climate*, **10**, 1769–1786.

HOUGHTON, R. W., 1996: Subsurface quasi-decadal fluctuations in the North Atlantic. *Journal of Climate*, **9**, 1363–1373.

HURRELL, J. W., 1995: Decadal trends in the North Atlantic Oscillation: Regional temperatures and precipitation. *Science*, **269**, 676–679.

JACOBS, G. A., HURLBERT, J. C., KINDLE, J. C., METZGER, E. J., MITCHELL, J. L., TEAGUE, W. J., and WALLCRAFT, A. J., 1994: Decade-scale trans-Pacific propagation and warming effects of an El Niño anomaly. *Nature*, **370**, 360–363.

JIANG, N., NEELIN, J. D., and GHIL, M., 1995: Quasi-quadrennial and quasi-biennial variability in the equatorial Pacific. *Climate Dynamics*, **12**, 101–112.

KAHYA, E. and DRACUP, J. A., 1993: U.S. Streamflow patterns in relation to the El Niño/Southern Oscillation. *Water Resources Research*, **29**, 2491–2503.

KAHYA, E. and DRACUP, J. A., 1994: The influences of type 1 El Niño and La Niña events on streamflows in the Pacific Southwest of the United States. *Journal of Climate*, **7**, 965–976.

KAYLOR, R. E., 1977: *Filtering and Decimation of Digital Time Series.* College Park: Institute of Physical Science and Technology, University of Maryland, Technical Report BN850, 14 pp.

KAZADI, S. N., 1996: Interannual and long-term climate variability over the Zaire River basin during the last 30 years. *Journal of Geophysical Research*, **101**, 21351–21360.

KILADIS, G. N. and DIAZ, H. F., 1989: Global climatic anomalies associated with extremes of the Southern Oscillation. *Journal of Climate*, **2**, 1029–1090.

KOCH, R. W., BUZZARD, C. F., and JOHNSON, D. M., 1991: Variation of snow water equivalent and streamflow in relation to the El Niño/Southern Oscillation. *Proceedings, Western Snow Conference*, **59**, 37.

KUMAR, A. and HOERLING, M. P., 1997: Interpretation and implications of the observed inter-El Niño variability. *Journal of Climate*, **10**, 83–91.

LATIF, M. and BARNETT, T. P., 1994: Causes of decadal climate variability over the North Pacific and North America. *Science*, **266**, 634–637.

LETTENMAIER, D. P., WOOD, E. F., and WALLIS, J. R., 1994: Hydro-climatological trends in the continental United States, 1948–88. *Journal of Climate*, **7**, 586–607.

LINS, H. F., 1985a: Streamflow variability in the United States: 1931–78. *Journal of Climate and Applied Meteorology*, **24**, 463–471.

LINS, H. F., 1985b: Interannual streamflow variability in the United States based on principal components. *Water Resources Research*, **21**, 691–701.

LOUGH, J. M., 1991: Rainfall variations in Queensland, Australia: 1891–1986. *International Journal of Climatology*, **11**, 745–768.

LOUGH, J. M., 1993: Variations of some seasonal rainfall characteristics in Queensland, Australia: 1921–1987. *International Journal of Climatology*, **13**, 391–409.

MANLY, B. F. J., 1986: *Multivariate Statistical Methods*. London: Chapman and Hall, 159 pp.

MARENGO, J., 1995: Variations and change in South American streamflow. *Climatic Change*, **31**, 99–117.

MARENGO, J. A., TOMSELLA, J., and UVO, C. R., 1998: Long-term streamflow and rainfall fluctuations in tropical South America: Amazonia, East Brazil, and Northwest Peru. *Journal of Geophysical Research*, **103**, 1775–1783.

McCABE, G. J., Jr. and DETTINGER, M. D., 1999: Decadal variability in the relations between ENSO and precipitation in the western United States. *International Journal of Climatology*, **19**, 1399–1410.

MOSS, M. E., PEARSON, C. P., and McKERCHAR, A. I., 1994: The Southern Oscillation index as a predictor of the probability of low streamflows in New Zealand. *Water Resources Research*, **30**, 2717–2723.

NEWMAN, M. and SARDESHMUKH, P. D., 1995: A caveat concerning Singular Value Decomposition. *Journal of Climate*, **8**, 353–360.

NICHOLLS, N., 1992: Historical El Niño/Southern Oscillation variability in the Australasian region. *In* Diaz, H. F. and Markgraf, V. (eds.), *El Niño: Historical and Paleoclimatic Aspects of the Southern Oscillation*. Cambridge: Cambridge University Press, 151–173.

NORTH, G. E., BELL, T. L., CAHALAN, R. F., and MOENG, F. J., 1982: Sampling errors in the estimation of empirical orthogonal functions. *Monthly Weather Review*, **110**, 699–702.

PARKER, D. E., FOLLAND, C. K., BEVAN, A., WARD, M. N., JACKSON, M., and MASKELL, K., 1995: Marine surface data for analysis of climatic fluctuations on interannual to century timescales. *In* Martinson, D. G., et al. (eds.), *Natural Climate Variability on Decade-to-Century Time Scales*. Washington: National Academy Press, 241–250.

PIECHOTA, T. C. and DRACUP, J. A., 1996: Drought and regional hydrologic variation in the United States: Associations with the El Niño–Southern Oscillation. *Water Resources Research*, **32**, 1359–1374.

PISCIOTTANO, G., DIAZ, A., and MECHOSO, C. R., 1994: El Niño–Southern Oscillation Impact on Rainfall in Uruguay. *Journal of Climate*, **7**, 1286–1303.

PROBST, J. L. and TARDY, Y., 1987: Long range streamflow and world continental runoff fluctuations since the beginning of this century. *Journal of Hydrology*, **94**, 289–311.

PROBST, J. L. and TARDY, Y., 1989: Global runoff fluctuations during the last 80 years in relation to world temperature change. *American Journal of Science*, **289**, 267–285.

QUINN, W. H., 1992: A study of Southern Oscillation-related climatic activity for A.D. 622–1900 incorporating Nile River flood data. *In* H. F. Diaz and V. Markgraf (eds.), *El Niño: Historical and Paleoclimatic Aspects of the Southern Oscillation*. Cambridge: Cambridge University Press, 119–149.

RAO, V. B. and HADA, K., 1990: Characteristics of rainfall over Brazil: Annual variations and connections with the Southern Oscillation. *Theoretical and Applied Climatology*, **42**, 81–91.

RASMUSSON, E. M. and CARPENTER, T. H., 1982: Variations in tropical sea surface temperature and surface wind fields associated with the Southern Oscillation/El Niño. *Monthly Weather Review*, **110**, 354–384.

RASMUSSON, E. M., WANG, X., and ROPELEWSKI, C., 1990: The biennial component of ENSO variability. *Journal of Marine Systems*, **1**, 71–96.

READ, J. F. and GOULD, W. J., 1992: Cooling and freshening of the subpolar North Atlantic Ocean since the 1960s. *Nature*, **360**, 55–57.

REDMOND, K. T. and KOCH, R. W., 1991: Surface climate and streamflow variability in the western United States and their relationship to large-scale circulation indices. *Water Resources Research*, **27**, 2381–2399.

RICHMAN, M. B., 1986: Rotation of principal components. *International Journal of Climatology*, **6**, 293–335.

RISBEY, J. S. and ENTEKHABI, D., 1996: Observed Sacramento Basin streamflow response to precipitation and temperature changes and its relevance to climate impact studies. *Journal of Hydrology*, **184**, 209–224.

ROGERS, J. C., 1988: Precipitation variability over the Caribbean and tropical Americas associated with the Southern Oscillation. *Journal of Climate*, **1**, 172–182.

ROPELEWSKI, C. F. and HALPERT, M. S., 1987: Global and regional scale precipitation patterns associated with the El Niño–Southern Oscillation (ENSO). *Monthly Weather Review*, **115**, 2352–2362.

ROPELEWSKI, C. F. and HALPERT, M. S., 1989: Precipitation patterns associated with the high-index of the Southern Oscillation. *Journal of Climate*, **2**, 268–284.

ROPELEWSKI, C. F. and HALPERT, M. S., 1996: Quantifying Southern Oscillation–precipitation relationships. *Journal of Climate*, **9**, 1043–1059.

SCHLOSSER, P., BONISCH, G., RHEIN, M., and BAYER, R., 1991: Reduction of deepwater formation in the Greenland Sea during the 1980s: Evidence from tracer data. *Science*, **251**, 1054–1056.

SLACK, J. R. and LANDWEHR, J. M., 1992: Hydro-climatic data network (HCDN): A U.S. Geological Survey streamflow data set for the United States for the study of climate variations, 1874–1988. *U.S. Geological Survey Open-File Report* 92–129, 193 pp.

STOCKER, T. F. and BROECKER, W. S., 1994: Observation and modeling of North Atlantic Deep Water formation and its variability: Introduction. *Journal of Geophysical Research*, **99**, 12317.

TANIMOTO, Y., IWASAKA, N., HANAWA, K., and TOBA, Y., 1993: Characteristic variations of SST with multiple time scales in the North Pacific. *Journal of Climate*, **6**, 1153–1160.

TRENBERTH, K. E., 1984: Signal versus noise in the Southern Oscillation. *Monthly Weather Review*, **112**, 326–332.

TRENBERTH, K. E., 1990: Recent observed interdecadal climate changes in the Northern Hemisphere. *Bulletin, American Meteorology Society*, **71**, 988–993.

TRENBERTH, K. E. and HOAR, T. J., 1996: The 1990–1995 El Niño–Southern Oscillation event: Longest on record. *Geophysical Research Letters*, **23**, 57–60.

TRENBERTH, K. E. and SHEA, D. J., 1987: On the evolution of the Southern Oscillation. *Monthly Weather Review*, **115**, 3078–3096.

TROUP, A. J., 1965: The 'southern oscillation.' *Quarterly Journal of the Royal Meteorological Society*, **91**, 490–506.

WAYLEN, P. R., QUESADA, M. E., and CAVIEDES, C. N., 1994: The effects of El Niño–Southern Oscillation on precipitation in San Jose, Costa Rica. *International Journal of Climatology*, **14**, 559–568.

XIE, P. and ARKIN, P. A., 1996: Analyses of global monthly precipitation using gauge observations, satellite estimates and model predictions. *Journal of Climate*, **9**, 840–858.

ZHANG, Y., WALLACE, J. M., and BATTISTI, D. S., 1997: ENSO-like interdecadal variability: 1900–93. *Journal of Climate*, **10**, 1004–1020.

ZORN, M. R. and WAYLEN, P. R., 1997: Seasonal response of mean monthly streamflow to El Niño/Southern Oscillation in North Central Florida. *The Professional Geographer*, **49**, 51–62.

5

El Niño/Southern Oscillation and the Seasonal Predictability of Tropical Cyclones

CHRISTOPHER W. LANDSEA

Hurricane Research Division, NOAA/AOML, Miami, Florida 33149

This chapter is dedicated to the memory of Mr. José Fernández-Partagás, hurricane forecaster and researcher, who passed away on 23 August 1997.

Abstract

Perhaps the most dramatic effect that El Niño has upon the climate system is in changing tropical cyclone characteristics around the world. This chapter reviews how tropical cyclone frequency, intensity, and areas of occurrence are altered in all of the cyclone basins by the phases of El Niño/Southern Oscillation (ENSO). In addition to ENSO, other global factors (such as the stratospheric Quasi-Biennial Oscillation) and local factors (such as sea surface temperature, monsoon intensity and rainfall, sea level pressures, and tropospheric vertical shear) can also help modulate tropical cyclone variability. Understanding how these various factors relate to tropical cyclone activity can be challenging owing to the fairly short record (on the scale of only tens of years) of reliable data. Despite this limitation, many of the variables that have been linked to tropical cyclones can be utilized to provide seasonal forecasts of tropical cyclones. Details of prediction methodologies that have been developed for the North Atlantic, Northwest Pacific, South Pacific, and Australian basin tropical cyclones are presented as well as the real-time forecasting performance for Atlantic hurricanes as issued by Prof. William Gray.

Introduction

Tropical cyclones are the costliest and deadliest natural disasters around the world, as the approximate 300,000 death toll in the infamous Bangladesh Cyclone of 1970 and the $26.5 billion (U.S.) in damages due to Hurricane Andrew in the southeast United States can attest (Holland 1993; Hebert et al. 1996). Pielke and Pielke (1997b) show that hurricane property losses (exceeding those due to earthquakes by a factor of 4) account for 40% of all insured losses in the United States for the period 1984 to 1993. Understanding and being able to predict how both tropical cyclone frequencies and intensities vary from year to year is obviously a topic of great interest to meteorologists, public and private decision makers, and the general public alike. Multidecadal-scale tropical cyclone variations and possible "greenhouse warming" effects have been reviewed in Landsea (1999). This chapter will explore the effects that the El Niño/Southern Oscillation (ENSO) and other phenomena have upon tropical cyclones around the world and the progress that has been made in utilizing such information to provide seasonal forecasting of these storms.

"Tropical cyclone" is the generic term for a nonfrontal synoptic-scale low-pressure system that develops over tropical or subtropical waters with organized convection and a well-defined cyclonic surface wind circulation. The energy source for tropical cyclones is primarily derived from evaporation and sensible heat flux from the sea in the presence of high winds and lowered surface pressure. These energy sources are tapped through condensation and fusion in convective clouds concentrated near the cyclone's "warm-core" center (Holland 1993). Tropical cyclones with maximum sustained surface winds of less than 18 m s^{-1} are called "tropical depressions." Once a tropical cyclone reaches winds of at least 18 m s^{-1}, it is typically called a "tropical storm" and assigned a name. Names are decided upon by representatives from countries in the basins affected at annual World Meteorological Organization regional meetings (Neumann 1993). If winds reach 33 m s^{-1}, the storm is then called a "hurricane" (in the North Atlantic Ocean, the Northeast Pacific Ocean east of the date line, or the South Pacific Ocean east of 160°E); a "typhoon" (in the Northwest Pacific Ocean west of the date line); a "severe tropical cyclone" (in the Southwest Pacific Ocean west of 160°E or Southeast Indian Ocean east of 90°E); a "severe cyclonic storm" (in the North Indian Ocean); and a "tropical cyclone" (in the Southwest Indian Ocean) (Neumann 1993). In addition, the category of "intense (or major) hurricane" has been utilized for the Atlantic basin for those tropical cyclones obtaining winds of at least 50 m s^{-1}, which corresponds to a category 3, 4, or 5 on the Saffir–Simpson hurricane intensity scale (Simpson 1974; Hebert et al. 1996).

It should be pointed out that such definitions are quite arbitrary ones and that nearly all intensity wind values at the surface are an estimation (by satellite pictures) or an extrapolation (from aircraft reconnaissance downward to the surface). Thus, by the nature of the tropical cyclone, by the limited data available, and by the way that meteorologists have defined intensity thresholds, the strength of an individual tropical cyclone can be difficult to pinpoint with certainty. Also, changes in observational platforms available to monitor tropical cyclones can produce as much or greater change in the cyclone record as can actual climate fluctuations. Studies of interannual (and especially interdecadal) changes in tropical cyclones must carefully consider both the relative arbitrariness of the intensity record of the storms and the dependency of intensity on the observations available.

Necessary (but Not Sufficient) Environmental Conditions

Before tropical cyclogenesis and development can occur, several precursor environmental conditions must be in place (Gray 1968, 1979):

(1) Warm ocean waters (of at least 26.5°C) throughout a sufficient depth (unknown how deep, but at least on the order of 50 m). Warm sea surface temperatures (SSTs) are necessary to fuel the heat engine of the tropical cyclone.[1]
(2) An atmosphere that cools fast enough with height such that it is potentially unstable to moist convection. It is the precipitating convection, typically in the form of thunderstorm complexes, that allows the heat stored in the ocean waters to be liberated for tropical cyclone development.
(3) Relatively moist layers near the midtroposphere. Dry mid-levels are not conducive to the continued development of widespread thunderstorm activity, because entrainment into the thunderstorms dries and cools the rising parcel, reducing buoyancy.
(4) A minimum distance of about 500 km from the equator. For tropical cyclogenesis to occur, nonnegligible amounts of the Coriolis force are required for near-gradient wind balance to occur. Without a substantial Coriolis force, inflow into the low pressure is not deflected to the right (to the left in the Southern Hemisphere) and the partial vacuum of the low is quickly filled.
(5) A preexisting near-surface disturbance with sufficient vorticity and convergence. Tropical cyclones cannot be generated spontaneously. To develop, they require a weakly organized system with sizable spin and low-level inflow.
(6) Low values (less than about 10 m s^{-1}) of vertical wind shear between the 850 and 200 mb levels. Vertical wind shear is the magnitude of wind change with height. Large values of vertical wind shear disrupt the incipient tropical cyclone and can prevent genesis, or, if a tropical cyclone has already formed, large vertical shear can weaken or destroy the tropical cyclone by interfering with the organization of deep convection around the cyclone center (DeMaria 1996).

Having these conditions met is necessary, but not sufficient, as many disturbances that appear to have favorable conditions do not develop. Recent work (Velasco and Fritsch 1987; Chen and Frank 1993; Emanuel 1993) has indicated that large thunderstorm systems (called mesoscale convective complexes [MCCs]) often produce an inertially stable, warm-core vortex in the trailing altostratus decks of the MCC. These mesovortices have a horizontal scale of approximately 100 to 200 km, are strongest in the midtroposphere, and have no appreciable signature at the surface. Zehr (1992) hypothesized that genesis of the tropical cyclones occurs in two stages: Stage one occurs when the MCC produces a mesoscale vortex, and stage two occurs when a second blowup of convection at the mesoscale vortex initiates the intensification process of lowering central pressure and increasing swirling winds.

[1] However, documented cases exist (e.g., Atlantic Hurricane Karl in 1980 [Lawrence and Pelissier 1982]) where this SST threshold of 26.5°C was not necessary. It may be instead that SSTs exceeding this amount are a general proxy for an environment that is conditionally unstable to moist convection. Conditions can – and apparently do – set up on occasion to allow for conditional moist instability in waters cooler than 26.5°C.

Variations in Environmental Conditions that Affect Tropical Cyclone Activity

Seasonal variations of tropical cyclone activity depend upon changes in one or more of the previous parameters. Many studies have focused upon the variations in these values both before and during the tropical cyclone season. While most of these studies have been focused on the Atlantic basin, all of the global basins have been analyzed to some degree for interannual predictability.

Globally, tropical cyclones are affected dramatically by the ENSO. ENSO is a fluctuation on the scale of a few years in the ocean–atmospheric system involving large changes in the Walker and Hadley Cells throughout the tropical Pacific Ocean region (Philander 1989). The state of ENSO can be characterized by, among other features, SST anomalies in the eastern and central equatorial Pacific: Warmings in this region are referred to as El Niño events and coolings are La Niña events. The Southern Oscillation Index (SOI), the standardized difference in sea level pressure (SLP) between Tahiti and Darwin, Australia, also describes the state of ENSO with high (low) pressures at Darwin and low (high) pressure at Tahiti corresponding to El Niño (La Niña) events. ENSO greatly alters global atmospheric circulation patterns, and it can affect tropical cyclone frequencies primarily by altering the lower tropospheric source of vorticity and by changing the vertical shear profile.

The various basins do not respond identically to ENSO. Some show changes in frequency of cyclogenesis, while others show shifts in the genesis locations. These variations are due to the time of year that the basin reaches its peak in activity versus the annual cycle of ENSO, the location of the basin with respect to the central equatorial Pacific, and the background climatological flow features within the basin. Basins within the Pacific can be partially forced by direct alterations of the SSTs in the genesis regions; however, most basins experience remote forcing through alteration of the tropospheric flow features. It is the combination of spatial, temporal, and climatological factors that determines how individual tropical cyclone basins will be altered by ENSO.

Nicholls (1979) first noticed that tropical cyclones in the vicinity of Australia (90°E to 165°E) are reduced in number during the warm (El Niño) phase of ENSO. Revell and Goulter (1986) and Hastings (1990) demonstrated that the reduction of tropical cyclones in the Australian region is compensated for by an increase in the South Pacific east of 165°E (Fig. 5.1), because of a shift in the center of action in tropical cyclone genesis. There is also a smaller tendency to have the tropical cyclones originate a bit closer to the equator (Revell and Goulter 1986). The opposite is observed in La Niña events. This difference appears to be due to a weakening of the Australian monsoon trough (e.g., the boundary between the cross-equatorial near-surface westerlies and the trade wind easterlies; McBride 1995) in the western portion of the basin and an extension of this trough well to the east of its usual location during an El Niño event; this changes the availability of lower tropospheric large-scale cyclonic circulation and convergence for the storms to develop (Fig. 5.2; Evans and Allen 1992). Evans and Allen (1992) also identified a regional change for the Northern Territory of Australia that is opposite to the general tendency for the entire basin. They found fewer tropical

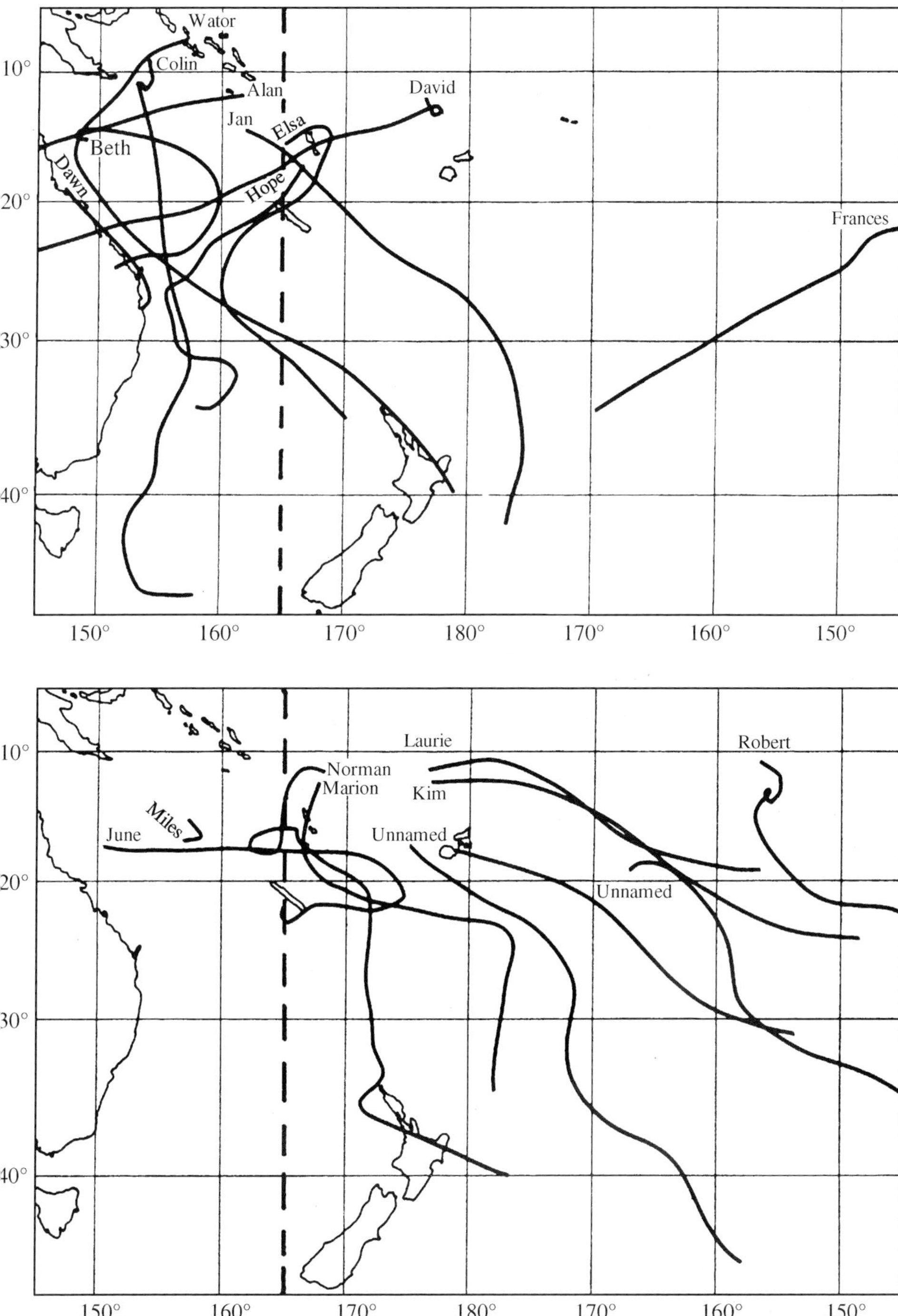

Fig. 5.1 Tracks of tropical cyclones in the South Pacific in two contrasting years: (top) 1976/77 season, warm phase of ENSO or El Niño and (bottom) 1975/76 season, cold phase of ENSO or La Niña. The dashed line indicates 165°E longitude. (Adapted from Revell and Goulter 1986.)

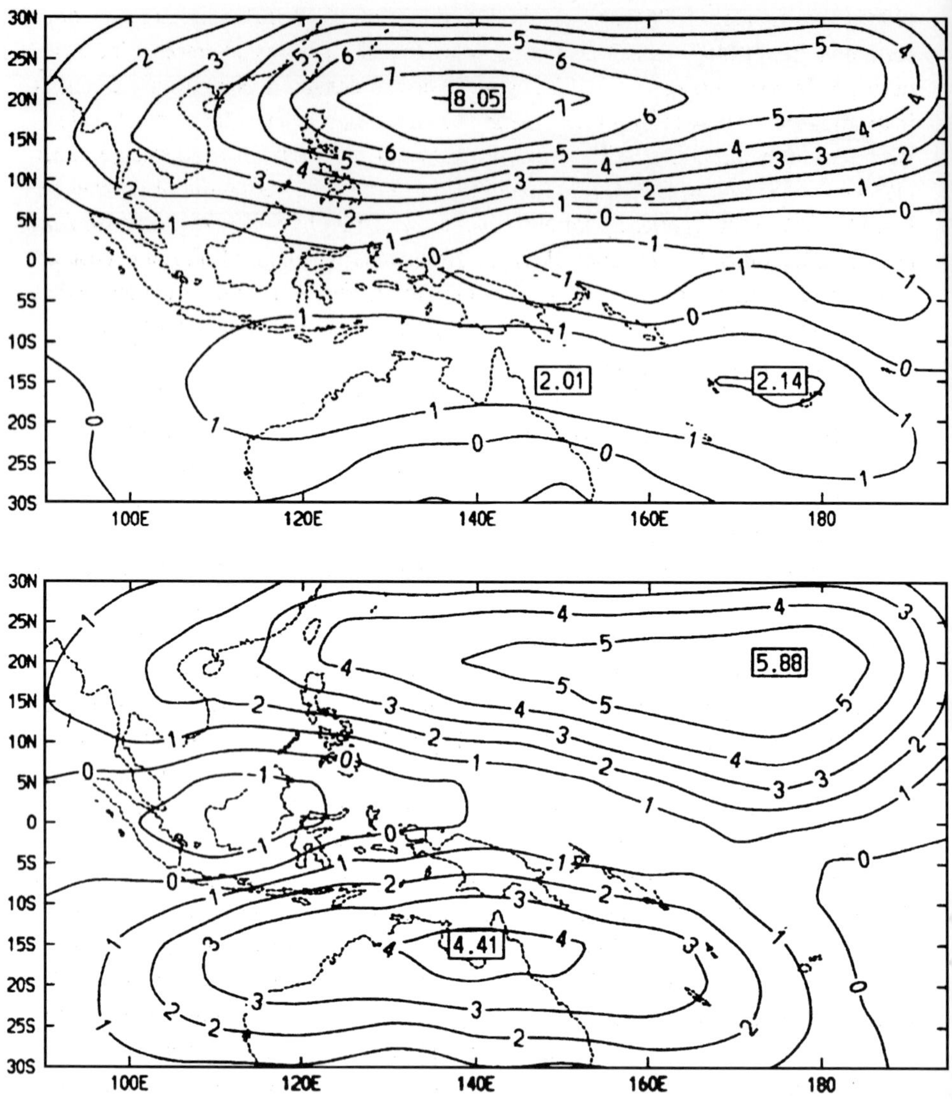

Fig. 5.2 January–February composite 850 mb streamfunction for (a) ENSO warm phases (1966, 1973, 1983, and 1987) and (b) ENSO cold phases (1971, 1974, 1976, and 1989). Positive values south of the equator indicate high streamfunction or large-scale cyclonic circulation. (From Evans and Allen 1992.)

cyclones (and fewer landfalls) during La Niña than in El Niño years because of a stronger – though landlocked – monsoon trough. Such an overland positioning of the monsoon trough, while allowing for large rainfall production over northern Australia, is not conducive to tropical cyclone formation, because genesis of tropical cyclones requires an oceanic moisture and heat source.

Likewise, the Northwest Pacific basin experiences a similar change in the location of tropical cyclone genesis without a total change in frequency. Pan (1981), Chan

(1985), and Lander (1994) have detailed that west of 160°E fewer tropical cyclones form and from 160°E to just east of the date line the amount of genesis increases during El Niño events (Fig. 5.3). The opposite occurs during La Niña events. Changes in the monsoon trough location and strength again appear to dictate the tropical cyclone variations, although this possible effect has not been documented. In addition, Lander (1994) uncovered a midseason increase in tropical cyclone formation in subtropical latitudes (20° to 30°N) during La Niña events, which he hypothesized to be tropical cyclogenesis forced by the Tropical Upper Tropospheric Trough (TUTT; a persistent, summer–autumn, "cold-core" trough with maximum amplitude at the tropopause that occurs primarily over the tropical and subtropical mid-oceans [see Fitzpatrick et al. 1995] within the trade wind belt).

The western portion of the Northeast Pacific basin near Hawaii (140°W to the date line) experiences more tropical cyclone genesis during an El Niño year and more tropical cyclones tracking into the subregion in the year following an El Niño (Schroeder and Yu 1995). The opposite effects of La Niña have yet to be analyzed, and the mechanism for such changes is unclear at this time.

Whereas previous studies have indicated that changes in local conditions are important in altering the tropical cyclogenesis frequencies, the Atlantic basin feels the effects of ENSO remotely through changes in the vertical shear wind profile. During El Niño events, the vertical shear increases primarily due to increases in the climatological westerly winds in the upper troposphere (Fig. 5.4), and reduced 200 mb westerlies and shear occur during La Niña events (Gray 1984a; Shapiro 1987). The larger (smaller) vertical shear accompanying El Niño (La Niña) events leads directly to decreased (increased) numbers of Atlantic hurricanes. Goldenberg and Shapiro (1996) found that the area between 10°N and 20°N from North Africa to Central America (hereby known as the Atlantic "main development region") shows the largest sensitivity toward changes in the vertical shear, with weakly opposite conditions occurring in the subtropical latitudes of 20°N to 35°N (Fig. 5.5). This tendency for weaker (stronger) vertical shear in the subtropical latitudes during El Niño (La Niña) events may account for increasing (decreasing) the number of subtropical-forming tropical cyclones, though these changes in the subtropical latitudes are weaker in magnitude than the changes occurring in the main development region. Additional impacts of ENSO on Atlantic climate are discussed in Enfield and Mayer (1997) and in the Enfield and Mestas-Nuñez chapter in this book.

The remaining basins – the eastern portion of the Northeast Pacific (the North Pacific Ocean from 140°W to North America), the Southwest Indian, and the North Indian – appear to have little ENSO-forced variations (i.e., Jury 1993; Dong and Holland 1994; McBride 1995), though there may be ENSO relationships produced in these areas that have not yet been identified.

Besides the ENSO, another global factor appears to force changes in tropical cyclones: the stratospheric Quasi-Biennial Oscillation (QBO), an east–west oscillation of stratospheric winds that encircle the globe near the equator (Wallace 1973). This oscillation has a distinct effect upon Atlantic (more activity in the west phase [Fig. 5.6]; Gray 1984a; Shapiro 1989), Southwest Indian (more activity in the east phase; Jury

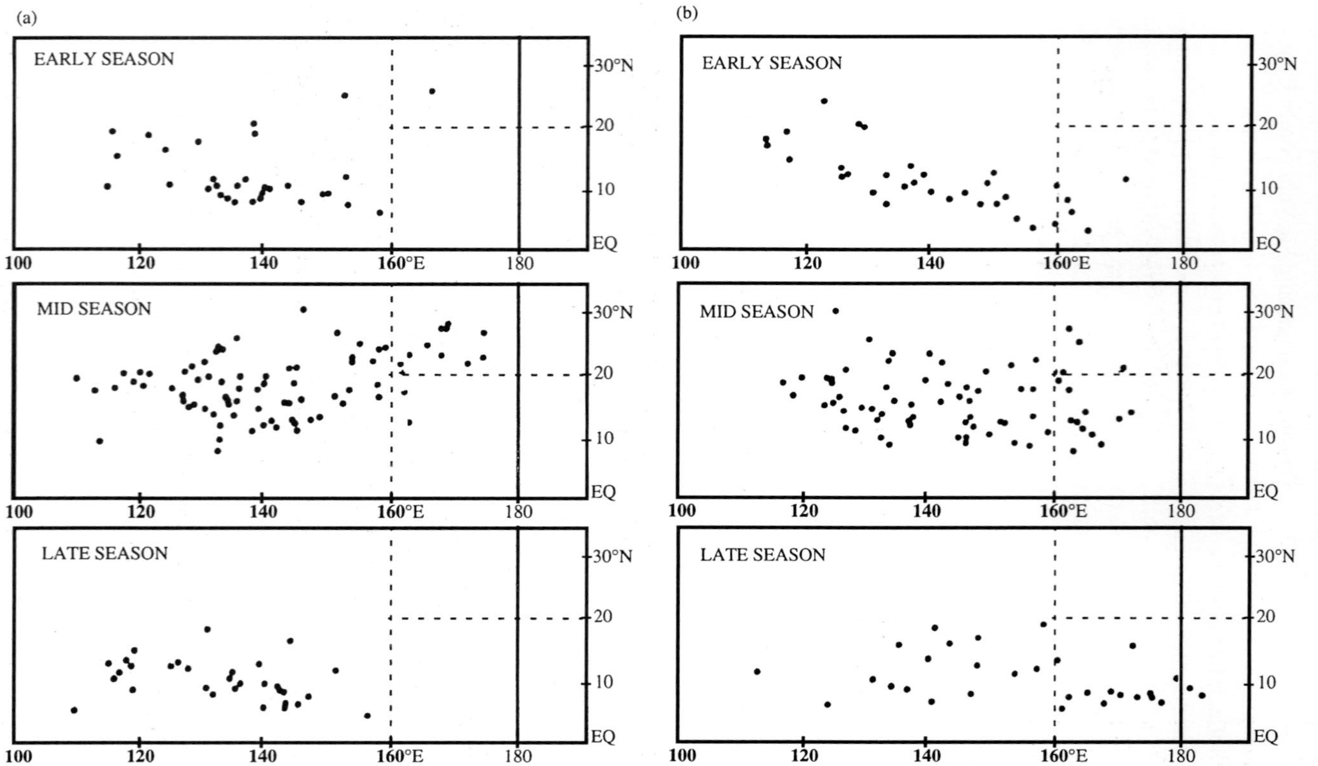

Fig. 5.3 Origins of tropical cyclones by season (early season, March to mid-July; midseason, mid-July to mid-October; and late season, mid-October to January) for the five years during the period 1970–91 with the five highest values of the March–January average of the Southern Oscillation Index (SOI) or La Niña [panel (a)] and for five years with the lowest values of the March–January average of the SOI or El Niño [panel (b)]. (From Lander 1994.)

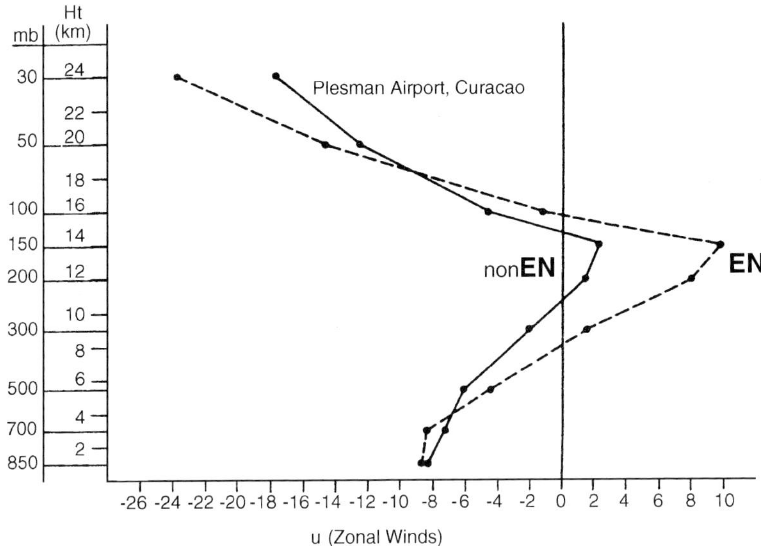

Fig. 5.4 Vertical profile of zonal wind during August and September at Curacao (12.2°N, 69.0°W) for an average of five El Niño years (1957, 1965, 1972, 1976, and 1982), denoted by EN and eighteen other non–El Niño years (non–EN). (From Gray 1984a.)

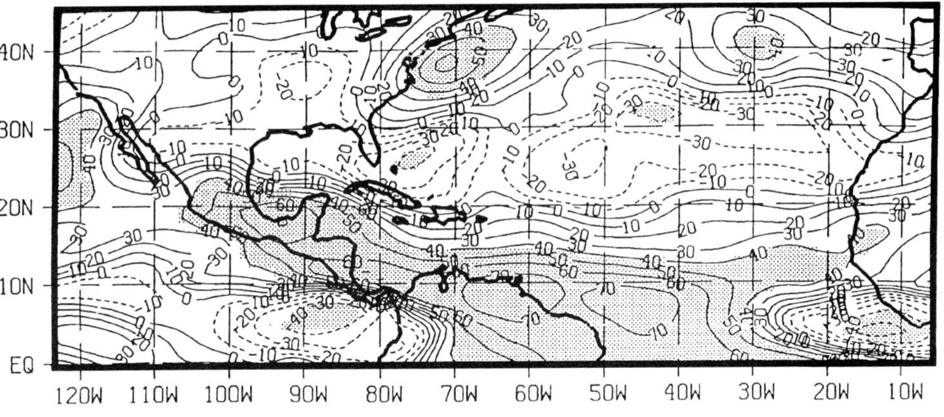

Fig. 5.5 Map of linear correlation coefficients between an index of the El Niño 1+2 SST anomalies (ENI) and the absolute value of the vertical shear between 700 and 200 mb for 1968–92. Magnitudes are multiplied by 100; the contour interval is 0.10. Negative contours are dashed. Regions where the correlations are significant at the 95% level are shaded. (From Goldenberg and Shapiro 1996.)

1993), and Northwest Pacific (more activity in the west phase; Chan 1995) tropical cyclones. While the exact mechanism of the stratospheric QBO's influence on tropical cyclones is uncertain, it has been hypothesized that upper tropospheric to lower stratospheric vertical shear variations (Gray et al. 1992b) and/or upper tropospheric static stability changes (Knaff 1993) may be responsible.

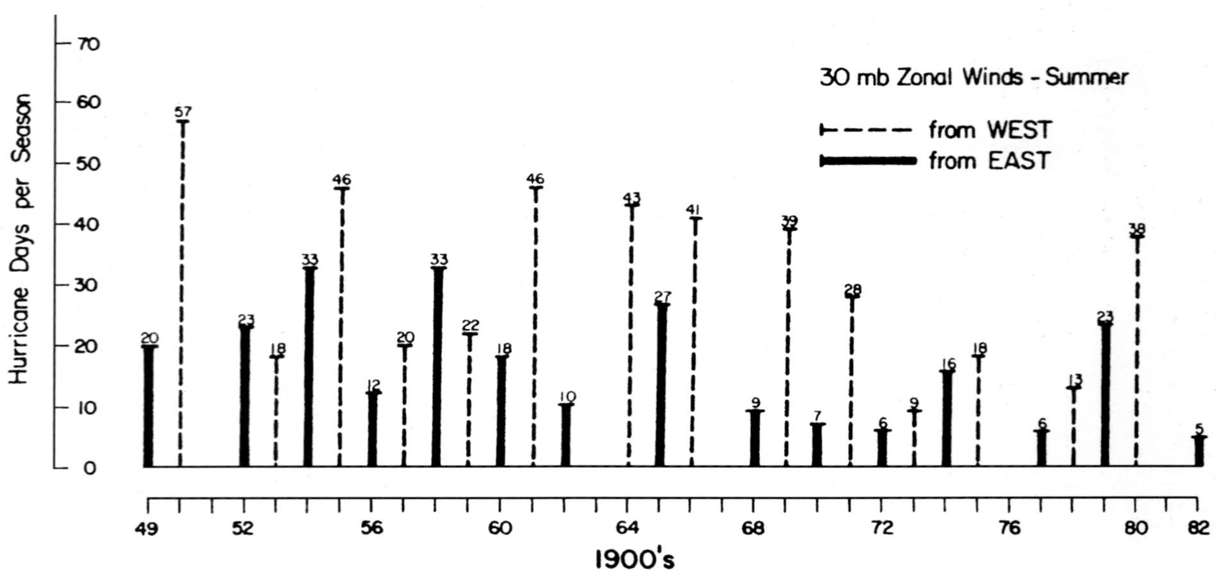

Fig. 5.6 Relationship between 30 mb stratospheric wind direction and seasonal number of hurricane days from 1949 to 1982. Years with no observation are those in which the 30 mb zonal wind is changing phase. (From Gray 1984a.)

In addition to the global effects of ENSO and QBO, there are also local effects that appear to directly impact tropical cyclone frequency within individual basins. These include variations of local SLPs, SSTs, and trade wind and monsoon circulations.

Sea level pressures act to directly impact the strength of the vertical wind shear. For example, in the Atlantic basin, because of a relatively invariant SLP field near the equator, above (below) normal SLP in the main development region from 10°N to 20°N between Africa and the Americas tightens (loosens) the local pressure gradient and strengthens (weakens) the easterly trade winds by 1 to 3 m s^{-1}, thereby contributing to increased (decreased) vertical shear (Gray et al. 1993, 1994). In addition, Gray et al. (1993) have suggested that abnormally low SLP indicates a poleward shift and/or a strengthening of the Intertropical Convergence Zone (ITCZ). Both situations contribute to less subsidence and drying in the main development region through which easterly waves move. Knaff (1997) indicates that low SLP is accompanied by a deeper moist boundary layer and a weakened trade wind inversion. Moreover, an enhanced ITCZ provides more large-scale, low-level cyclonic vorticity to incipient tropical cyclones, thereby creating an environment that is more favorable for tropical cyclogenesis (Gray 1968). In contrast, above normal SLP tends to be associated with opposite conditions that are unfavorable for tropical cyclogenesis. Ray (1935), Brennan (1935), Shapiro (1982), Gray (1984b), and Gray et al. (1993, 1994) have discussed the relationship between SLP anomalies and Atlantic basin activity, while Nicholls (1984) has analyzed Australian tropical cyclones and local pressure values.

Sea surface temperatures in the genesis regions of tropical cyclone basins have a direct thermodynamic effect on tropical cyclones through their influence on moist static stability (Malkus and Riehl 1960). Sea surface temperature also indirectly influences the vertical shear through its strong inverse relationship with surface pressures in some regions (Shapiro 1982; Gray 1984b; Nicholls 1984). (These direct and indirect effects of local SST variations are considered separately from the remote forcings of the SST modulations directly due to ENSO.) In particular for the Atlantic basin, warmer than average waters are usually accompanied by lower than average surface pressures, and thus weaker trade winds and reduced shear. Cooler than average waters are usually accompanied by higher pressure, stronger trade winds, and increased shear. Somewhat surprisingly, interannual SST variations have relatively small or negligible contributions toward increasing the tropical cyclone frequency in most basins. Only the Atlantic, Southwest Indian, and Australian regions have significant though small, positive associations in the months directly before the tropical cyclone seasons begin (Raper 1992; Shapiro and Goldenberg 1998). However, Saunders and Harris (1997) provide substantial evidence that both preceding and during the hurricane season, Atlantic SSTs in the main development region contribute a large percentage of the explained variance (over 30% during the height of the season) of the number of hurricanes generated in that area. Indeed, they argue through a partial correlation analysis that these Atlantic SSTs are the *dominant* physical modulator of tropical Atlantic hurricanes. In addition to these studies, Ray (1935), Carlson (1971), Wendland (1977), and Shapiro (1982) have also examined the Atlantic basin; Jury (1993) has investigated the Southwest Indian; and Nicholls (1984) and Basher and Zheng (1995) have analyzed the Australian/Southwest Pacific for SST associations.

One aspect that has recently been uncovered is the association of a tropical cyclone basin with its generating (or nearby) monsoon trough. As was previously discussed, Evans and Allen (1992) found that variations in the Australian monsoonal flow can be associated with changes in tropical cyclone activity such that a strong (weak) monsoon circulation during a cold (warm) phase of ENSO is accompanied by many (few) tropical cyclones. Bate et al. (1989) also suggested that variations in the Australian monsoon could alter the tropical cyclone activity, independent of any pronounced ENSO events. Over the Atlantic basin, June through September monsoonal rainfall in Africa's western Sahel has shown a very close association with intense hurricane activity (Fig. 5.7; Reed 1988; Gray 1990; Landsea and Gray 1992; Landsea et al. 1992). Wet years in the western Sahel (e.g, 1988 and 1989) are accompanied by dramatic increases in the incidence of intense hurricanes, while drought years (e.g., 1990 through 1993) are accompanied by a decrease in intense hurricane activity. Variations in tropospheric vertical shear and African easterly wave intensity have been hypothesized as

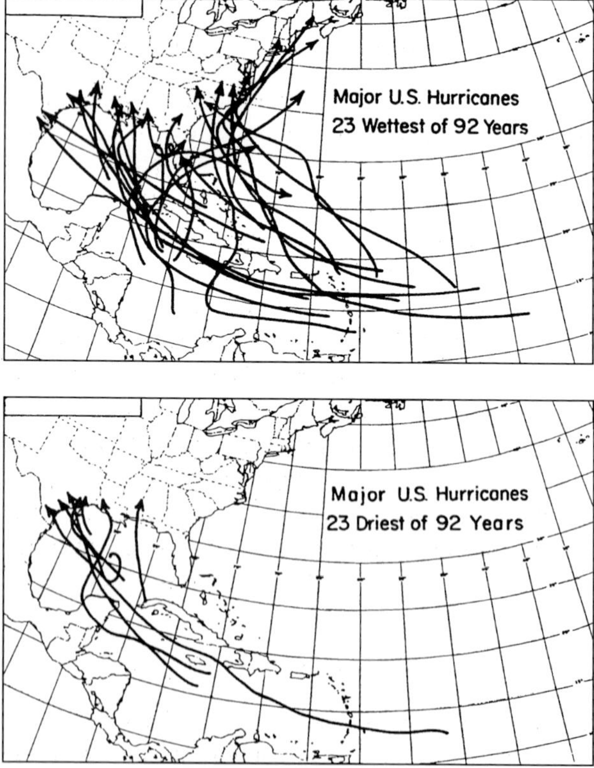

Fig. 5.7 Hurricane tracks of the tropical cyclones that made landfall as an intense hurricane (category 3, 4, or 5) along the U.S. coastline during the twenty-three wettest years (upper panel) and the twenty-three driest years (lower panel) from 1899 to 1990 in the western Sahel. (From Landsea et al. 1992.)

the physical mechanisms that link the two phenomena (Gray 1990; Landsea and Gray 1992), although Goldenberg and Shapiro (1996) have demonstrated that changes in the vertical shear probably dominate. They note that wet (dry) years are associated with reduced (increased) wind shear, due to both weaker (stronger) than average lower tropospheric trade winds and upper tropospheric westerlies throughout the main development region.

A final factor that has been considered for forcing interannual variations of tropical cyclone activity is changes in the "steering flow" in which the storms are embedded. (To a first approximation, tropical cyclones can be considered to be steered by the surrounding deep layer [the ocean surface to 100 mb] atmospheric flow features [Franklin et al. 1996].) Namias (1955) and Ballenzweig (1959) first suggested that interannual variations in the midtropospheric flow fields could help account for variations both in Atlantic basin tropical cyclogenesis and in the tracks of the storms once formed. Although their ideas regarding genesis have not borne out, the hypothesis regarding changes in steering have held up. Shapiro (1982) confirmed that midtropospheric flow features can account for subregions within the Atlantic basin experiencing more or less activity in any particular year.

Predicting Seasonal Variations of Tropical Cyclones

Currently, the only feasible methodology for seasonal tropical cyclone forecasting involves the use of statistical regression models. Eventually, the use of numerical models (or global circulation models, GCMs) to produce seasonal forecasts may also be possible. Indeed, a couple of encouraging steps forward (e.g., Wu and Lau 1992; Watterson et al. 1995) has shown that, either directly through the number of tropical cyclone-like vortices or indirectly through measurements of crucial environmental fields, there may someday be a skill with such models. However, real-time skill today is unattainable because of (1) the inability in some GCMs to produce a realistic representation of tropical cyclones in the coarse grid spacing available; (2) the complete lack of a stratospheric QBO – shown earlier to be a crucial component in the tropical cyclone variability of many regions – in the GCMs; and (3) the inability to forecast the oceanic boundary conditions, including the timing, phase, and magnitude of the ENSO phenomena as well as local SST anomalies. However, as is detailed below, statistical forecasting schemes have already provided and are continuing to provide skilled and useful predictions of tropical cyclone activity around the world.

Atlantic Basin

With the completion of the 1996 hurricane season, Prof. William Gray and colleagues at Colorado State University have issued real-time seasonal hurricane forecasts for thirteen years. The original forecasting procedures are described in Gray (1984a,b) but have since been substantially redeveloped and improved. Forecast techniques have been developed from the analysis of data going back to 1950. Instead of an ordinary least squares (OLS) regression technique, Gray et al. (1992a, 1993, 1994) have utilized

Table 5.1 *Various predictive groups utilized in Gray et al.'s (1992a, 1993, 1994) 1 December, 1 June, and 1 August forecasts of Atlantic seasonal tropical activity. The numbers in parentheses indicate the number of predictors used in each group.*

1 December	1 June	1 August
QBO (3)	QBO (3)	QBO (3)
African rainfall (2)	African rainfall (2)	African rainfall (2)
	Current ENSO conditions (4)	Current ENSO conditions (2)
	Caribbean pressures and 200 mb winds (2)	Caribbean pressures and 200 mb winds (2)
	African temperature and pressure gradients (2)	

a linear regression model based upon the least absolute deviations (LAD). The LAD method creates regression lines that are fitted to the data by minimizing the actual distance between hindcasted values and the observations. This differs from the more traditional OLS regression approach that is based upon the unphysical square of the same distance. Thus all observations are weighted equally in LAD rather than an undue emphasis being placed on the outliers as is seen with OLS. Complementary with LAD is the use of the agreement coefficient, ρ, which provides a measure of the fit of hindcasted and observed tropical cyclone values. The agreement coefficient (Mielke 1991) measures skill by comparing the absolute differences between hindcasted and observed values versus a random assortment of these absolute differences: A $\rho = 0$ indicates absolutely no agreement between hindcasted and observed values, and a $\rho = 1$ indicates perfect agreement between the two. Values of ρ that range from 0 to 1 can be considered the amount of variability that the hindcasts can explain in the observations.

Forecasts issued at the end of the previous year's hurricane season are a fairly recent endeavor. The 1 December forecast is based upon five predictors (Gray et al. 1992a). These predictors include those based upon the extrapolated state of the stratospheric QBO through the zonal winds at 50 mb and 30 mb and the vertical shear of the zonal winds between the two levels and previously measured North African rainfall – August and September precipitation within the western Sahel and August through November precipitation along the Gulf of Guinea. Table 5.1 lists these predictive groupings, and Figure 5.8 shows the location of these various predictors.

Because of the consistency of the QBO, successful long-range extrapolations of the mean stratospheric zonal winds can be made almost a year in advance. For this 1 December forecast time, mean QBO conditions for September of the following year are extrapolated based upon November information. The two West African rainfall indices are needed for Atlantic tropical cyclone forecasting because of the intimate link between concurrent seasonal amounts of intense hurricane activity and seasonal rainfall in the Sahel of West Africa (Landsea and Gray 1992). Gray et al. (1992a) found that rainfall along the Gulf of Guinea and in the Sahel itself provides a somewhat

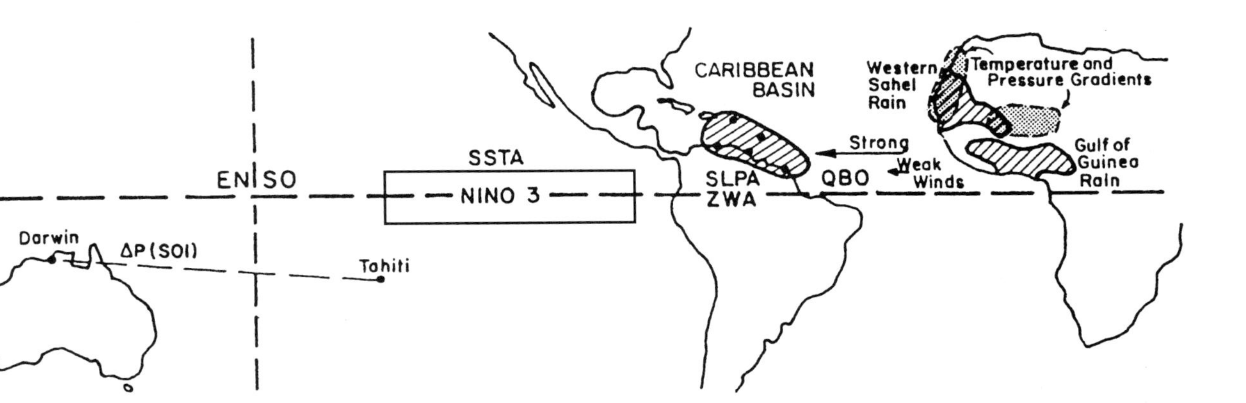

Fig. 5.8 Locations of meteorological and oceanographical parameters used in the Atlantic seasonal forecasts by Gray et al. (1992a). See the text for details.

dependable indication of future Sahel rainfall (and thus Atlantic hurricane activity). The Sahel rainfall correlation to the previous year's rainfall is reflected in the strong tendency for anomalies of precipitation to continue from year to year. This persistence is likely due to a combination of global SST forcing (Lamb 1978; Folland et al. 1986) and changes in the land surfaces, including desertification, which may reinforce drought conditions (Nicholson 1988; Xue and Shukla 1993). The positive feedback between the Gulf of Guinea rainfall in August through November to Sahel rainfall/Atlantic hurricanes the following year appears to result from changes in available moisture for the North African monsoon through long-term storage in the soil and biosphere (Gray et al. 1992a). While the previous year's Sahel rainfall can be used to forecast for only about 5% of the intense hurricane variability, the Gulf of Guinea rainfall anomalies provide a much stronger predictor of around a third of the variability hindcasted in the intense hurricane activity.

Overall, the 1 December hindcasts were able to explain about 40% to 50% of the variability of the tropical cyclone activity. Because of the tendency to overfit statistical regressions with large numbers of predictors relative to the number of data points (e.g., greater than around 1 to 10) in a non-cross-validated approach (Elsner and Schmertmann 1994), true independent forecasts will have a substantial degradation in skill. Thus the skill estimated to be available for future independent predictions is at the level of 20–35% of the variability according to methodology described in Mielke et al. (1996). This can be compared to climatology, which provides none of the variance by definition, and to year-to-year persistence (i.e., an autoregressive model with a 1-year lag), which can only account for about 5% of the variability. Figure 5.9 demonstrates the observed differences in intense hurricanes for the ten hindcasts for the most active tropical cyclone seasons and the ten hindcasts for the quietest seasons. Note the very large differences in observed intense hurricane tracks, indicating a substantial amount of skill present in these hindcasts. This is an impressive result considering that this forecast is issued six months before the start of the "official" hurricane season and eight months before the active portion of the hurricane season. The latter forecasts of early June and early August make substantial use of physical parameters that affect the Atlantic hurricanes (e.g., ENSO conditions, SLP anomalies, upper tropospheric zonal winds, etc.) and that also have the tendency to persist from the forecast date through the peak of the season. This is not feasible for the early December forecasts with such a long lead time, especially for ENSO's upcoming state because of the difficulty in obtaining skill across the March–May "predictability barrier"[2] (Wright 1985; Wright et al. 1988).

The 1 June seasonal tropical cyclone forecast incorporates elements from the 1 December forecast as well as more timely information from the most recent few months

[2] Some forecasting groups have claimed ENSO predictive skill through the spring season (e.g., Chen et al. 1995; Penland and Sardeshmukh 1995) primarily through hindcast runs on dependent data. However, independent tests show that no ENSO mode exists – statistical or numerical – that exhibits skill in real-time predictions relative to a simple ENSO climatology and persistence (ENSO-CLIPER) model (Knaff and Landsea 1997). Thus until truly skillful ENSO models for lead times of several seasons become available, improvements in forecasts issued in early December for Atlantic hurricanes the following year may be slow to occur.

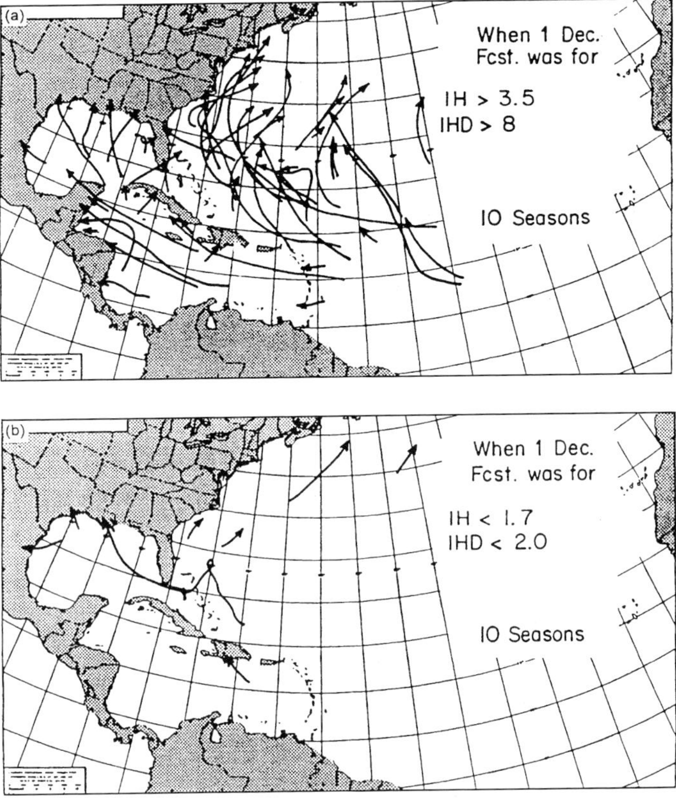

Fig. 5.9 Contrast of observed intense hurricane tracks between 1950 and 1990 from the ten most active hindcast (>3.5 intense hurricanes [IH] and >8.0 days of intense hurricanes [IHD] occurring) seasons versus the ten calmest hindcast seasons (<1.7 IH and <2.0 IHD) from an initial forecast date of 1 December of the previous year. The ratio of observed IHD between the two composites is 9.5 to 1. (From Landsea et al. 1994.)

(Gray et al. 1994), the most important being an indication of ENSO's evolving state. There are thirteen predictors in five groups as listed in Table 5.1 used in this forecast. Figure 5.8 shows the locations of these various predictors. As with the 1 December forecast, three of the predictors are for extrapolating the state of the QBO expected during September zonal winds at 50 mb and 30 mb and for vertical shear between the two layers. Four predictors involve North African surface parameters. Two of these, the Gulf of Guinea and western Sahel rainfall, were described in the previous section. The other two North African predictors are the anomalous surface temperature and sea level pressure gradients from February through May of the current year. The remaining six predictors involve conditions over the Caribbean Sea (April to May sea level pressure anomalies and 200 mb zonal wind anomalies) and current information regarding the strength and trend of ENSO.

The new predictors include two North African surface predictors that relate to the pre-rainy season conditions over sub-Saharan North Africa. When zonal surface temperature and sea level pressure gradients during February through May are relaxed during monsoon onset, the Sahel rainfall and Atlantic hurricane activity become stronger than normal. Conversely, when the surface temperature and sea level pressures have tightened gradients from the west coast to the interior, Sahel rainfall is reduced and Atlantic hurricane activity is quieter than normal. These surface conditions act to alter the strength of the southwesterly monsoon flow into the Sahel. Over the Caribbean, April and May sea level pressure and 200 mb zonal wind anomalies (reliable measures of crucial vertical wind shear variations) are utilized as predictors for the hurricane season. The preseason sea level pressure anomalies and the 200 mb zonal winds over the Caribbean have a tendency to persist into the heart of the hurricane season and thus are useful as predictors of hurricane activity. The last four predictors give indications of the current strength of ENSO and its trend in the previous few months: the April and May equatorial eastern Pacific SSTs and the SOI and their changes between January/February to April/May. These values provide reliable indications of how ENSO will likely behave during August through October, the crucial peak Atlantic basin hurricane months.

With the use of these thirteen predictors, the hindcast testing can anticipate between 50% and 70% of the variability by 1 June. This should degrade to 25–55% of the variability in independent real-time (operational) forecasts, demonstrating a substantial improvement over the skill levels that are suggested for our 1 December forecasts. If these atmospheric and oceanic relationships are stable, then substantial independent real-time forecast skill is available.

For the final initial time forecast of 1 August, information is utilized that extends right up to the start of the active portion of the hurricane season (Gray et al. 1993). This forecast may appear to be more of a "nowcast" than a prediction when one recalls that the "official" Atlantic hurricane season extends from June through November. However, an inspection of the seasonal variation of named storms and hurricanes reveals that only 11% and 6% of the annual named storm and hurricane activity (as measured by days in which these cyclones are present), respectively, occur before 1 August on average (Landsea 1993). Less than 2% of the intense hurricane activity is observed on average before 1 August, and 95% occurs just in the three months of August through October. In addition, the small amount of activity that does occur in June or July has shown no predictive value for the entire season: A busy (e.g., two or three named storms) June and July can precede a very active year (such as 1990, when fourteen named storms occurred) or a very quiet year (such as 1986 when only six named storms were observed). Alternatively, quiescent (e.g., with no named storms observed) June and July years can either precede very active years (such as 1988, with twelve named storms) or very quiet seasons (such as 1983, with only four named storms observed).

Nine predictors in four predictor groupings (listed in Table 5.1; locations are shown in Fig. 5.8) are used in the 1 August forecast (Gray et al. 1993); all but one of these are simply updates of predictors described earlier. The QBO measures of 50 mb and 30 mb zonal winds and the vertical shear between the two levels through July are extrapolated two months forward to September. The Caribbean Sea sea level pressure anomalies and 200 mb zonal wind anomalies are again utilized, but now updated for the months of June and July. In addition, the June and July values of SST anomaly (SSTA) and SOI are used for a current indication of ENSO's phase and strength. The Caribbean Sea and ENSO predictors are useful because of their strong tendency to persist through the remainder of the hurricane season. Rainfall for the Gulf of Guinea for August through November of the previous year is utilized, but in combination with one additional predictor – the rainfall anomaly in the western Sahel during June and July. Since the rainy season usually commences during these two months, this rainfall index provides a reliable idea of the early summer strength of the monsoon in its effect on the Sahel. Typically, the use of June and July rainfall provides a useful indication of how rainy the remaining two months, August and September, of the rainy season will be (Bunting et al. 1975; Gray et al. 1994). Because of the strong concurrent correlation between Atlantic tropical cyclone activity and seasonal Sahel rainfall, June and July rainfall for the western Sahel provides an excellent precursor signal for hurricane activity from August until the end of the hurricane season, particularly for the expected intense hurricane activity. Note that these more recent rainfall measurements replace the western Sahel rainfall anomalies for August and September of the previous year.

The skill levels based upon hindcast testing range between 45 and 60% of the variability explained by 1 August. In real-time predictions, the amount of skill likely to be available will be in the range of 25–40%. While this is an improvement over the hindcast skill available by 1 December, it is somewhat lower than what may be possible by 1 June. This is because the 1 June forecast model (Gray et al. 1994) – developed after the 1 August model (Gray et al. 1993) – is believed to be a superior, improved product despite having an initial time two months earlier. Work is currently under way to reduce the number of predictors for all of the lead times, to include the years of the early 1990s, and to only select those predictors that contribute a reasonable amount of variability toward the regression equation. One particular change will to be to utilize the Niño 3.4 region (5°N–5°S, 120°W–170°W) in place of both the SOI and the original ENSO SST index (Niño 3), which was farther to the east. The Niño 3.4 index has been identified as the SST region having the strongest concurrent association with midlatitude and tropical ENSO-forced circulation variations (Barnston et al. 1997).

Regardless of the exact performance of the published regression schemes in Gray et al. (1992b, 1993, 1994) in the future, thirteen years of forecasts have now been issued in real time by Prof. Gray and his collaborators at Colorado State University. As in any real-time forecasting situation, the seasonal forecasts have not relied solely upon the quantitative regression results in Gray (1984b) and Gray et al. (1992a, 1993, 1994). The forecasts issued also give some weight to consistency between predictands, predictive factors not explicitly in the regression model, and forecaster intuition. Thus the forecast results presented below are the final "official" forecasts and are not strictly the regression model results. A full independent verification of LAD regression results presented in Gray et al. (1992a, 1993, 1994) will need to wait until a larger sample (at least ten years of data) is available.

In most of the prior real-time forecasts from 1984–96, predictions have beaten those of climatology and persistence, which were previously the only ways to estimate future hurricane activity. Table 5.2 presents the real-time (operational) seasonal forecasts for named storms, hurricanes, intense hurricanes, and hurricane days (a measure of the duration of the season) from various starting times and their verification. The eight early June seasonal forecasts for 1985, 1986, 1987, 1990, 1991, 1992, 1994, and 1995 were more accurate in general than climatology for both named storms and hurricanes (1950–90 mean values of 9.3 named storms and 5.8 hurricanes). The forecasts for 1984, 1988, and 1996 were about as successful as climatology, while the two seasonal forecasts for 1989 and 1993 were failures. To quantify the amount of skill available, the agreement coefficient, ρ, is utilized to compare the real-time forecasts against the observations. Table 5.2 shows that the early June predictions have explained about 25% of the variability for named storms and hurricanes, significantly greater than that available by persistence (2% and 15%, respectively) and by climatology (0%). The early June intense hurricane forecasts have yet to show significant skill, however; seven years is too small a test database to say anything definitive. The early December forecasts have too small a sample size (five years) of independent data to reveal any conclusions as of yet. However, the early August forecasts for all of the tropical cyclone parameters show increased, significant skill over and above the early June predictions, with up to 55% of the named storms and 40% of the hurricane variability. These values of variability explained for the early June and early August named storm and hurricane forecasts are in the range, and even higher than for the 1 August forecasts, of the expected independent skill discussed earlier. Real-time forecast and verification reports for all of these forecast dates during the past several years are now available via the World Wide Web: *http://www.tropical.atmos.colostate.edu/*.

In addition to the contributions by Gray et al., substantial progress has been made toward seasonal forecasting of Atlantic basin tropical cyclones, including U.S. landfalling hurricanes, in research led by Prof. James Elsner at Florida State University. Elsner and Schmertmann (1993) utilized the predictors from Gray et al. (1992a) to derive a fully cross-validated Poisson regression model that outperforms the LAD model for intense hurricanes. In Hess et al. (1995), a subjective stratification is performed to remove those hurricanes that had a midlatitude baroclinic influence sometime during their genesis or development to hurricane force. With these removed from the database, it was found that the predictors from Gray et al. (1992a, 1993) show a stronger relationship with the "tropical-only" hurricanes (Fig. 5.10), though there are concerns regarding the subjectivity inherent in removing the baroclinically influenced hurricanes from the entire database. The baroclinically influenced hurricanes were not found to be predictable by available predictors. Hess et al. (1995) then employed an OLS multiple linear regression to provide for tropical-only hurricane (to which they added a climatological value of baroclinically influenced hurricanes) forecasts at both 1 December and 1 August initial times. Again this methodology shows a modest, but significant, improvement over that of Gray et al. (1992a, 1994) in the hindcast data set.

In an attempt to go from the entire Atlantic basin to more regional scales including landfalling U.S. hurricanes, Lehmiller et al. (1997) split portions of the Atlantic basin up into four threat regions: the U.S. northeast coast (from northern North Carolina to New England), the U.S. southeast coast (from eastern Florida to southern North Carolina), the Gulf of Mexico, and the Caribbean Sea. For the first two regions, they considered

Table 5.2 *Real-time Gray et al. forecasts and results for the Atlantic tropical cyclone season during the years 1984–96. Predictions and verifications that were significantly above the climatological average (11–19 for named storms, 7–12 for hurricanes, 4–7 for intense hurricanes, and 31–60 for hurricane days) are indicated by underlined numbers. Those that were below average (4–7 named storms, 2–4 hurricanes, 0–1 intense hurricanes, and 4–17 hurricane days) are noted by boldface numbers. Skill of forecasts and of predictions of year-to-year persistence is assessed by ρ, the agreement coefficient (Gray et al. 1992a). Significance for the ρ values is given by: "*" significant at the 0.10 level, "**" significant at the 0.05 level, and "***" significant at the 0.01 level.*

Year	Early December forecast	Early June forecast	Early August forecast	Observed
Named Storms:	1950–90	Mean = 9.3		
1984	—	10	10	12
1985	—	11	10	11
1986	—	8	**7**	**6**
1987	—	8	**7**	**7**
1988	—	11	11	12
1989	—	**7**	9	11
1990	—	11	11	14
1991	—	8	**7**	8
1992	8	8	8	**6**
1993	11	11	10	8
1994	10	9	**7**	**7**
1995	12	12	16	19
1996	8	10	11	13
ρ	+.038	+.270**	+.558***	Persistence:+.023
Hurricanes:	1950–90	Mean = 5.8		
1984	—	7	7	5
1985	—	8	7	7
1986	—	**4**	**4**	**4**
1987	—	5	**4**	**3**
1988	—	7	7	5
1989	—	**4**	**4**	7
1990	—	7	6	8
1991	—	**4**	3	**4**
1992	**4**	**4**	**4**	**4**
1993	6	7	6	**4**
1994	6	5	**4**	**3**
1995	8	8	9	11
1996	5	6	7	9
ρ	+.189	+.251*	+.403**	Persistence:+.150

Table 5.2 *(cont.)*

Year	Early December forecast	Early June forecast	Early August forecast	Observed
Intense Hurricanes:	1950–90	Mean = 2.3		
1990	—	3	2	**1**
1991	—	**1**	**0**	2
1992	**1**	**1**	**1**	**1**
1993	3	2	2	**1**
1994	2	**1**	**1**	**0**
1995	3	3	3	5
1996	2	2	3	6
ρ	+.107	+.115	+.239*	Persistence: +.255
Hurricane Days:	1950–90	Mean = 23.7		
1984	—	30	30	18
1985	—	35	30	21
1986	—	**15**	**10**	**11**
1987	—	20	**15**	**5**
1988	—	30	30	21
1989	—	**15**	**15**	32
1990	—	30	25	27
1991	—	**15**	**8**	**8**
1992	**15**	**15**	**15**	**16**
1993	25	25	25	**10**
1994	25	**15**	**12**	**7**
1995	35	35	30	60
1996	20	20	25	45
ρ	+.129	+.139	+.247**	Persistence: +.175

U.S. landfalling hurricanes and intense hurricanes, and for the second two regions, they considered hurricanes and intense hurricanes that occurred anywhere within these water boundaries. The predictors utilized are those from Gray et al. (1992a) for a 1 December forecast of the previous year and from Gray et al. (1993), with some additional ones for use in a 1 August forecast. The additional 1 August predictors (July 700–200 mb vertical shear at Miami/West Palm Beach [U.S.], July sea level pressure in Cape Hatteras [U.S.], and July averaged U.S. East Coast sea level pressure) assist in determining the regional vertical shear and the steering flow strength. Similarly to the other 1 August predictors, these are suggested to work through a persistence of anomalous conditions through the height of the hurricane season. With a multivariate discriminant analysis, Lehmiller et al. (1997) were able to successfully hindcast the occurrence or nonoccurrence of storms at least three quarters of the time (versus a climatological accuracy of nearly 50%) for the following forecasts: 1 December Caribbean Sea hurricanes, 1 August Gulf of Mexico intense hurricanes, 1 August Caribbean Sea intense hurricanes, and 1 August

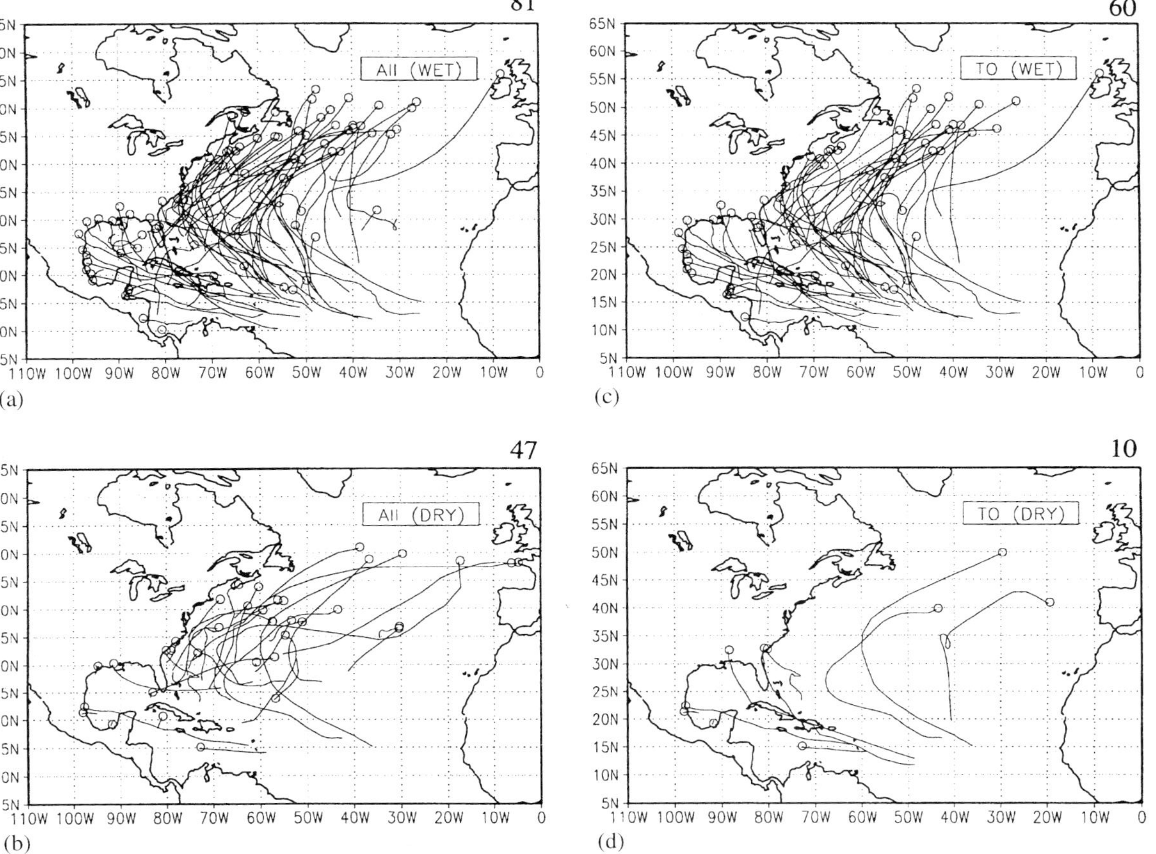

Fig. 5.10 Comparison of hurricane tracks for wet versus dry years based on Gulf of Guinea rainfall. (a) The ten wettest years and all hurricanes; (b) ten driest years and all hurricanes; (c) ten wettest years and tropical-only hurricanes; (d) ten driest years and tropical-only hurricanes. Number of hurricanes for each category is shown in the upper right corner of each panel. (From Hess et al. 1995.)

U.S. southeast coast hurricanes. The other hindcasts, including all efforts for the U.S. Northeast, were unable to significantly improve upon results from climatology. The predictions of regional hurricanes, as well as the basinwide measures of Atlantic hurricane activity, issued by J. Elsner and colleagues can be found in issues of the *Experimental Long-Lead Forecast Bulletin* (Barnston 1996).

Northwest Pacific

The use of steering flow variations is also key for predictions of year-to-year tropical cyclone movement over regions of the Northwest Pacific Ocean (Chan 1994). In this forecasting methodology, Chan utilizes the zonal winds at both 850 and 500 mb before the season commences in a cross-validated linear OLS regression model to forecast the tropical cyclone numbers moving through designated regions. This technique allows for predictions of the annual number of westward moving tropical cyclones through the Philippines by early April (Table 5.3) and the annual number of northward moving

Table 5.3 *Prediction of the annual number of occurrences of westward tropical cyclones in the region bounded by 7.5°N, 117.5°E and 17.5°N, 142.5°E (i.e., the area surrounding and just to the east of the Philippines) using 850 mb zonal winds in January and March. The "predicted" values are obtained via a cross-validated regression. The error of prediction is in parentheses. RMS error=root-mean-square error.*

Year	Observed	Predicted
1975	15	18(+3)
1976	20	22(+2)
1977	21	24(+3)
1978	31	25(−6)
1979	32	33(+1)
1980	32	25(−7)
1981	36	32(−4)
1982	34	29(−5)
1983	24	27(+3)
1984	21	32(+11)
1985	25	30(+5)
1986	35	31(−4)
1987	39	40(+1)
1988	33	38(+5)
Mean	28	
Standard Deviation	7.3	
RMS Error		5.1

Source: Chan (1994).

tropical cyclones through the region just south of Japan by early March. Both methods appear to provide skill in predicting the number of tropical cyclones substantially above that of climatology. However, the physical link – presumably persistence of steering flow conditions before the season begins in March to the end of the season in late autumn – has not yet been convincingly shown.

Australia/South Pacific Basin

Research into seasonal tropical cyclone forecasting in the Australian region (5°S–32°S, 105°E–165°E) began with Nicholls's (1979) report on the usage of an index of ENSO to predict the upcoming October–May cyclone season. Nicholls utilized the Darwin surface pressure (one half of the SOI) to hindcast about 40% of the variance of activity, with skill coming primarily through the early and midseason tropical cyclones. In Nicholls (1984), this work was extended to show that local SSTs and SSTs in the equatorial eastern Pacific (a direct measure of ENSO) could also be utilized in a forecast mode (Fig. 5.11). Holland et al. (1988) showed that most of the local SST predictive signal was due directly to ENSO effects.

Recently, Nicholls (1992) noticed a temporal change in the behavior of the tropical cyclone numbers versus the ENSO predictors. The predicted values have been consistently too high compared to what has been observed since the season of 1986/1987. The values in the correlation for independent data dropped to only about 20% of the variability. He suggests that most of this bias is artificial being caused by changes in tropical cyclone monitoring policies. Instead, utilizing the year-to-year differences in cyclone numbers and the SOI gives a predictive association explaining about 65% of the variability. Thus with just one predictor (the SOI, or more recently, the 1-year change in SOI), a regression scheme is able to capture almost two thirds of the interannual variability in Australian basin tropical cyclone activity by early December and slightly less by early September. Nicholls's real-time forecasts are posted in November of each year and can be found at the following World Wide Web site: *http://www.bom.gov.au/bmrc/mrlr/nnn/climtcfc.htm.*

While Nicholls has utilized El Niño's effect of reducing tropical cyclone activity in the Australian basin, Basher and Zheng (1995) use the tendency for El Niño increases of South Pacific tropical cyclones for forecasting purposes in that region. Again making use of an OLS multiple regression, Basher and Zheng use SOI and local SSTs during September to forecast the number of cyclone occurrences in 20° longitude by 12° latitude boxes extending across the South Pacific. Using just the September SOI, they find that they can explain 52% of the variance for the entire region extending from 170°E to 130°W. The local SSTs do not contribute significantly above this value. It is just in the region from 150°E to 170°E where the local SSTs are able to forecast 35% of the variance, and the SOI provides no additional information. This is because ENSO's influence reverses from inhibiting to enhancing tropical cyclogenesis (for El Niño events) around the longitude of 160°E.

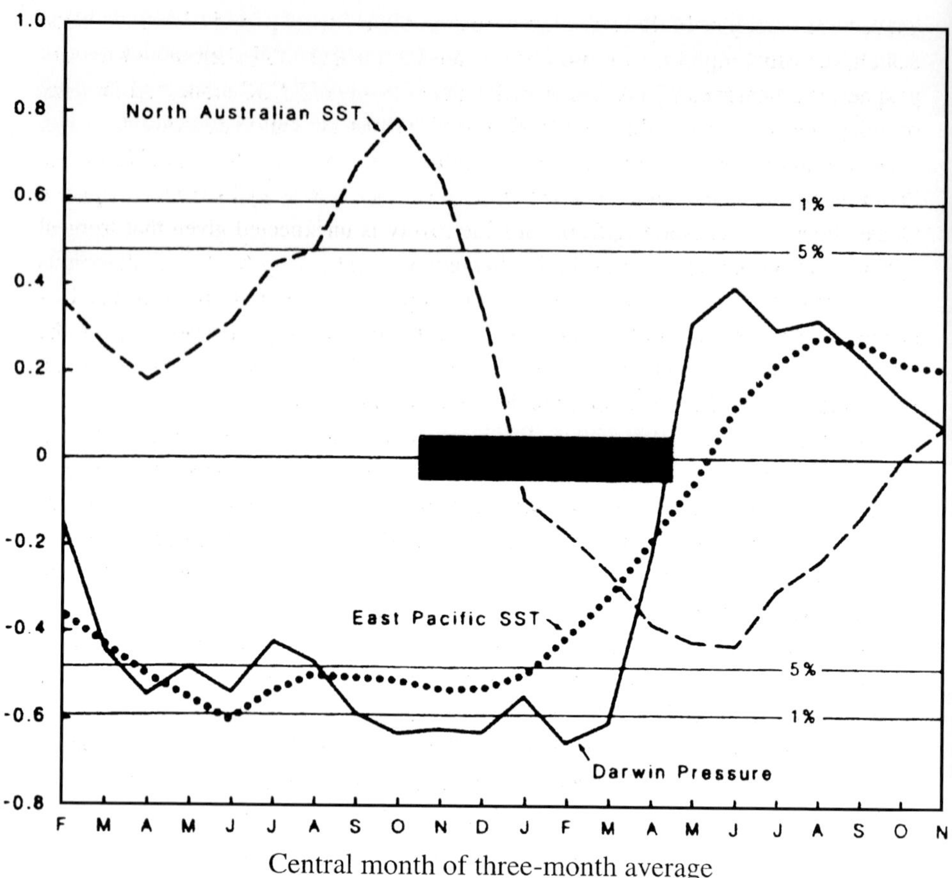

Fig. 5.11 Correlations between number of Australian tropical cyclones observed in the total cyclone season and three-month averages of three variables associated with ENSO. The thick horizontal bar indicates the extent of the cyclone season. Central month of three-month averages of the ENSO variables is indicated on the horizontal axis. Data are from 1964–82. All correlations are calculated on at least eighteen pairs of observations. (From Nicholls 1984.)

Discussion and Conclusions

In the section "Necessary (but Not Sufficient) Environmental Conditions," discussion centered on the need for favorable synoptic-scale variations of the environment to allow genesis and development of tropical cyclones to proceed. Since the El Niño/Southern Oscillation (ENSO) produces such large, wide-scale changes in tropical circulation, perhaps it is not surprising that tropical cyclones are strongly altered by ENSO. However, these changes are not uniform throughout the global tropics. In some regions, an El Niño event would bring increases in tropical cyclone formation (e.g., the South Pacific and the North Pacific between 140°W to 160°E) while others would see decreases (e.g., the North Atlantic, the Northwest Pacific west of 160°E, and the Australian region). La Niña typically brings opposite conditions. These alterations in tropical cyclone activity are due to a variety of ENSO effects: modulation of the intensity of the local monsoon trough, repositioning of the location of the monsoon trough, and alteration of the tropospheric vertical shear. Thus, based on the global nature of ENSO alone, assumptions of independence between different tropical cyclone basins would be incorrect.

In addition to ENSO, three basins (the Atlantic, Southwest Indian, and Northwest Pacific) show systematic alterations of tropical cyclone frequency by the stratospheric Quasi-Biennial Oscillation (QBO). This intuitively is unexpected given that tropical cyclones are primarily a tropospheric phenomenon, but it may be due to alterations in the static stability and dynamics near the tropopause. Certainly more research is needed to provide a thorough explanation of these relationships. However, given the robustness of these alterations in tropical cyclone activity that match the QBO phases, it appears unlikely that the association is purely a chance correlation.

Interannual tropical cyclone variations have also been linked to more localized, basin-specific features such as sea surface temperatures (SSTs), monsoon strength and rainfall, sea level pressures (SLPs), and tropospheric vertical changes. These regional factors can be as large as the forcing due to ENSO, though most are not. Together with ENSO and the QBO, these factors produce changes in the frequency, intensity, formation region, and track of tropical cyclones in all basins. However, understanding how tropical cyclone variability relates to the surrounding environmental conditions is hampered because we have only a few decades worth of reliable tropical cyclone records. The emerging field of "paleotempestology" – the study of prehistoric tropical cyclones (e.g., Liu and Fearn 1993; Keen and Slingerland 1993) – may be able in coming years to assist in analyzing how and why tropical cyclones change from year to year.

Despite the current lack of accurate long-term tropical cyclone records, some of the tropical cyclone–environmental associations have led to methodologies that can be used in seasonal forecasting by the onset of the tropical cyclone season. Given a typical life cycle for a particular phase of ENSO of a year or greater, one can utilize this as a basis for seasonal predictions with lead times of up to several months. These are now being done (or can be performed) for the Atlantic basin tropical cyclones, those near Australia, and those in the South Pacific basin. These forecasts assume no knowledge of the future state of ENSO, except that the current ENSO phase will persist for at least the next few months.

Over the past fifteen years, seasonal forecasting for various basins has evolved to the point that up to 50% (or more) of the tropical cyclone variability can be predicted at the start of the cyclone season. In addition to the frequency and intensity of the storms, statistical methodologies have also been introduced to forecast track frequency and the likelihood of hurricane landfall in certain coastal zones. Currently, using such statistical models is the only feasible methodology for seasonal tropical cyclone forecasting, because of the lack of skill in global circulation models (GCMs). Because of the complex difficulties faced in utilizing these numerical models, it may take a decade or two of concerted research effort before such GCM forecasting is feasible, if ever. However, for the time being, creative use of statistical regression schemes can provide and will continue to provide skilled and useful predictions of tropical cyclone activity around the world.

A sensible question would be: How can seasonal forecasts of tropical cyclones be utilized when they are for large geographic regions such as the entire North Atlantic Ocean, Caribbean Sea, and Gulf of Mexico? Note that to "utilize" a prediction would also include when it is considered in a particular decision process of an individual or group. This allows for the possibility that someone considers a forecast but does not necessarily change his or her behavior. "Value" of a forecast is typically associated with changes in behavior, but this is not necessarily so (e.g., increased confidence about the future) (Pielke and Pielke 1997a). There are a number of reasons for issuing such predictions. Practically, most people in the general public cannot (and should not in most cases) utilize the forecast directly. As an example, it would be foolish if John Q. Public decided to ignore hurricane preparedness and mitigation plans because the year was one predicted to be below average. The storms of 1992 serve as an excellent warning against such actions: The Atlantic hurricane season was very successfully forecasted by Prof. Gray (see Table 5.2) as a quiet year with only four hurricanes, yet one of those was Hurricane Andrew, the most destructive U.S. hurricane on record (Mayfield et al. 1994). Strong wording should be added to any seasonal hurricane forecast to discourage individuals from misusing the forecasts.

Corporations and governments, however, because of their size and scope of operations, are starting to make reasonable use of such forecasts each year. One large, unnamed, private manufacturing company with interests all along the U.S. coastline serves as an example of the many positive uses that can be made with these predictions: decisions on the amount of "hurricane" liability insurance that covers preparations, damages, and repair costs; determinations of annual budgets and preventative maintenance schedules; an aid in production and inventory storage planning; plans for data processing disaster recovery; and schedules for workers' shifts in the upcoming year. Such usage can be better enhanced by increases in skill, by providing longer-range accurate predictions, and by regionalizing the forecast for smaller locales (such as is being pursued by Lehmiller et al. 1997). Of course, different users have differing thresholds of skill needed to derive value from the forecasts. It has been found that the more climate sensitive the decision, the smaller the improvement over climatology needed to gain benefits.

Another justification for such forecasting efforts is that they lead to further advances in our understanding of linkages in the climate system. It was the failure of the 1989 seasonal Atlantic hurricane forecast (Table 5.2) that led to the discovery that the West African monsoon is intimately tied to the occurrence of Atlantic hurricanes on an interannual basis. Successes in predicting seasonal tropical cyclone activity may also provide insight into forecasts of other tropical phenomena such as droughts and flooding associated with anomalous changes in the strength and location of the Intertropical Convergence Zone (ITCZ) and the monsoons (e.g., Hastenrath 1995).

Finally, the seasonal tropical cyclone forecasts at times generate considerable public and media interest. The effect of such interest is a qualitative, rather than quantitative, use in that it heightens the public awareness to the danger of hurricanes and, it is hoped, prompts more people to take precautions and make preparations.

Acknowledgments Stan Goldenberg and John Kaplan at the NOAA Hurricane Research Division, Todd Kimberlain of Colorado State University, Henry Diaz at the NOAA Climate Diagnostics Center, Roger Pielke, Jr., at the National Center for Atmospheric Research, and two anonymous reviewers provided quite detailed, helpful comments on an earlier draft of this paper. Prof. Bill Gray of Colorado State University has, as always, sparked many useful and enlightening discussions on the topic. Finally, the author thanks the Bermuda Biological Research Station's Risk Prediction Initiative for providing financial support through a grant on the topic.

References

BALLENZWEIG, E., 1959: Relation of long-period circulation anomalies to tropical cyclone formation and motion. *Journal of Meteorology*, **16**, 121–139.

BARNSTON, A. G. (ed.), 1996: *Experimental Long-Lead Forecast Bulletin, 5.1.* Climate Prediction Center, NOAA, Washington, D.C., 54 pp.

BARNSTON, A. G., CHELLIAH, M., and GOLDENBERG, S. B., 1997: Documentation of a highly ENSO-related SST region in the equatorial Pacific. *Atmosphere-Ocean*, **35**, 367–383.

BASHER, R. E. and ZHENG, X., 1995: Tropical cyclones in the Southwest Pacific: Spatial patterns and relationships to Southern Oscillation and sea surface temperature. *Journal of Climate*, **8**, 1249–1260.

BATE, P. W., GARDEN, G. S., JACKSON, G. E., CHEANG, B. K., and SANKARAN, P., 1989: The tropical circulation in the Australian/Asian region – November 1987 to April 1988. *Australian Meteorological Magazine*, **37**, 201–216.

BRENNAN, J. F., 1935: Relation of May–June weather conditions in Jamaica to the Caribbean tropical disturbances of the following season. *Monthly Weather Review*, **63**, 13–14.

BUNTING, A. H., DENNETT, M. D., ELSTON, J., and MILFORD, J. R., 1975: Seasonal rainfall forecasting in West Africa. *Nature*, **253**, 622–623.

CARLSON, T., 1971: An apparent relationship between sea surface temperature of the tropical Atlantic and the development of African disturbances into tropical storms. *Monthly Weather Review*, **99**, 309–310.

CHAN, J. C. L., 1985: Tropical cyclone activity in the Northwest Pacific in relation to the El Niño/Southern Oscillation phenomenon. *Monthly Weather Review*, **113**, 599–606.

CHAN, J. C. L., 1994: Prediction of the interannual variations of tropical cyclone movement over regions of the western North Pacific. *International Journal of Climatology*, **14**, 527–538.

CHAN, J. C. L., 1995: Tropical cyclone activity in the western North Pacific in relation to the stratospheric quasi-biennial oscillation. *Monthly Weather Review*, **123**, 2567–2571.

CHEN, S. A. and FRANK, W. M., 1993: A numerical study of the genesis of extratropical convective mesovortices. Part I: Evolution and dynamics. *Journal of Atmospheric Science*, **50**, 2401–2426.

CHEN, D., ZEBIAK, S. E., BUSALACCHI, A. J., and CANE, M. A., 1995: An improved procedure for El Niño forecasting: Implications for predictability. *Science*, **269**, 1699–1702.

DeMARIA, M., 1996: The effect of vertical shear on tropical cyclone intensity change. *Journal of the Atmospheric Sciences*, **53**, 2076–2087.

DONG, K. and HOLLAND, G. J., 1994: A global view of the relationships between ENSO and tropical cyclone frequencies. *Acta Meteorology Sinica*, **8**, 19–29.

ELSNER, J. B. and SCHMERTMANN, C. P., 1993: Improving extended-range seasonal predictions of intense Atlantic hurricane activity. *Weather and Forecasting*, **8**, 345–351.

ELSNER, J. B. and SCHMERTMANN, C. P., 1994: Assessing forecast skill through cross validation. *Weather and Forecasting*, **9**, 619–624.

EMANUEL, K. A., 1993: The physics of tropical cyclogenesis over the Eastern Pacific. *In* Lighthill, J., Zhemin, Z., Holland, G. J., and Emanuel, K. (eds.), *Tropical Cyclone Disasters*. Beijing: Peking University Press, 136–142.

ENFIELD, D. B. and MAYER, D. A., 1997: Tropical Atlantic sea surface temperature variability and its relation to El Niño–Southern Oscillation. *Journal of Geophysical Research*, **102**, 929–945.

EVANS, J. L. and ALLEN, R. J., 1992: El Niño–Southern Oscillation modification to the structure of the monsoon and tropical activity in the Australian region. *International Journal of Climatology*, **12**, 611–623.

FITZPATRICK, P. J., KNAFF, J. A., LANDSEA, C. W., and FINLEY, S. V., 1995: Documentation of a systematic bias in the Aviation model's forecast of the Atlantic tropical upper-tropospheric trough: Implications for tropical cyclone forecasting. *Weather and Forecasting*, **10**, 433–446.

FOLLAND, C. K., PALMER, T. N., and PARKER, D. E., 1986: Sahel rainfall and worldwide sea temperatures, 1901–1985. *Nature*, **320**, 602–607.

FRANKLIN, J. L., FEUER, S. E., KAPLAN, J., and ABERSON, S. D., 1996: Tropical cyclone motion and surrounding flow relationships: Searching for beta gyres in Omega dropwindsonde datasets. *Monthly Weather Review*, **124**, 64–84.

GOLDENBERG, S. B. and SHAPIRO, L. J., 1996: Physical mechanisms for the association of El Niño and West African rainfall with Atlantic major hurricane activity. *Journal of Climate*, **9**, 1169–1187.

GRAY, W. M., 1968: Global view of the origins of tropical disturbances and storms. *Monthly Weather Review*, **96**, 669–700.

GRAY, W. M., 1979: Hurricanes: Their formation, structure and likely role in the tropical circulation. *In* Shaw, D. B. (ed.), *Meteorology Over the Tropical Oceans*. Royal Meteorology Society, James Glaisher House, Grenville Place, Bracknell, Berkshire, RG12 1BX, 155–218.

GRAY, W. M., 1984a: Atlantic seasonal hurricane frequency: Part I: El Niño and 30 mb quasi-biennial oscillation influences. *Monthly Weather Review*, **112**, 1649–1668.

GRAY, W. M., 1984b: Atlantic seasonal hurricane frequency: Part II: Forecasting its variability. *Monthly Weather Review*, **112**, 1669–1683.

GRAY, W. M., 1990: Strong association between West African rainfall and US landfall of intense hurricanes. *Science*, **249**, 1251–1256.

GRAY, W. M., LANDSEA, C. W., MIELKE, P. W., Jr., and BERRY, K. J., 1992a: Predicting Atlantic seasonal hurricane activity 6–11 months in advance. *Weather and Forecasting*, **7**, 440–455.

GRAY, W. M., LANDSEA, C. W., MIELKE, P. W., Jr., and BERRY, K. J., 1993: Predicting Atlantic basin seasonal tropical cyclone activity by 1 August. *Weather and Forecasting*, **8**, 73–86.

GRAY, W. M., LANDSEA, C. W., MIELKE, P. W., Jr., and BERRY, K. J., 1994: Predicting Atlantic basin seasonal tropical cyclone activity by 1 June. *Weather and Forecasting*, **9**, 103–115.

GRAY, W. M., SHEAFFER, J. D., and KNAFF, J. A., 1992b: Influence of the stratospheric QBO on ENSO variability. *Journal of the Meteorological Society of Japan*, **70**, 975–995.

HASTENRATH, S., 1995: Recent advances in tropical climate prediction. *Journal of Climate*, **8**, 1519–1532.

HASTINGS, P. A., 1990: Southern Oscillation influences on tropical cyclone activity in the Australian/Southwest Pacific region. *International Journal of Climatology*, **10**, 291–298.

HEBERT, P. J., JARRELL, J. D., and MAYFIELD, M., 1996: The deadliest, costliest, and most intense United States hurricanes of this century (and other frequently requested hurricane facts). NOAA Tech. Memo., NWS TPC-1, Miami, Florida, 75–104.

HESS, J. C., ELSNER, J. B., and La SEUR, N. E., 1995: Improving seasonal predictions for the Atlantic basin. *Weather and Forecasting*, **10**, 425–432.

HOLLAND, G. J., 1993: Ready reckoner, Chapter 9 of *Global Guide to Tropical Cyclone Forecasting*, WMO/TC-No. 560, Report No. TCP-31, World Meteorological Organization, Geneva.

HOLLAND, G. J., McBRIDE, J. L., and NICHOLLS, N., 1988: Australian tropical cyclones and the greenhouse effect. In *Greenhouse: Planning for Climate Change*. Leiden, The Netherlands: E. J. Brill, 438–455.

JURY, M., 1993: A preliminary study of climatological associations and characteristics of tropical cyclones in the SW Indian Ocean. *Meteorology and Atmospheric Physics*, **51**, 101–115.

KEEN, T. R. and SLINGERLAND, R. L., 1993: Four storm-event beds and the tropical cyclones that produced them: A numerical hindcast. *Journal of Sedimentary Petrology*, **63**, 218.

KNAFF. J. A. 1993: Evidence of a stratospheric QBO modulation of tropical convection. Atmospheric Science Paper No. 520, Colorado State University, Ft. Collins, CO 80523, 91 pp.

KNAFF, J. A., 1997: Implications of summertime sea level pressure anomalies in the tropical Atlantic region. *Monthly Weather Review*, **10**, 789–804.

KNAFF, J. A. and LANDSEA, C. W., 1997: An El Niño-Southern Oscillation CLImatology and PERsistence (CLIPER) forecasting scheme. *Weather and Forecasting*, **12**, 633–652.

LAMB, P. J., 1978: Large-scale tropical Atlantic surface circulation patterns associated with sub-Saharan weather anomalies. *Tellus*, **30**, 240–251.

LANDER, M., 1994: An exploratory analysis of the relationship between tropical storm formation in the Western North Pacific and ENSO. *Monthly Weather Review*, **122**, 636–651.

LANDSEA, C. W., 1993: A climatology of intense (or major) Atlantic hurricanes. *Monthly Weather Review*, **121**, 1703–1713.

LANDSEA, C. W., 1999: Climate variability of tropical cyclones: Past, present and future. *In* Pielke, R. A., Jr. and Pielke, R. A., Sr. (eds.), *Storms*. London: Routledge Press, in press.

LANDSEA, C. W. and GRAY, W. M., 1992: The strong association between Western Sahel monsoon rainfall and intense Atlantic hurricanes. *Journal of Climate*, **5**, 435–453.

LANDSEA, C. W., GRAY, W. M., MIELKE, P. W., Jr., and BERRY, K. J., 1992: Long-term variations of Western Sahelian monsoon rainfall and intense U.S. landfalling hurricanes. *Journal of Climate*, **5**, 1528–1534.

LANDSEA, C. W., GRAY, W. M., MIELKE, P. W., Jr., and BERRY, K. J., 1994: Seasonal forecasting of Atlantic hurricane activity. *Weather*, **49**, 273–284.

LAWRENCE, M. B. and PELISSIER, J. M., 1982: Atlantic hurricane season of 1981. *Monthly Weather Review*, **110**, 852–866.

LEHMILLER, G. S., KIMBERLAIN, T. B., and ELSNER, J. B., 1997: Seasonal prediction models for North Atlantic basin hurricane location. *Monthly Weather Review*, **125**, 1780–1791.

LIU, K. B. and FEARN, M. L., 1993: Lake-sediment record of late Holocene hurricane activities from coastal Alabama. *Geology*, **21**, 793–796.

MALKUS, J. S. and RIEHL, H., 1960: On the dynamics and energy transformations in steady-state hurricanes. *Tellus*, **12**, 1–20.

MAYFIELD, M., AVILA, L., and RAPPAPORT, E. N., 1994: Atlantic hurricane season of 1992. *Monthly Weather Review*, **122**, 517–538.

McBRIDE, J. L., 1995: Tropical cyclone formation. *In* Elsberry, R. L. (ed.), *Global Perspectives on Tropical Cyclones*. WMO/TC-No. 693, Geneva: World Meteorological Organization, 63–105.

MIELKE, P. W., Jr., 1991: The application of multivariate permutation methods based upon distance functions in the earth sciences. *Earth-Science Review*, **31**, 55–71.

MIELKE, P. W., Jr., BERRY, K. J., LANDSEA, C. W., and GRAY, W. M., 1996: Artificial skill and validation in weather forecasting. *Weather and Forecasting*, **11**, 153–169.

NAMIAS, J., 1955: Secular fluctuations in vulnerability to tropical cyclone activity in and off New England. *Monthly Weather Review*, **83**, 155–162.

NEUMANN, C. J., 1993: Global overview. Chapter 1 of *Global Guide to Tropical Cyclone Forecasting*. WMO/TC-No. 560, Report No. TCP-31, Geneva: World Meteorological Organization.

NICHOLLS, N., 1979: A possible method for predicting seasonal tropical cyclone activity in the Australian region. *Monthly Weather Review*, **107**, 1221–1224.

NICHOLLS, N., 1984: The Southern Oscillation, sea surface temperature, and interannual fluctuations in Australian tropical cyclone activity. *Journal of Climatology*, **4**, 661–670.

NICHOLLS, N., 1992: Recent performance of a method for forecasting Australian seasonal tropical cyclone activity. *Australian Meteorological Magazine*, **40**, 105–110.

NICHOLSON, S. E., 1988: Land surface-atmosphere interaction: Physical processes and surface changes and their impact. *Progress in Physical Geography*, **12**, 36–65.

PAN, Y., 1981: The effect of the thermal state of eastern equatorial Pacific on the frequency of typhoons over western Pacific. *Acta Meteorological Sinica*, **40**, 24–32.

PENLAND, C. and SARDESHMUKH, P. D., 1995: The optimal growth of tropical sea surface temperature anomalies. *Journal of Climate*, **8**, 1999–2024.

PHILANDER, S. G. H., 1989: *El Niño, La Niña, and the Southern Oscillation*. New York, Academic Press, 293 pp.

PIELKE, R. A., Jr. and PIELKE, R. A., Sr., 1997a: Vulnerability to hurricanes along the U. S. Atlantic Gulf Coasts: Consideration of the use of long-term forecasts. *In* Diaz, H. F. and Pulwarty, R. S., (eds.), *Hurricanes: Climate and Socioeconomic Impacts*. New York: Springer-Verlag, 147–184.

PIELKE, R. A., Jr. and PIELKE, R. A., Sr., 1997b: *Hurricanes: Their Nature and Impacts on Society*. New York: John Wiley & Sons, 279 pp.

RAPER, S., 1992: Observational data on the relationships between climate change and the frequency and magnitude of severe tropical storms. *In* Warrick, R. A., Barrow, E. M., and Wigley. T. M. L. (eds.), *Climate and Sea Level Change: Observations, Projections and Implications*. Cambridge: Cambridge University Press, 192–212.

RAY, C. L., 1935: Relation of tropical cyclone frequency to summer pressures and ocean surface-water temperatures. *Monthly Weather Review*, **63**, 10–12.

REED, R. J., 1988: On understanding the meteorological causes of Sahelian drought. *Pontificiae Academiae Scientarvm Scripta Varia*, **69**, 179–213.

REVELL, C. G. and GOULTER, S. W., 1986: South Pacific tropical cyclones and the Southern Oscillation. *Monthly Weather Review*, **114**, 1138–1145.

SAUNDERS, M. A. and HARRIS, A. R., 1997: Statistical evidence links exceptional 1995 Atlantic hurricane season to record sea warming. *Geophysical Research Letters*, **24**, 1255–1258.

SCHROEDER, T. A. and YU, Z. 1995: Interannual variability of central Pacific tropical cyclones. *Preprints of the 21st Conference on Hurricanes and Tropical Meteorology*, Miami: American Meteorological Society, 437–439.

SHAPIRO, L. J., 1982: Hurricane climatic fluctuations. Part II: Relation to large-scale circulation. *Monthly Weather Review*, **110**, 1014–1023.

SHAPIRO, L. J., 1987: Month-to-month variability of the Atlantic tropical circulation and its relationship to tropical storm formation. *Monthly Weather Review*, **115**, 2598–2614.

SHAPIRO, L. J., 1989: The relationship of the quasi-biennial oscillation to Atlantic tropical storm activity. *Monthly Weather Review*, **117**, 2598–2614.

SHAPIRO, L. J. and S. B. GOLDENBERG, 1998: Atlantic sea surface temperatures and tropical cyclone formation. *Journal of Climate*, **11**, 578–590.

SIMPSON, R. H., 1974: The hurricane disaster potential scale. *Weatherwise*, **27**, 169 and 186.

VELASCO, I. and FRITSCH, J. M., 1987: Mesoscale convective complexes in the Americas. *Journal of Geophysical Research*, **92**, 9561–9613.

WALLACE, J. M., 1973: General circulation of the tropical lower stratosphere. *Reviews of Geophysics and Space Physics*, **11**, 191–222.

WATTERSON, I. G., EVANS, J. L., and RYAN, B. F., 1995: Seasonal and interannual variability of tropical cyclogenesis: Diagnostics from large-scale fields. *Journal of Climate*, **8**, 3052–3066.

WENDLAND, W., 1977: Tropical storm frequencies related to sea surface temperature. *Journal of Applied Meteorology*, **16**, 477–481.

WRIGHT, P. B., 1985: The Southern Oscillation: An ocean–atmosphere feedback system? *Bulletin of the American Meteorological Society*, **66**, 398–412.

WRIGHT, P. B., WALLACE, J. M., MITCHELL, T. P., and DESER, C., 1988: Correlation structure of the El Niño/Southern Oscillation phenomenon. *Journal of Climate*, **1**, 609–625.

WU, G. and LAU, N., 1992: A GCM simulation of the relationship between tropical-storm formation and ENSO. *Monthly Weather Review*, **120**, 958–977.

XUE, Y. and SHUKLA, J., 1993: The influence of land surface properties on Sahel climate. Part I: Desertification. *Journal of Climate*, **6**, 2232–2245.

ZEHR, R. M., 1992: *Tropical Cyclogenesis in the Western North Pacific*. NOAA Technical Report NESDIS 61, Washington, DC: U.S. Department of Commerce, 181 pp.

6

Climate and ENSO Variability Associated with Vector-Borne Diseases in Colombia

GERMÁN POVEDA

Postgrado en Recursos Hidráulicos, Universidad Nacional de Colombia, Facultad de Minas, Carrera 80 Calle 65, Bloque M2-300 Medellín, Colombia

NICHOLAS E. GRAHAM

Scripps Institution of Oceanography, University of California, San Diego, California, U.S.A.

PAUL R. EPSTEIN

Center for Health and the Global Environment, Harvard Medical School, Boston, Massachusetts, U.S.A.

WILLIAM ROJAS

Corporación para Investigaciones Biológicas (CIB), Carrera 72 A No. 78 B 141, Medellín, Colombia

MARTHA L. QUIÑONES AND IVÁN DARÍO VÉLEZ

Programa de Control de Enfermedades Tropicales (PECET), Universidad de Antioquia, Calle 62 No. 52-19, Medellín, Colombia

WILLEM J. M. MARTENS

International Centre for Integrative Studies, Maastricht University, P.O. Box 616, 6200 MD, Maastricht, The Netherlands

Abstract

Climatic factors are associated with the incidence of diverse vector-borne diseases (VBDs). Colombia, located in tropical South America, witnesses high precipitation rates and temperatures, varying with elevation over the Andes. We show how temperatures are linked to malaria incidence throughout the country, and we compare those results with those obtained via simple mathematical expressions that represent indices associated with malaria transmission as a function of temperature. Interannual climatic variability in tropical South America is strongly associated with El Niño/Southern Oscillation (ENSO). Most of the region, including Colombia, experiences prolonged dry periods and above normal air temperatures during El Niño, and generally opposite conditions during La Niña. Through correlation analysis, we show that during El Niño events there are outbreaks of malaria and dengue fever in Colombia. These outbreaks could be explained in terms of a decrease in precipitation and an increase in air temperature, which favor the ecological, biological, and entomological components of these diseases.

We illustrate the ability to predict malaria cases in Colombia by using an epidemiological model based on the concept of vectorial capacity (see Martens et al. 1997). This transmission potential model is driven with surface air temperatures derived from an atmospheric general circulation model (ECHAM3 model, Max Planck Institute for Meteorology) with a spatial resolution of about 300 km. The malarial model produces peaks in *Plasmodium vivax* vectorial capacity during El Niño years and an upward trend with time, in agreement with the Colombian malarial historical record.

These statistical correlations and modeling results may be used for developing health early warning systems of climate conditions conducive to outbreaks, which may facilitate early public health interventions to control and mitigate the incidence of these VBDs.

Introduction

Outbreaks and spreading of vector-borne diseases (VBDs) in human populations are clearly multifactorial, involving social, biological, and environmental factors, of which climatic variability has long been recognized as important (Gill 1920, 1921). El Niño/Southern Oscillation (ENSO) is the major forcing mechanism of climatic and hydrological variability at interannual timescales, with especially strong impacts on northern South America. El Niño events are related to strong perturbations in global atmospheric circulation and associated anomalies in seasonal-to-interannual weather patterns that can trigger profound societal, economical, and environmental consequences.

Some of the relevant biological/climatic factors that interact to affect human health are (Epstein and Stewart 1998) (1) the quality and distribution of surface water and insect breeding sites (Dobson and Carper 1993); (2) temperature and humidity, affecting the life cycles of disease vectors and of the parasites within the vector (Patz et al. 1996; Martens et al. 1997);. and (3) impacts on the ecosystems of predators of insects (Unninayar and Sprigg 1995; Epstein and Chikwenhere 1994). There is diverse evidence relating outbreaks of several waterborne diseases and VBDs with climatic anomalies forced by ENSO events. Malaria outbreaks have been found to be associated with ENSO in Pakistan (Bouma and van der Kaay 1994) and Venezuela (Bouma and Dye 1997), and in Colombia (Poveda and Rojas 1996, 1997). Diseases involving mosquitoes and rodents may cluster after extreme events (especially flooding) in association with the ENSO phenomenon (Epstein et al. 1995); and in Ecuador and Peru epidemics of malaria appear related to flooding associated with El Niño occurrences (Epstein and Stewart 1998). Related research is found in Epstein (1995), Patz et al. (1996), and McMichael and Haines (1997).

In the next section, general features of Colombia's climate and the ENSO phenomenon and its influence on the hydroclimatology of Colombia and tropical South America are discussed. Next, climatic factors associated with the incidence of malaria over Colombia are illustrated, and simple mathematical relationships are used to represent three malaria transmission indices as a function of average temperature. Then we analyze the annual record of malaria and dengue fever cases in Colombia at the national and regional levels, for the period 1959–94, and show the linkages between those cases and ENSO, through correlation analysis with sea surface temperatures (SSTs) in the equatorial Pacific Ocean. We then present results of a malaria model driven by the results of an atmospheric general circulation model (AGCM), with promising predictability capacity. Finally, we discuss the influence of predictability of ENSO and Colombia's hydroclimatology with respect to malaria and dengue prevention and control campaigns.

Colombia's Climate and ENSO

Diurnal, semi-annual, annual, and interannual cycles strongly characterize weather and climate variability in Colombia. The annual distribution of rainfall is primarily influenced by the position of the Inter-Tropical Convergence Zone (ITCZ), while its spatial variability is controlled by the Andes mountains, the eastern Pacific and western Atlantic oceans, the atmospheric circulation over the Amazon basin, and vegetation and soil moisture contrasts. Mean annual precipitation ranges from less than 300 mm in a small region of the Caribbean coast to regions with 10,000–13,000 mm along the Pacific coast (see Snow 1976, p. 371). Over the three branches of the Andes and the Sierra Nevada de Santa Marta, temperature and precipitation vary with altitude in the range 0–5,800 m, influenced by the strength of the trade winds and local circulation. Over the Andes, precipitation increases with height from the valley floor up to an altitude where the so-called Pluviometric optimum occurs, and decreases upwards. Location of this precipitation maximum over the mountains depends on the valley floor height, the air absolute humidity, and local circulation. The relatively cooler and superficial winds that penetrate from the Pacific Ocean into Colombia form a low-level westerly jet (referred to hereafter by the acronym CHOCO jet; see Poveda et al. 1999) that is permanently centered around 5°N throughout the year. The CHOCO jet interacts with the warmer easterlies, thus causing high atmospheric instability that triggers enormous amounts of precipitation over western and central Colombia and producing large meso-scale complexes that penetrate into Colombia and interact with the ITCZ (Velasco and Frisch, 1987). Northeastern Colombian rainfall is highly influenced by the easterly trade winds coming from the Caribbean Sea, as well as by summertime tropical easterly waves. Southeastern Colombian rainfall is mostly associated with the strength of the southeasterly trade winds from the Amazon basin, and therefore the eastern slope of the Andes experiences high precipitation rates (around 5,000 mm per year) due to orography.

El Niño refers to the unusual warming of SSTs in the eastern and central tropical Pacific. Important components of this anomaly in the global climate system are the deepening of the oceanic thermocline in the eastern Pacific and the weakening of the dominant surface easterly trade winds. During an El Niño event, there is a shift in the center of convection from the western to the central Pacific. The accompanying Southern Oscillation, the "seesaw" of the atmospheric mass that produces a pressure gradient between the western and the eastern equatorial Pacific, is quantified by the Southern Oscillation Index (SOI), defined as the standardized difference between Tahiti and Darwin sea level pressures (SLPs). Negative values of the SOI are associated with warm events (El Niño), while positive values accompany cold events (La Niña). El Niño/Southern

Oscillation is an aperiodic oscillation with a recurrence varying from 2 to 10 years, for an average of about every 4 years (Trenberth 1991). Kiladis and Diaz (1989) recognize the following years for the occurrence of El Niño events during this century: 1902, 1904, 1911, 1913, 1918, 1923, 1925, 1930, 1932, 1939, 1951, 1953, 1957, 1958, 1963, 1969, 1972, 1973, 1976, 1977, 1982, 1983, 1986, 1987, and 1991/1992. The onset of El Niño events occurs during the Northern Hemisphere spring, exhibiting a strong locking with the annual cycle (Webster, 1995). El Niño events that comprise two calendar years are generally characterized by SST positive anomalies that increase during the Northern Hemisphere spring and fall of the first year (year 0), with the maximum SST anomalies occurring during the winter of the following year (year +1) and SST anomalies receding during the spring and summer of the year +1. Recently, El Niño occurred during 1994–95 and 1997–98. Trenberth and Hoar (1996) argue that during the period 1991–95 there was a five-year single El Niño event, while Goddard and Graham (1997) suggest that, although that period was marked by generally warmer than normal SSTs in parts of the eastern and central equatorial Pacific, analyses of ocean thermodynamics argue against its characterization as a single extended episode. The physics of ENSO and its climatic consequences can be found in Horel and Wallace (1981), van Loon and Madden (1981), Glantz et al. (1991), Rasmusson and Carpenter (1982), Ropelewski and Halpert (1987), Rasmusson (1991), Diaz and Markgraf (1992), Diaz and Kiladis (1992), and Battisti and Sarachick (1995). El Niño/Southern Oscillation disrupts the normal patterns of the global atmosphere–oceanic circulation and the land surface hydrology affecting weather events and climate. The associated extreme weather events, including floods, droughts, and heat waves, produce severe socioeconomic and environmental impacts, including crop and fishery failures and food shortages, infrastructure disruption, forest fires, reduced hydropower generation, electricity shortages, harmful algal blooms, and epidemics.

In general, there is a coherent pattern of climatic and hydrological anomalies in tropical South America during extreme phases of ENSO (Poveda and Mesa 1997). Overall, negative anomalies in rainfall, soil moisture, and river discharges, as well as positive air temperature anomalies, occur during El Niño. The reverse picture is generally valid for the cold phase (La Niña). Figure 6.1 (top row) shows the annual cycle of air temperature for the difference between El Niño events of 1982–83, 1976–77, and 1991–92, as compared to the 1988–89 cold event in the Pacific Ocean, according to data from the National Centers for Environmental Prediction/National Center for Atmospheric Research (NCEP/NCAR) Reanalysis Project (Kalnay et al. 1996). We conclude that there is an overall increase in air temperature during El Niño in Colombia as compared with the La Niña situation, in particular during December–January–February (DJF) of the mature ENSO phase (referred to as DJF+1). Also, precipitation decreases (Fig. 6.1, bottom row) during El Niño as compared with "normal" and La Niña years throughout the country. The first empirical orthogonal function of standardized monthly rainfall over Colombia, for the period 1958–90, which explains more than 30% of rainfall variability (see Figure 6 of Poveda and Mesa 1997), exhibits the same sign over the entire region, confirming the generalized negative rainfall anomalies associated with El Niño.

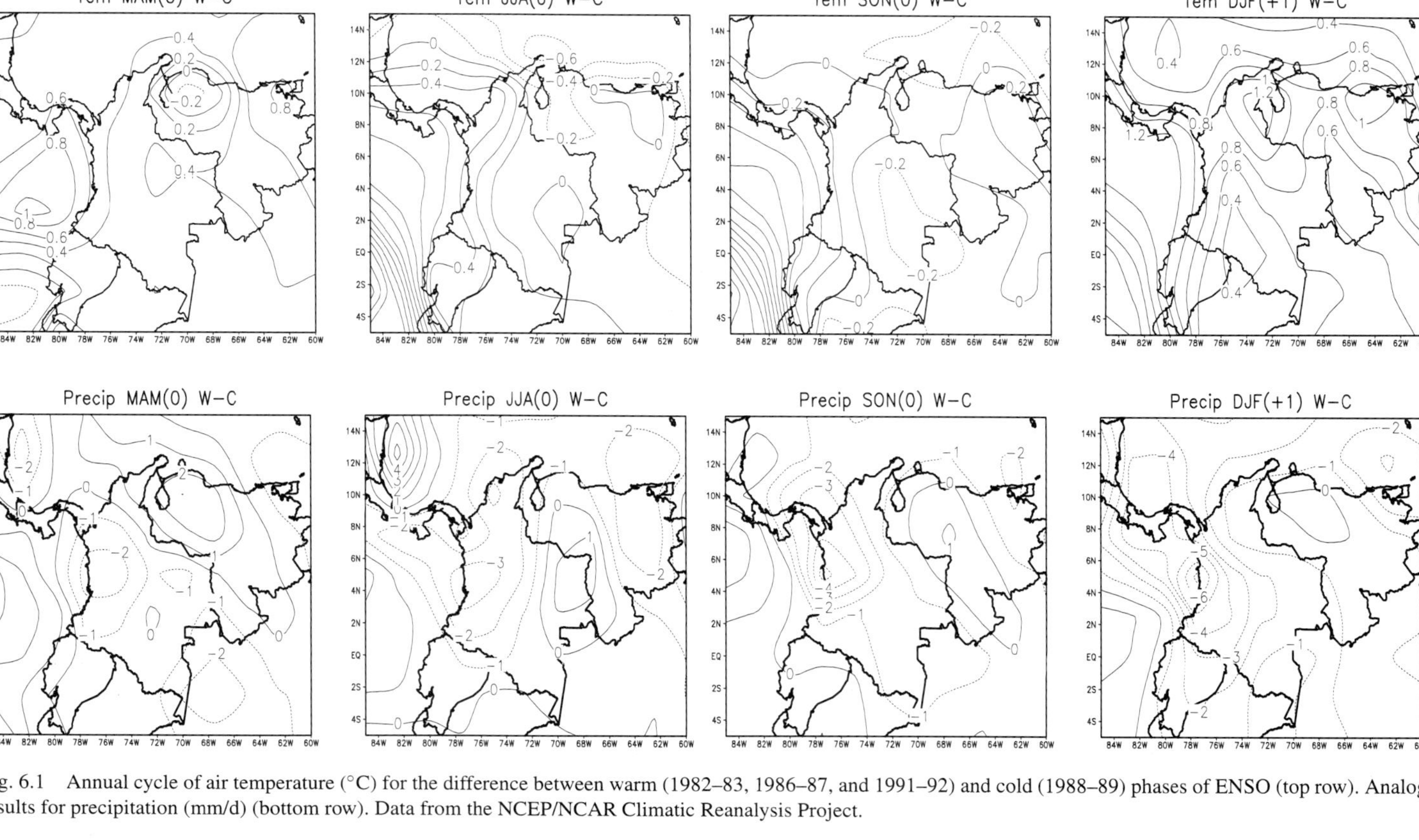

Fig. 6.1 Annual cycle of air temperature (°C) for the difference between warm (1982–83, 1986–87, and 1991–92) and cold (1988–89) phases of ENSO (top row). Analog results for precipitation (mm/d) (bottom row). Data from the NCEP/NCAR Climatic Reanalysis Project.

Western Colombia and the Andes are the regions of Colombia most strongly affected by El Niño, in particular during DJF+1, September–October–November of the ENSO year (referred to as SON-0), and June–July–August of the ENSO year (JJA-0), in decreasing order. March–April–May of both years 0 and +1 (MAM-0 and MAM+1) are the least affected by ENSO (Poveda et al. 1997 1999). In Figure 6.1 (bottom row), one can see that the Colombian Orinoco River basin might exhibit an increase in precipitation during JJA-0. Positive anomalies in precipitation during El Niño may also appear in the southernmost fringe of the Pacific coast along the Ecuadorian border. Observations show that negative anomalies associated with ENSO's effect on river discharges occur progressively later for rivers toward the east in Colombia and northern South America. El Niño impact is also felt in the Caribbean plains and in the Orinoco and Amazonian regions of Colombia.

The main mechanisms of the ocean–land–atmosphere system that interact to reduce rainfall during El Niño are (1) the weakening of the low-level westerly jet ("Chocó Jet") that penetrates from the Pacific Ocean to inland Colombia at 5°N (Poveda et al. 1997 and 1999); (2) the reduction of the 700 hPa equatorial easterly jet; (3) the reduction of moisture advection from the Caribbean Sea; (4) the reduction in number and intensity of easterly waves along the tropical North Atlantic; (5) the displacement of the Intertropical Convergence Zone (ITCZ) to the southwest of its normal position (Pulwarty and Diaz 1993), due to an anomalous Hadley cell over tropical South America (Rasmusson and Mo 1993), and (6) land–atmosphere feedback (Poveda and Mesa 1997).

Figure 6.2 shows the correlation map between SSTs in the Pacific and Indian Oceans and the first principal component (PC) of Colombia's monthly precipitation field for the 1958–90 period (see Figure 7 of Poveda and Mesa 1997). Figure 6.2 confirms the strong inverse association between the Pacific Ocean SSTs anomalies and rainfall in Colombia. Regions of greatest negative correlation are the central and eastern equatorial Pacific and the area of the Indian monsoon. In fact, Colombian precipitation is most highly correlated with SSTs at the Niño-4 region (5°N–5°S, 160°E–150°W), the

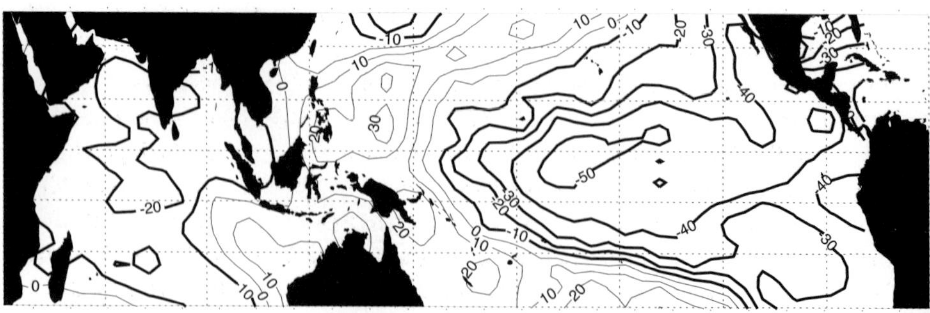

Fig. 6.2 Correlation map (%) between SSTs at the Pacific and Indian Oceans and the first PC of the Colombian monthly precipitation records (three-month running averages). Note the high values of the correlation coefficients over the Niño-4 and Niño-3 regions, as well as over the Indian monsoon region.

Niño-3 region (5°N–5°S, 150°W–90°W), and the Niño-1+2 region (0°–10°S, 90°–80°W) of the Pacific Ocean. Interestingly, the Niño-4 region in the Central Pacific returns higher correlations than other regions closer to South America.

Climate and ENSO Variability versus Malaria and Dengue Fever

Malaria

As the result of climatic and topographic features, two thirds of Colombia's population live in malaria endemic areas (Rojas et al. 1992). The most important malaria vectors in the country are *Anopheles albimanus*, *A. darlingi*, and *A. nuñeztovari* (Quiñones et al. 1987), transmitting *Plasmodium falciparum* (46.5%) and *P. vivax* (53.5%) and rare cases (8–10 per year) of *Plasmodium malariae* (Haworth 1988). The geographical distribution of the disease in Colombia is associated with prevalent climatic conditions. As has been stated, temperature and precipitation are related to elevation, in particular over the Andes and the Sierra Nevada de Santa Marta. We use the association existing between altitude and temperature to represent diverse indices associated with malaria transmission in Colombia as follows. Figure 6.3 shows the map for the extrinsic incubation period (EIP, or the incubation period of the parasite inside the mosquito) for *P. vivax*, estimated as

$$n = \frac{D_{\mathrm{m}}}{T - T_{\mathrm{min},\, n}}, \tag{1}$$

(MacDonald 1957), where n is the incubation period of the parasite inside the vector (in days), D_{m} is the number of degree-days required for the development of the parasite (105°C for *P. vivax*), T is the actual average temperature, and $T_{\mathrm{min},\, n}$ is the maximum temperature required for parasite development (14.5°C for *P. vivax*). The temperature map of Colombia (not shown) considers the regional variability associated with elevation and rainfall. According to Equation (1), the Amazon and Orinoco basins and the lowlands of the Pacific and Caribbean coasts exhibit the shortest EIP, while the warmer zones of the Andean region and the Sierra Nevada de Santa Marta in northeastern Colombia (red colors) exhibit the longest EIP for *P. vivax*. In Figure 6.4 we show a map for the daily survival probability of the *Anopheles* vector, as a function of mean temperature, T. The functional relationship we have used is $p = \exp[-1/(-4.4 + 1.31T - 0.03T)]$ (Martens 1998). Again, almost the same regions of the Andes and the Sierra Nevada de Santa Marta exhibit the highest daily survival probabilities in the country. If we assume that the daily survival probability is an independent event, then p^n can be interpreted as a "survival rate" (SR) of the parasite, as a function of temperature. Figure 6.5 presents the map of SR for Colombia. The Amazon and Orinoco basins and the lowlands of both the Pacific and the Caribbean coasts exhibit the highest parasite survival rates for *P. vivax*. These modeling results agree with estimations of malarial risk in Colombia for 1996 (see Fig. 6.6, adapted from a map kindly provided by Julio C. Padilla of the Colombian Ministry of Health), confirming that the Colombian Amazon (southeastern), the Orinoco basin (eastern),

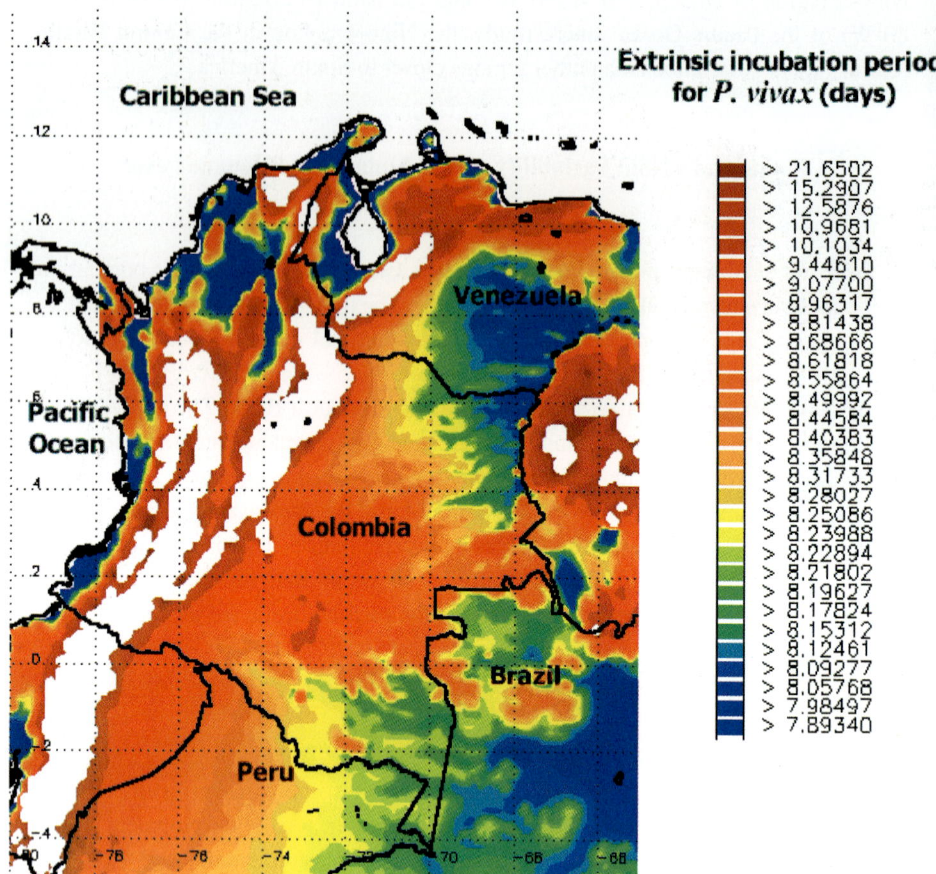

Fig. 6.3 Distribution of the estimated extrinsic incubation period (Equation 1) for *P. vivax* over Colombia. Regions in white do not support *P. vivax* development because of temperature constraints.

and the Pacific coast are the regions more prone to malaria in the country. These maps illustrate the association between some indices associated with malaria transmission and temperature, which is related to altitude and temperature, as well as to precipitation and humidity in Colombia.

Now, we examine the temporal variability of malaria cases. In Figure 6.7 we present the evolution of the Annual Parasitic Incidence index (API), defined as the ratio between the number of cases reported and the population at risk per 1,000 inhabitants, for the period 1959–94. The API index is computed as the total of cases of both *P. vivax* (represented by the AVI index in Fig. 6.7) and *P. falciparum* (AFI index), from data reported by the Colombian Ministry of Health. Arrows in Figure 6.7 indicate El Niño years according to the classification given by Kiladis and Diaz (1989). Three facts emerge from the figure: First, outbreaks of malaria are identified during 1961, 1968, 1972, 1977, 1983, 1987, and 1991/1992. Second, there is a clear increasing

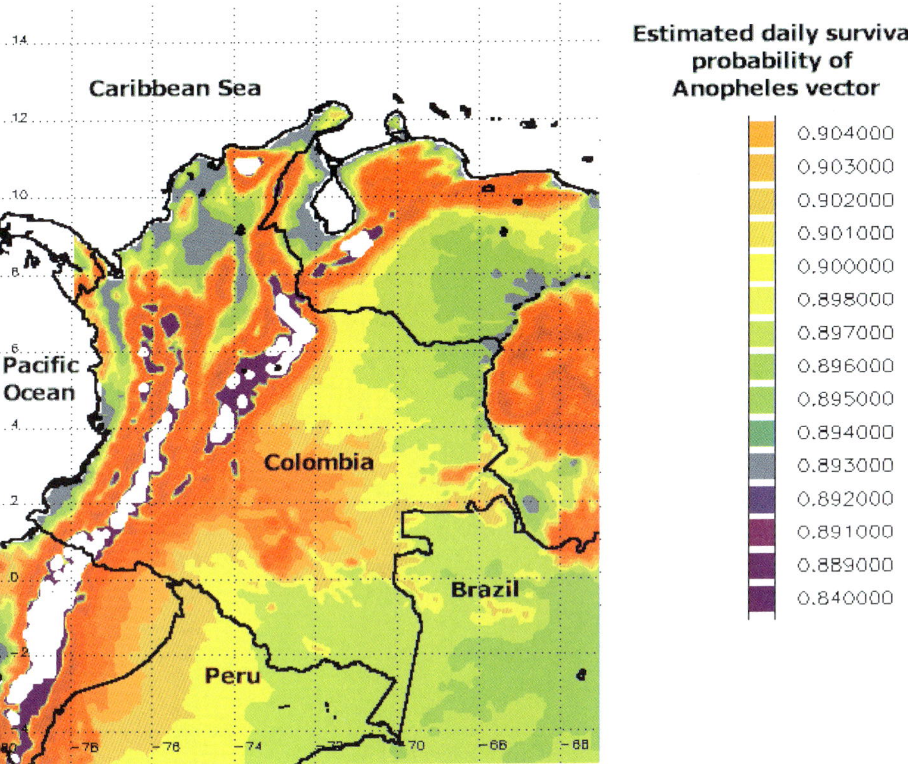

Fig. 6.4 Distribution of the estimated daily survival probability ($p = \exp[-1/(-4.4 + 1.31T - 0.03T)]$, Martens 1998) of the *Anopheles* vector over Colombia. Regions in white are associated with null probability.

trend in the number of malaria cases. And third, since the mid-1970s, *P. vivax* has become more predominant than *P. falciparum* nationally. With respect to the first point, it is worth noting that all years but 1961 correspond to ENSO warm events. However, the anomalous dry period from 1957 through 1961 in Colombia was influenced by the 1957–58 El Niño event (Poveda 1994; Poveda and Mesa 1996, 1997). Figure 6.8 shows the time evolution of the detrended national API index, together with the series of SST anomalies over the Pacific Ocean Niño-4 region, which exhibits the highest correlation with the Colombian hydroclimatology (Poveda 1994; Poveda and Mesa 1997), as well as the series of the (annually averaged) first Principal Component of monthly precipitation in Colombia. A simple detrending procedure for the API series consisted of removing a linear trend fitted to the historical record. Simultaneous correlation between the API and the SST anomalies over Niño-4 is 0.50, statistically significant at the 95% level, while lag-1-year cross-correlation is 0.49. This result indicates that interannual climatic variability could explain a very important part of the variance of the malaria record in Colombia. The correlation coefficient between the Principal Component No. 1 of precipitation and the detrended API series is 0.42, statistically significant at the

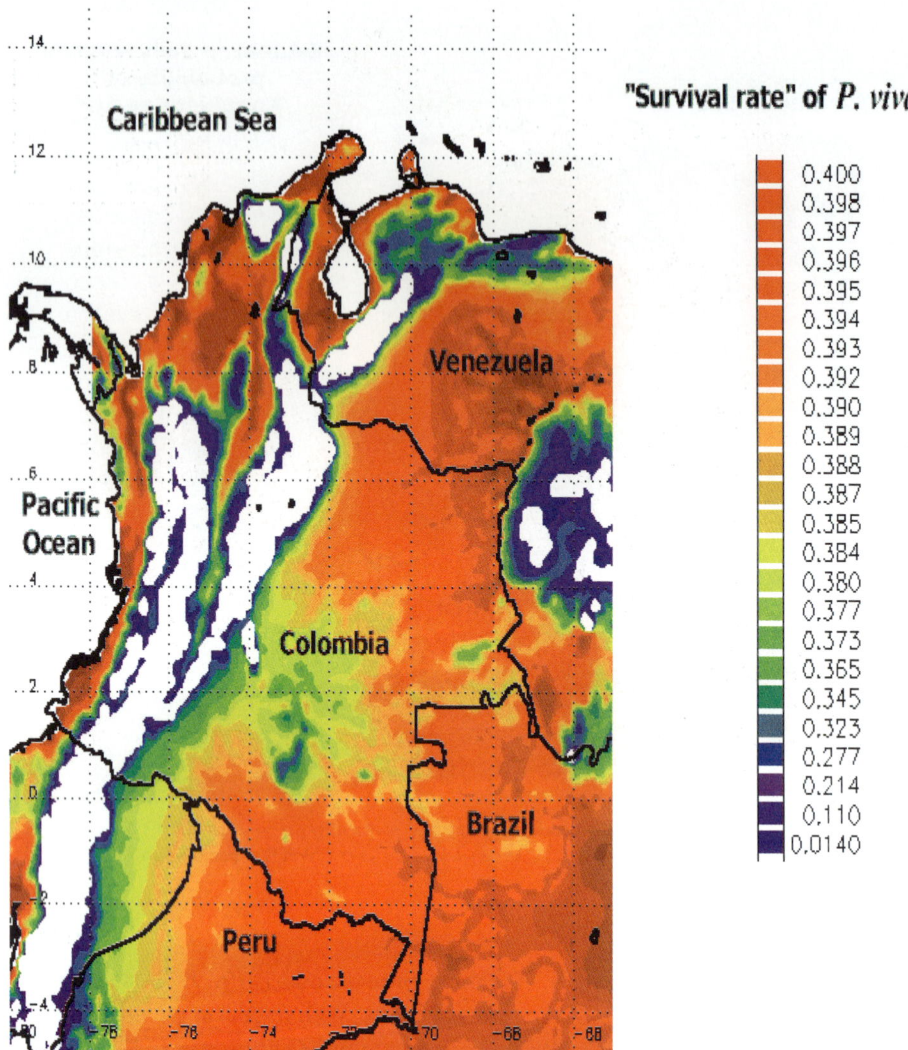

Fig. 6.5 Distribution of the "survival rate" of *P. vivax*, defined as p^n, over Colombia. Regions in white are associated with null probability.

95% level. Analysis performed with regional API series (not shown here) indicates that the association between El Niño occurrences and regional malaria outbreaks is very high, particularly in the Andean, Pacific, and Orinoco regions. Figure 6.9 shows the evolution of the API index for the *Departamentos* (States) of Antioquia and Putumayo.

Dengue Fever

Dengue fever (DF), spread by mosquito vectors of the genus *Aedes*, has a worldwide distribution in the tropics. Outbreaks are clearly multifactorial, involving compounding

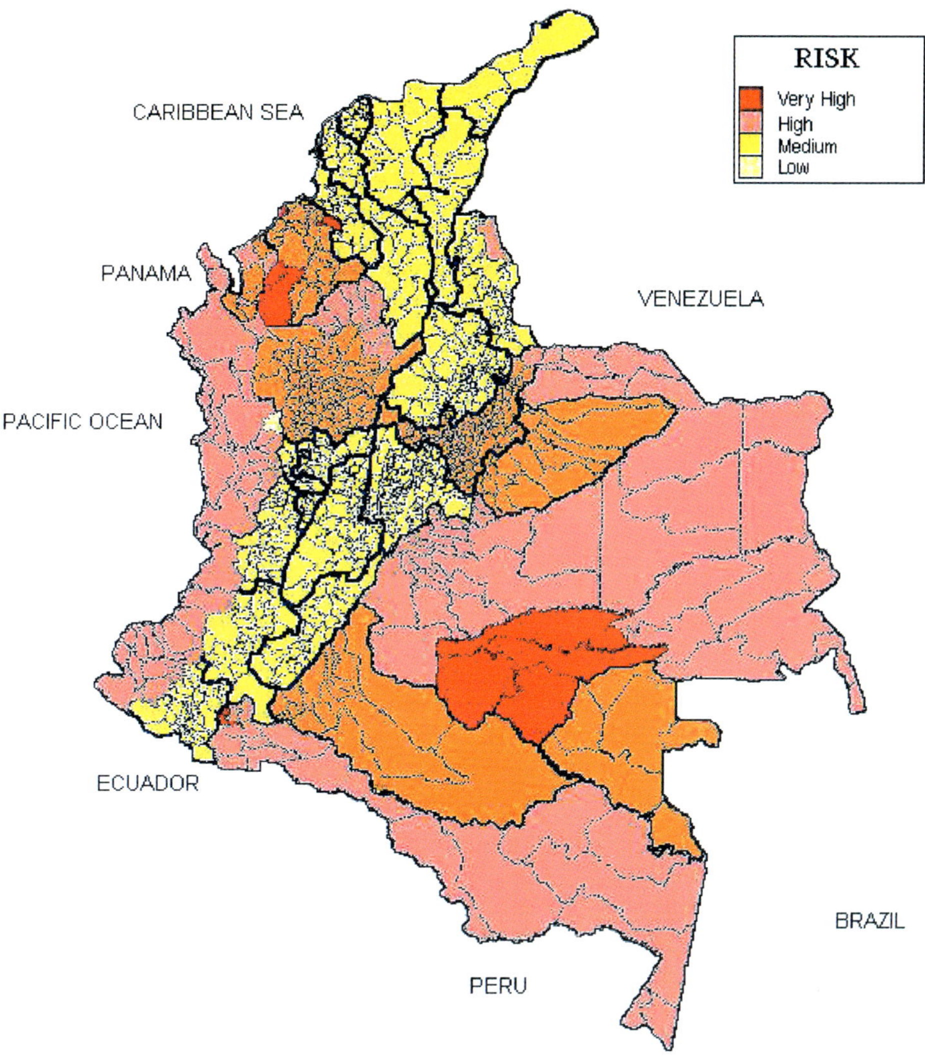

Fig. 6.6 Distribution of malaria risk over Colombia. Adapted from a map kindly provided by Dr. Julio César Padilla of the Colombian Ministry of Health.

social, biological, and environmental factors. Peri-urban sprawl, poor sanitation, and the proliferation of nonbiodegradable (and other) water containers, and social inequities in general provide the setting for the resurgence of dengue fever in Latin America. But meteorological factors also play a role. In general terms, climate circumscribes the range at which VBDs can occur, while weather influences the timing and intensity of outbreaks (Dobson and Carper 1993). Two characteristics of climate change are germane to the distribution: (1) gradual warming and (2) the disproportionate rise in minimum temperatures (TMINs) (i.e., nighttime and winter), in relation to the gradual

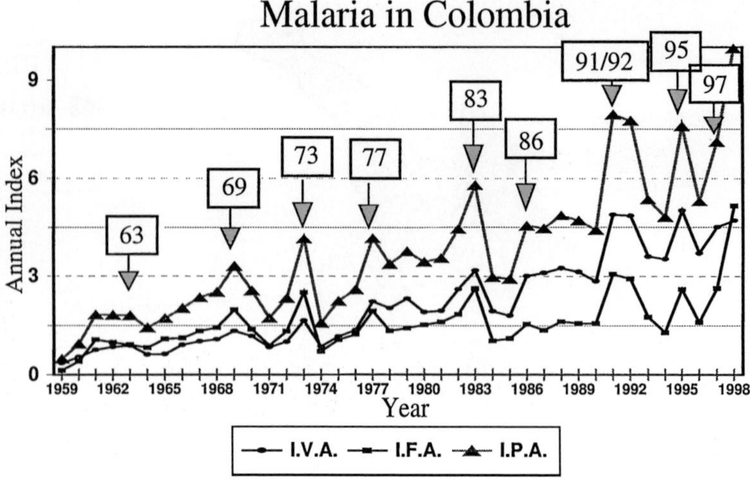

Fig. 6.7 Evolution of the API, defined as the ratio between the number of cases reported and the population at risk per 1,000 inhabitants, computed as the total of cases of both *P. vivax* (represented by the AVI index) and *P. falciparum* (AFI index). Data reported by the Colombian Ministry of Health.

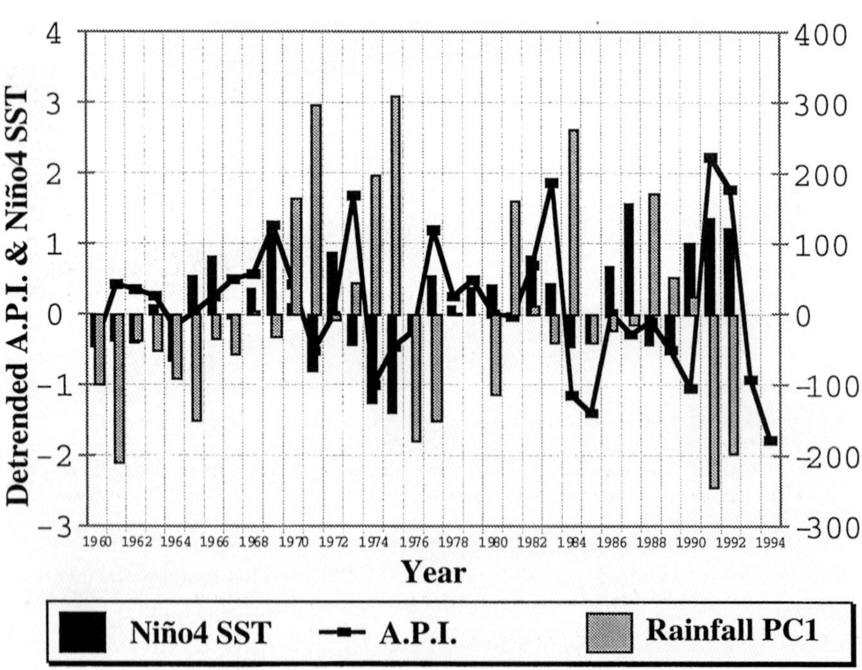

Fig. 6.8 Series of the detrended API index for Colombia SST anomalies over the Pacific Ocean Niño-4 region (5°N–5°S, 160°E–150°W) and first Principal Component of Colombia precipitation (scale on the right). The simultaneous correlation is 0.50, and the lag-1-year correlation is 0.49, both statistically significant at the 95% level.

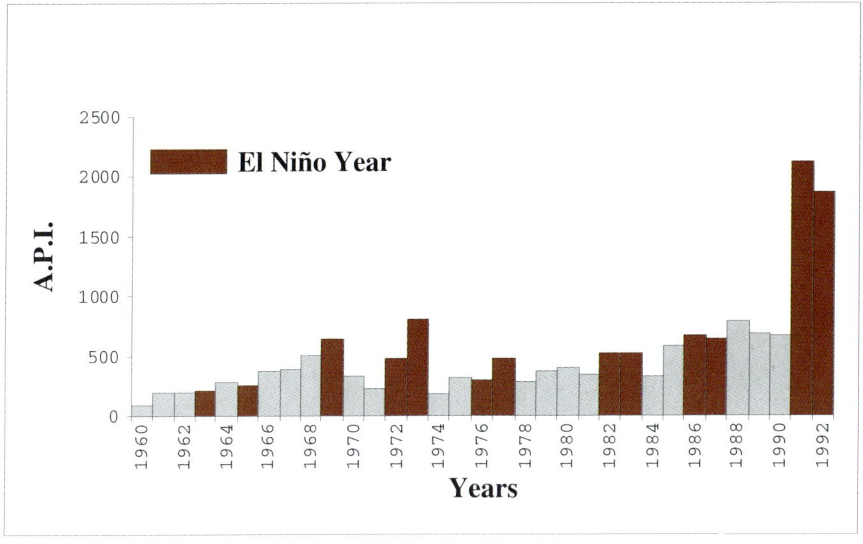

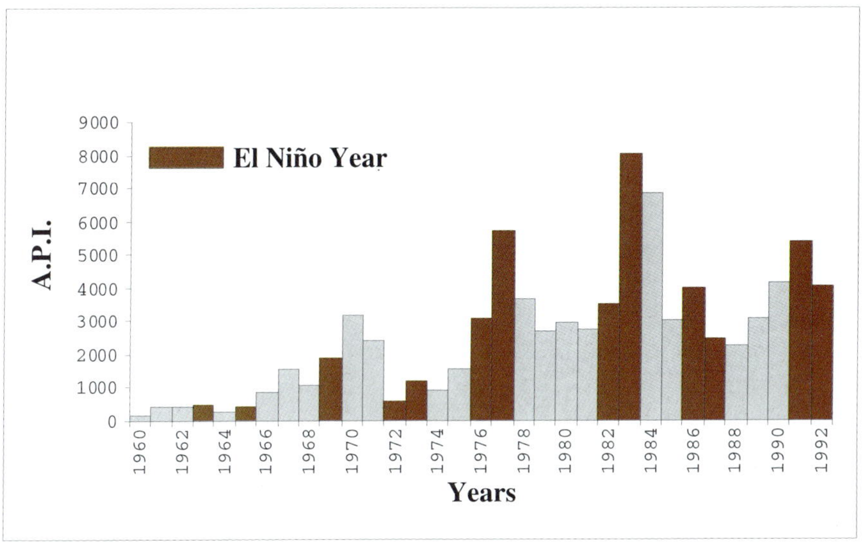

Fig. 6.9 Evolution of regional API index for (top graph) Antioquia (northwest, Andean region) and (bottom graph) Putumayo (Amazon region).

rise in average global temperatures (Karl et al. 1993; Easterling et al. 1997). Several investigations have linked dengue fever transmission to temperature (Hales et al. 1996; Donald Burke, unpublished data, 1997). The mosquito larvae can develop only at temperatures above a crucial threshold, and freezing kills eggs. Moderately high temperature can also hasten the larval stage, leading to smaller mosquitoes, which require more frequent blood meals. Also, the EIP shortens at higher temperatures. Dengue type 2 virus has an EIP of 12 days at 30°C but only 7 days at 32–35°C

(Focks et al. 1995). Koopman et al. (1991) found that decreasing the incubation period by 5 days can lead to a threefold higher transmission rate of dengue and that raising the temperature from 17°C to 30°C increases dengue transmission fourfold. Higher temperatures may increase the amount of feeding within the gonotrophic cycle (MacDonald 1957), given the smaller body size and enhanced metabolism with increasing temperature.

Dengue fever upsurges in the islands of the South Pacific are associated with ENSO events (Hales et al. 1996). Also, the incidence of dengue fever in Colombia, much of Central America, and several Caribbean nations (and in Rio de Janeiro) is strongly influenced by ENSO events. Figure 6.10 (top) shows the evolution of dengue fever

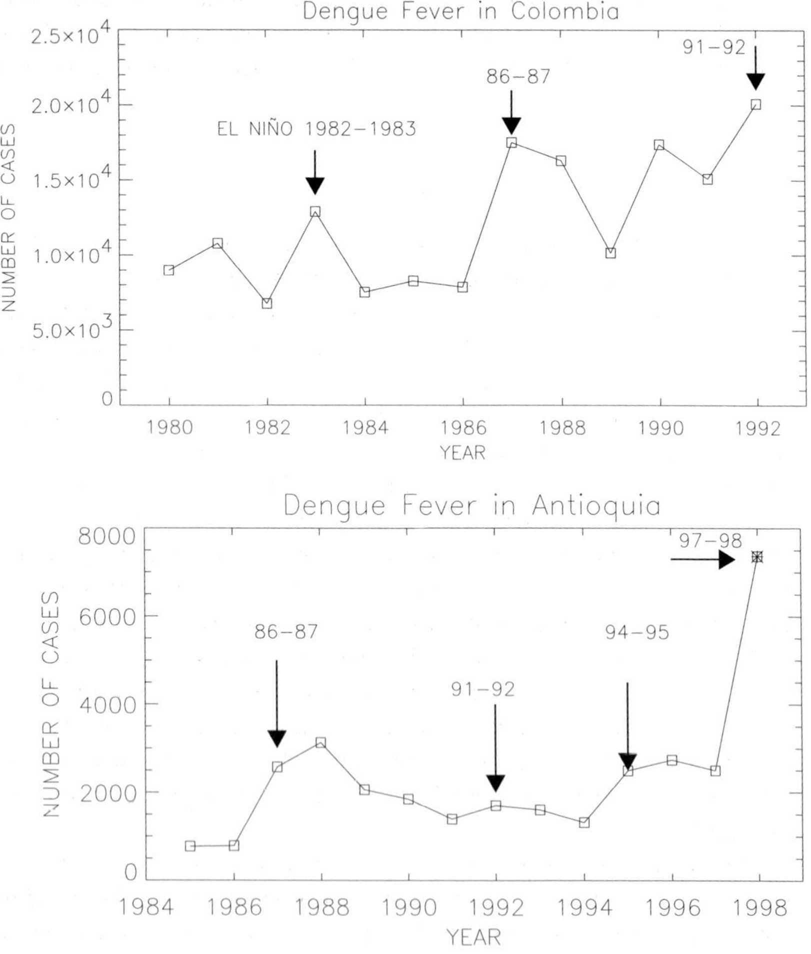

Fig. 6.10 Annual reported dengue cases for Colombia (missing data in 1993–94) (top) and hemorrhagic dengue fever cases in Antioquia, northwest Colombia (bottom).

cases in Colombia during 1980–92. Interestingly, peaks are evident for 1983, 1987, and 1992, all of them being El Niño years (+1). Figure 6.10 (bottom) shows the number of dengue hemorrhagic fever (DHF) cases in Antioquia (northwestern Colombia), for the period 1985–98, with most of the peaks corresponding to El Niño (+1) years. Notice the extraordinary peak in 1998 (7,368 cases until September 12th). During 1997, there were 3,950 cases of DHF in Colombia (41 deaths), equivalent to 9.82 per 100,000 inhabitants, as compared to the 1,750 cases in 1996 (14 deaths), equivalent to 4.44 per 100,000 inhabitants, as reported by the Colombian Ministry of Health. In Antioquia there were 67 cases of DHF during 1997, and up to this writing (09/1998) there have been 208 cases of DHF. Again, temperature increases and available stagnant waters may account for the upsurges. Water supply for human consumption becomes a serious problem during El Niño in Colombia, owing to the prolonged drought. Many rural towns require the storage of water in cans and tanks, thus creating more breeding sites for the *Aedes* mosquito and favoring the spread of dengue fever in the country.

Discussion

Among the reasons that could help explain the strong relationship between El Niño and malaria and dengue fever outbreaks in Colombia are the increased air temperatures and the reduced precipitation during El Niño events. Temperature, precipitation, and humidity affect the epidemiology of malaria (Bouma 1995). Temperature impacts the dynamics of the vector population (reproductive rates, longevity, and biting rates), as well as the duration of the EIP or sporogony of the malaria parasites inside the mosquitoes. Warmer temperatures increase reproductive and biting rates and decrease the EIP. Very warm (above about 35°C) and, to a lesser extent, very cold (below 10°C) temperatures decrease mosquito longevity. Rainfall and humidity impact these dynamics, as well as mosquito breeding sites. For some areas, excessive precipitation can increase breeding sites. In other areas with many mountain-derived rivers such as Sri Lanka (Bouma 1995) and Colombia, decreased rainfall may help create ponds and stagnant waters along the riverbanks and in the lower valleys (that otherwise would overflow during a non–El Niño year), thus providing adequate breeding sites for the mosquito. The increase in water temperature associated with ENSO years may further favor anopheline mosquito and parasite maturation. Thus, temperature increases and availability of adequate aquatic conditions could be the main factors that explain the observed relationship.

The apparent linkages between the trends in climate variables and the incidence of malaria in Colombia are clear. In particular, the association between increased mean and minimum temperatures (shown in Colombia by Mesa et al. 1997) and increased numbers of malaria cases are at least consistent with what is known about the biological factors involved in transmission of this disease. In the following section, we use a simple epidemiological model of malaria transmission to test one of these hypothesized linkages.

Climate–Malaria Modeling: Results for Colombia

We modeled the spread of malaria as a function of important meteorological factors with the Epidemic (or Transmission) Potential Model (Martens et al. 1995a,b; Martens 1998), which is based on the concept of vectorial capacity (MacDonald 1957; Garret-Jones 1964). Epidemic (or transmission) potential is the reciprocal of the critical mosquito density threshold and summarizes how climate change would affect the mosquito population directly. This value is obtained by computing the effect of temperature on feeding frequency and longevity and the incubation period of the parasite in the vector. Within the model, the relation between ambient temperature and latent period is calculated by using a temperature sum as described by Macdonald (1957). The frequency of feeding depends mainly on the rapidity with which a blood meal is digested, which increases as temperature rises, and can also be calculated by means of a thermal temperature sum (Detinova et al. 1962). Between certain temperature thresholds, the longevity of a mosquito decreases with rising temperature (Molineaux 1988). The optimal temperature for mosquito survival lies in the 20–25°C range. The concept of vectorial capacity (VC) involves other factors, including density of mosquitoes (which in turn depends on temperature and precipitation), blood meals and biting proportions necessary for transmission, and daily survivability, which also depend on temperature and humidity. In these terms, incubation period (probably the most important), mosquito density and survivability, and even host mosquito biting activity are dependent on meteorological conditions.

This epidemiological model was driven with surface air temperatures derived from an AGCM (ECHAM3 model of the Max Planck Institute for Meteorology in Hamburg) with a spatial resolution of about 300 km. The AGCM was run for 1970–92 using observed SSTs as boundary forcing. Previous analyses have shown that the surface air temperature data produced by this AGCM agree well with large-scale spatial averages of observed surface air temperatures and precipitation over Colombia and much of northern South America. In particular, the model correctly shows warmer and drier than usual conditions over much of this region during El Niño years and an upward trend in surface air temperatures during the period since the mid-1970s. Given these temperature signals, the malaria model produces the peaks in vectorial capacity (VC) during El Niño years and an upward trend with time. The qualitative agreement of the model for *P. vivax* and the observed number of cases of malaria in Colombia are also apparent (see Fig. 6.11), suggesting the possibility that the malaria model is correctly showing the important role of temperature and precipitation variability in modulating malaria transmission in this region. Nevertheless, one needs to interpret the results with caution and realize that the Epidemic Potential Model may serve as an index "for early warning" in combination with climate forecasts.

Several remarks must be made concerning the modeling results. The patterns in vectorial capacity are not straightforwardly linked with patterns of incidence, given the complexity of the malaria transmission dynamics that include immunity, loss of infection, land use and cover changes, human interventions, etc. We note that the climate data used (on a very large grid level) contain uncertainties that should be borne in

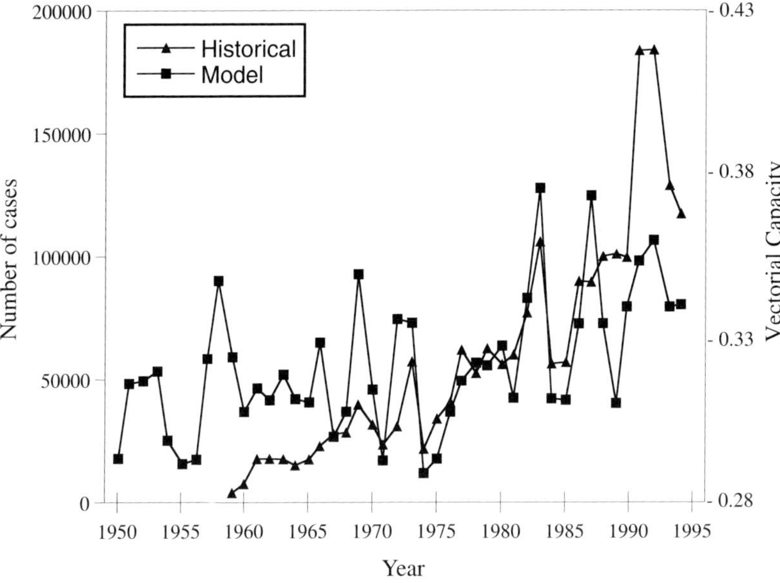

Fig. 6.11 Comparison between modeled vectorial capacity for *P. vivax* and the historical number of malaria cases in Colombia.

mind when dealing with smaller spatial scales. The model's broad geographical spatial scale makes it unfeasible for it to take into account the different *Anopheles* species, although different parts of the country have different species of anopheline vectors (Quiñones et al. 1987). Caution is necessary also in the interpretation of modeling results using surface air temperature data, since mosquitoes can select local microhabitats with different but more suitable temperatures to rest. In these microhabitats, other climatic variables can be "ideal" for mosquito populations despite apparent adverse climate conditions. This is the case for humidity, which can play a major role in the determination of population longevity, a variable with a higher impact on VC (MacDonald 1957). Also, the model results do not prove that climate is involved with the variability in malaria activity, but they are quite consistent with that idea. Indeed, the role of temperature is perhaps overstated in the model and other climatic factors may be involved; nonetheless, the model aptly reproduces the peaks and the increasing trend existing in the Colombian malarial record. If one considers including precipitation projections in the model one should remember that these are far less certain than temperature projections, particularly in tropical regions, and recall that some of the largest malaria outbreaks have occurred during very dry years. In these terms, we would consider rainfall not to be a limiting factor.

Conclusions

There is a strong association between climatic conditions and malarial risk throughout Colombia. Consistently, we have found that the cases of malaria (*P. vivax* and

P. falciparum) in Colombia exhibit (1) peaks during El Niño events, which holds true for dengue fever as well; (2) upward trends for the 1959–93 period; and (3) since the mid-1970s, a greater predominance of *P. vivax* over *P. falciparum* as a national average. We associate the outbreaks in malaria and dengue fever in Colombia during El Niño to an increase in temperature (minimum and mean) and a decrease in precipitation throughout the country, although changes in other climatic variables such as humidity may play a role, too. Warmer temperatures increase reproductive and biting rates and decrease the EIP. Diminished rainfall can lead to the formation of ponds and stagnant pools, which may create more mosquito breeding sites. The increase in temperature in ponds and stagnant waters associated with El Niño events may increase temperatures and provide breeding sites that further favor anopheline mosquito and parasite maturation. With regard to DF, diminished precipitation causes water shortages in many regions of the country, leading to the storage of water in receptacles and tanks, which constitute preferred breeding sites for *Aedes aegypti*.

Epidemics of these diseases are the result of multiple factors, including socioeconomic determinants, migratory patterns, demographic features, and local environmental constraints. But climatic variability is also an important incidence factor. Thus the statistical correlations may be helpful for developing health early warning systems (HEWS) that inform the public of meteorological conditions conducive to outbreaks. The coupling of epidemiological and climate models produces results that replicate the historical peaks and trends in the Colombian malaria record, thus giving a promising tool for forecasting the disease. Institutional support for HEWS may help to facilitate early, environmentally sound public health interventions. These measures include increased surveillance and plans for mobilizing public health responses, information dissemination to the public, environmental cleanups for DF, distribution of pesticide-impregnated bed nets and medications for malaria control, targeted spraying of pesticides, stocking of drugs, and applications of biological control measures to mosquito breeding sites. Use of chemical interventions and biological controls can be planned when an El Niño episode is expected or once it is under way.

In addition to ENSO, other large-scale ocean–atmospheric phenomena affect the hydroclimatology of Colombia at interannual timescales. The North Atlantic Oscillation and the Quasi-Biennial Oscillation exhibit a significant influence on Colombia's climatology (Poveda and Mesa 1996, 1997). These findings bode well for developing adequate climatic predictive models specifically designed for Colombia (Salazar et al. 1994a,b; Poveda and Penland 1994; Carvajal et al. 1998). The improved predictive capability can enhance the planning and decision processes for control of malaria, dengue fever, and other VBDs, helping to prevent disease and the associated social and economic troubles.

The Colombian authors are conducting an ongoing research effort to refine the regional and local aspects of the malaria incidence and to better understand the relationships with climatic variability at different timescales in Colombia. That research includes the response in the field and in the laboratory of vector and parasite life cycles to the environmental changes associated with extreme phases of ENSO. Further research on climatic patterns and weather parameters may be useful for developing public health priorities throughout Colombia and in other regions. This work can provide a framework for future modeling exercises to investigate the relationships among climatic factors, climate variability, and human health.

Acknowledgments The Colombian team of researchers is grateful to the Colombian Council for Science and Technology, COLCIENCIAS, for partially funding this research. Discussions with J. C. Padilla and O. J. Mesa have been highly valuable. Part of the work done by W. J. M. Martens has been funded by the Dutch National Research Program on Global Air Pollution and Climate Change (NOP), Project Number 952257. The Colombian Ministry of Health and the Health Service of Antioquia have provided the malaria and dengue data. Mr. Ricardo Mantilla help to prepare Figures 6.3 through 6.5. The authors thank the anonymous reviewer for helpful suggestions, which led to an improved paper.

References

BATTISTI, D. S. and SARACHIK, E. S., 1995: Understanding and predicting ENSO, U.S. National Report to IUGG, 1991–1994, American Geophysical Union. *Review of Geophysics*, **33** Suppl.

BOUMA, M. J., 1995: *Epidemiology and Control of Malaria in Northern Pakistan. With Reference to the Afghan Refugees, Climate Change, and El Niño Southern Oscillation.* Dordrecht: ICG Printing.

BOUMA, M. J. and DYE, C., 1997: Cycles of malaria associated with El Niño in Venezuela. *Journal of the American Medical Association*, **278**, 1772–1774.

BOUMA, M. J. and van der KAAY, H.J., 1994: Epidemic malaria in India and El Niño Southern Oscillation. *Lancet*, 344, 1638–1639.

CARVAJAL, L. F., MESA, O. J., SALAZAR, J. E., and POVEDA, G., 1998: Singular spectral analysis applied to hydrological time series in Colombia (in Spanish). *Ingeniería Hidráulica en México*, **XIII**(1), 7–16.

DETINOVA, T. S., BEKLEMISHEV, W. N., and BERTRAM, D. S., 1962: Age-grouping methods in diptera of medical importance. WHO Monograph 47, World Health Organization, Geneva.

DIAZ, H. F. and KILADIS, G. N., 1992: Atmospheric teleconnections associated with the extreme phases of the Southern Oscillation. *In* Diaz, H.F. and Markgraf, V. (eds.), *El Niño: Historical and Paleoclimatic Aspects of the Southern Oscillation.* Cambridge: Cambridge University Press, 7–28.

DIAZ, H. F. and MARKGRAF, V. (eds.), 1992: *El Niño: Historical and Paleoclimatic Aspects of the Southern Oscillation.* Cambridge: Cambridge University Press, 476 pp.

DOBSON, A. and CARPER, R., 1993: Biodiversity. *Lancet*, **342**, 1096–1099.

EASTERLING, D. R., HORTON, B., JONES, P. D., PETERSON, T. C., KARL, T. R., PARKER, D. E., SALINGER, M. J., RAZUVAYEV, V., PLUMMER, N., JAMASON, P., and FOLLAND, C. K., 1997: Maximum and minimum temperature trends for the globe. *Science*, **277**, 363–367.

EPSTEIN, P. R., 1995: Emerging diseases and ecosystem instability: New threats to public health. *American Journal of Public Health*, **85**, 168–172.

EPSTEIN, P. R. and CHIKWENHERE, G. P., 1994: Biodiversity questions (letter). *Science*, **265**, 1510–1511.

EPSTEIN, P. R., PEÑA, O. C., and RACEDO, J. B. 1995: Climate and disease in Colombia. *Lancet*, **346**, 1243.

EPSTEIN, P. R. and STEWART, M., 1998: Saving scarce public health resources and saving lives: Health sector applications of climate forecasting. Preprint.

FOCKS, D. A., DANIELS, E., HAILE, D. G., and KEESLING, L. E., 1995: A simulation model of the epidemiology of urban dengue fever: Literature analysis, model development, preliminary validation, and samples of simulation results. *American Journal of Tropical Medicine and Hygiene*, **53**, 489–506.

GARRET-JONES, C., 1964: Prognosis for interruption of malaria transmission through assessment of the mosquito's vectorial capacity. *Nature*, **204**, 1173–1175.

GILL, C. A., 1920: The relationship between malaria and rainfall. *Indian Journal of Medical Research*, **7**, 618–632.

GILL, C. A., 1921: The role of meteorology and malaria. *Indian Journal of Medical Research*, **8**, 633–693.

GLANTZ, M., KATZ, R., and NICHOLLS, N. (eds.), 1991: *Teleconnections Linking Worldwide Climate Anomalies*. Cambridge: Cambridge University Press, 535 pp.

GODDARD, L. and GRAHAM, N. E., 1997: E1 Niño in the 1990s. *Journal of Geophysical Research*, **102** (C5), 10423–10436.

HALES, S., WEINSTEIN, P., and WOODWARD, A., 1996: Dengue fever in the South Pacific: Driven by El Niño Southern Oscillation? *Lancet*, **348**, 1664–1665.

HAWORTH, J., 1988: The global distribution of malaria and the present control effort. *In* Wersdorfer, W. H. and McGregor, I. (eds.), *Malaria*. Edinburgh: Churchill Livingstone.

HOREL, J. D. and WALLACE, J. M., 1981. Planetary scale atmospheric phenomena associated with the Southern Oscillation. *Monthly Weather Review*, **109**, 813–829.

KALNAY, E., KANAMITSU, M., KISTLER, R., COLLINS, W., DEAVEN, D., GANDIN, L., IREDELL, M., SAHA, S., WHITE, G., WOOLLEN, J., ZHU, Y., CHELLIAH, M., EBISUZAKI, W., HIGGINS, W., JANOWIAK, J., MO, K. C., ROPELEWSKI, C., WANG, J., LEETMAA, A., REYNOLDS, R., JENNE, R., and JOSEPH, D., 1996: The NCEP/NCAR 40-Year Reanalysis Project. *Bulletin of the American Meteorological Society*, **77**, 437–471.

KARL, T. R., JONES, P. D., KNIGHT, R. W., KUKLA, G., PLUMMER, N., RAZUVAYEV, V., GALLO, K. P., LINDSAY, J., CHARLSON, R. J., and PETERSON, T. C., 1993: A new perspective on recent global warming: Asymmetric trends of daily maximum and minimum temperature. *Bulletin of the American Meteorological Society*, **74**, 1007–1023.

KILADIS, G. and DIAZ, H. F., 1989: Global climatic anomalies associated with extremes in the Southern Oscillation. *Journal of Climate*, **2**, 1069–1090.

KOOPMAN, J. S., PREVOTS, D. R., MARIN, M. A. V., DANTES, H. G., AQUINO, M. L. Z., LONGINI, I. M., Jr., and AMOR, J. S., 1991: Determinants and predictors of dengue infection in Mexico. *American Journal of Epidemiology*, **133**, 1168–1178.

MacDONALD, G., 1957: *The Epidemiology and Control of Malaria*. London, UK: Oxford University Press, 201 pp.

MARTENS, W. J. M., 1998: *Health and Climate Change: Modelling the Impacts of Global Warming and Ozone Depletion*. London: Earthscan Publications Ltd.

MARTENS, W. J. M., JETTEN, T. H., and FOCKS, D. A., 1997: Sensitivity of malaria, schistosomiasis and dengue to global warming. *Climatic Change*, **35**, 145–156.

MARTENS, W. J. M., JETTEN, T. H., ROTMANS, J., and NIESSEN, L. W., 1995a: Climate change and vector-borne diseases: A global modelling perspective. *Global Environmental Change*, **5**, 195–209.

MARTENS, W. J. M., NIESSEN, L. W., ROTMANS, J., JETTEN, T. H., and McMICHAEL, A. J., 1995b: Potential impact of global climate change on malaria risk. *Environmental Health Perspectives*, **103**, 458–464.

McMICHAEL, A. J. and HAINES, A., 1997: Global climate change: The potential effects on health. *British Medical Journal*, **315**, 805–809.

MESA, O. J., POVEDA, G., and CARVAJAL, L. F., 1997: *Introducción al Clima de Colombia*. [Introduction to the Climate of Colombia]. Bogotá: Universidad Nacional de Colombia Press, 390 pp.

MOLINEAUX, L., 1988: The epidemiology of human malaria as an explanation of its distribution, including some implications for its control. *In* Wernsdorfer, W. H. and McGregor, I. (eds.), *Malaria, Principles and Practice of Malariology* (vol. 2). New York: Churchill Livingstone, 913–998.

PATZ, J. A., EPSTEIN, P. R., BURKE, T. A., and BALBUS, J. M., 1996: Global climate change and emerging infectious diseases. *Journal of the American Medical Association*, **275**, 217–223.

POVEDA, G., 1994: Empirical orthogonal functions to study the relationship between river discharges and sea surface temperatures in the Pacific and Atlantic Oceans (in Spanish). *Proceedings XVI Latin-American Hydraulics and Hydrological Meeting*, IAHR, Santiago, Chile, **4**, 131–144.

POVEDA, G. and MESA, O. J., 1996: Extreme phases of ENSO (El Niño and La Niña) and their influence on the hydrology of Colombia (in Spanish). *Ingenieria Hidráulica en México*, **XI**(1), 21–37.

POVEDA, G. and MESA, O. J., 1997: Feedbacks between hydrological processes in tropical South America and large scale oceanic–atmospheric phenomena. *Journal of Climate*, **10**, 2690–2702.

POVEDA, G. and PENLAND, C., 1994: Forecasting monthly mean river discharges using linear inverse modeling (in Spanish). *Proceedings XVI Latin-American Hydraulics and Hydrology Meeting*, IAHR, Santiago, Chile, **4**, 119–129.

POVEDA, G. and ROJAS, W., 1996: Impact of El Niño phenomenon on malaria outbreaks in Colombia (in Spanish). *Proceedings XII Colombian Hydrological Meeting*, Bogatá: Colombian Society of Engineers, 647–654.

POVEDA, G. and ROJAS, W., 1997: Evidences of the association between malaria outbreaks in Colombia and the El Niño–Southern Oscillation (in Spanish). *Revista Academia Colombiana de Ciencias*, **XXI** (81), 421–429.

POVEDA, G., GIL, M. M., and QUICENO, N., 1997: Impact of ENSO and NAO on the annual cycle of the Colombian hydrology (in Spanish). *Proc. International Workshop on the Climatic Impacts of ENSO at Regional and Local Scales*, ORSTOM-INHAMI, Quito, Ecuador.

POVEDA, G., GIL, M. M., and QUICENO, N., 1999: Associations between ENSO and the annual cycle of Colombia's hydro-climatology. *Proceedings 10th Symposium on Global Change*, 79th American Meteorological Society Meeting, Dallas, Texas, 1999.

PULWARTY R. S. and DIAZ, H. F., 1993: A study of the seasonal cycle and its perturbation by ENSO in the tropical Americas. *IV International Conference on Southern Hemisphere Meteorology and Oceanography*, American Meteorological Society, 262–263.

QUIÑONES, M. L., SUAREZ, M. F., and FLEMING, G. A., 1987: Estado de la susceptibilidad al DDT de los principales vectores de malaria en Colombia y su implicación epidemiológica. *Biomedica*, **7**, 81–86.

RASMUSSON, E. M., 1991: Observational aspects of ENSO cycle teleconnections. *In* Glantz, M., Katz, R. W., and Nicholls, N. (eds.), *Teleconnections Linking Worldwide Climate Anomalies. Scientific Basis and Societal Impacts*, Cambridge: Cambridge University Press, 309–343.

RASMUSSON, E. M. and CARPENTER, T. H., 1982: Variations in tropical sea surface temperature and surface wind fields associated with the Southern Oscillation. *Monthly Weather Review*, **110**, 354–384.

RASMUSSON, E. M. and MO, K., 1993: Linkages between 200 mb tropical and extratropical circulation anomalies during the 1986–1989 ENSO cycle. *Journal of Climate*, **6**, 595–616.

ROJAS, W., PEÑARANDA, F., and ECHAVARRIA, M., 1992: Strategies for malaria control in Colombia. *Parasitology Today*, **8**, 141–144.

ROPELEWSKI, C. F. and HALPERT, M. S., 1987: Global and regional scale precipitation associated with El Niño/Southern Oscillation. *Monthly Weather Review*, **115**, 1606–1626.

SALAZAR, J. E., MESA, O. J., POVEDA, G., and CARVAJAL, L. F., 1994a: Application of a continuous non-linear model to hydrological series (in Spanish). *Proceedings XVI Latin-American Hydraulics and Hydrological Meeting*, IAHR, Santiago, Chile, **4**, 169–180.

SALAZAR, J. E., MESA, O. J., POVEDA, G., and CARVAJAL, L. F., 1994b: Modeling ENSO impacts on Colombian hydrology through Regime Dependent Autoregressive Models (in Spanish). *Proceedings XVI Latin-American Hydraulics and Hydrological Meeting*, IAHR, Santiago, Chile, **4**, 181–191.

SNOW, J. W., 1976: The climate of northern South America. *In* W. Schwerdtfeger (ed.), *Climates of Central and South America*. Amsterdam: Elsevier, 295–403.

TRENBERTH, K., 1991: General characteristics of El Niño–Southern Oscillation. *In* Glantz, R. M., Katz, R., and Nicholls, N. (eds.), *Teleconnections Linking Worldwide Climate Anomalies*. Cambridge: Cambridge University Press, 13–42.

TRENBERTH, K. and HOAR, T., 1996: The 1990–1995 El Niño Southern Oscillation event: Longest on record. *Geophysical Research Letters*, **23**, 57–60.

UNNINAYAR, S. S. and SPRIGG, W., 1995: Climate and the emergence and spread of infectious diseases. *EOS*, **76** (47), 478.

van LOON, H. and MADDEN, R. A., 1981: The Southern Oscillation, I, Global associations with pressure and temperature in northern winter. *Monthly Weather Review*, **109**, 1150–1162.

VELASCO, I. and FRISCH, M., 1987: Mesoscale convective complexes in the Americas. *Journal of Geophysical Research*, **92**, 9591–9613.

WEBSTER, P. J., 1995: The annual cycle and the predictability of the tropical coupled ocean–atmosphere system, *Meteorology and Atmospheric Physics*, **56**, 33–55.

SECTION B

Long-Term Changes in ENSO: Historical, Paleoclimatic, and Theoretical Aspects

7

The Documented Historical Record of El Niño Events in Peru: An Update of the Quinn Record (Sixteenth through Nineteenth Centuries)

LUC ORTLIEB

ORSTOM, Institut Français de Recherche Scientifique pour le Développement en Coopération, Programme Paléoclimatologie et Variabilité Climatique Tropicale (UR 1), Centre Ile de France, 32 Avenue Henri-Varagnat, F-93143 Bondy-Cedex, France

Abstract

The áclassical chronology of El Niño events for the past four and a half centuries proposed by Quinn et al. (1987) was primarily based upon indications of anomalous meteorological and hydrological phenomena observed in Peru and neighboring areas, as described by various authors and anonymous sources. This sequence of reconstructed El Niño events, later improved and modified by Quinn (1992, 1993; Quinn and Neal 1992), became the major reference for proxy calibrations and for most studies on climate variability related to El Niño/Southern Oscillation (ENSO) during historical, pre-instrumental, times. Precisely because global and regional records of interannual climate variability are becoming more diversified and accurate, there is an urgent need to reevaluate and consolidate the documentary record of El Niño manifestations, particularly in southwestern South America, a key area for ENSO studies.

A preliminary revision of some of the sources used by Quinn et al. (1987) to elaborate on their record (Hocquenghem and Ortlieb 1992b) showed that some of the El Niño events were actually poorly documented and simply may not have occurred. For instance, some events had been reconstructed exclusively from evidence of Rímac River floods at Lima, while no clear relationship has been established between these floods and ENSO manifestations. Another question concerns the significance of anomalous rains in southern Peru: Do they correspond to El Niño situations, as inferred by Quinn et al., or rather to conditions associated with the opposite phase of the Southern Oscillation (La Niña)? Furthermore, a previous analysis of documentary sources on rainfall excess in central Chile during the sixteenth through nineteenth centuries (Ortlieb 1994) revealed many discrepancies with respect to the regional El Niño record of Quinn. The lack of coincidence (especially in the sixteenth and seventeenth centuries) may reflect inaccuracies in the Chilean and Peruvian records, but it may also indicate a different regime, during the Little Ice Age (LIA), of the teleconnection pattern as observed nowadays in the precipitation excess anomalies in northwestern Peru and central Chile.

This study thus focuses on the sources provided by Quinn et al. and involves a detailed critical analysis of the source reliability, the interpretations of the strength of the events, and the significance of the data with regard to the reconstruction of past El Niño events. For each event, the nature, location, and sometimes the date of the meteorological anomalies that support Quinn interpretations (information not given in the 1987 paper) are included. Additional data on historical rainfall excess (or drought) provided by recent studies are also integrated into the overview covering the 1525–1900 period. For some particular (so-called) El Niño events, the reliability of the references, some transcription problems, and internal contradictions within the sources are reviewed. A major case is made for the need for evidence of rainfall in the coastal region of northern Peru in the assessment of El Niño event reconstruction. Conversely, it is assumed that drought episodes in northern Peru should be coeval with non–El Niño situations.

With respect to the Quinn et al. (1987) and Quinn and Neal (1992) sequences, the resulting compilation of El Niño manifestations in Peru and southernmost Ecuador puts into question the occurrence of some 42 events and suggests the exclusion of 25 previously identified El Niño years. New sources support the inclusion of 7 previously unrecognized El Niño years. The new revised chronological sequence of historical Peruvian El Niños is then compared with other compiled documentary records from the western Pacific region (Whetton and Rutherfurd 1994) and with the coral reef proxy record from the Galapagos Islands (Dunbar et al. 1994). These comparisons lead to the conclusion that a more reliable, consolidated, El Niño record for the past few centuries is still needed. More precise reconstructions of the historical climatology of some key areas of South America, a better assessment of the teleconnections through time, and multiproxy studies that associate documentary records should help researchers to reach this objective.

Introduction

The Quinn Record(s) of Historical El Niño Events

Without question, the late William Quinn was a true pioneer in the study of variability of El Niño manifestations through time. After his work on historical reconstructions of Peruvian river floods and anomalous rainy events, Quinn is rightfully considered as the "father" of the past few centuries' record of the El Niño/Southern Oscillation (ENSO) phenomenon. Quinn et al. (1987; henceforth "QNA") established the strength scale of the El Niño events that has been generally adopted by the large community of scientists working on ENSO and climate variability. Quinn's list of past El Niño events recorded in the eastern Pacific during the past four and a half centuries has been viewed as the major reference for any long-term analysis of the ENSO mode. Practically all the centennial/decadal studies within the past decade that used dendroclimatology, coral reef sequences, annually layered tropical ice cores, or other proxy sequences were compared to, if not calibrated with, Quinn's El Niño chronologies (Quinn and Neal 1983a, b, 1992; QNA; Quinn 1992, 1993).

The key paper for the historical chronology of El Niño events, which included the sources of the data on which Quinn based his interpretations, was the one published in 1987 (QNA). In the early 1990s, Quinn extended his reconstructions of past El Niño (ENSO) events, both geographically and chronologically. With the purpose of strengthening the historical ENSO chronology, he began to correlate the documentary record from South America with data from India, China, and Nile floods (Quinn 1992, 1993). In this process, he was thus led to distinguish two chronological records of climatic anomalies: one considered to be of global meaning, based on all the available data from East Africa, and the Indian and Pacific Oceans, and referred to as the "ENSO chronology," and another one called the "regional El Niño chronology," which was established from eastern Pacific and western South American data. With respect to the original 1987 work, the regional El Niño chronology differed in the extension to several years of some events, a shift to the following (or preceding) year, or modifications in the evaluated strength of some events. Some moderate events that had not been qualified for the sixteenth through eighteenth centuries in QNA were included in Quinn (1992, 1993) and Quinn and Neal (1992; henceforth referred to as "Q&N") records. To obtain his latest regional El Niño chronologies, Quinn reinterpreted some data and/or revised some previous interpretations. However, he did not plainly discuss the old or new sources of information that led him to the 1992 and 1993 papers. The list of documentary sources published by QNA in 1987 was modified and completed by Q&N: These two lists constitute the basic reference for Quinn's reconstruction of El Niño events in South America.

Previous Work

After a short note on the most improbable occurrence of an El Niño event in 1531–32, during the conquest of Peru by Pizarro, Hocquenghem and Ortlieb, in 1991, took advantage of a relatively easy access in Peru to most of the original sources of information cited by QNA to critically reexamine the historical documents and references used by QNA (Hocquenghem and Ortlieb 1992a,b). They located the proper information in many of the references cited by QNA (the 1987 work did not mention page references) and considered it useful to quote the significant sentences of the relevant data that had led to the interpretation of El Niño occurrences (Hocquenghem and Ortlieb 1992b; hereafter noted as "H&O"). Naturally, in almost all the cases, the original information had been written in Spanish, as was the H&O paper. As a result of their critical analysis, H&O questioned the occurrence of some of the events and cast some doubts on the intensities of others.

Among a body of references that mainly concerned evidence from Peru, QNA included a source (Taulis 1934) that deals with the variation of annual precipitation in central Chile during the past few centuries. Later, Q&N added two other sources from Chile (Vicuña Mackenna 1877; Vidal Gormaz 1901). The inclusion of rainfall data from 30°S in the QNA record posed a problem of teleconnection within the South

American region. As was shown by Quinn and Neal (1983a,b), Deser and Wallace (1987), Aceituno (1987, 1988), and Ruttlant and Fuenzalida (1991), there is a very close relationship between the negative phase of the Southern Oscillation (warm El Niño events) and precipitation excess in central Chile. Ortlieb (1994) thus tried to consolidate the historical sequence of rainy years in central Chile through an analysis of Taulis's (1934) work, by comparing it to two other records (Vicuña Mackenna 1877; Urrutia de Hazbún and Lanza Lazcano 1993). This study showed that aside from the fact that Taulis's work was not fully reliable, there is no satisfactory correlation with the QNA and Quinn (1993) records, specifically for the sixteenth through eighteenth centuries. The lack of coincidence between El Niño manifestations in Peru and Chile could mean that the documentary records were still substantially inaccurate and incomplete, but it might also imply that during the Little Ice Age (LIA) a different teleconnection pattern may have existed between northern Peru and central Chile. This interesting conclusion regarding a possible variation of the ENSO mode during the larger scale climatic variations of the past few centuries calls for a more precise study of the historical climate variability in southwestern South America.

Among other recent studies worth mentioning on the relationship between the El Niño system and climate variability of the past centuries are several papers presented at the 1992 international symposium on "Former ENSO phenomenon in western South America: Records of El Niño events" (Hisard 1992; Huertas 1993; Macharé and Ortlieb 1993; Mabres et al. 1993). For northern Peru, a doctoral dissertation in history brought out previously unpublished material from national and regional archives, some of which is of major interest for the reconstruction of climate variability in Piura province (Schlüpmann 1994). A dissertation (Minaya 1994) was focused on the correlation between the precipitation regime in southwestern Peru and El Niño occurrences during the past forty years. Conclusions of this work were examined in relation to the QNA record (Ortlieb et al. 1995) and with regard to the link between the El Niño phenomenon and the exceptional rainfalls in the extremely arid Atacama Desert of northern Chile (Ortlieb 1995).

It is timely to synthesize the data accumulated during the past few years and to verify how they combine with the published Quinn records (QNA; Q&N; Quinn 1993). It might also be useful to recapitulate and revisit the reservations regarding the reconstruction of some El Niño events previously expressed by H&O in the light of newly available information.

A Need for Reevaluation of Quinn's Records

In the recent studies dealing with climatic variability of the past few centuries, particularly those referring to the ENSO mode, it is striking to note how the El Niño chronologies proposed by Quinn are accepted without discussion. In particular, it is seldom mentioned that QNA, Q&N, and Quinn (1993) had ranked the confidence in their reconstruction of past El Niño events. As it commonly happens in such cases, Quinn himself was more cautious with his own ENSO chronological sequence than authors who used his records to compare or calibrate their data. Actually, there has been a general tendency to lean upon Quinn's work – to consider his records as a "black box" that need not be opened and scrutinized. No one (to my knowledge) questioned the fact that QNA, Q&N, or Quinn (1992, 1993) mixed documentary data for Peru, Chile, Bolivia, and Brazil, while much remains to be understood regarding the El Niño teleconnection pattern within southwestern South America and its possible evolution in the course of the past few centuries.

Another justification for this overview is provided by the recent publication of ENSO chronologies for the Indo-Pacific region (Whetton and Rutherfurd 1994; Whetton et al. 1996; Allan et al. 1996), which also reveal some discrepancies with the QNA chronology of El Niño events and with the records of larger scale, global ENSO records of Quinn (1992, 1993) and Q&N. Some of these discrepancies should vanish through a closer look into the original sources used by QNA, Q&N, and Quinn and through a critical reevaluation of some of the criteria used over a decade ago in the reconstruction of former El Niño events.

Methodological Problems

As may be expected, the elaboration of a historical sequence of El Niño events from documentary sources is not an easy task and faces problems of various kinds. Some of these problems, as for any historical work, concern the availability and diversity of written reports and other sources, the appropriate selection of original observations, the evaluation of their reliability, the detection and elimination of distorted or spurious information, etc. Another kind of difficulty, more specific to paleo-ENSO studies, involves the link between the detectable effects of a meteorological anomaly (flood, drought, destruction) and the El Niño phenomenon. Because earthquakes and anomalous rainfall have been considered as closely associated during colonial times in Peru, it can be expected that some reports on natural disasters may have led observers to erroneously attribute building destruction to unusual meteorological conditions rather than to seismic activity or other causes.

The determination of the intensity of former ENSO events is particularly difficult to assess. This task is hampered by the extreme heterogeneity of the written sources, the variable degree of exaggeration of the chronicles, the intrinsic difficulty of quantifying an atmospheric phenomenon through its effects on the environment (which may itself have changed significantly in the course of the past centuries), and finally by the known fact that "normal weather" is not news.

Historical Data Analysis

The sources of information used and cited by QNA and Q&N consist of documents of varied origin: published books and articles, newspaper articles, review studies, and a few unpublished archives. Obviously, the role of Antunez de Mayolo, distinguished Peruvian geographer and third coauthor of the QNA chronology, was essential in the data selection and analysis of the sources. The compilation made by QNA can be considered as rather complete, as far as published material is concerned. Not many important new sources have been found since the QNA work. Any improvement of the El Niño chronology, be it for data consolidation or for inclusion of new evidence, should come from time-consuming research into unpublished (regional or national) archives in Lima and other Peruvian towns, especially in Trujillo, Lambayeque, Piura, and Tumbes. A clear example of such fruitful research is the doctoral study of Schlüpmann (1994), which dealt with agrarian socioeconomic structures in Piura (northern Peru) in the sixteenth through eighteenth centuries.

One of the fundamental criticisms that may be raised about the tables published by QNA is that the sources are presented as of equivalent value. Hocquenghem and Ortlieb (1992b) stressed that eyewitness reports and compiled works or journal articles must certainly not be placed at the same level. Historical data analysis consists of evaluating the trustworthiness of written reports. The fact that an item of information is repeated in several successive compilations cannot grant more veracity to the data per se. Actually, QNA detected and commented upon some errors in the date of one particular event that had been wrongly repeated in several documents. Another problem, also taken into consideration by QNA and Q&N, deals with the reports of authors who tended to find periodicities in the meteorological manifestations (e.g., 35-year Bruckner or 11-year sunspot cycles). However, QNA relied heavily upon some authors (Labarthe 1914; Taulis 1934) who might have been influenced by such cyclical theories and who, additionally, did not fully acknowledge the precise sources of the data that they used. The indiscriminate use of data from compilers who do not give information on their original sources may seriously weaken the value of the El Niño reconstructions.

El Niño Event Reconstruction

The reconstruction of paleometeorological situations from documentary sources is necessarily speculative. The destruction of a bridge produced by a river flood, an exceptional thunderstorm, and a single shower in the coastal desert of Peru are pieces of information that have been used to infer the occurrence of former El Niño events. In other more favorable cases, independent reports of climate anomalies or meteorological conditions from different regions of Peru (and neighboring countries) are available and provide much more satisfactory and precise criteria for the reconstruction of El Niño (or La Niña) conditions. As can be easily understood, the strongest former events are those which are most likely to have been commented upon as catastrophic phenomena. Of course, this is why the QNA record dealt only with the strong and very strong events of the sixteenth, seventeenth, and eighteenth centuries. Moderate and weak El Niño events had been identified, by QNA, only for the nineteenth and twentieth centuries, when more written data were at hand. It was only when Quinn (1992, 1993) incorporated the Indian drought sequence and the Nile River flood record, produced on an annual basis, that it became possible to evaluate the strengths of moderate events for the previous centuries. Quinn and Neal (1992) thus provided a series of new sources to strengthen the original QNA sequence of El Niño occurrences in South America. The additional evidence then provided by Q&N included more data on anomalies in Chile, northeastern Brazil, and Bolivia.

Paleo–El Niño studies are seriously hampered by the fact that no two events are alike. Recent El Niño events present large variability in terms of intensity, location, and season of occurrence (Philander 1991). Events of similar strength may show large variations in their impacts. Huertas (1987, 1993) stressed that during very strong events, the maximum effects of the El Niño phenomenon could be located either in the Trujillo area (as in 1578 or 1728) or in the Piura region (as in 1983). This spatial variability of the effects constitutes another obstacle for the assessment of the strength of former events.

The El Niño phenomenon, as defined a century ago (Carranza 1891; Carrillo 1893; Eguiguren 1894), is characterized by anomalous rains in the coastal desert of northern Peru. Based on twentieth-century observations, it can be added that these exceptional rains in the arid coastal region of Peru do not normally extend southward to the latitude of Lima. During the most recent strong or very strong events, particularly in 1982–83, the coastal area of southern Peru, as well as the cordilleran region of southeastern Peru and Bolivia, suffered from severe droughts. The nature of El Niño impacts on southern Peru constitutes one of the major problems raised by H&O with respect to a series of historical events identified by QNA on the basis of flood evidence. These cases will be discussed below.

Confidence Rating

The "confidence rating" attributed by QNA to every event exemplifies the difficulty of reaching consistency within the sequence of reconstructed occurrences. On a theoretical basis, and as expressed by QNA (and Q&N), such a rating is determined by the number of different sources that lead to the interpretation of a former El Niño event. The validity of such a confidence rating in the cases where the sources are not independent of each other, or where the sources have not been previously submitted to a critical evaluation, was already discussed. As was stressed by H&O, as well as by Ortlieb (1994) regarding the Chilean record, there is an intrinsic difference in the quality of an original source or a contemporaneous witness report, on one hand, and newspaper articles or compiled studies written two or three centuries after the fact, on the other hand. Quinn and Neal (1992), who referred to this aspect of happenstance, were aware of the problem and actually revised the confidence ratings for many events listed by QNA, but they did not qualifiy their sources accordingly. As no information was given by QNA and Q&N regarding the nature of the evidence leading to the reconstruction of every El Niño event (only references are listed), the reader must rely heavily upon the indicated confidence rating. In some cases, major discrepancies with respect to the values expressed by QNA and Q&N may be justified.

Quinn et al. (1987) explained (p. 14,454) that their published record did not include events with confidence ratings of 1 (meaning a single source) because they had required at least one confirmation of any single piece of information. If this requirement were strictly applied, and if only independent sources were selected, this criterion would be too drastic: Many reconstructed events of the sixteenth and seventeenth centuries would be excluded from the present records! A more appropriate solution would seem to be to perform a stricter data analysis, to rely more on reliable informants, and to minimize the importance of all references that constitute only repetitions of previously published data.

A Contribution to the Revision of the Quinn Record for Peru

Presentation of the Revised Peruvian Record (Sixteenth through Nineteenth Centuries)

Table 7.1 presents an analysis of the main data upon which the Quinn records (QNA; Q&N; and Quinn 1992, 1993) were based, along with the revised chronological sequence that I propose. Unlike the tables published in 1987, Table 7.1 includes information on the location (Fig. 7.1) and time of year of the phenomena or anomalies that led to the interpretation of every event. It eventually includes some comments regarding the accuracy, relevance, and reliability of the sources. The table also includes the

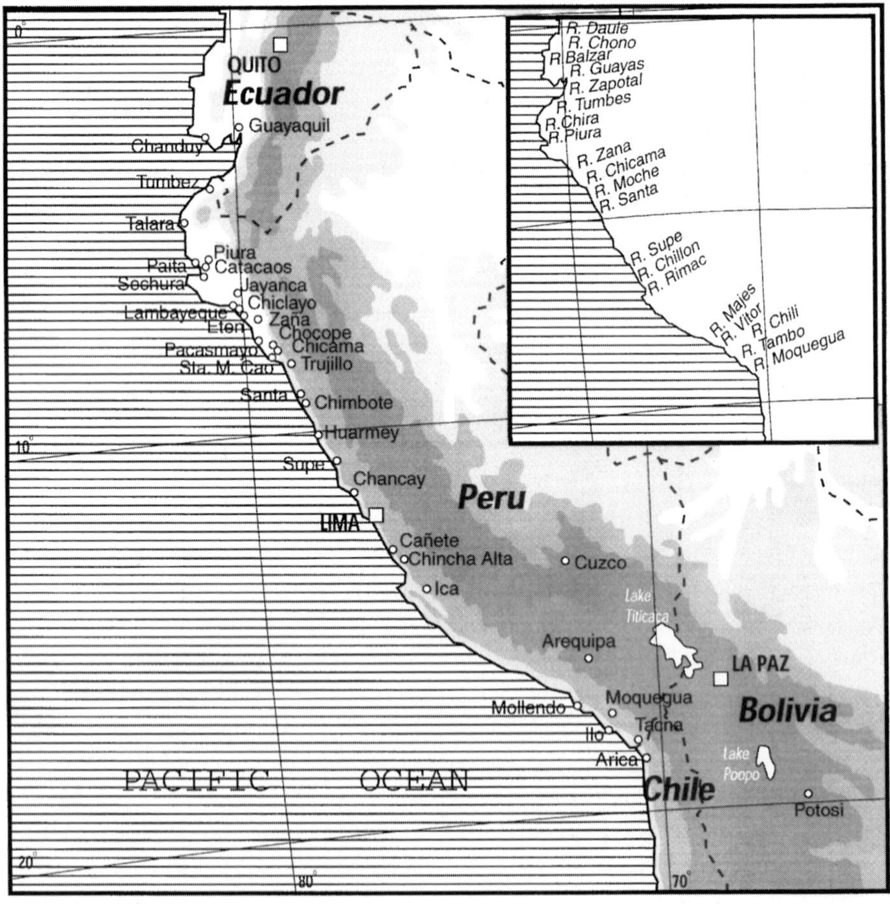

Fig. 7.1 Map of the Peruvian coastal region, with most of the localities mentioned in the text and Table 7.1.

Table 7.1 *Compilation of main available historical documentary data from Peru on which can be based reconstructions of El Niño anomalies between 1525 and 1900. Indicated documentary sources are those used by QNA and/or Q&N to which were added new references (in italics and shaded areas), partly taken from H&O. Not all the sources referred to by QNA and Q&N are indicated: Some were eliminated because they were mere repetitions of original information (e.g., Portocarrero 1926), and others were not included because they deal specifically with central Chile anomalies (Taulis 1934, Vicuña Mackenna 1877, Vidal Gormaz 1901) or with northeast Brazil droughts (Andrade 1948, Brooks 1971). The last column summarizes the proposed updated interpretations as to the occurrences and strengths of El Niño events (lack of event occurrence and new event occurrences are underlined). The sign § (fifth column) designates the reproduction of quotes of original information in H&O. EN = El Niño. Strength of events: VS = very strong, S = strong, M = moderate, W = weak.*

Years	Event intensity in QNA (*Q&N)	Confidence rating in QNA (*Q&N)	Major original sources in QNA and (*) in Q&N	Precise location of relevant quote (§: in H&O)	Location of climatic/ oceanographic anomaly	Phenomenon/effects leading to the reconstruction of EN event	Remarks	Proposed interpretation
1525–1526	S	3	Xerez 1534	QNA, pp.197–198 §	Eastern equatorial Pacific	Thunderstorms and heavy squalls off Colombia and Ecuador	Insufficient data to assess EN conditions	<u>No EN</u> ?
1531–1532	S	4	Xerez 1534	QNA, p. 200 §	Eastern Pacific	Sailing time (only 13, or 7?, days) from Panama to Ecuador in a 1531 trip . . .	. . . but other route than in 1525 (H&O)	No EN
			Prescott 1892	p. 175 (in Spanish ed.) §	Piura, N Peru	"Flooded" rivers in N Peru (actually perennial)	Unreliable source (H&O)	
			*Murphy 1926				Ref. not seen	
1539–1540	M/S	3	Montesinos 1642	See QNA: 14454	Cuzco, SE Peru	Death of 30,000 Indians due to drought (?) in 1539	Real cause of those deaths?	<u>No EN</u> ?
					Cuzco, SE Peru	Storm and hail in South Peru Andes in 1540	Not a clear EN signal	

Table 7.1 (*cont.*)

Years	Event intensity in QNA (*Q&N)	Confidence rating in QNA (*Q&N)	Major original sources in QNA and (*) in Q&N	Precise location of relevant quote (§: in H&O)	Location of climatic/oceanographic anomaly	Phenomenon/effects leading to the reconstruction of EN event	Remarks	Proposed interpretation
1541			Cobo 1653	(1): 90 §	Lima	Rain and flood in 1541	Rainfall in Lima	
			QNA	p. 14454	Lima (?)	Red tide ("Aguaje") on 12 July 1540	Which original source?	
			*Raimondi 1876				Ref. not seen	
*1544	Not in QNA		*Albenino 1549				Ref. not seen	
	*M+	*4	*Montesinos 1642	(1): 140–158		Data not found in ref.		?
*1546–	Not in QNA	—	*Benzoni ("1565") 1572	p. 57	Guayaquil, S Ecuador	Rio Chiono flood and reconstruction of Guayaquil (?)	Possible EN conditions	M?
1547	*S	*4	*Albenino 1549				Ref. not seen	
			*Raimondi 1876				Ref. not seen	
			Moreno 1804, in Palma 1894, and in Unanue 1806	p. 1151 §	Lima	Two lightning bolts and a single thunderstroke (no rain) in Lima on 13 July 1552	Very poor (and only) evidence for El Niño manifestation!	
				p. 38 §				
1552	S	4	*Humboldt 1804	p. 11	Lima	Same source as Moreno, Unanue, and Palma		No EN

			Actas Cabildo Trujillo 1549–60 (in H&O)	*p. 213*	*Trujillo, N Peru*	*No mention of any rain between 1549 and 1560; see also Lizárraga 1609*	*No EN conditions in N Peru?*	
*1558–	Not in QNA		*Montesinos 1642	p. 158	Central Peru	Epidemic diseases	EN conditions?	
*1559– *1560– *1561	*M/S	*3	*Martinez y Vela 1702 (=Arzans de Orsúa y Vela 1965)	(1): 115	Potosí, Bolivia	Drought from October 1560 to January 1561 (see Table 7.2)	No clear relation with EN conditions (see text)	No EN ?
			*Garcia Rosell 1903			Data not found in ref.		
*1565	Not in QNA *M+	*2	*Montesinos 1642	(2):18	Ayacucho, central Peru	Famine in Huamanga (no explicit reason given)	Single source and poor evidence!	No EN ?
1567–			Oliva 1631	See QNA, p. 14453	Eastern Pacific	Panama-Lima trip in 26 days (March 1568) . . .	. . . instead of 6 (?) months	
1568	S+	5	Cobo 1639, as cited by Labarthe 1914	p. 307	Lima	Destruction of a bridge: See text (erroneous quote!)	Misreading: "1567" (=1607)	No EN ?
			*Montesinos 1642			Data not found in ref.		
1574	S	4	García Rosell 1903	(3): 334 §	Piura, N Peru	Strong rains which led to emigration of Piura population toward Paita	Single source	M?
			Acosta 1590	p. 82 §	Trujillo, N Peru	Heavy rains and floods		
			Cobo 1639	p. 311 §	Lima	Rímac River flood		
			Cobo 1653	(1): 90 §	Trujillo, N Peru	Second record of rainy episode in N Peru (after 1541)	The first well-documented (very) strong EN event	

Table 7.1 (*cont.*)

Years	Event intensity in QNA (*Q&N)	Confidence rating in QNA (*Q&N)	Major original sources in QNA and (*) in Q&N	Precise location of relevant quote (§: in H&O)	Location of climatic/ oceanographic anomaly	Phenomenon/effects leading to the reconstruction of EN event	Remarks	Proposed interpretation
1578	VS	5	García Rosell 1903	(3): 334 §	Piura, N Peru	Heavy rains		VS
			*1580 anon. ms. in Brüning 1922–23	pp. 13, 119, 180	N Peru coast	Very strong rains in February–March with much destruction and food shortage		
			*Huertas 1984			(See Huertas 1987 below)		
			*Cabello Valboa 1586	pp. 223–224 §	N Peru coast	Rainfall, weakening of trade winds, and strong northerly winds	Print mistake: "1576" = 1578 (see H&O)	
			Lizárraga 1603–1609	*(17): 14–15 §*	*Chicama-Trujillo, N Peru*	*Very strong rainfalls, "never seen before"*		
			Anon. ms. published by Huertas 1987	*pp. 39–40 §*	*Lambayeque, Trujillo (N Peru)*	*Rains and floods (24 Feb – 6 Apr.)*	*Detailed eyewitness report*	
			Rostworowski, in Peralta 1985	*pp. 122–124*	*Lambayeque and N of Peru*	*Compilation of socio-economical impacts*		
*1582	Not in QNA *M	*3	*Montesinos 1642	(2): 86	Ayacucho, central Peru	Drought in the Andes	EN conditions?	?
*1585	Not in QNA *M+	*2	*Montesinos 1642			Data not found in ref.		?

*1589–	1589 & 1590: Not in QNA		*Montesinos 1642	(2): 111	Cuzco, central Peru	Epidemic diseases in 1590	Relation with EN conditions?	
*1590–	*M/S	*3	*Barriga 1951	p. 47	Arequipa, S Peru	Only data found refers to lack of rain in December 1589	No (?) evidence for EN conditions	No EN ?
1591–1592	S 1592: Not in Q&N	2	Martinez y Vela 1702 (=Arzans de Orsúa y Vela 1965)	(1): 217–218	Potosí, Bolivia	Drought in Potosí in late 1591–early 1592 (see Table 7.2)	No clear relation with EN conditions (see text)	
1593	*Not in QNA and Q&N*	—	*Lizárraga 1603–09*	*pp. 14–15 §*	*Trujillo, N Peru*	*Heavy rainstorm, but less strong than in 1578*	Single (but reliable) source	M?
	Not in QNA		*Montesinos 1642	(2): 130–131	Central S Peru	Cauca and Magdalena River floods; heavy rains	EN conditions?	
*1596	*M+	*3	*Ocaña and Alvarez 1969*	*p. 38 §*	*Paita, N Peru*	*Heavy rainfall in Paita (destruction) and floods*	Possibly strong EN conditions	S
*1600	Not in QNA *S	*3	*Barriga 1951		Arequipa, S Peru	Data not found in ref.	Confusion with volcanic effects?	No EN
*1604	Not in QNA *M+	*3	*Montesinos 1642	(2): 168–169	Huamanga, central S Peru	Miraculous rainfall under a blue sky (after a drought)	Untrustworthy data	No EN
			Cobo 1639	(1): 313 §	Lima	Rímac River flood, and bridge destruction in February 1607	(See 1567 and 1671)	
1607*–1608	S	5	Alcedo y Herrera 1740	pp. 122–123	Eastern Pacific	Unusually (?) easy travel between Panama and Lima in December 1607	NE winds, possibly En related	M?

Table 7.1 (*cont.*)

Years	Event intensity in QNA (*Q&N)	Confidence rating in QNA (*Q&N)	Major original sources in QNA and (*) in Q&N	Precise location of relevant quote (§: in H&O)	Location of climatic/ oceanographic anomaly	Phenomenon/effects leading to the reconstruction of EN event	Remarks	Proposed interpretation
	1608: Not in QNA		*Martinez y Vela 1702 (= Arzans de Orsúa y Vela 1965)	(1): 265	Potosí, Bolivia	Snowfalls and rains in late 1607 (see Table 7.2)	Not clear manifestation of EN	
1614	S	5	Cobo 1653	(1): 90 §	Chancay, central Peru	Single (and local?) rainfall event in March	Poor evidence for EN	NO EN ?
			*Haenke 1799			Data not found in ref.		
1618–1619			Vásquez de Espinoza 1629		S Peru	Data not found in ref.		
1618–1619	S	4	Cobo 1653	(1): 90 §	Ilo, S Peru	Lightning, thunderstorm, and rain on 12 June 1619	Winter rainstorm in 1619	M?
			Anon. ms. cited by Huertas 1992	*p. 105 §*	*Zaña. N Peru*	*Strong rainfall, but casualties possibly more related to 1619 earthquake*	No precise EN evidence (single rainfall?)	
1622	*Not in QNA and Q&N*	—	*H. Brüning, cited by Gorbitz 1978*	*p. 36*	*Jayanca, N Peru*	*Rains and floods in 1622 "and previous years"*	Single source	M?
1624	S+	4	Cobo 1653	(1): 90 §	Zaña and Trujillo, N Peru	"Copious" rainfall and floods	Possibly EN conditions	M?
		*5	*Montesinos 1642	(2): 228	Central Andes	Drought between Cajamarca and Huamachuco!	EN conditions?	
			Puente 1885	p. 38	Lima	Rímac River flood	Not sufficient	1634:

1634–			Palma 1894	p. 42 §	Lima	Rímac River flood	evidence	Not in Q93
*1635	S	4	*Montesinos 1642	(2): 249		Data not found in ref.		No EN
*1634– *1635*	*1635: Not in QNA*	*3	*Suardo 1634*	*(2): 13–15 §*	*Lima and coast of S Peru*	*Rain in Lima; floods in S Peru (February–March 1634)*	EN (or La Niña) evidence?	?
*1640–1641	Not in QNA S*	*2	*Martinez y Vela 1702 (=Arzans de Orsúa y Vela 1965)		Potosí, Bolivia	Data not found in ref.	No source from Peru	No EN
*1647	Not in QNA *M+	*3				Only data on central Chile	No source from Peru	No EN
1652	S+	4	Cobo 1653	(1): 90 §	Lima	Single rainfall in February	Very poor EN evidence	No EN ?
	Not in QNA		*Alcedo y Herrera 1740	p. 164	Eastern Pacific	Easy trip from Colombia to Lima in early 1655	Data reliability?	No EN
*1655	*M	*3	*Actas Cabildo Guayaquil 1650–57*	*p. 154*	*Guayaquil, S Ecuador*	*Drought in 1654–55*	La Niña conditions?	
1660	S	3	Anon, ms. cited by Labarthe 1914	p. 309	Supe, central Peru	Supe River flood	Insufficient data (and from central Peru)	Not in Q93 No EN
1671	S	3	Anon. ms. cited by Labarthe 1914	p. 309	Supe and Lima, central Peru	Rímac and Supe River floods	Reliable data? (see text)	No EN
			*Martinez y Vela 1702 (=Arzans de Orsúa y Vela 1965)	(2): 259	Potosí, Bolivia	Drought in October–December 1671	No clear manifestation of EN	?
1678	*Not in QNA and Q&N*	—	*Gorbitz 1978*	*p. 30*	*Jayanca, N Peru*	*Complete destruction of Jayanca Vieja*	Single source (to confirm)	M?

Table 7.1 (*cont.*)

Years	Event intensity in QNA (*Q&N)	Confidence rating in QNA (*Q&N)	Major original sources in QNA and (*) in Q&N	Precise location of relevant quote (§: in H&O)	Location of climatic/ oceanographic anomaly	Phenomenon/effects leading to the reconstruction of EN event	Remarks	Proposed interpretation
1681	S	3	Rocha 1681	(2): 168–169	Lima	Two thunderstrokes E of Lima (no rain) on 3 July 1680 (not 1681!)	Phenomenon probably not related to EN	No EN
*1684	Not in QNA *M+	*2	*Martinez y Vela 1702 (=Arzans de Orsúa y Vela 1965)	(2): 316	Potosí, Bolivia	Drought in late 1683–early 1684	No clear manifestation of EN	No EN ?
1686	*Not in QNA and Q&N*	—	*1780 anon. ms. cited by Schlüpmann 1988*	*p. 40 §*	*Yapatera (E Piura), N Peru*	*Abundant rains that caused destruction of the hacienda Yapatera*	Rains possibly EN related	1686– 1687–1688: M?
1687–1688	S+	4	Juan and Ulloa 1748	(2, 1): 20 §	Zaña, N Peru	Erroneous mention of the destruction of Zaña (see 1720)	Reference suppressed in Q&N	
			Unanue 1806			Data not found in ref. (quote of Juan and Ulloa?)	Repetition of Juan and Ulloa	
			Melo 1913	p. 152	Zaña, N Peru	Reproduction of Juan and Ulloa misinterpretation	Unreliable source!	
			Remy 1931, in Petersen 1935	(2): 36 §	Lima	Single rainstorm (2 Dec. 1687) with destruction	Evidence for EN conditions?	

*1692– *1693	Not in QNA *S	*3	*Martinez y Vela 1702 (=Arzans de Orsúa y Vela 1965)	(2): 368, 393	Potosí, Bolivia	Drought in January–March 1693 (+1694 and 1695!)	Not reliable data from Bolivia	No EN
1696–	S	3	Palma 1894	p. 42 §	Lima	Rímac River flood on 11 Feb. 1696	Poor evidence for EN conditions	1696: Not in Q93
*1697	*M+ 1697: Not in QNA		*Actas Cabildo Guayaquil, in Estrada Ycaza 1977*	*pp. 111–112*	*Guayaquil, S Ecuador*	*Guayas River flood in May 1696*	Possibly EN conditions in S Ecuador	1696: M? 1697: No EN
1701	S+	4 *5	Feijoo de Sosa 1763	1: 158 §	Trujillo, N Peru	Abundant rainfalls	Concordant sources for EN conditions in N-central Peru, but no information for Piura	S
			Bueno 1763	P. 50 §	Trujillo, N Peru	Anomalous rainstorms in 1701 (as in 1720 and 1728)		
			Haenke 1799	p. 234	Trujillo, N Peru	Unusually strong rains		
			Anon. ms., cited by Labarthe 1914	p. 309	Zaña, N Peru	Zaña River flood and impacts on crops		
			*Humboldt 1804	p. 12	Lima and N Peru coast	Strong rainfalls (secondhand information taken from Bueno 1763)		
			*Unanue 1806	pp. 38–39 §	Trujillo, N Peru			
			1706 ms. cited by Huertas 1992	*pp. 105–106*	*Zaña, N Peru*	*Strong rains with severe destruction of crops*		
			Cook 1712				Ref. not seen	

Table 7.1 (*cont.*)

Years	Event intensity in QNA (*Q&N)	Confidence rating in QNA (*Q&N)	Major original sources in QNA and (*) in Q&N	Precise location of relevant quote (§: in H&O)	Location of climatic/ oceanographic anomaly	Phenomenon/effects leading to the reconstruction of EN event	Remarks	Proposed interpretation
1707– 1708–	S	3	Alcedo y Herrera 1740	pp. 228–230	Eastern Pacific	Easy navigation between Panama and Lima, in June–July 1707	Reliable EN oceanographic conditions?	No EN
1709	Not in QNA *M/S		*Schlüpmann 1994*	*p. 61*	*Piura*	*Drought in Piura in 1706–15*	Well-documented study	
1714–	1714: Not in Q&N		Le Gentil 1728	p. 88	Cañete, S Lima	Flood of Cañete River in September 1715	Not clear EN indication	1714–1715:
1715–	S	4	Odriozola ms. cited by Labarthe 1914	p. 309	Arequipa, S Peru	Chili River flood in 1714, but “no data from other Peruvian rivers”	EN or La Niña evidence?	No EN 1716:
*1716	1716: Not in QNA	*3	*Schlüpmann 1994*	*p. 61*	*Piura*	*Drought in Piura in 1706–15*	1716: normal (or EN) year?	No EN?
	Not in QNA		*Bueno 1763			Data not found in ref.		
			*Barriga 1951			Data not found in ref.		
*1718	*M+	*3	*1718 ms. cited by Schlüpmann 1994*	*p. 61*	*Piura*	*Flood of Piura River that caused destruction*	Sierra rains, or EN conditions?	M?
1720			Shelvocke 1726	p. 103 §	Paita, N Peru	“Wet rainy weather” in Paita (March)		

1720	S+ *VS	4 *5	Feijoo de Sosa 1763	1 (12): 158–161 §	Zaña and Lambayeque, N Peru	Consistent rains (weaker rains than in 1728) which led to Zaña destruction	Manifestations of a strong EN event in N Peru	S
			Bueno 1763	pp. 50, 53 §	Trujillo, Zaña, N Peru	Heavy rains and flooding; destruction of Zaña on 15 March 1720	(Not as strong as 1728, but stronger than 1701?)	
			Haenke 1790	pp. 234, 245	Trujillo, Zaña, Lambayeque, N Peru	Unusual heavy rains, stronger than in 1701; alternating NE and S winds		
			Alcedo 1786–89			Data not found in ref.		
			Moreno 1804, cited by Palma 1894	p. 1151 §	Lima	Second thunderstorm noted after 1552 in Lima	Only source on Lima	
			Bachmann 1921	p. 14	Zaña, N Peru	Zaña destruction in 1720	Secondhand (compiled) data	
			*Humboldt 1804	pp. 11, 12	Coast of N Peru	Destructive rainfalls in January 1720, with thunder		
			*Unanue 1806	pp. 29–30 §	Coast of Peru	Quote of Bueno 1763		
			*Raimondi 1876			Zaña destruction in 1720?		
			*Adams 1905	p. 97	Zaña, N Peru	Zaña destruction in 1720		
			*Huertas 1984 (1987)	p. 16	Zaña, N Peru	Zaña destruction in 1720		

Table 7.1 (*cont.*)

Years	Event intensity in QNA (*Q&N)	Confidence rating in QNA (*Q&N)	Major original sources in QNA and (*) in Q&N	Precise location of relevant quote (§: in H&O)	Location of climatic/ oceanographic anomaly	Phenomenon/effects leading to the reconstruction of EN event	Remarks	Proposed interpretation
			Rubiños y Andrade 1782, in H&O	*pp. 228–230 §*	*Zaña, N Peru*	*Zaña River flood, and rains during 2 weeks in March*	Detailed report by eyewitness	
*1723	Not in QNA *M+	*3				Only data concerning central Chile and Brazil		<u>No EN</u>
1728	VS	5	Feijoo de Sosa 1763	1 (12): 158–160 §	Trujillo, N Peru	In February–March 1728 stronger rains than in 1720 (not 1726)	Feijoo de Sosa data more precise than Bueno (see H&O)	VS
			Bueno 1763	p. 50	Trujillo, N Peru	"Copious rains," less strong than in 1720		
			Alcedo 1786–89	(3): 344 § (4):16, 490	Paita and Zaña, N Peru	Zaña ruined (see 1720!); damage in Paita	Very strong EN conditions in northern Peru region	
			Spruce 1864			Data not found in ref.		
			Eguiguren 1894	p. 247 §	Piura	Rainfall and river flood at Piura (Távara 1854 data)		
			*Anson 1748	p. 178 §	Paita; N Peru	Great destruction in Paita		
			*Humboldt 1804	p. 12	Coast of Peru	Cites Feijoo de Sosa data	Secondhand (compiled) data	

1728			*Unanue 1806	pp. 29–30	Coast of Peru	Quote of Bueno 1763		
			*Palma 1894			Data not found in ref.		
			*Garcia Rosell 1903	p. 427 §	Paita and Piura	Destruction due to 1728 rains		
			Esquivel y Navia 1746, in Huertas 1993	*p. 366*	*Cuzco, SE Peru*	*Strong rainfall in Cuzco*	Rains in SE Peru Andes	
			Juan and Ulloa 1748	*p. 22 §*	*Chocope, Chicama, N Peru*	*Rainfall during 40 nights in "1726" (in fact 1728)*	Examples of misquotation of the 1728 event	
			Stevenson 1825	*(2): 177–178 §*	*Chocope, N Peru*	*Rains during 34 nights in "1746" (= 1728: H&O)*		
			1752 anon. ms. cited by Schlupmann 1994	*C 2330 §(see maps)*	*Catacaos and Sechura, N Peru*	*Transformation of Piura valley, after flooding*	Eyewitness descriptions of manifestations of a very strong EN event in N Peru	
			1778 and 1809 ms., in Cruz Villegas 1982	*pp. 138–140 §*	*Piura and Catacaos, N Peru*	*Destruction of the banks of Piura River, after flooding*		
			Lequanda 1793	*pp. 168–169 §*	*Piura*	*Destruction in the city and effects of flooding*		
*1736	Not in QNA *S	*2				Only data from NE Brazil		Not in Q93 No EN
*1740	Not in QNA *M	*3	*Juan and Ulloa 1748	1(1): 20	Lambayeque, N Peru	Lambayeque River limited flood in November 1740	Not an indication of EN conditions	Not in Q93 No EN

Table 7.1 (*cont.*)

Years	Event intensity in QNA (*Q&N)	Confidence rating in QNA (*Q&N)	Major original sources in QNA and (*) in Q&N	Precise location of relevant quote (§: in H&O)	Location of climatic/ oceanographic anomaly	Phenomenon/effects leading to the reconstruction of EN event	Remarks	Proposed interpretation
*1744	Not in QNA *M+	*3				Only data from NE Brazil		<u>No EN</u>
1747	S *S+	5	Llano y Zapata 1748	pp. 2–3	Moquegua and Abancay, S Peru; Lima	1747: heavy rainstorms and destruction (March) in S Peru; rainfall in Lima on 1 July	Rains in S and central Peru	1747–1748: S
			Feijoo de Sosa 1763	(1): 163 §	Trujillo, N Peru	Two rainstorms in a single day in Trujillo	Only available data on N Peru	
			Moreno 1804, in Unanue 1806	p. 38 §	Lima	Thunderstorm in Lima, like in 1552, 1720, and 1803	Thunderstorm in Lima as evidence for EN conditions?	
			Palma 1894	p. 1151 §	Lima	Same source (Moreno 1804) as Unanue		
			*Humboldt 1804	p. 11	Lima			
1748	*Not in QNA and Q&N*	—	*Anon. ms. cited by Schlüpmann 1994*	*pp. 62, 241 §*	*Sancor (Piura), N Peru*	*Heavy rainfalls and floods E of Piura in 1748 (47?)*	Unconfirmed data (1748?)	
			Stevenson 1825	*(2): 178 §*	*Chocope, N Peru*	*Local (?) exceptional rains during 11 nights*	Unreliable source? (H&O)	
*1750	M *M+	— *4	QNA	p. 14455	?	No source given by QNA	No source	<u>No EN</u> Not in Q93
			*Cerdán, cited by Puente 1885	p. 43	Lima area	Rímac River flood	No En conditions?	

*1755– *1756	Not in QNA *M	*3	*Garcia Rodriguez 1779				Ref. not seen	<u>No EN</u>
			*Garcia Rosell 1903	p. 453	Paita, Piura	Smallpox epidemic	EN conditions?	
1761	S	5	Bueno 1763	p. 39 §	Santa, N-central Peru	Several mentions of the same story: a flood of Santa River that produced severe damage in the small town of Santa	No data on the origin of the flood; possibly not related to En conditions	M?
			Alcedo 1786–89	(4)				
			Haenke 1790 Ruschenberger 1835	p. 185 (2): 309				
			*Garcia Rodriguez 1779				Ref. not seen	
			Cicala 1994	*p. 547*	*Piura*	*Strong rainfalls in 1761 (or 1760?)*	EN conditions?	
*1764	Not in QNA *M	*2				Only data from central Chile		<u>No EN</u>
*1768	Not in QNA *M	*2	*Garcia Rodriguez 1779				Ref. not seen	?
1775	S	4 *3	*Cerdan, cited by Puente 1885	p. 43	Lima area	Rímac River flood (same source)	EN conditions?	<u>No EN</u>
			Cerdan, cited by Labarthe 1914	p. 311	Lima area			
			Estrada Ycaza 1977	*p. 112*	*S Ecuador*	*Drought in 1775–80*	No EN conditions	
			Schlüpmann 1994	*p. 62*	*N Peru*	*Drought in N Peru between 1766 and 1776*	No strong EN in the decade	Not in Q93

Table 7.1 (*cont.*)

Years	Event intensity in QNA (*Q&N)	Confidence rating in QNA (*Q&N)	Major original sources in QNA and (*) in Q&N	Precise location of relevant quote (§: in H&O)	Location of climatic/ oceanographic anomaly	Phenomenon/effects leading to the reconstruction of EN event	Remarks	Proposed interpretation
1778–	M?	—	QNA	p. 14455		No source given by QNA		No EN ?
			*Garcia Rodriguez 1779				Ref. not seen	
1779	*M+	*4	*Cerdan, cited by Puente 1885	p. 43	Lima area	Rímac River flood in 1779		1779: Not in Q93
1783	M?	—	QNA	p. 14455		No source given by QNA		No EN ?
			Q&N	p. 630		Only data from central Chile and NE Brazil		
	*S	*3	*Ms. cited by Huertas 1993*	*p. 366*	*Ica, central Peru*	*Drought in 1783–84–85 in S-central Peru*	Data from S Peru difficult to interpret (EN conditions?)	
			1785 ms., in Galdós 1988, in Huertas 1993	*p. 365*	*Arica, N Chile*	*Drought in 1783–85 at Azapa (Arica)*		
1784	*Not in QNA and Q&N*	—	*Castillo 1931 (p. 219), cited by Hamerly 1973*	*pp. 68, 105*	*S Ecuador*	*Daule River flood; destruction of banana and tobacco fields in 1784–85*	EN conditions in Ecuador in December 1784–January 1885	1784–1785:
1785–	1785: Not in Q&N	4	Cerdan, cited by Labarthe 1914	pp. 311–312	Lima	Rímac River flood on 26 Feb. 1786	(No?) EN conditions in 1786?	M

1786	S *M+	*3	Estrada Ycaza 1977	p. 122	S Ecuador	Strong rainfalls and Daule and Balzar River floods in 1785	1784–85: EN conditions	1786: <u>No EN</u>
1791	VS	5	Unanue 1806	pp. 29–30 §	Peru coast	Abundant summer rains, like in 1701, 1720, 1728	No precise data	S
			Ruschenberger "1834" (=1835)	(2): pp. 354–355	Lambayeque, N Peru	Great destruction and loss of lives in March 1791, because of snowmelt	No evidence for rainfall in Lambayeque; only snowmelt in the Andes	
			Hutchinson 1873	p. 211	Lambayeque, N Peru	Great destruction by floods from the sierra		
			Spruce 1864	p. 29 §	Piura, N coast	Major flood of Piura River, as remembered in 1864	No rain in Piura but upstream rainfall led to the catastrophic flood	
			Távara 1854, in Eguiguren 1894	p. 247 §	Piura	Violent Piura River flood and bridge destruction in summer 1790–91 (upstream rains)		
			Diario de Lima, in Labarthe 1914	p. 312	Lima area	On 13 Feb., rainstorm in Caraballo valley	Single event?	
			Bachmann 1921			Data not found in ref.		
			*Garcia Rosell 1903	(4): 461	Piura	Piura River flood story		
			*Adams 1905	p. 97	Lambayeque	Lambayeque River flood		

Table 7.1 (*cont.*)

Years	Event intensity in QNA (*Q&N)	Confidence rating in QNA (*Q&N)	Major original sources in QNA and (*) in Q&N	Precise location of relevant quote (§: in H&O)	Location of climatic/ oceanographic anomaly	Phenomenon/effects leading to the reconstruction of EN event	Remarks	Proposed interpretation
			*Leguia y Martinez 1914	p. 182	Piura	Comparison (?) with the exceptional rainfall of 1891	Inaccurate comparison!	
			Garcia Rosell 1904	*p. 102*	*Piura*	*Last rains before 1804*		
			Mercurio Peruano 1791 (7Aug.) [1964]	*(2): 253 §*	*Lambayeque, N Peru*	*Destructive floods on 1–3 March 1791*	Snowmelt in the Andes?	
1803–1804	S+	5	Moreno 1804, in Palma 1894	p. 1150 §	Lima	Rainstorm and high air temperature in 1803	Warm air temperature in central Peru	S
			Unanue 1806	pp. 34, 36–38, 39 §	Lima and central Peru	Warm temperature in January and February 1803 and early 1804; rainfall on 1 April 1803		
			Stevenson 1829			Data not found in ref.		
			Spruce 1864	p. 29 §	Piura	In 1804, first rains after 1791	No strong EN conditions?	
			Eguiguren 1894	pp. 250–251	Piura	1803: weak rainfall 1804: abundant rainfalls	EN conditions in central and N Peru in summer 1803–04	
			Labarthe 1914	p. 313	Chiclayo, N Peru; Lima	Floods in February 1804; for Lima, cites Unanue 1806		

			Petersen 1935	(2): 4, 35–36	Lima	Quotes of Moreno 1804 and Palma 1894		
			*Humboldt 1804	p. 11	Lima	Thunderstroke in Lima on 19 April 1803		
			*Lastres 1937			No full ref. given by Q&N	Ref. not seen	
			Humboldt 1804	*p. 22*	*Callao (Lima)*	*Positive (3°C) SST anomaly in January 1803*	Oceanographic EN conditions	
			Garcia Rosell 1904	*pp. 102–103*	*Piura*	*Abundant rains in February 1804*	EN conditions in N Peru	
			1804 ms. cited by Hamerly 1973	*p. 68*	*Guayaquil, S Ecuador*	*Heavy rainfall and floods for 7 months in 1803–04*	EN conditions in S Ecuador	
1806–1807	M *S+	3 *5	Unanue "1815" (=1806)		Lima?	Data not found in ref. (written in 1806!)		No EN
			Stevenson 1829			Data not found in ref.		
			Remy 1931 in Petersen 1935	(2): 36	Lima	Single rainfall (15 Dec. 1806)	Poor evidence for EN	
1806–1807			*Eguiguren 1894*	*pp. 250–251 §*	*Piura*	*Drought in 1806 and 1807*	No EN conditions in northern Peru	
			Schlüpmann 1994	*pp. 63–64*	*Piura*	*1805–14: drought in Piura, affecting cattle*		
			Palma 1894			Date not found in ref.		
	M	4	González 1913		Lima	Data (which begin in 1832) not found in ref.		

Table 7.1 (*cont.*)

Years	Event intensity in QNA (*Q&N)	Confidence rating in QNA (*Q&N)	Major original sources in QNA and (*) in Q&N	Precise location of relevant quote (§: in H&O)	Location of climatic/ oceanographic anomaly	Phenomenon/effects leading to the reconstruction of EN event	Remarks	Proposed interpretation
1812	*M+	*3	*Eguiguren 1894*	*pp. 250–251 §*	*Piura*	*Drought in Piura*	Drought in N Peru: no EN conditions!	No EN
			Garcia Rosell 1904	*p. 102*	*Piura*	*No rain in 1811–13*		
			Schlüpmann 1994	*pp. 63–64*	*Piura*	*1805–14: drought in Piura, affecting cattle*		
1814	S	4	Spruce 1864	p. 29 §	Piura	Piura River flood (but Chira River not swollen)	Only moderate EN conditions?	M
			Eguiguren 1894	pp. 250–251 §	Piura	Exceptional (?) rainfall (1 Feb.) after earthquake	Single rainfall event?	
			Anon. ms. cited by Schlüpmann 1994	*p. 64*	*Trujillo, N Peru*	*First rainfalls after years of drought*	Only moderate EN conditions?	
1817	M+	5	Eguiguren 1894	pp. 250–251 §	Piura	Abundant rainfalls	EN conditions in N Peru and S Ecuador in 1817 and 1818	
			Labarthe 1914	p. 313	N Peru	Floods in N Peru rivers		
			1825 anon. ms. in Hamerly 1973	*pp. 68, 131*	*Guayaquil, S Ecuador*	*Strong rains in 1817–18*		1817–
			Eguiguren 1894	pp. 250–251 §	Piura	1819: abundant rainfalls	EN conditions in N Peru and S Ecuador in 1819 but also in 1818	1818–

1819	M+	4	*Ms. in Seminario Ojeda 1994*	*p. 62*	*Coast of N Peru*	*Strong rains that limited communications in May*		1819: M
			1825 anon. ms. in Hamerly 1973	*pp. 68, 131*	*Guayaquil, S Ecuador*	*Strong rains in 1818–19*		
			1820 anon. ms. in Schlüpmann 1994	*p. 64*	*Piura*	*Rainfall that destroyed Lancones church in 1819*		
1821	M	5	Eguiguren 1894	pp. 250–251 §	Piura	Abundant rainfalls		M
			Fuchs 1925	p. 524	Trujillo, N Peru	Strong rainfalls during 3 months "near 1821"	Accurate data? (see Petersen)	
			Remy 1931		Lima?		Ref. not seen	
			1825 anon. ms. in Hamerly 1973	*pp. 68, 131*	*Guayaquil, S Ecuador*	*Strong rains in 1820–21*	Moderate (?) EN conditions in N Peru and S Ecuador	
			Garcia Rosell 1907	*p. 91*	*Piura*	*Abundant rainfalls which favored 1822 crops*		
			Petersen 1935 (=1956)	*(2): 33*	*Trujillo, N Peru*	*Doubts about data on Strong rains in Trujillo*	Unreliability of Fuchs data?	
1824	M	5	Spruce 1864	p. 29 §	Piura	Moderate rainy season	Moderate EN?	M?
			Basadre 1884	p. 3	Ilo region, S Peru	Uncommon winter grass vegetation on the coast	EN conditions in S Peru?	
	*M+		Eguiguren 1894	pp. 250–251 §	Piura	Abundant rainfalls	(See Spruce!)	
			1825 anon. ms. in Hamerly 1973	*p. 68*	*Guayaquil, S Ecuador*	*Intense rains in 1824–25; Zapotal River floods*	Moderate (?) EN conditions	

Table 7.1 (*cont.*)

Years	Event intensity in QNA (*Q&N)	Confidence rating in QNA (*Q&N)	Major original sources in QNA and (*) in Q&N	Precise location of relevant quote (§: in H&O)	Location of climatic/ oceanographic anomaly	Phenomenon/effects leading to the reconstruction of EN event	Remarks	Proposed interpretation
1828	VS	5	Ruschenberger “1834” (=1835)	(2): pp. 354–355	Lambayeque, N Peru	Floods and destruction of Lambayeque hospital	Snowmelt (like in 1791?)	
			Spruce 1864	p. 29 §	Piura and S Ecuador	Chira River flood; rainfalls in N Peru and Ecuador	Strong (VS) EN manifestations in N and N-central Peru	VS
			Hutchinson 1873	p. 211	Lambayeque	Large river flood		
			Eguiguren 1894	pp. 248, 251	Piura	Exceptional rainfall with thunderstrokes in Piura		
			Paredes n.d., cited by Eguiguren 1894	pp. 247–248 §	Piura and Trujillo, N Peru	14-day rainfall with thunderstorms in N Peru, in March; Sechura flood		
			Sievers 1914				Ref. not seen	
			Bachmann 1921			Data not found in ref.		
			*Middendorf 1894				Ref. not seen	
			*Adams 1905	p. 97	Lambayeque	Lambayeque River flood		
			1868 ms. cited in Schlüpmann 1994	*p. 64*	*Piura*	*Changes in the course of flooded rivers*	Very strong EN conditions	

			Garcia Rosell 1907	*p. 107*	*Piura, Santa, N Peru*	*Rains in N Peru; very strong flood of Chira River*		
1832	M *M+	5	Spruce 1864	p. 29 §	Tumbes and Piura, N Peru	Limited rains N of Piura and in Chira area	Moderate EN conditions?	M
			Eguiguren 1894	pp. 250–251 §	Piura	Abundant rainfalls		
			*Basadre 1884	p. 3	Ilo region, S Peru	Abundant winter vegetation in 1831 (not in 1832)	Not pertinent data (1831)	
			Távara 1854, cited in Mabres et al. 1993	*p. 398*	*Piura*	*Abundant rains, but less strong than in 1828*	Moderate EN conditions?	
1837	M *M+	5	Eguiguren 1894	pp. 250–251 §	Piura	Abundant rainfalls	Moderate EN conditions?	M
			Labarthe 1914	p. 314	Piura	Floods, less strong than in 1828		
1844– 1845– 1846	S+ *M/S+	5	Spruce 1864	p. 29 §	N Peru and S Ecuador	Heavier rains in 1845 than 1828 at Guayaquil	Strong rains in S Ecuador	1844–1845: S?
			Eguiguren 1894	pp. 250–251 §	Piura	1844: abundant rainfalls 1845: exceptional (?) rains 1846: regular rains	Moderate (?) EN effects in N Peru in 1844–45	
1844– 1845– 1846	1846: Not in QNA	*4	*Basadre 1884	p. 3	Ilo region, S Peru	Winter vegetation on the coast of S Peru in 1846	EN (or La Niña?) conditions in 1846?	1846: No EN?
			*Adams 1905			Data not found in ref.		
1850	M	5 *4	Eguiguren 1894	pp. 250–251 §	Piura	Regular rainfalls		W?
			Fuchs 1925	p. 524	Peru coast	Strong (?) rains	No detail given	
			Petersen 1935	*(2): p. 36*	*Peru coast*	*Questions Fuchs data*	No evidence!	

Table 7.1 (*cont.*)

Years	Event intensity in QNA (*Q&N)	Confidence rating in QNA (*Q&N)	Major original sources in QNA and (*) in Q&N	Precise location of relevant quote (§: in H&O)	Location of climatic/ oceanographic anomaly	Phenomenon/effects leading to the reconstruction of EN event	Remarks	Proposed interpretation
			Spruce 1864	*p. 29*	*N Peru and S Ecuador*	*No strong rains in 1845–64*	Less than moderate intensity?	
*1852	Not in QNA	*4	*Spruce 1864	p. 29	N Peru and S Ecuador	No strong rains in 1845–64	Not more than weak EN conditions	W?
	*M		*Eguiguren 1894	pp. 250–251	Piura	Regular rainfalls		
1854	W/M	4	Spruce 1864	p. 29	N Peru and S Ecuador	No strong rains in 1845–64	Not more than weak EN conditions	W?
			Eguiguren 1894	pp. 250–251 §	Piura	Regular rainfalls		
1857–	M+	5	Eguiguren 1894	pp. 250–251 §	Piura	1857: regular rainfalls 1858: drought	No EN conditions in 1858	1857: M?
			Labarthe 1914	p. 315	Piura, Moquegua, S Peru	1857: rain in Piura and large floods in Moquegua	1857: rains in N and S Peru	
	*M		Gaudron 1925	pp. 362, 365	Peru coast	Strong rains in 1858	Trustworthy?	
1858			Zegarra 1926	p. 23	Trujillo, N Peru	Exceptional rains in 1858	"Exceptional"?	1858:
			León Barandiarán 1938, in Huertas 1993	*pp. 356, 380*	*Lambayeque, N Peru*	*Flood and destruction in Lambayeque in 1857*	Moderate EN conditions?	No EN?

1860	M	4	Labarthe 1914	pp. 315–316	Lima, coast of S Peru	Rímac River flood; rains in Ica, Moquegua, Arequipa	Rains in S Peru: La Niña conditions?	No EN
			**El Comercio* 1860 (7 Jan.)		Lima	(Very small) Rímac River "flood"	No relevant data	
			idem (4 Feb.)		Lima?		Ref. not seen	
			Eguiguren 1894	*pp. 250–251 §*	*Piura*	*Drought in Piura*	No EN conditions!	
1861	*Not in QNA or Q&N*	—	*Ramirez Zenon 1888, in San Cristoval 1938*	*p. 424*	*Piura and Paita*	*Strong rains in March 1861*	Reliability of this single source?	1861–(?)
*1862	Not in QNA *M–	*4	*Spruce 1864	p. 30 §	Chanduy, S Ecuador; Piura	Heavy rains with thunder in February–March, but only two short showers in Piura	Only weak (?) EN conditions for both 1861 and 1862?	–1862: W?
			*Eguiguren 1894	pp. 250–251 §	Piura	Regular rainfalls		
1864	S	5	Spruce 1864	pp. 25–30 §	Piura, Tumbes, and S Ecuador	Warm temperatures (January–February) and rain in March (Chira)	Strong or moderate EN conditions?	S?
			Eguiguren 1894	pp. 248, 251 §	Piura	Exceptional rainfalls		
1866	M *M+	4 *5	Eguiguren 1894	pp. 250–251 §	Piura	Regular rainfalls	Weak or moderate EN conditions?	M?
			Labarthe 1914	p. 316	Lambayeque, N Peru	Rainfall and floods with destruction		
			Bachmann 1921			Data not found in ref.		
			*Adams 1905	p. 97	Lambayeque	Lambayeque River flood		
			El Comercio 1872 (10 Jan.)			Data not found in ref. (wrong reference?)		

Table 7.1 (*cont.*)

Years	Event intensity in QNA (*Q&N)	Confidence rating in QNA (*Q&N)	Major original sources in QNA and (*) in Q&N	Precise location of relevant quote (§: in H&O)	Location of climatic/ oceanographic anomaly	Phenomenon/effects leading to the reconstruction of EN event	Remarks	Proposed interpretation
1867–1868	M *M+	4	Eguiguren 1894	pp. 250–251 §	Piura	1867: drought in Piura 1868: weak rainfalls	No EN conditions in 1867 and 1868!	No EN
			Raimondi 1897, in Schweigger 1964	p. 151 §	Guañape and Sta Magdalena de Cao, N Peru	November 1867: thunder (?) and rainfall; warm SST, red tide; yellow fever epidemic	Interpreted as submarine volcanic eruption!	
			*Bachmann 1921			Data not found in ref.		
			Raimondi 1874	*p. 363 §*	*Piura*	*Chira River flood*	Sierra rains and no EN conditions?	
			El Amigo del Pueblo 1906, in Mabres et al. 1993	*p. 398*	*Piura*	*Drought in 1867–70*		
1871	S+	5	Hutchinson 1873	(2): 147, 211–212 §	Trujillo and N Peru coast	Large flood (and locust plague) in 1870 (not 1871)	Date confusion 1870/1871	S
			Eguiguren 1894	pp. 250–251 §	Piura	Exceptional rainfalls		
			Tizón y Bueno 1907				Ref. not seen	
			Sievers 1914			Data not found in ref.		
			Labarthe 1914	p. 316	Piura, Lambayeque, and Lima	Floods and destruction in February–March in Lambayeque; 450 m^3/s in Rímac River	Coincidence of Rímac flood and rains in N Peru	

			Bachmann 1921			Data not found in ref.		
			Gaudron 1925	p. 365	Peru coast	Strong rains	33-year solar cycle	
			*Middendorf 1894				Ref. not seen	
			*Adams 1905	p. 97	Lambayeque	Lambayeque River flood		
			*Leguia y Martinez 1914	pp. 45, 211	Piura and Lambayeque	Very strong rainfall, like in 1891 (?); change of course of Chira River		
			*Anonymous 1925	p. 238	Peru coast	Only mention of strong rains (but less than in 1925)	No firsthand data	
			León B. 1938, cited by Huertas 1993	*p. 380*	*Lambayeque, N Peru*	*Flooding of Lambayeque in March*		
1874	M	4	*La Patria*, 9 Feb. 1874	p. 2	Cañete, central Peru (S Lima)	Thunder and rainstorm in February on coast S of Lima	EN, or La Niña, evidence?	No EN
			Bravo 1903	p. 14	Santa, N-central Peru	Santa River blocked by a landslide; no rain excess	Source suppressed in Q&N	
			Bachmann 1921			Data not found in ref.		
			*Adams 1905			Data not found in ref.		
1874			*Eguiguren 1894*	*pp. 250–251 §*	*Piura*	*Drought*	No EN conditions!	
			Eguiguren 1894	pp. 250–251 §	Piura	Exceptional rainfalls in 1877 and 1878		

Table 7.1 (*cont.*)

Years	Event intensity in QNA (*Q&N)	Confidence rating in QNA (*Q&N)	Major original sources in QNA and (*) in Q&N	Precise location of relevant quote (§: in H&O)	Location of climatic/ oceanographic anomaly	Phenomenon/effects leading to the reconstruction of EN event	Remarks	Proposed interpretation
1877–1878	VS	5	Palma 1894	p. 1150 §	Lima	On 31 Dec. 1877, strong rainfall with exceptional thunderstrokes	Impressive meteorological phenomenon in Lima, but not typical of EN conditions	VS
			Remy 1931, in Petersen 1935	(2): 37 §	Lima	Thunderstorm and 18-minute rainfall (31 Dec.)		
			Portal 1932, in Petersen 1935	(2): 3, 35 §	Lima	Eyewitness report on 31 Dec. 1877 thunderstorm		
			Melo 1913	p. 156	Mollendo, S Peru	Rains that lasted 14 (?) months and floods	Rains also in S Peru	
			Sievers 1914	p. 276 §	Piura	1877–78 rainfalls compare with 1884 and 1891		
			Labarthe 1914	p. 317	Pacasmayo, N-central Peru	Floods and casualties on railway in 1877	Heavy rains in N-central Peru	
				p. 317	Chimbote, N-central Peru	Santa River flood, with railway destruction in 1878		
			Bachmann 1921			Data not found in ref.		

			Murphy 1926	p. 53 §	Piura and N Peru	Heavy rainfalls compare with 1884, 1891, and 1918, but are less than in 1925	Possibly very strong intensity EN	
			Kiladis and Diaz 1986			Global comparisons, and similarities with 1982–83		
			*Basadre 1884	p. 44	Tarapaca, N Chile	Strong rains and floods in Pampa del Tamarugal	N Chile evidence	
			*Adams 1905	p. 97	Lambayeque	Lambayeque River flood in 1878	Strong intensity of EN in 1878	
			*Leguia y Martinez 1914	pp. 73, 77	Chira River area (Piura)	Chira River flood with destruction in 1878		
			*Anonymous 1925	p. 238	Peru coast	Only mention of strong rains (but less than 1925)	No firsthand data	
			El Amigo del Pueblo. 1906, in Mabres et al. 1993	*p. 398*	*Piura*	*Very abundant rains in 1878*		
1880	M	4	Eguiguren 1894	pp. 250–251 §	Piura	Regular rainfalls	Weak EN conditions?	W?
			Puls 1895				Ref. not seen	
			Eguiguren 1894	p. 250–251 §	Piura	Exceptional rainfalls	Strong EN manifestations in N Peru	
			Sievers 1914	p. 276 §	Piura	Comparison between 1877–78, 1884, and 1891		

Table 7.1 (*cont.*)

Years	Event intensity in QNA (*Q&N)	Confidence rating in QNA (*Q&N)	Major original sources in QNA and (*) in Q&N	Precise location of relevant quote (§: in H&O)	Location of climatic/ oceanographic anomaly	Phenomenon/effects leading to the reconstruction of EN event	Remarks	Proposed interpretation
1884			Labarthe 1914	pp. 317–318	Peru coast, from N to S	Rains and floods all along the coast		
1884	S+	5	Bachmann 1921 Murphy 1925	pp. 169–170	Eten, N Peru	Data not found in ref. High fish mortality (QNA: 14457)	Impacts on marine resource	S
			*Anonymous 1925	p. 238	Peru coast	Only mention of strong rains (but less than 1925)	No firsthand data	
			Weberbauer 1914, in Petersen 1956	*(1): 91*	*Paita, N Peru*	*Vegetation linked to heavy rains as in 1891*		
			Murphy 1926	*p. 53 §*	*Piura*	*Heavy rainfalls compare with 1878, 1891, and 1918, but are less than in 1925*	As strong EN as in 1878 and 1891?	
1887–			Eguiguren 1894	pp. 250–251 §	Piura	1887 and 1888: regular rains 1889: weak rains	No EN conditions in 1889?	
			Labarthe 1914	pp. 318–319	Lima	Rímac River flood and bridge destruction in 1889	1889 Rímac River flood not necessarily related to EN	
1888	W/M	5	*Bravo 1903	p. 14	Verrugas, N-central Peru	Landslide with bridge destruction in March 1889		W?

1889	*M/M+	*4	*El Amigo del Pueblo 1906, in Mabres et al. 1993*	*p. 398*	*Piura*	*Rains more abundant in 1888 than in 1887 and 1889*	Not consistent data (see also Eguiguren 1894)	
			Ramos Seminario n.d., cited by H&O	*pp. 267–268 §*	*Piura*	*Some rains in March 1887 and in February 1889*		
			Carranza 1891, in Schweigger 1964	p. 59 §	Offshore N Peru coast	Combination of oceanic and climatic effects on the S Ecuador/N Peru coast		
			Eguiguren 1894	pp. 248–249 §	Piura and Paita, N Peru	60-day rains (February–April) stronger than in 1828, 1871, 1877–78, and 1884		
			Fuchs 1907	p. 288	Huarmey and Chimbote, N-central Peru	Elevated temperature and strong rainfalls coming from the sea		
			Labarthe 1914	p. 319	Peru coast, from Piura to Lima	Floods in Piura, Lambayeque, Pacasmayo, Santa, Supe, and Lima	Very strong EN event which led to the concept of a combined climatic and oceanographic El Niño phenomenon (Carranza 1891)	
			Sievers 1914	p. 276 §	Piura	Comparison between 1877–78, 1884, and 1891		

Table 7.1 (*cont.*)

Years	Event intensity in QNA (*Q&N)	Confidence rating in QNA (*Q&N)	Major original sources in QNA and (*) in Q&N	Precise location of relevant quote (§: in H&O)	Location of climatic/ oceanographic anomaly	Phenomenon/effects leading to the reconstruction of EN event	Remarks	Proposed interpretation
1891	VS	5	Bachmann 1921	pp. 40–43, 46	N Peru coast	Comments on Carranza 1891	To be noted: no rain excess in central or S Peru!	VS
			1922 ms. in Murphy 1926	p. 36 §	Talara, N Peru	Very strong rainfalls in February		
			Petersen 1935	(2): 37 §	Tumbes and Zorritos, N Peru	Strong rainfalls with electrical storms		
			Zegarra 1926	pp. 23, 34	Trujillo, N Peru	Exceptional rains		
			*Adams 1905	p. 97	Lambayeque	Lambayeque River flood		
			*Leguia y Martinez 1914	pp. 43–44, 45, 51, 71, 288–289	Chira River area (Piura)	Large floods of Chira and Piura rivers, with destruction; Chira flow: 5,400 m^3/s		
			*Anonymous 1925	p. 238	N Peru coast	Strong rains from Piura to Huarmey (not Lima)		
			Weberbauer 1914, in Petersen 1956	*(1): 91*	*Paita*	*Vegetation linked to heavy rains, as in 1884*		
			López Martinez n.d., in Peralta 1985	*pp. 128–130*	*Lima and N coast of Peru*	*El Comercio articles on impacts of 1891 rains*		
	1896: Not in Q&N		*El Comercio* 1897 22 Feb., not 3 Feb.)		Chiclayo, N Peru	Strong rains on 12–13 Feb. 1897 in Chiclayo	1897: single episode of rain?	1896: Not in Q93

			Bravo 1903	p. 18	Lima region	Large landslide in Rímac River valley in 1897	Rainfall E of Lima	1896:
1896–			Bachmann 1921			Data not found in ref.		No EN
	M+	4	*Jones 1933	p. 18	Piura	Abundant rains in 1897 (and drought in 1896)	No EN conditions in 1896	
1897			*El Comercio 1897 (30 Jan. and 1 Feb.)*		*Trujillo, Pacasmayo, N Peru*	*Chicama and Moche River floods in 1897*	EN conditions in N Peru in 1897	1897: M
			El Comercio 1897 (11 Feb.)		*Lima, Cuzco*	*Rain at Lima (January 1897) and flood in Cuzco (February)*	Rains in central and SE Peru	
			El Amigo del Pueblo 1906, in Mabres et al. 1993	*p. 398*	*Piura*	*Drought in 1892–96; some rains in 1897*	No EN conditions in 1896	
			El Comercio 1899 (10 Feb.)			Data not found in ref. (apparently wrong ref.)		
			Labarthe 1914	p. 320	N-central and S coast of Peru	Floods in Ferreñafe, Lima, and Moquegua in 1900	Rains in central and S Peru	
			Bachmann 1921		Trujillo, N Peru	Strong rains in 1899		
			Murphy 1923				Ref. not seen	
			Hutchinson 1950				Ref. not seen	
1899–			*Jones 1933	p. 18	Piura	Drought in 1899 and abundant rains in 1900	Contradicts data from other refs.!	1899: S?
			*Schott 1938				Ref. not seen	

Table 7.1 (*cont.*)

Years	Event intensity in QNA (*Q&N)	Confidence rating in QNA (*Q&N)	Major original sources in QNA and (*) in Q&N	Precise location of relevant quote (§: in H&O)	Location of climatic/ oceanographic anomaly	Phenomenon/effects leading to the reconstruction of EN event	Remarks	Proposed interpretation
1900	S	5	*El Comercio 1899 (15 and 21 Feb.)*		*Lima, Piura, and Paita*	*1899: Chillon River flood and rainfalls in Piura area*	Strong rainfalls in S, central, and N Peru in 1899 (including Lima)	1900: No EN
			El Comercio 1899 (17 Feb.)		*Mollendo and Moquegua*	*Tambo and Moquegua River floods in 1899*		
			El Comercio 1899 (22, 24, 25 Feb.)		*Arequipa, Moquegua, S Peru*	*1899: Tambo and Vitor River floods; heaviest rains since 1884 in Arequipa*		
			Bravo 1903	*p. 20*	*Lima*	*Local rainfall in April 1899 in Rímac valley*		
			El Amigo del Pueblo 1906, in Mabres et al. 1993	*p. 398*	*Piura*	*Regular rains in 1899; drought in 1900*	1899: only moderate (?) EN conditions 1900: no EN!	
			Leguia y Martinez 1914	*pp. 9, 45, 180–181*	*Piura*	*(Relative) drought in the 22 years following 1891*		

page references of the sources, with the aim of making further research easier. It also indicates when excerpts of the original text were reproduced in H&O (unfortunately, it was not possible to reproduce here all the excerpts; these should soon be gathered in a database on a web site). In the last column, updated interpretations are proposed for the occurrence and strength of El Niño events during the studied period.

Table 7.1 recapitulates the sources listed by QNA and Q&N, some sources indicated by H&O, and some recently found references. All the sources not mentioned by QNA or Q&N are indicated in italics and shaded areas. For the sake of conciseness, and to reduce any unnecessary "noise," I eliminated a series of references originally listed by QNA or Q&N from authors who merely repeated previously available data, without adding any relevant information (e.g., Portocarrero 1926) and from an author (Taulis 1934), cited twenty times for the 1525–1900 period by QNA, who does not qualify as a reliable source.

In the following section, a series of cases are discussed. They concern interpretations at odds with those proposed by QNA and/or Q&N, and they illustrate problems of text interpretation, unreliability of some sources, fragility of the evidence of inferred events, and validity of the teleconnected manifestations of the El Niño phenomenon between Peru and the Bolivian altiplano. The problem of the evidence of rainfall anomalies limited to southern Peru and of the Rímac floods, with respect to El Niño reconstruction, will then be presented.

Critical Analysis of the Foundation of Some of the Earliest Historical Events

Definitely, the Second "Event" of the QNA Record (1531–32) Did Not Occur!

The first two El Niño events identified by Quinn (QNA; Q&N; Quinn 1992, 1993) would have occurred during the years 1525–26 and 1531–32. The arguments in favor of such an interpretation, developed by QNA, bear upon the duration of ship time between Panama and Ecuador and the crossing of rivers (supposedly swollen by heavy rainfall). These arguments were extensively and specifically discussed by Hocquenghem and Ortlieb (1990, 1992a,b), and it may be considered as well established now that, at least during the years 1531–32, and probably also in 1525–26, no excess precipitation occurred in northern Peru. All the available written information from the earliest "cronistas" of the Peruvian history supports that interpretation.

Quinn et al.'s (1987) misinterpretation of the 1531–32 "event" is due to the confidence that Quinn and collaborators had in a text written at the end of the nineteenth century (Prescott 1892). Prescott (1892) tried to explain that the conquest of Peru benefited from anomalous climatic conditions in the coastal desert of northern Peru. Thus, Prescott (1892) wrote, for instance, that on 24 September 1532, Pizarro left San Miguel de Piura and crossed "the smooth water of the Piura River." Quinn et al. (1987) took this information at face value and noted that this river, normally dry, is known to be flooded only during rainy (El Niño) episodes. This indication is misleading in several ways. In 1532, the village of San Miguel de Piura was built on the banks of the Chira (not Piura) River (the village of Piura was moved to the Piura

River banks much later; Raimondi 1876; Eguiguren 1894; Schweigger 1959). Thus, the river crossed by Pizarro and his party was the perennial Chira River, the second most important of the country. In September (certainly not a typical period for El Niño manifestation, by the way), the Chira River is normally fed by water from the cordilleran winter rainfall (the "smooth waters" do not indicate any anomalous excess of precipitation).

Another argument used by QNA in favor of an El Niño event reconstruction in 1531–32 is that Pizarro's party was blocked in Puna Island (southern Ecuador) by heavy rains. However, one of the conquistadores, Xerez (1534), indicated that the party had been exhausted and clearly stated that they remained on the island because they needed some rest. Then, according to QNA, the party is said to have had difficulties in crossing the Tumbes River (located at the present-day boundary between Ecuador and Peru), which is normally a large, flooded river in winter. Another argument presented by QNA is that the Zaña River was also flooded. Actually, the same "cronista" (Trujillo 1571), who wrote that the river was swollen, explained (in the same sentence) that it was because the Indians had intentionally directed all the water from their agricultural diversion canal system.

In unambiguous contradiction to Prescott's theory, all the documents left by the participants in the conquest – namely, Ruiz de Arce (1545), Estete (1535), Xerez (1534), Cieza de Leon (1553), and Trujillo (1571) – insist on the fact that at the end of 1532 the conquistadores crossed a warm desert, without enough water supply, where it "never rains." From this unanimous observation, it may be inferred that, even in the year preceding the conquest (1531), it had not rained. These eyewitness reports should be given much more weight than a historian's adventurous interpretation proposed three and a half centuries afterward!

In fact, for more than a century, several authors have been discussing the idea that anomalous rains helped the Spaniards in their rapid conquest of Peru (Raimondi 1876; Schweigger 1959; Hamilton and Garcia 1986), and they all reached the same conclusion and dismissed this hypothesis. This "romantic" theory had likely been developed to flatter the national Peruvian pride, at a time when repeated El Niño heavy rainfalls struck the north of the country (1871, 1877, 1884, and 1891; see Table 7.1). It is surprising that QNA overlooked it. Finally, it may be noted that Quinn (1993) had downgraded his confidence rating from 4 (in QNA and Q&N) to 2 for the 1531–32 so-called event, and altogether modified its strength evaluation from "strong" to "moderate."

For the supposedly "strong" event of 1525–26, the original sources of information are much less abundant than for 1532. The most important available information (actually the only original source, despite the confidence rating of 3 noted in QNA) was provided by Xerez (1534), who indicated that, off the coast of (present-day) Colombia and Ecuador, in 1525–26, the sea was rough, with northerly winds, thunderstorms, and lightning. Nothing is known about the onshore weather. Such oceanic conditions per se can hardly be considered as depicting typical El Niño manifestations. As was previously indicated by H&O, it will be difficult to assess whether these were (or were not) El Niño years. In 1993, Quinn considered the event as a moderate one, with a confidence rating of only 2.

A Transcription Problem: The 1567–68, 1607, and 1671 Cases

As an example of trivial, but real, problems sometimes met in the historical reconstruction of El Niño records, it is interesting to examine in some detail the consequences of a typographical error in a seventeenth-century manuscript. This work is an important book on the early history of Lima, by Father Bernabé Cobo (1639), which was extensively referred to by most of the authors who subsequently wrote about climate features at Lima, including many of those mentioned by QNA for the sixteenth through seventeenth centuries.

The story starts with the mention of a Rímac River flood that destroyed a pillar that supported one of the six or seven arches of the first bridge built in stone and bricks in Lima. The bridge fell down in February of 1607. Cobo then explains that the Virrey Montesclaro decided that it would be more convenient to build another bridge than to repair the old one. The new bridge was finished in 1610. The problem arose from the fact that in the original manuscript, Cobo wrote "167" instead of "1607" (see the note in the 1964 edition of the Cobo work). From the context (p. 313, Cobo 1964), and thanks to the well-written date of the construction of the new bridge (1610), there is no question that the old bridge had been destroyed in 1607. The river flood that occurred in 1607, known to us (through Cobo) because of its consequences, may (or may not) be interpreted as evidence of an El Niño manifestation (see below).

In a general historical study on river floods in Peru, Labarthe (1914, p. 307) refers to Cobo's work and tells the story of the river flood and its consequences on the first stone bridge of Lima, but he incorrectly states that it occurred in the year 1567. Labarthe, who read a previous edition of Cobo's work (without the note of the 1964 editor), misinterpreted the "167" mention. For their interpretation of an El Niño event in 1567–68, QNA explicitly referred to four sources: Cobo (1639), Labarthe (1914), Portocarrero (1926), and Oliva (1631). The first source (Cobo) had been taken from the erroneously cited text of Labarthe. The third one (Portocarrero) only repeated in a condensed way Labarthe's data. Thus, the only remaining acceptable source is that of Oliva, who mentions that in 1568 a Father Geronimo Ruiz Portillo sailed from Panama to Lima in only (?) 26 days, "a trip which usually took six months" (quote from QNA, p. 14,453). Quinn et al. (1987) add, "An accomplishment such as this in a sailing vessel would indicate the presence of highly favourable winds and currents during their journey southward." Hocquenghem and Ortlieb expressed reservations regarding the interpretation of the normal and unusual (?) travel time between Panama and Peru, and they stressed that neither provincial archives (Actas del Cabildo de Trujillo, 1566–71, 1969) nor Lizárraga (1603–09, 1969) mentioned heavy rainfalls in these years in northern Peru. Considering the Labarthe misreading (1567 instead of 1607), the erroneous reference to Cobo's work, and the weak argument on sailing times, I conclude that neither for 1567 nor for 1568 is there enough information to reconstruct an El Niño event (Table 7.1). Additional information for 1568, given by Labarthe (1914, who himself refers to Montesinos 1642), concerns a river flood in Cuzco, in the southeastern Peruvian Andes. These data rather imply that La Niña conditions were prevailing in 1568.

The transcription problem may also affect a hypothetical El Niño reconstruction for 1671. Quinn et al. (1987) inferred a "strong" event from only two references for this year: Labarthe (1914) and Portocarrero (1926). Portocarrero, as has already been said, only reproduced Labarthe data and thus cannot be viewed as a relevant informant. For 1671, Labarthe mentioned floods of the Supe and Rímac Rivers. He provided some details (but no reference) from an anonymous manuscript concerning the Supe River but did not say from which source he interpreted a Rímac flood. I wonder if he did not consider that "167" may also be interpreted as 1671! This suspicion is supported by the fact that Labarthe (p. 308) did repeat for 1607, with other words than those used for the year 1567 (!), the same story of the destruction of the first stone bridge built in Lima.

Quinn (1993) maintained the confidence rating of 3 and the assignation of a "strong" intensity for the 1671 event. Because of a lack of confidence in Labarthe and for additional reasons dealt with below, I express serious doubts regarding the occurrence of an El Niño event in 1671.

Events Related to Single Rainfalls (or Thunderstrokes): 1552, 1614, 1619, 1652, and 1687

For the earliest two centuries, for which the documentary sources are naturally much less abundant than for later on, QNA were led to identify El Niño events on what may be viewed as particularly weak evidence. In several cases, the mention of a single rainstorm is the unique information that supports the recognition of an event (furthermore qualified as a "strong" event). Such were the cases for the years 1614, 1619, and 1652. In 1552, it was not rainfall but a couple of lightning bolts that constituted the evidence for a "strong" event (confidence rating of 4, although only two references are given by QNA and three by Q&N, with the latter ones being a mere repetition of the earlier one). The interpretation of the 1614 event relies upon the occurrence of one rainfall at some distance north of Lima (Cobo 1653), information that was repeated and exaggerated by Labarthe (1914); the latter was subsequently repeated by Portocarrero (1926). In 1619, a winter (12 June) thunderstorm with lightning was reported at Ilo (coast of southern Peru) by the same Cobo (1653). The 1652 rainfall, also reported by Cobo (with subsequent repetitions by Labarthe and then Portocarrero), occurred in February in Lima.

After H&O, I tend to conclude that the years 1552, 1614, 1619, and 1652 should not be considered as El Niño years until more data are found in each case. Up to now an additional source has been found only for the 1619 event (anonymous manuscript cited by Huertas 1992), mentioning a rainfall in Zaña, in northern Peru (Table 7.1).

In 1687, several anomalies were reported (Table 7.1), but as is shown below, the only one that endures the analysis is a single shower on 2 December (Remy 1931). So, in this case also, detecting the occurrence of an El Niño event is based on a single rainfall event in Lima.

About the 1591–92 Case: The Teleconnection with Bolivia

In spite of an apparent confidence rating of 2, QNA relied on a single source (Martínez y Vela 1702) to reconstruct a strong event for 1591–92. Later, Quinn (1992, 1993) and Q&N modified and extended the duration of the event (1589–91) and downgraded its strength (M/S). In QNA, no information other than a mention of a drought in the Potosí area of Bolivia was thus available for an El Niño event reconstruction for these years. This posed a problem of internal consistency in the QNA record, since no other data for the Bolivian altiplano were considered in the rest of the historical sequence. If the 1591–92 (or 1589–91) event were confirmed, relying more on Bolivian historical data to consolidate the El Niño record would be justified. Conversely, if no correlation can be established between dry years in Bolivia and identified El Niño events, it would not be justified to include a 1591–92 event in the record. Later on, in 1992, Q&N added two other sources for a 1589–91 (not 1591–92) event, the first one (Montesinos 1642) referring to epidemic diseases in 1590, and a second one (Barriga 1951) that mentions a lack of rains in December 1589 in southern Peru. Both sources provide rather weak evidence for El Niño event reconstruction. Furthermore, Q&N relied more heavily on the same source to infer the occurrence of six more El Niño events (1558–61, 1607–08, 1640–41, 1671, 1684, and 1692–93).

The source used by QNA and Q&N is a reliable chronicle of the Potosí mining district that covers the 1545–1737 period. Several versions of this major source for the history of Bolivia have been edited, and apparently QNA did not refer to the most complete one (Arzans de Orsúa y Vela 1965). A peculiarity of this work is that it has been published under several names: Bartolomé Martínez y Vela was also known as B. Arzáns de Orsúa y Vela.

Through a preliminary analysis of the huge work of Arzans de Orsúa y Vela, I looked for climatic data that might be significant and useful for paleo–El Niño studies. In Table 7.2, I recapitulate the major drought and heavy rainfall episodes counted for the 1545–1737 period and compare them to different records (QNA; Q&N; Quinn 1993; the present study). Table 7.2 indicates that among the twenty-three detected episodes of drought (of variable intensity), only four would coincide with QNA El Niño events (1591–92, 1671, 1714–15?, and 1728), ten coincide with the Q&N record, and up to twelve of them would be coeval (at least partly) with events of the Quinn (1993) record. At the same time, it can be noted that three strong (or M/S) El Niño events of the Q&N and Quinn (1993) records are coeval with rainfall excess in Potosí (1600, 1607–08, and 1707–09) and that none of these three events are documented as rainy years in northern Peru (see Table 7.1).

With respect to evidence of northern Peru rainfall anomalies, which I tend to consider as primary criteria for assessment of El Niño reconstruction, only two coincidences with Potosí drought were found: 1678–79 and 1728 (Table 7.2). Two coincidences between northern Peru rainfall and heavy rainfall in Potosí (1593 and 1607) must also be noted. Until ongoing studies (by M. R. Prieto, at Mendoza, and A. Gioda, at Cochabamba, in collaboration with the author; see also Prieto 1994; Prieto et al., 1999) on historical teleconnections between Peru, Chile, northern Argentina, and Bolivia are completed, it is difficult to formally conclude that drought evidence for Potosí can be used straightforwardly to reconstruct El Niño past occurrences.

Table 7.2 *Compilation of indications of drought (bold) and rainfall excess (italics) anomalies in Potosí (Bolivia) during the period 1560–1737, as recorded by Arzans de Orsúa y Vela (1965, written in 1705–37). Quinn et al. (1987) had based their interpretation of a strong El Niño event, for 1591–92 (shaded), on the only reference of Arzans de Orsúa y Vela (work referred to by QNA as "Martínez y Vela 1702"). Later, Q&N used the "Martínez y Vela 1702" source to reconstruct El Niño events in 1558–61, 1589–91 (instead of 1591–92), 1607–08, 1640–41, 1671, 1684, and 1692–93 (shaded areas). Actually, a poor correlation is observed between dry years in Potosí and reconstructed El Niño events according to QNA, Q&N, and Quinn (1993). El Niño events generally occurred coeval with droughts in Potosí (but not always). However, notice that many dry years in Potosí were not interpreted as El Niño years. EN = El Niño; VS = very strong; S = strong; M = moderate strength.*

Year & months (I = January)	Drought in Potosí	Rainfall excess in Potosí	Quote in Arzans de Orsúa y Vela 1965	QNA	Q&N	Quinn 1993	EN anomaly recorded in N Peru (Table 7.1)	Interpretation of EN occurrence (Table 7.1)
1557 (VIII–IX)		*Snowfall and icy cold*	*(1): 102–103*	—	—	—	—	—
1560 (X–XII)–1561 (I)	**Drought**		**(1): 115**	—	**1558–61 M/S**	**1558–61 M/S**	**No**	**No EN?**
1582–83	**Drought**		**(1): 192**	—	**1582 M**	**1581–82 M+**	**No**	**?**
1588 (I–III)		*Disastrous rainfalls*	*(1): 203*	—	—	—	—	—
1591 (X–XII)–1592 (I)	**Drought**		**(1): 217, 218**	**1591–92 S**	**1589–91 M/S**	**1589–91 M/S**	**No**	**No EN?**
1593 (I–IV)		*Abundant rains*	*(1): 218*	—	—	—	*Yes*	*M event?*
1600		*Abundant rains*	*(1): 244*	—	*1600 S*	*1600 S*	*No*	*No EN*
1605 (X–XII)–1606 (I–III)	**Strong drought**		**(1): 263**	—	—	—	—	—

1607 (VIII–IX)		*Snow (VIII) and rains (IX)*	*(1): 265*	*1607 S*	*1607–08 S*	*1607–08 S*	*No*	*M event?*
1609 (?)	**Strong drought**		**(1): 263 (Note 1)**	—		—	—	—
1671 (X–XII)	**Drought**		**(2): 259**	**1671 S**	**1671 S**	**1671 S**	**No**	**No EN?**
1677		*Abundant rains*	*(2): 285*	—	—	—	—	—
1678 (X–XII)–1679 (I)	**Strong drought**		**(2): 293–294**	—	—	—	**Yes**	**M event?**
1683 (X–XII)–1684 (I)	**Drought**		**(2): 316**	—	**1684 M+**	**1684 M+**	**No**	**No EN?**
1693 (I–III)–1694–95	**Drought**		**(2): 368, 393**	—	**1692–93 S**	**1695 M**	**No**	**No EN**
1698–1699	**Regional drought**		**(2): 393, 394**	—	—	—	—	—
1705 (X–XII)	**Drought**		(2): 435	—	—	—	—	—
1709 (XII–1710 (I–II)		*Abundant rains*	*(2): 479*	—	1707–09 M/S	1707–09 M/S		*No EN*
1712 (II–III & X–XI)	**Drought**		**(2): 495, 501**	—	—	—	**Drought in Piura between 1706 & 1715 (Schlüpmann 1994: 61)**	***No EN***
1713 (II–III & X–XII)	**Strong drought**		**(3): 3, 12**	—	—	**1713 M**		***No EN***
1714 (I–II)		*Abundant rains*	*(3): 14*	*1714–15* S	—	—		*No EN*
1715 (I)	**Limited drought**		**(3): 26**		**1715–16**	**1715–16**		**No EN**

Table 7.2 (*cont.*)

Year & months (I = January)	Drought in Potosí	Rainfall excess in Potosí	Quote in Arzans de Orsúa y Vela 1965	QNA	Q&N	Quinn 1993	EN anomaly recorded in N Peru (Table 7.1)	Interpretation of EN occurrence (Table 7.1)
1716 (I–III)	**Regional drought**		**(3): 43**	—	**S**	**S**	**No**	**No EN**
1719 (I–III)		*Abundant rains*	*(3): 78*	—	—	—	—	—
1721 (I)	**Drought**		**(3): 124, 163**	—	—	—	—	—
1722 (I & X–XII)	**Drought**		**(3): 138, 150–151, 163**	—	—	—	—	—
1723 (I–VII)	**Drought**		**(3): 153, 156, 163**	—	**1723 M+**	**1723 M+**	**No**	**No EN**
1724 (I–II)		*Abundant rains*	*(3): 162*	—	—	—	—	—
1728 (X-XII)	**Drought**		**(3): 287**	**1728 VS**	**1728 VS**	**1728 VS**	**Yes**	**Very strong event**
1729 (I–II)	**Drought**		**(3): 289**	—	—	—	—	—
1732 (X)	**Drought**		**(3): 349**	—	—	—	—	—
1732 (XII)–1733 (I–II)		*Disastrous rainfall*	*(3): 350*	—	—	—	—	—
1734 (I–II & X–XI)	**Drought**		**(3): 363, 378**	—	—	**1734 M**	—	—
1736 (X–XII)–1737 (I)	**Drought**		**(3): 411–412**	—	**1736 S**	**1737 S**	**No**	**No EN**

The Date of the Destruction of Zaña by a Huge Flood: 1720, Not 1687–88 or 1728

Large floods that lead to the destruction of a city and the emigration of its inhabitants may constitute reliable indicators of anomalous rains, and hence of El Niño past occurrences in coastal northern Peru. Quinn et al. (1987) were apparently confused by misleading references to the destruction of Zaña (or Saña) by different sources that mention that it happened in 1687–88, or in 1720, or else in 1728. As was shown by H&O (pp. 225–226, 228–231), who reproduced a text including eyewitness reports of Zaña destruction (Rubiños y Andrade 1782), and as indicated by Bueno (1763, p. 53), the large flood that caused the complete destruction of the city occurred on 15 March 1720, after several days of uninterrupted rainfall. This disaster occurred only once, so authors who erroneously mention other dates for the same destruction episode should not be trusted.

For the two years 1687 and 1688, QNA based their interpretation of a "strong +" El Niño event on four sources (excluding Taulis 1934): Juan and Ulloa (1748), Unanue (1806), Melo (1913), and Remy (1931). The latter reference deals with data for Lima (a single rainfall on 2 December 1687). I did not find the information concerning 1687–88 in Unanue's work (1806). The latter two sources actually correspond to a single one, because Melo (1913, p. 152) only repeated information from Juan and Ulloa, a century and a half later. Juan and Ulloa (1748) mentioned that Zaña was pillaged in 1685 by the English pirate Edward David and was then completely destroyed "some years later" by a formidable flood. Quinn et al. (1987) concluded that this flood occurred in 1687 or 1688. As we know that this actually happened in 1720, and as no mention of heavy rainfall or flood was found for the years 1687–88 by Huertas (1987, 1992) in his studies on the history of Zaña (H&O), it is confidently inferred that no such flood occurred in these years. Actually, Q&N suppressed the Juan and Ulloa (1748) source in their revised record but still included the Unanue (1806) and Melo (1913) sources. Another indication of the destruction of Zaña (by a large flood) was given by Alcedo (1786–89, p. 344) for 1728. Quinn et al. (1987) cite this source as an additional reference among those that support the 1728 El Niño event. Curiously, QNA also used the Alcedo work as a source for the 1720 event. Actually, abundant information confirms the occurrence of El Niño events in both 1720 and 1728 (Table 7.1).

Meteorological Anomalies Restricted to Central and Southern Peru

One of the major points made by the earlier H&O analysis of the QNA record concerns the evidence for climatic anomalies restricted to Lima and the southern coast of Peru. As rainfall and thunderstrokes are exceptional in Lima, in the arid coastal fringe of Peru, and because more precise and abundant information is available from the capital, it is easily understandable that such phenomena were recorded in colonial times. Anecdotes and comments on these matters from many authors were then used by QNA. However, the problem of the relationship between Rímac River floods, or Lima showers, and El Niño manifestations was scarcely tackled. In their interpretations, QNA apparently relied much on the exceptional character of rainfall in Lima and did not question whether these phenomena were related to El Niño, or La Niña, conditions. It must be noted, for instance, that they did not address the fact that the typical "garúa" of Lima is winter precipitation, and as such this may rather be linked to La Niña meteorological conditions. Quinn et al. (1987) may have amalgamated the anomalous rainfall signals of northern Peru, central Peru, and southern Peru. These premises may be wrong.

El Niño Impacts along the Peruvian Coast

From a climatological point of view, the El Niño phenomenon was defined in the Piura–Paita area, far north of Lima. There is no doubt whatsoever that the Sechura Desert, which is different from the narrow coastal desert that borders the whole country, constitutes the core of the El Niño "land" (see Eguiguren 1894; Petersen 1935; Ortlieb and Macharé 1993). The amount of anomalous rainfall in the coastal region of extreme northern Peru remains the most reliable indicator of the strength of the El Niño events. While a clear relationship links river flooding in the northern reaches of the country and the precipitation that falls within the wide coastal area of the Sechura Desert, there are uncertainties as to the significance of river floods in the central coast of Peru. In Lima and in central Peru, the river floods imply upstream rainfalls that occur either at the foot of the nearby Andes or within the 4,000 m high cordillera. A Rímac River flood is never produced by rainfall in Lima, and thus the report of a past flood should not be interpreted as evidence for rainfall in the coastal desert of central Peru. In fact, not even rainfall in Lima can be linked directly to El Niño episodes.

The amount of annual rainfall at Lima varies between a few millimeters and less than 10 cm (Fig. 7.2). It must be stressed that unlike northern Peru or central Chile, the hyperarid coastal desert that stretches between 6° and 25°S latitude (including the Atacama Desert of northern Chile) never registers "heavy" rainfalls. The exceptional showers that may fall in this coastal desert do not exceed a few centimeters of precipitation, while the Sechura Desert may receive hundreds of centimeters (up to 400 cm locally in 1982–83) of precipitation during El Niño years (Huaman Solis and Garcia Peña 1985; Woodman 1985). But what is most important is that even the small amounts of precipitation that occur in the coastal desert of central and southern Peru do not seem to be related to El Niño conditions. Figure 7.2 shows that during the past forty years (1950–91) it did not rain more during El Niño years than in "normal" years or La Niña years. The amount of precipitation during the very strong 1982–83 event does not depart from the overall mean of the past thirty years. This observation for the second half of the twentieth century must be taken into consideration when one looks at the historical climatic record of Lima.

The coastal region of southern Peru and the southern Peruvian Andes are known to suffer deficits of precipitation during El Niño years (Huaman Solis and Garcia Peña 1985; Francou and Pizarro 1985; Garcia Peña and Fernández 1985; Ropelewski and Halpert 1987). In a study on the relationship between precipitation in the coast of southern Peru and the well-established occurrences of El Niño events during the past forty years, Minaya (1994) showed that for Lima there is no direct and unequivocal link between rainfall (or drought) and El Niño (or La Niña) conditions (Fig. 7.2; Table 7.3). At Tacna, in the coastal area close to the Chilean border, as well as in Arequipa on the high inland plateau, no relationship can be established between the strengths of the

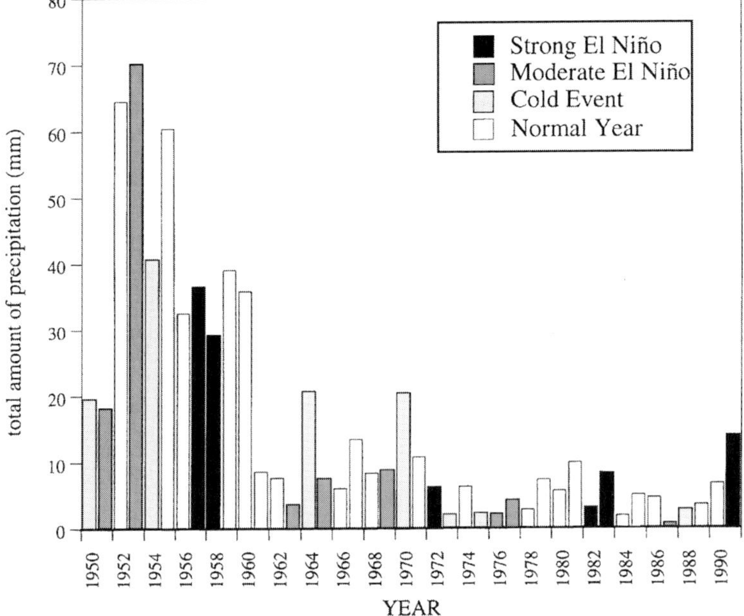

Fig. 7.2 Annual rainfall variation in Lima for the 1950–91 period. No evidence is seen for a straightforward relationship with recent El Niño events (data from Corporación Peruana de la Aviación Civil [CORPAC] compiled by Minaya 1994). Neither strong nor moderate events are characterized by rainfall more abundant than the decadal mean. If extrapolated to the past few centuries, this observation leads one to question the occurrence of a series of events as proposed by QNA, Q&N, and Quinn (1993).

events and the amounts of precipitation. The very strong 1982–83 El Niño event was characterized by a total drought in Arequipa and a strong deficit of the Majes River flow (Table 7.3), but during strong events (such as in 1972–73) exceptional rainfall at Arequipa and maximum flows of the Majes River were registered (Minaya 1993). Moderate events also correspond to opposite extremes, in Tacna for instance: total drought in the 1965 and 1969 events and maximum annual rainfall in the 1953 event (Table 7.3).

Based on instrumental data of the past decades, it thus appears that neither droughts, nor anomalous precipitation episodes, nor river floods in the southern half of the country can be used to predict El Niño conditions. This conclusion has serious implications for the elaboration of the historical record of El Niño events.

Rímac River Floods and El Niño Events

Quinn et al. (1987) and Q&N often refer to evidence of floods of the Rímac River as an indication of anomalous rainfall, and hence of El Niño conditions. We saw that actually the floods of this river, like others in the central part of the Peruvian coastal desert, do not reflect properly rainfall excess in the coastal region; this observation, however, does not preclude the hypothesis that precipitation on the western flank of the Andes is, in some way, related to El Niño circulation patterns. A careful study of the

Table 7.3 *Amounts of annual precipitation in southern Peru localities during the last El Niño events of the past forty years, from Corporación Peruana de la Aviación Civil (CORPAC) data (Minaya 1994). No clear relationship can be established between the rainfall in southern Peru and the occurrence of El Niño events: Some events are characterized by strong deficits and others by rainfall excess. Rio Majes flow data (from Dirección General de Aguas, Ministerio de Agricultura, Lima, in Minaya 1993) show the same extreme variability with the recent El Niño events.*

ENSO events 1950–90		Annual precipitation (mm)				Streamflow (m^3)
Year	Strength	Lima	Pisco	Arequipa	Tacna	Rio Majes
1951	M−	18	0	72	25	3,304
1953	M+	70	0	248	114	2,855
1958	S	29	4	55	63	3,177
1965	M+	8	1	33	0	1,079
1969	M−	9	0	30	0	1,801
1972–	S	7	6	253	81	3,337
1973	S	2	0	95	7	4,227
1976	M	2	1	112	53	n.d.
1983	VS	9	0	0	34	426
1987	M	8	0	49	6	1,393
Mean 1950–90		16	2	93	26	2,391

twentieth-century Rímac River floods in regard to El Niño events is hampered by the inadequacy of the instrumental record of precipitation in the first half of the century (precise locations of the rainfalls) and by the intense development of human activities (hydroelectric power station, water supply plant) upstream in the Rímac valley. Otherwise, it should be useful to try to determine in the historical record how tight the relationship is between Rímac floods and El Niño events that were unambiguously identified.

Table 7.4 recapitulates eighteen cases of Rímac floods known to have occurred between 1567 and 1900, sixteen of which are mentioned in the QNA record. Practically all of these floods had been identified by Labarthe (1914), a source that was not entirely reliable, as has been mentioned (e.g., the 1567/1607/1671 problem). In several cases, the original sources of information of Labarthe could not be verified. In other instances, as for the four eighteenth-century cases of Rímac River floods, Labarthe indicated that he relied upon a newspaper (*Mercurio Peruano*) review article written by a journalist, Ambrosio Cerdán. It can be noted that several of the so-called events identified by QNA on the basis of Rímac River floods (1634, 1696, 1750, 1755, and 1779) were not confirmed in the 1993 Quinn record (Table 7.4).

Among the sixteen events that QNA correlated with Rímac River floods, only seven are confirmed as El Niño episodes (Table 7.1) by assessing evidence of rainfall in northern Peru (indicated in bold in Table 7.4). These reconfirmed events coincided with floods of other rivers in northern, central, and eventually southern Peru. It is certainly

Table 7.4 *Chronological list of El Niño years (according to QNA and Quinn 1993) that were (partially or exclusively) identified on the basis of evidence for Rímac River floods, at Lima, in the sixteenth through nineteenth centuries. In some cases (in boldface), Rímac River floods appear to be coeval with rainfall excess in northern Peru, or southern Ecuador, and thus seem to have occurred during El Niño events. In other instances (shaded areas), Rímac floods are rather correlated with southern Peru rains. It is deduced that by themselves floods at Lima should not be used to reconstruct El Niño events.*

Years	Rímac River flood (month of occurrence)	Original source and quote in Labarthe 1914 (L.)	El Niño intensity according to QNA (confidence)	El Niño intensity according to Quinn 1993	Evidence for N Peru and S Ecuador rains	Proposed interpretation (see Table 7.1)	Remarks
1567	Yes	Cobo 1639 L., p. 307	1567–68 S+ (5)	1567–68 S+ (5)	No	No EN event?	Transcription problem: 1567/1607 (see text)
1578	**Yes**	**Cobo 1639 L., p. 308**	**1578 VS (5)**	**1578–79 VS (5)**	**Yes**	**Very strong event**	**Large floods at Lima and in N Peru**
1607	February	Cobo 1639 L., p. 308	1607 S (5)	1607–08 S (5)	No	Moderate event?	No other data available
1634	February/ March	Source? Palma 1894, p. 42	1634 S (4)		No	No EN event?	Floods at Lima and in S Peru
1671	Yes	Source? L., p. 309	1671 S (3)	1671 S (3)	No	No EN event?	Transcription problem: 1671/1607 (see text)
1696	**11 Feb.**	**Source? Palma 1894, p. 42**	**1696 S (3)**		**Yes**	**Moderate event?**	**Flood in S Ecuador**
1750	Yes	A. Cerdan L., p. 311	1750 M (p. 14455)		No	No EN event	Only source for QNA?
1775	Yes	A. Cerdan L., p. 311	1775 S (4)		No	No EN event	No report on destruction; drought in N Peru
1779	Yes	A. Cerdan L., p. 311	1778–79 M		No	No EN event?	No report on destruction

Table 7.4 (*cont.*)

Years	Rímac River flood (month of occurrence)	Original source and quote in Labarthe 1914 (L.)	El Niño intensity according to QNA (confidence)	El Niño intensity according to Quinn 1993	Evidence for N Peru and S Ecuador rains	Proposed interpretation (see Table 7.1)	Remarks
1786	Yes	A. Cerdan L., p. 311	1785–86 S (4)	1785–86 M+ (2)	No	No EN event	No other data for 1786
1804	**February/ March**	**Guía de Forasteros L., p. 313**	**1803–04 S (5)**	**1803–04 S + (5)**	**Yes**	**Strong event**	**Limited destruction but warm temperature**
1860	8 Mar.	L., p. 315	1860 M (4)	1860 M (4)	No	No EN event	Coeval with rains in S Peru (drought in N Peru)
1871	**February**	**L., p. 316**	**1871 S + (5)**	**1871 S + (5)**	**Yes**	**Strong event**	**Large flood coeval with N Peru rainfalls**
1872	January & 28 Feb.	L., p. 316			No		Rains in southern and central Peru
1884	**7 Jan.**	**L., p. 317**	**1884 S (5)**	**1884 S+ (5)**	**Yes**	**Strong event**	**Rains in N, S, and central Peru**
1889	**12 Mar.**	**L., p. 318**	**1887–89 M (4)**	**1887–89 M (4)**	**Some rain**	**Weak event?**	**Bridges destroyed E of Lima**
1891	**20 Mar.**	**L., p. 319**	**1891 VS (5)**	**1891 VS (5)**	**Yes**	**Very strong event**	**Bridges destroyed E of Lima**
1900	February	L., p. 320	1899–1900 S (5)	1899–1900 S (5)	No	No EN event?	Rains in S and central Peru; drought in N Peru

significant that the few Rímac floods that occur only along with floods in southern Peru do not seem to be associated with El Niño events (Table 7.1).

As was hypothesized by H&O, some of the events proposed by QNA on the basis of climatic anomalies and river floods in southern Peru might actually be manifestations of La Niña conditions. This may apply to the years 1540, 1634, 1714, 1775, 1806–07, 1812, 1860, 1874, and 1900. The best candidates for cold event (La Niña) years are those for which there are combined indications of drought in northern Peru and above average rainfall in the southern half of the Peruvian coast (1714–15, 1775, 1806–07, 1812, 1860, and 1874). Further historical studies planned for the Piura area and aimed to complement the century-old Eguiguren (1894) work should help to discriminate El Niño, La Niña, and normal years of the past few centuries.

New Data from Southernmost Ecuador

Among the new data gathered to assess the reconstruction of former manifestations of El Niño events, some information relative to northernmost Peru and the southern part of Ecuador is included here. Quinn et al. (1987) previously referred to two informants who reported information from southern Ecuador: Spruce (1864) and Estrada Ycaza (1977). The former provided trustworthy data, especially for the first half of the nineteenth century (Table 7.1). Estrada Ycaza's (1977) work was referred to by QNA in a single case (1785–86), although the book contains a series of relevant data on flooding and particularly heavy rainfalls that occurred in the past centuries in the Guayaquil region. Quinn and Neal (1992) cited a 1565 (or 1572) book, by Benzoni, in which is mentioned that the first settlement of Guayaquil suffered from a large flood of the Chiono River to the point that the town was reconstructed some distance to the south. This flood occurred in 1546 and is probably related to an El Niño manifestation.

Table 7.1 includes relevant additional data on climatic anomalies reported in South Ecuador. Evidence for heavy rainfalls and floods of several rivers in southern Ecuador is presented for the following years: 1696, 1760, 1784–85, 1804, 1817, 1819, 1821, 1824–25, and 1850 (Table 7.1). This information suggests or reconfirms the occurrence of El Niño events. In a few cases, reports on drought in southern Ecuador may be used to infer that no El Niño event occurred (1654–55, 1775–80).

Certainly further investigations into the documentary record of southern Ecuador should be encouraged. Unlike the central Chilean data, there is no question that meteorological anomalies that occurred in northern Peru and those reported in southern Ecuador are closely related and (most often) directly linked with the El Niño phenomenon. In fact, there are much closer similarities between El Niño manifestations within an area that encompasses southern Ecuador and northern Peru, than between northern and southern Peru.

The Northern Peru–Central Chile Teleconnection

The El Niño Record for Central Chile

For several reasons explained previously, the Taulis (1934) reference was suppressed from the critical analysis of the QNA record (Table 7.1). This source, extensively used by QNA, Q&N, and Quinn (1992, 1993), consists of a mere chronological table, in which every year between 1535 and 1933 is graphically depicted as either normal, dry, very dry, wet, or very wet. No precise indication of documentary sources is given by Taulis, a major inconvenience for this kind of work. As was demonstrated in a previous work (Ortlieb 1994), many indications for rainy or anomalously rainy years as reported by Taulis were compiled from a well-documented work written by a respected historian and national figure, Benjamin Vicuña Mackenna (1877). A close correspondence between Taulis and Vicuña Mackenna records is observed for the period 1723–1877 (Table 7.5). It is assumed that Taulis used instrumental records of precipitation for the period 1877–1933, but we totally ignore his sources for the sixteenth and seventeenth centuries.

In his most recent papers, Quinn relied even more heavily on Taulis as well as on Vicuña Mackenna (1877) and an informant on wreck occurrences related to storminess in central Chile (Vidal Gormaz 1901). As a result, the sequences presented by Q&N and Quinn (1992, 1993) include some fourteen additional events (with respect to the QNA chronology) that were partly inferred from evidence that came from central Chile.

Through a comparison between Taulis's and Vicuña Mackenna's records, which also includes a third chronicle of past climatic anomalies extracted from a historical review of natural disasters in Chile (Urrutia de Hazbún and Lanza Lazcano 1993; hereafter U&L), Ortlieb (1994) intended to consolidate the chronological sequence of rainfall excess in central Chile. The proposed sequence of rainy years thus included Taulis's data only when additional confirmation was obtained in Vicuña Mackenna or U&L (Table 7.5). Precipitation excesses were qualified as regular, strong (S), and very strong (VS) (bold characters in Table 7.5). In this way, only two rainy years were assessed in the sixteenth century, and eight in both the seventeenth and eighteenth centuries, while up to twenty years (with varying amounts of excess rainfall) were counted in the nineteenth century (Table 7.5). The much larger number of rainy events recognized during the nineteenth century should be related primarily to the major accessibility of documentary sources. But the possibility cannot be excluded that the nineteenth and twentieth (see Ruttlant and Fuenzalida 1991) centuries were actually more "rainy" than the previous centuries.

The comparison of available instrumental records of precipitation at Santiago for the past century and a half and El Niño sequences (Kiladis and Diaz 1989; Ruttlant and Fuenzalida 1991) suggests that some proportionality exists between the amount of winter rainfall in central Chile and the strength of El Niño events. Therefore, it can be expected that the record of the major historical rainfall anomalies at Santiago may correspond to the strongest events of the past few centuries. How could this hypothesis be tested? The record of well-assessed rainy years in pre–nineteenth century times (Ortlieb 1994) can be compared neither with QNA nor with the last published regional chronologies by Quinn, since these were developed with central Chile data. Verifying the evolution through time of the relationship between precipitation excess in central Chile and El Niño manifestations cannot be performed if the historical record of El Niño events was built (at least partially) upon rainfall data for central Chile.

Table 7.5 *Historical reconstruction of rainfall excess anomalies in central Chile compared to Quinn's records of El Niño events (QNA; Quinn 1993). The sequences of rainy years were deduced from analysis of reports from Vicuña Mackenna (1877), Taulis (1934), and Urrutia de Hazbún and Lanza Lazcano (1993). The sequence of Ortlieb (1994) synthesized the three previous studies, after respective evaluation. Strength of El Niño events: VS = very strong, S = strong, M = moderate, W = weak. Years with the strongest rainfall excess are indicated in boldface and shaded; "No" means: No evidence for rainfall excess.*

Rainy years in central and north-central Chile				QNA EN chronology	Revised EN chronology
Vicuña M. 1877	Taulis 1934	U&L 1993	Ortlieb 1994	Quinn et al. 1987	Quinn 1993
No data	No data	No data	No data	**1525** **S** **1526**	1525 M 1526
				1531 **S** **1532**	1531 M 1532
	1535?		?		1535 M+
1536			?		
				1539 1540 M/S 1541	1539 1540 M/S 1541
1544	**1544**		**1544** **S**		1544 M+
					1546 **S** **1547**
	1548 1550 1551		? ? ?		
				1552 **S**	**1552** **S**
	1559		?		1558 1559 M/S 1560 1561
	No data				1565 M+
				1567 **S+** **1568**	**1567** **S+** **1568**
	1574	**1574**	**1574** **S**	**1574** **S**	**1574** **S**
	1575		?		
				1578 **VS**	**1578** **VS** **1579**

Table 7.5 (*cont.*)

Rainy years in central and north-central Chile					QNA EN chronology		Revised EN chronology	
Vicuña M. 1877	Taulis 1934	U&L 1993	Ortlieb 1994		Quinn et al. 1987		Quinn 1993	
	1581	1581	**?**				1581	M+
							1582	
							1585	M+
							1589	
							1590	M/S
					1591	**S**	1591	
					1592			
							1596	M+
	No data	**1597**	**?**					
							1600	**S**
							1604	M+
	1607	1607	1607		**1607**	**S**	**1607**	**S**
							1608	
1609	**1609**	**1609**	**1609**	**S**				
					1614	**S**	**1614**	**S**
1618	1618		1618		**1618**	**S**	**1618**	**S**
					1619		**1619**	
							1621	M+
					1624	**S+**	**1624**	**S+**
							1630	M
					1634	**S**		
							1635	**S**
							1640	M
							1641	
1647	**1647**	**1647**	**1647**	**S**			1647	M+
	1648		**?**					
	1650	1650	1650				1650	M
					1652	**S+**	**1652**	**S+**
	1655		**?**				1655	M
		1657	**No**					

Table 7.5 (*cont.*)

Rainy years in central and north-central Chile				QNA EN chronology		Revised EN chronology	
Vicuña M. 1877	Taulis 1934	U&L 1993	Ortlieb 1994	Quinn et al. 1987		Quinn 1993	
		1660	**No**	**1660**	**S**		
						1661	**S**
				1671	**S**	**1671**	**S**
		1679	**No**				
				1681	**S**	**1681**	**S**
	1683		**?**				
						1684	M+
	1686		**?**				
?	1687	1687	1687	**1687**	S+	**1687**	S+
?	1688	1688	1688	**1688**			
						1692	**S**
		1694	**No**				
	1695		**?**			1695	M
				1696	**S**		
1697	**1697**		**1697** **S**			1697	M+
	1698		**?**				
				1701	S+	**1701**	S+
						1704	M
	1705		**?**				
				1707	**S**	1707	
				1708		1708	M/S
						1709	
						1713	M
				1714	**S**		
				1715		**1715**	**S**
						1716	
						1718	M+
				1720	S+	**1720**	**VS**
		1722	**No?**				
1723	1723	1723	1723			1723	M+
				1728	**VS**	**1728**	**VS**
						1734	M
						1737	**S**

Table 7.5 (*cont.*)

Rainy years in central and north-central Chile					QNA EN chronology		Revised EN chronology	
Vicuña M. 1877	Taulis 1934	U&L 1993	Ortlieb 1994		Quinn et al. 1987		Quinn 1993	
		1743	**No?**					
1744	1744	1744	1744				1744	M+
		1745	**No?**					
1746	**1746**		**1746**	**S**				
					1747	**S**	**1747**	**S+**
1748	1748	1748	**1748**	**S**				
	1751	1751	1751				1751	M+
							1754	M
							1755	
							1758	M
					1761	**S**	**1761**	**S**
1764	**1764**	1764	1764					
							1765	M
1768	1768		1768				1768	M
							1772	M
					1775	**S**		
							1776	
							1777	**S**
							1778	
1779?		1779	**No**					
		1779	**No**					
							1782	**S**
1783	**1783**	**1783**	**1783**	**VS**			**1783**	
					1785	**S**	1785	M+
					1786		1786	
1790?			**No**					
					1791	**VS**	**1791**	**VS**
					1803	**S+**	**1803**	**S+**
					1804		**1804**	
					1806	M	1806	M
					1807		1807	

Table 7.5 (*cont.*)

Rainy years in central and north-central Chile				QNA EN chronology	Revised EN chronology
Vicuña M. 1877	Taulis 1934	U&L 1993	Ortlieb 1994	Quinn et al. 1987	Quinn 1993
					1810 M
				1812 M	1812 M+
	1813		?		
				1814 S	**1814 S**
1817	**1817**		**1817 S**	1817 M+	1817 M+
1819	1819		1819	1819 M+	1819 M+
1820	1820		1820		
1821	1821		1821	1821 M	1821 M
		1823	**No?**		
				1824 M	1824 M+
		1826	**No**		
1827	**1827**	1827	**1827 S**		
1828	1828	1828	1828	**1828 VS**	**1828 VS**
1829	**1829**		**1829 S**		
					1830 M
				1832 M	1832 M+
1833	**1833**	**1833**	**1833 S**		
		1835	**No?**		
		1836	**No**		
1837	1837	1837	1837	1837 M	1837 M+
1841	1841		1841		
1843	**1843**		**1843 S**		
				1844 S +	**1844**
1845	**1845**		**1845 VS**	**1845**	**1845 S**
					1846
		1848	**No**		
1850	1850	**1850**	**1850 S**	1850 M	1850 M
1851	**1851**		**1851 S**		
					1852 M
					1853
1854			**No**	1854 W/M	1854 M
1855	1855	1855	1855		
1856	1856	1856	1856		
		1857	**No**	1857 M+	1857 M
1858	1858		1858	1858	1858

Table 7.5 (*cont.*)

Rainy years in central and north-central Chile				QNA EN chronology	Revised EN chronology
Vicuña M. 1877	Taulis 1934	U&L 1993	Ortlieb 1994	Quinn et al. 1987	Quinn 1993
1860 1862	1860		1860 **No**	1860 M	1860 M 1862 M−
1864	**1864**	1864	**1864 S**	**1864 S**	**1864 S**
				1866 M 1867 M	1866 M+ 1867 M+
1868	**1868**		**1868 S**	1868	1868
				1871 S+	**1871 S+**
1873			**No?**		
1874		1875	**No?** **No**	1874 M	1874 M
1877	**1877**	**1877**	**1877 VS**	**1877 VS**	**1877 VS**
No data		1878	1878	**1878**	**1878**
	1880	**1880**	**1880 S**	1880 M	1880 M
		1884	**No**	**1884 S+**	**1884 S+**
	1887		**?**	1887	1887
	1888	**1888**	**1888 S**	1888 W/M	1888 M
				1889	1889
	1891	**1891**	**1891 S**	**1891 VS**	**1891 VS**
No data		1896	**No**	1896 M+ 1897	1897 M+
	1899 **1900**	**1899** **1900**	**1899 VS** **1900 VS**	**1899 S** **1900**	**1899 S** **1900**

Revised Peruvian Record versus Revised Chilean Record

Previous studies (Ortlieb 1994, 1995; Ortlieb et al. 1995) evaluated the consistency of reconstructed sequences of El Niño events for central and northern Chile with different chronologies of regional events (QNA; Quinn 1993; H&O). The new Peruvian sequence proposed here constitutes a more internally consistent reference, since it excludes all data from central Chile and is, furthermore, supposedly better consolidated. As has been discussed, it is inferred here that the southern Ecuador historical data, and maybe that of the high Andes of Peru or Bolivia, are open to being more directly linked to the regional climate of coastal northern Peru than to the climate system of central Chile. The so-called Peruvian record derived from the interpretation of Table 7.1 is summarized in the second column of Table 7.6.

Table 7.6 *Western South American El Niño records compared to summarized Indo-Pacific ENSO records. The Chile (Ortlieb 1994, and Table 7.5) and Peru (Table 7.1) records are based on documentary records, while the Galapagos sequence (Dunbar et al. 1994) is deduced from SST reconstructions based on ^{18}O composition of an emerged coral reef. The eastern Pacific record of Quinn (1993), indicated in the fourth column, is his last published sequence of El Niño events (modified from the QNA record). The fifth column, also reproduced from Quinn (1993), represents a global combination of ENSO manifestations in Egypt, India, China, and South America. The sequence of India droughts and the synthetic eastern ENSO chronology were compiled by Whetton and Rutherfurd (1994). Legend: (a) El Niño events (based on documentary records): In bold and shaded = strong rainfall anomaly (underlined = very strong); italics = small anomaly; ? = insufficient data. (b) ENSO reconstruction from ^{18}O data from UR-86 coral record; strongest events in bold. (c) Quinn records with ranking of El Niño event intensity: W = weak; M = moderate; S (+shaded) = strong; VS (+shaded and underlined) = very strong. Period of occurrence in the year: E = early (January–March); L = late (September–December). (d) Droughts in India according to several sources (see Whetton & Rutherfurd 1994), considered as coeval with ENSO events. Less well assessed data in italics. (e) ENSO years determination based on coincidence of at least three indicators from the Nile region, Java, North China, India, and Peru (Quinn's data); years in bold are the best correlated (four coincidental indicators within the five areas; underlined: five coincidences).*

South America El Niño records			Quinn's (1993) revised chronology and ranking		Eastern Hemisphere ENSO compilation		India droughts	South America El Niño records
Chile	Peru	Galapagos	E Pacific El Niño events		Global ENSO events			
Ortlieb 1994	Present work	Dunbar et al. 1994	Quinn 1993		Quinn 1993		Whetton & Rutherfurd 1994	
(a)	(a)	(b)	(c)		(c)		(d)	(e)
							1520–21?	
	No?		1525–	M	1525–	M		
	No?		E1526		E1526			
	No		1531–	M	1531–	M		
	No		E1532		E1532			

Table 7.6 (*cont.*)

South America El Niño records			Quinn's (1993) revised chronology and ranking		Eastern Hemisphere ENSO compilation		India droughts	South America El Niño records
Chile	Peru	Galapagos	E Pacific El Niño events		Global ENSO events			
Ortlieb 1994	Present work	Dunbar et al. 1994	Quinn 1993		Quinn 1993		Whetton & Rutherfurd 1994	
(a)	(a)	(b)	(c)		(c)		(d)	(e)
?			1535	M+	1535	M+		
	No?		1539–		**1539–**			
	No?		1540–	M/S	**1540–**	**S**	1540–	
	No?		1541		**1541**		1541–	
							1542–	
							1543	
1544	?		1544	M+	1544	M+		
	1546–		**1546–**	**S**	**1546–**	**S**		
	1547		**1546**		**1547**			
	No	No	**1552**	**S**	**1552–**	**S**		No
					1553			
							1554–56?	
	No?	Data	1558–		**1558–**			Data
?	No?		1559–	M/S	**1559–**	**S**		
	No?		1560–		**1560–**			
	No?		1561		**E1561**			
	No?		1565	M+	1565	M+		
	No?		**1567–**	**S+**	**1567–**	**S+**		

	No?		**1568**		**1568**			
1574	1574		**1574**	**S**	**1574**	**S**		
							1576–77?	
	1578		**1578–**	**VS**	**1578–**	**S**		
			E1579		**E1579**			
?			1581–	M+	1581–	M+		
	?		1582		1582			
	?		1585	M+	1585	M		
	No?		1589–		**1589–**			
	No?		1590–	M/S	**1590–**	**S**		
	No?		1591		**1591**			
	No?						*1592?*	
	1593						*1594–1995–*	
	1596		1596	M+	1596	M	*1996–1997–*	
	?		**1600**	**S**	**1600–**	**S**	*1998?*	
					1601			
	No		1604	M+	**1604**	**S**		
1607	1607–	1607	**1607–**	**S**	**1607–**	**S**		
	1608		**1608**		**1608**			
1609								
	No?		**1614**	**S**	**1614**	**S**	*1613–1915?*	
1618	1618–		**1618–**	**S**	1618–	M	*1618–*	
	1619		**1619**		1619		*1619?*	
		1621	1621	M+	**1621**	**S**		
	1622							
		1623					*1623?*	
	1624		**1624**	**S+**	1624	M+		
							1629–	
			1630	M	**1630–**	**S+**	1630–	
					1631		1631	
		1633						No

Table 7.6 (*cont.*)

South America El Niño records			Quinn's (1993) revised chronology and ranking		Eastern Hemisphere ENSO compilation		India droughts	South America El Niño records
Chile	Peru	Galapagos	E Pacific El Niño events		Global ENSO events			
Ortlieb 1994	Present work	Dunbar et al. 1994	Quinn 1993		Quinn 1993		Whetton & Rutherfurd 1994	
(a)	(a)	(b)	(c)		(c)		(d)	(e)
	No?		**1635**	**S**	1635	M		Data
	No		1640–	M	**1640–**	**S+**		
	No		1641		**1641**			
1647	No		1647	M+	1647	M		
							1648	
?								
1650			1650	M	**1650**	**S+**	*1650?*	
	No?	**1652**	**1652**	**S+**	1652	M		
?	No		1655	M	1655	M		
							1659–	
							1660–	
	No							
			1661	**S**	**<u>1661</u>**	**<u>VS</u>**	1661	
		1670						
	No?		**1671**	**S+**	1671	M+		
		1674						
	1678							
	No		**1681**	**S**	**1681**	**S**		
?					1683–	M+		
	No?		1684	M+	1684			
							1685?	

?	1686–							
1687	1687–	1687	**1687**	**S+**	**1687–**	**S**	*1687?*	
1688	1688				**1688**			
	No		**1692**	**S**	1692	M+		
					1694–	**VS**		
?			1695	M	**1695**			
	1696							
1697	No		1697	M+	1697	M		
	1701		**1701**	**S+**	1701	M		No?
							1702–	No
		1703			**1703–**	**S**	*1703–*	**1703**
			1704	M	**1704**		*1704?*	**1704**
	No		1707–		1707–			
	No		1708–	M/S	1708–	M		
	No		1709		1709		*1709?*	1709?
								1711
			1713	M	1713–	M+		
	No				1714			
	No		**1715–**	**S**	**1715–**	**S+**		
	No?		**1716**		**1716**			
	1718		1718	M+	1718	M	*1718?*	
	1720		**1720**	**VS**	1720	M+		1720
1723	No		1723	M+	**1723**	**S**		1723
					1725	M		
	1728		**1728**	**VS**	1728	M		
								1729
					1731	M+		1731
								1732
		1733					*1733?*	
			1734	M	1734	M		1734
			1737	**S**	**1737**	**S**	*1737?*	**1737**
							1739?	

Table 7.6 (*cont.*)

South America El Niño records			Quinn's (1993) revised chronology and ranking		Eastern Hemisphere ENSO compilation		India droughts	South America El Niño records
Chile	Peru	Galapagos	E Pacific El Niño events		Global ENSO events			
Ortlieb 1994	Present work	Dunbar et al. 1994	Quinn 1993		Quinn 1993		Whetton & Rutherfurd 1994	
(a)	(a)	(b)	(c)		(c)		(d)	(e)
?								**1743**
1744	No		1744	M+	1744	M+	*1744?*	1744
1746							*1746–*	1746
	1747–		**1747**	S+	**1747–**	**S**	*1747?*	
1748	**1748**				**1748**			1748
1751			1751	M+	1751	M+		No
							1752?	1752
			1754–	M	**1754–**	**S**		
	No		1755		**1755**			
	No		1758	M	1758	M		
	1761	1761	**1761**	**S**	**1761–**	**S**		
		1762			**1762**			1762
		1763						
1764	No							
			1765	M	1765–	M+		1765
					1766			
1768	?		1768	M	1768–	M+		
					1769		*1769–*	**1769**
							1770	1770

			1772	M	1772–	M		
				M	1773			
			1776–		1776–			
			1777–	**S**	1777–	M+		
	No?		**E1778**		E1778			1778
	No?	1779						No
			1782–	**S**	**1782–**		1782	**1782**
1783	No?		**1783**		**1783–**	**VS**	1783	**1783**
	1784–				**1784**			1784
	1785		1785–	M+	1785–	M+		
	No		1786		1786			
					1790–			
	1791	1791	**1791**	**VS**	**1791–**	**VS**	1791	1791
					1792–			
					1793			
					1794–			
		1795			1795–	M+		
					1796–			
					1797			
					1799	M		1799
		1800						
					1802–		1802–	1802
	1803–		**1803–**	**S+**	**1803–**	**S+**	1803	1803
	1804		**1804**		**1804**			
	No		1806–	M	1806–	M	1806	
	No	1807	1807		1807			
			1810	M	1810	M		
	No		1812	M+	1812	M+	1812	
	1814		**1814**	**S**	**1814**	**S**		1814

Table 7.6 (*cont.*)

South America El Niño records			Quinn's (1993) revised chronology and ranking		Eastern Hemisphere ENSO compilation		India droughts	South America El Niño records
Chile	Peru	Galapagos	E Pacific El Niño events		Global ENSO events			
Ortlieb 1994	Present work	Dunbar et al. 1994	Quinn 1993		Quinn 1993		Whetton & Rutherfurd 1994	
(a)	(a)	(b)	(c)		(c)		(d)	(e)
1817	1817–		1817	M+	1817	M+		1817
	1818–							1818
1819	1819		1819	M+	1819	M+		
1820								
1821	1821		1821	M	1821	M		
		1823					1823–	
	1824		1824	M+	**1824–**	**S**	1824	**1824**
					1825			**1825**
1827					**1827–**	**S+**		
1828	**1828**		**1828**	**VS**	**1828**			
1829								
			1830	M	1830	M		
		1831						
	1832		1832	M+	**1832–**	**S+**	1832–	1832
1833					**1833**		1833	1833?
					1835–	M		1835
					1836			

1837	1837		1837	M+	**1837–**		1837–	1837
					1838–	**S**	1838	
					1839			
		1840						
1841								
								1842
1843								No?
	1844–		**1844–**		**1844–**		1844	
1845	**1845**		**1845–**	**S**	**1845–**	**VS**		
	No?		**E1846**		**E1846**			1846
1850	1850		1850	M	**1850**	**S**		1850
1851								
	1852		1852	M	1852–	M		
		1853			E1853		1853	1853?
	1854		1854	M	**1854–**	**S**		No?
1855					**1855**			1855
1856								
	1857		1857–	M	1857–			
1858	No?		1858		1858–	M+		
					E1859			
1860	No		1860	M	1860	M	1860	1860?
	1861–							
	1862		1862	M–	1862	M–		1862
1864	**1864**		**1864**	**S**	**1864**	**S+**	1864–	1864
		1865			L1865–	M+	1865–	
	1866		E1866	M+	E1866		1866	1866?
	No		L1867–	M+	**L1867–**			1867
1868	No		1868		**1868–**	**S+**	1868	**1868**
					E1869			No?
	1871		**1871**	**S+**	1871	M	1871	
					1873–	M+		1873

Table 7.6 (*cont.*)

South America El Niño records			Quinn's (1993) revised chronology and ranking		Eastern Hemisphere ENSO compilation		India droughts	South America El Niño records
Chile	Peru	Galapagos	E Pacific El Niño events		Global ENSO events			
Ortlieb 1994	Present work	Dunbar et al. 1994	Quinn 1993		Quinn 1993		Whetton & Rutherfurd 1994	
(a)	(a)	(b)	(c)		(c)		(d)	(e)
	No		1874	M	1874			No
					L1876–		1876–	1876
1877	**1877**		**1877–**	**VS**	**1877–**	**VS**	1877	**1877**
1878	**1878**		**1878**		**1878**			
1880	1880		1880	M	1880–	M+		
					1881			No?
								1882
	1884		**1884**	**S+**	1884–	M+		
					1885			
								1886
1887?	1887–	1887	L1887–		**L1887**			No?
1888	1888–	**1888**	1888–	M	**1888–**	**S**		**1888**
	1889		E1889		**E1889**			No?
1891	**1891**		**1891**	**VS**	1891	M	1891	**1891**
	No				1896–	M+		
	1897		1897	M+	1897			
1899	**1899**		**1899–**	**S**	**1899–**	**VS**	1899	**1899**
1900	No		**E1900**		**M1900**			

For the sixteenth century, only one possible coincidence is observed (1574; Table 7.6). For the seventeenth century, there are only three possible coincidences (1607, 1618, and 1687–98). So, in these two early centuries, none of the strongest events in either Chile or Peru correspond to each other (1544, 1574, 1609, 1647, and 1697 in Chile; 1578 and 1596 in Peru). For the eighteenth century, the correlation is scarcely better. One strong event (1748) was identified in both regions (1747–48 in Peru), while four strong events that were identified in Peru (1701, 1720, 1728, and 1791) have no counterpart in central Chile. For the much better documented nineteenth century, the coincidences are more numerous (fifteen episodes). Six strong/very strong events were clearly recognized as such in both areas: 1828 (including 1827 and 1829), 1845, 1864, 1877 (1877–78), 1891, and 1899. Two strong events in Peru have no counterparts in Chile (1871 and 1884). Several strong or very strong events in Chile (1817, 1845, 1850, 1880, 1888, and 1899) are identified with an apparently weaker relative intensity in Peru (strong, moderate, or weak), but this comparison obviously relies upon scales that should be adjusted.

On the whole, it may thus be concluded that a fair correlation between the Peruvian and Chilean records existed only since the early nineteenth century. During the sixteenth, seventeenth, and eighteenth centuries, the best assessed indications of El Niño manifestations seldom coincide in time in both regions (only seven cases total). This observation, which is based on data that are better assessed than they were previously (e.g., Ortlieb 1994), reinforces the hypothesis that the teleconnection pattern that presently links the precipitation regimes in northern Peru and central Chile during El Niño events was different before the early nineteenth century. Such a hypothesis merits further investigation, especially of climatology in northern Peru for the sixteenth through eighteenth centuries. To establish, or reject, the nonsynchronicity of the meterological manifestations assigned to the El Niño system in both regions, we need to exclude the possibility that it is due to the inaccuracy of the documentary data.

A complementary approach would be to validate one or both records from western South America by correlation with other records from the eastern Pacific or from the other rim of the Pacific Ocean.

Comparisons with Other Historical ENSO Records

Among the high-resolution natural records that can be used to establish the El Niño chronological sequence for South America, one of the most favorable is a coral reef sequence of the Galapagos. The other proxy records are provided by high Andes ice caps (see chapter by Thompson et al., this book) and tree rings from subtropical Chile and Argentina (see chapter by Cook et al., this book). The coral record presents the advantage that it reflects more directly oceanic perturbations than the ice sequence (which is linked more to the Atlantic/Amazonian system) or the dendroclimatic records of Chile and Argentina (which are influenced by southern South American circulation patterns). A yearly resolved coral sequence from the Galapagos Islands, covering the past four centuries (1607–1953), has been published by Dunbar et al. (1994). Oxygen isotope measurements on annual growth layers provide information on paleotemperature variations that relate to El Niño conditions. The isotope record shows a satisfactory correspondence with the Quinn record, although a number of warmer annual episodes may need to be shifted by one year (after or before) with respect to the events defined by QNA (Dunbar et al. did not use the more recently published papers of Quinn). Some episodes do not have counterparts in the QNA chronology and are supposed, according to Dunbar et al. (1994), to represent El Niño events that might not have been identified in the Quinn record. The chronological sequence of the Galapagos coral reef is constructed from the thirty largest negative $\delta^{18}O$ excursions (with the strongest sea surface temperature [SST] anomalies, indicated in bold, Table 7.6).

The comparison between the coral record and both the Chilean and Peruvian records is disappointing because the largest isotope anomalies do not correlate with the strongest events as recognized in either Peru or Chile (Table 7.6). In only three instances (1607, 1687, and 1888) are there coincidences between the three records, and it is only in 1888 that the coincidence concerns a strong event (in Galapagos and in Chile). If the intensities of the events/episodes of elevated SST are not taken into consideration, and if quasi-coincidences ($\pm$1-year shift) are accepted, about a dozen fits are observed between the Galapagos data and either the Chilean or the Peruvian records (Table 7.6). The coincidences do not favor one of the two (Chile and Peru) records. No systematic temporal shift with the Peruvian or Chilean records is observed. This situation is somewhat puzzling, to the point that one may wonder if the chronological control is as tight as it seemed to Dunbar et al. (1994), or alternatively whether some bias may explain the general lack of correspondence between the strongest events/episodes of the Galapagos, Chilean, and Peruvian records. Ongoing research on seasonal variations in reconstructed paleotemperatures for the past two centuries on another coral sequence from the Galapagos (R. Dunbar and colleagues, in preparation) should bring new light on this problem.

Table 7.6 also shows chronological series of ENSO events as they were determined in India (Whetton and Rutherfurd 1994), and by way of correlation of different records (India, Java, Nile, North China, and Peru–Chile). The table also includes the global ENSO chronology of Quinn (1993). The two synthetic chronological sequences based on the India and Eastern Hemisphere data are not independent from QNA and regional El Niño chronologies from Quinn (1993), since Quinn also integrated part of this information. Comparison of both sequences – regional evidence of El Niño and global-scale features associated with the large-scale ENSO phenomenon – shows that global ENSO events are more numerous than the regionally experienced El Niño events, and that the former tend to last longer than the eastern Pacific events, particularly in the eighteenth and nineteenth centuries. It may also be noted that a number of strong events identified by Quinn (1993), in both the regional and global El Niño patterns during the sixteenth, seventeenth, and eighteenth centuries, have not been identified, either in the Chilean record (Ortlieb 1994) or in the revised Peruvian record (this work, second column, Table 7.6). For the nineteenth century, the larger number of reconstructed events in all records may be attributed to the increased availability of data, including instrumental data. However, there is also a possibility that the ENSO system behaved differently at the end of and shortly after the LIA and that the teleconnection pattern was modified at the LIA–post-LIA transition. If this is the case, the better correlation observed since the beginning of the nineteenth century would not be an artifact of the documentary records.

Intrinsically, the records from Chile, Peru, and India are based on the same kind of documentary data and include comparable numbers of El Niño/ENSO manifestations. The Galapagos coral record presents a lower number of events, but this is due to the threshold fixed by Dunbar et al. (1994) for the stable isotope excursions. It should be clear that if the records of El Niño/ENSO events proposed by Quinn (1993) are denser than the other ones in Table 7.6, it is basically because Quinn compiled information from Chile, Peru, and India (among other regions) to elaborate these records. Quinn's chronological sequences thus appear much more "complete" than any of the regional series, but, clearly, this does not imply that the former are more accurate than the latter.

Another problem that arises from the comparison shown in Table 7.6 deals with the validity of the reconstruction of the event intensities. As was stressed previously, the Chilean and Peruvian records generally do not present chronological coincidences of the strongest events (except in three instances, at the end of the nineteenth century). The intensity ratings of the events used in Quinn's compilation do not necessarily represent an integration of the widely spaced proxies but seem to rely upon one or another regional record, according to the events. This is how Quinn's record of "regional" El Niño events includes the strongest events documented in Peru and those registered in central Chile. This seems to be true for the first three centuries of the historical sequence.

Conclusions

From QNA and Q&N Chronologies to the "Peruvian" Record of El Niño Events

One of the aims of this chapter was to give an insight into the large body of documents that constitute the background of Quinn's work. After many years of general uncritical acceptance of the QNA chronology, it seemed useful to reexamine critically the nature and quality of the data that support the determination of the event occurrences in the past few centuries. The summary Table 7.1 provides for each of the events of the sixteenth through nineteenth centuries another kind of "confidence rating" than the indices given by Quinn.

Although this critical analysis of the sources used in the QNA record should certainly not be considered as definitive, it takes into consideration a number of sources that had not been previously available to H&O. Table 7.1 also includes some important new data. Of particular relevance are the new sources and unpublished archives revealed by historical studies in northern Peru (Schlüpmann 1988, 1994; Huertas 1987, 1992, 1993).

At this stage of a long-range, ongoing study it might be relevant to state a few points regarding the reliability of the sources used in the reconstruction of El Niño events, the evaluation of the event strengths, the number of historical events, and the concept of a "regional El Niño" record.

Reliability of Sources

As is shown by a few examples in the critical analysis of the Quinn records, evaluation of the source reliability is of major importance in this kind of study. It seemed useful to comment on the trustworthiness of particular sources in Table 7.1. Since the question of the discrimination between reliable and less reliable sources was not given particular emphasis in the QNA paper, one may conclude that they considered as equally reliable all the sources cited by them. It was shown here that several sources are actually unreliable. As a result of the H&O work and this study, a ranking of the trustworthiness of the sources may be proposed.

For the conquest period, the best informants are those who accompanied Pizarro during his rapid march toward Cuzco (i.e., Xerez 1534; Estete 1535; Mena 1534; Trujillo 1571; and Ruiz de Arce 1545). For the rest of the sixteenth century, the most reliable sources proved to be Zárate (1545), Cieza de Leon (1553), Benzoni (1572), Cabello Valboa (1586), Acosta (1590), Ocaña and Alvarez (1596), and Lizárraga (1603–09). For the seventeenth century, trustworthy information was given by Suardo (1629–39) and Cobo (1639, 1653), but not precisely by Montesinos (1642). For the eighteenth century, Anson (1748), Feijoo de Sosa (1763), Bueno (1763), Rubiños y Andrade (1782), Haenke (1790), and Lequanda (1793) are to be considered among the reliable authors, while Juan and Ulloa (1748) and Alcedo (1786–89) committed several errors in their writings and induced several misinterpretations. For the nineteenth century, the most reliable and accurate informants were Unanue (1806), Helguero (1802–03), Spruce (1864), and Eguiguren (1894), while Stevenson (1825), Ruschenberger (1835), Paz Soldán (1862), and Palma (1894) should be classified as compilers who were not always critical enough. Finally, for the twentieth century, it must be noted that, unfortunately, the three authors most frequently cited by QNA and Q&N should not be fully trusted: Labarthe (1914), who provided some erroneous data; Portocarrero (1926), who did not bring any new information with respect to Labarthe; and Taulis (1934), who did not cite any of his sources. Among the twentieth-century authors to be trusted, one may distinguish those from the beginning of the century, such as Garcia Rosell (1903, 1904, 1907), Fuchs (1907), Murphy (1925, 1926), Remy (1931), or Petersen (1935), from those who investigated in a modern way national, provincial, and municipal archives of the past centuries, such as Hamerly (1973), Estrada Ycaza (1977), Huertas (1987, 1993), or Schlüpmann (1988, 1994).

Event Strength Reevaluation

Determination of the strengths, or intensities, of past El Niño manifestations is a demanding endeavor, even when instrumental records are available. Without precise knowledge of the spatial extension of El Niño manifestations, it is hazardous to attempt to make a fair determination of the strengths of the events. And, clearly, this information is seldom available in the documentary records. Consequently, it is natural that subjective elements are involved in the classification of events.

The tendency of H&O to propose a reduction of the strengths of many events, with respect to the QNA evaluation, is generally confirmed here. In many cases for the sixteenth and seventeenth centuries, QNA tended to exaggerate the intensities of El Niño events (Table 7.1). Typical cases are those for which there is only an indication of a single shower, a thunderstroke, or a river flood, and which were related to "strong" events by QNA. Actually, in his latest papers (1992, 1993, Q&N), Quinn downgraded the intensity of a number of events with respect to the original QNA record (compare the last two columns of Table 7.5).

Because of the intrinsically fragmentary information provided by the documentary sources, it may be concluded that the evaluation of the intensities of former events should involve a wider regional analysis of meteorological impacts that include the countries neighboring Peru. This task, beyond the scope of the present study focused on the Peruvian (and southernmost Ecuador) record, should incorporate not only documentary data from Ecuador, Bolivia, and Argentina but also proxy records, particularly dendroclimatic records from southern South America.

Number of Events

This revision of the QNA and Q&N sequences led me to propose the suppression of some events and the addition of some new events (Table 7.1). Four newly proposed events and three cases of extension to a previous or succeeding year of a previously defined El Niño year are supported by evidence of rainfall occurrences in northern Peru or southern Ecuador (Table 7.7). In most cases, the interpretation of an El Niño occurrence relies on a single source and may not be fully accepted until a confirmation is obtained.

For some forty-two events of the Quinn records (QNA and Q&N), the reevaluation of the sources and combination with new sources led me to question the occurrence of El Niño conditions. Two situations were found: Either the available data were precise enough to determine that no El Niño occurred that year or the information at hand was not sufficient to preclude the occurrence of an El Niño event. In some twenty-five cases, it was possible to deny the occurrence of El Niño conditions, most often because drought conditions seem to have been prevailing in northern Peru at those times (e.g., 1531–32, 1552, 1655, 1707–08–09, 1714–15, 1775, 1806–07, 1812, 1860, 1867–68, 1874, and 1900). In some cases, it was because there was a unique, poorly reliable source (e.g., 1600, 1604, 1660, 1681, 1740, 1750) or because data from central Chile, eastern Bolivia, and/or northeast Brazil were provided as the only evidence (e.g., 1640–41, 1647, 1692–93, 1723, 1736, 1744, and 1764). In seventeen other cases (indications "?" or "No EN?" in the last column of Table 7.1), the analysis of the sources on which QNA and Q&N had based their interpretation showed that the data were irrelevant, or insufficient, to support the occurrence of an event, but it could not be demonstrated unequivocally that no event occurred that year.

Finally, there are some instances in which the reconstruction of an El Niño event was confirmed but its duration was restricted, on the basis that no positive evidence was confirmed for the first or the second year, in Peru at least (e.g., 1697, 1786, 1896, and 1900).

Table 7.7 *List of El Niño years not identified by QNA, Q&N, or Quinn (1992, 1993) and for which new evidence has been obtained from northern Peru and southern Ecuador anecdotical records. See Table 7.1 for sources of the records.*

Year	Reference	Location of anomaly	Interpretation of El Niño event occurrence and strength
1593	H&O 1990	Trujillo, N Peru	Moderate (?) event
1622	This work (Table 7.1)	Jayanca, N Peru	Moderate (?) event
1678	This work (Table 7.1)	Jayanca Vieja, N Peru	Moderate (?) event
1686 (–1687–1688)	H&O 1990	Yapatera (Piura), N Peru	Extension to 1686 of the 1687–88 Moderate (?) event
(1747–) 1748	H&O 1990	Sancor (Piura) and Chocope, N Peru	Extension to 1748 of the 1747 Strong event
1784 (–1785)	This work (Table 7.1)	Daule River, S Ecuador	Moderate event
1861–(1862)	This work (Table 7.1)	Paita and Piura, N Peru	Extension to 1861 of the 1862 Weak event?

The "Regional El Niño Record" Concept

The general lack of correlation between the local records in Chile and Peru in pre-nineteenth century times casts some doubts on the concept of a "regional El Niño chronology" in the sense of Quinn (1992, 1993). It appears that the so-called regional character of the chronological sequence of events might have actually resulted from the amalgamation of data from both regions.

The careful revision of the QNA and other sequences of Quinn, and the poor correlation finally observed between the Chilean and Peruvian (revised) records, suggest modifications in the teleconnection pattern of El Niño manifestations during the sixteenth through eighteenth centuries versus the twentieth-century situation. If this were confirmed, it may be due to some interactions between the LIA climatic system and the ENSO mode. Some previous studies, which had relied entirely upon QNA data, had concluded that the frequency of El Niño events did not show variations with respect to the LIA–post-LIA climate change (Enfield 1988, 1992; Enfield and Cid 1991). Since the present study modified part of the database of these statistical studies, it would not be surprising that a reprocessing would produce a different conclusion. It might even be possible that such a treatment could provide some assessment of the teleconnection problem brought up here.

Acknowledgments This work was supported by ORSTOM, L'Institut Français de Recherche Scientifique pour le Développement en Coopération (successively: UR 12, Programme AIMPACT, Analyses Intégrées des Marqueurs Paléoclimatiques Continentaux et littoraux Tropicaux and UR 1, Programme PVC, Paléoclimatologie et Variabilité Climatique) in the framework of cooperative scientific agreements with the Instituto Geofísico del Perú (Lima), the Universidad de Chile, and the Universidad de Antofagasta (Antofagasta, Chile). The author thanks many colleagues for fruitful discussions, and particularly A.-M. Hocquenghem, P. Aceituno, J. Ruttlant, H. Fuenzalida, J. Macharé, M. R. Prieto, A. Gioda, and G. Vargas. Special thanks are due to M. Soto (IFEA, Lima), who succeeded in finding in Lima most of the sources cited by W. Quinn and collaborators.

The author sincerely thanks H. Diaz and V. Markgraf for their kind invitation to contribute a chapter to this book. An anonymous reviewer helped to improve a preliminary version of the manuscript.

This paper is dedicated to Paul Ortlieb, who helped his son in many ways during the long-lasting preparation of the manuscript, and who died before the book was printed.

References

ACEITUNO, P., 1987: On the interannual variability of South American climate and the Southern Oscillation. Ph.D. thesis, University of Wisconsin, Madison, 128 pp.

ACEITUNO, P., 1988: On the functioning of the Southern Oscillation in the South American sector. Part 1: Surface climate. *Monthly Weather Review*, **116**(3), 505–524.

ACOSTA, J., 1590 [1954]: Historia Natural de las Indias. *In Obras del Padre José de Acosta.* Biblioteca de Autores Españoles, Madrid:

ESTRADA YCAZA, J., 1977: *Regionalismo y Migración*. Archivo Histórico del Guayas, Guayaquil, 296 pp.

FEIJOO de SOSA, M., 1763 [1984]: *Relacion Descriptiva de la Ciudad y Provincia de Trujillo del Perú*, 2 vol., Banco Industrial del Perú, Lima, 293 pp.

FRANCOU, B. and PIZARRO, L., 1985: El Niño y la sequía en los altos Andes centrales (Perú y Bolivia). *Bulletin de l'Institut Français d'Etudes Andines*, **14**(1–2), 1–18.

FUCHS, F. C., 1907: Zonas Iluviosas y secas del Perú. *Memorias y Boletín de la Sociedad de Ingenieros del Perú*, t. **VII**, 270–297.

FUCHS, F. C., 1925: Las Ùltimas Lluvias y la Ciencia Meteorológica. *La Vida Agricola*, Lima, pp. 521–524.

GALDÓS, G., 1988: La sequía en el sur de la Intendencia. *Diario El Pueblo*, 15 Diciembre 1988, Arequipa.

GARCIA PEÑA, A. and FERNANDEZ, I., 1985: Analisis climático de las regiones de la sierra y selva en períodos que ocurren fenómenos El Niño. *In Proceedings of the Seminario Regional Ciencia, Tecnologia y Agresión Ambiental, El Fenómeno El Niño*. Lima: CONCYTEC, pp. 521–550.

GARCIA RODRIGUEZ, V. J., 1779: *Geografía de el Perú*, Tomo II. Lima.

GARCIA ROSELL, R., 1903: Monografía histórica del Departamento de Piura. *Boletín de la Sociedad Geográfica de Lima*, **13**(2), 193–242; **13**(3), 310–351; **13**(4), 419–462.

GARCIA ROSELL, R., 1904: Monografía histórica del Departamento de Piura. *Boletín de la Sociedad Geográfica de Lima*, **15**(1), 96–112.

GARCIA ROSELL, R., 1907: Monografía historica sobre el Departamento de Piura. *Boletín de la Sociedad Geográfica de Lima*, **21**, 86–120.

GAUDRON, J., 1925: Las Lluvias en la Costa y la Periodicidad de los Fenómenos Meteorológicos. *La Vida Agricola*, Lima, pp. 361–368.

GONZÁLEZ, B., 1913: *Ligero Estudio Sobre la Meteorologia de los Vientos en Lima*. Tesis de doctorado en Ciencias, Facultad de Ciencias, Universidad Nacional Mayor de San Marcos, Casa Editorial Sanmarti y Cia, Lima, 38 pp. + tables.

GORBITZ, G., 1978: Las alteraciones del clima de verano en la costa norte del Perú. *Boletín de la Sociedad Geográfica de Lima*, **97**, 32–42.

HAENKE, T., 1799 [1901]: *Descripción del Perú* (reprint of a British Museum manuscript by Imprenta El Lucero), Lima, 320 pp.

HAMERLY, M. T., 1973: *Historia Social y Económica de la Antigua Provincia de Guayaquil, 1763–1842*. Publicaciones del Archivo histórico del Guayas, Guayaquil, 209 pp.

HAMILTON, K. and GARCIA, R. R., 1986: El Niño/Southern Oscillation events and their associated midlatitude teleconnections, 1531–1841. *Bulletin of the American Meteorological Society*, **67**, 1354–1361.

HELGUERO, J., 1802–03 [1984]: *Informe Económico de Piura 1802*. Lima: Editorial CIPCA.

HISARD, PH., 1992: Centenaire de l'observation du courant côtier El Niño, Carranza, 1892: Contributions de Krusenstern et de Humboldt à l'observation du phénomène El Niño. *In* Ortlieb, L. and Macharé, J. (eds.), *Paleo ENSO Records International Symposium Extended Abstracts*. Lima: ORSTOM/CONCYTEC, pp. 133–141.

HOCQUENGHEM, A. M. and ORTLIEB, L., 1990: Pizarre n'est pas arrivé au Pérou durant une année El Niño. *Bulletin de l'Institut Français d'Études Andines*, **19**(2), 327–534.

HOCQUENGHEM, A. M. and ORTLIEB, L., 1992a: Historical records of El Niño events in Peru (XVI–XVIIIth centuries): The Quinn et al. (1987) chronology revisited. *In* Ortlieb, L. and Macharé, J. (eds.), *Paleo ENSO Records International Symposium Extended Abstracts*. Lima: ORSTOM/CONCYTEC, pp. 143–149.

HOCQUENGHEM, A. M. and ORTLIEB, L., 1992b: Eventos El Niño y lluvias anormales en la costa del Peru: Siglos XVI–XIX. *Bulletin de l'Institut Français d'Etudes Andines*, **21**(1), 197–278.

HUAMAN SOLIS, F. and GARCIA PEÑA, A., 1985: Condiciones meteorológicas en el Perú durante el fenómeno "El Niño" 1982–83. *In Proceedings of the Seminario Regional Ciencia, Tecnologia y Agresión Ambiental, El Fenómeno El Niño*. Lima: CONCYTEC, pp. 333–353.

HUERTAS, J., 1984: *Tierras, diezmos y tributos en el obispado de Trujillo*. Lima: Universidad Mayor de San Marcos.

HUERTAS, L., 1987: *Ecologia e Historia. Probanzas de Indios y Españoles Referentes a las Catastróficas lluvias Producidas en 1578 en los Corregimientos de Trujillo y Zaña (Francisco Alcocer, Escribano Receptor)*. Chiclayo: Editorial CES Solidaridad, 208 pp.

HUERTAS, L., 1992: Diluvios, terremotos y sequias: Factores disturbadores del orden económico y social. *Revista Rimaq*, **1**(1), 99–118.

HUERTAS, L., 1993: Anomalías cíclicas naturales y su impacto en la sociedad: El fenómeno El Niño. *Bulletin de l'Institut Français d'Etudes Andines*, **22**(1), 345–393.

HUTCHINSON, G. E., 1950: Survey of existing knowledge of biogeochemistry, 3. The biogeochemistry of vertebrate excretion. *Bulletin of the American Museum of Natural History*, **96**, 554 pp.

HUTCHINSON, T. J., 1873: *Two Years in Peru with Exploration of Its Antiquities*, 2 vol. London: Samson Low, Marston & Searle.

JONES, H. H., 1933: Notas sobre la meteorologia y agricultura del Departamento de Piura. *Boletín de la Sociedad Geográfica de Lima*, **50** (1), 16–18.

JUAN, J. and ULLOA, A. de, 1748 [1978]: *Relación Histórica del Viaie a la America Meridional*, 2 vol. Fundación Universitaria Española, Madrid, 877 pp.

KILADIS, G. N. and DIAZ, H. F., 1986: An analysis of the 1877–78 ENSO episode and comparison with 1982–83. *Monthly Weather Review*, **114**, 1035–1047.

KILADIS, G. N., and DIAZ, H. F., 1989: Global climatic anomalies associated with extremes in the Southern Oscillation. *Monthly Weather Review*, **2**, 1069–1090.

La PATRIA, 1874: Crónica interior (Chincha Alta, Cañete, Pisco, Ica), Lunes 9 de Febrero 1874, p. 2, Lima. [Same text published on 10 February 1874 in *El Comercio*, p. 4.]

LABARTHE, P. A., 1914: Las avenidas extraordinarias en los rios de la costa. *Informes y Memorias de la Sociedad de Ingenieros del Perú*, **16**(11–12), 301–329.

LASTRES, 1937: (Citation referred to, but not given, in Q&N Table 32.1.)

Le GENTIL, L., 1728: *Nouveau Voyage Autour du Monde*. Amsterdam: P. Mortier. (Cited as "Gentil" by QNA.)

LEGUIA Y MARTINEZ, G., 1914: *Diccionario Geográfico, Histórico, Estadístico, etc. del Departamento de Piura*. Lima: Tipografía El Lucero.

LEÓN BARANDIARÁN, A., 1938: *Mitos, Leyendas y Tradiciones Lambayecanas*. Lima, 280 pp.

LEQUANDA, J. I., 1793 [1965]: Descripción geográfica del Partido de Piura perteneciente a la Intendencia de Truxillo. *In Mercurio Peruano*, 11 de Julio de 1793, **VIII**, 167–181.

LIZÁRRAGA, R. de, 1603–09 [1968]: *Descripción Breve de Toda la Tierra del Perú, Tucumán, Río de la Plata y Chile*. Biblioteca de Autores Españoles, Madrid, pp. 1–213.

LLANO Y ZAPATA, J. E., 1748: *Observación Diaria Critico-Histórico Meteorológica, Contiene Todo lo Acaecido en Lima Desde Marzo de 1747 Hasta 28 de Octubre del Mismo Año*, etc. Manuscript letters, Biblioteca Nacional de Lima, Lima, 50 pp.

LÓPEZ MARTINEZ, H., n. d: LAS LLUVIAS de 1891. (Manuscript reproduced in Peralta 1985.)

MABRES, A., WOODMAN, R., and ZETA, R., 1993: Algunos apuntes históricos adicionales sobre la cronología de El Niño. *Bulletin de l'Institut Français d'Études Andines*, **22**(1), 395–406.

MACHARÉ, J. and ORTLIEB, L., 1993: Registros del Fenómeno El Niño en el Perú. *Bulletin de l'Institut Français d'Études Andines*, Lima, **22**(1), 35–52.

MARTÍNEZ Y VELA, B., 1702 [1939]: *Anales de la Villa Imperial de Potosí* (Reprinted Artística), La Paz. (See Arzans de Orsúa y Vela, same author, same work, later edition.)

MELO, R., 1913: Hidrografía del Perú. *Boletín de la Sociedad Geográfica de Lima*, **29**(1–2), 141–159.

MENA, C. de, 1534 [1968]: La Conquista del Perú. *In El Perú a Través de los Siglos*. Biblioteca Peruana, Tomo 1, 133–169. Lima: Editores Técnicos Asociados.

MERCURIO PEROANO, 1791 [1964]: Ed. Facsimilar, Tomo II. Biblioteca Nacional del Perú, Lima.

MIDDENDORF, E. W., 1894: *Peru-Beobachtungen und studien uber das land und seine bewohner. Band II, Das kustenland von Peru.* Berlin: Rober Oppenheim (Gustav-Schmidt).

MINAYA, N. A., 1993: El Niño/Oscilción del Sur y las descargas del Rio Camana-Majes. *Espacio y Desarrollo*, **5**, 53–71.

MINAYA, N. A., 1994: El Niño/Oscilación del Sur y las precipitaciones en la Costa central y sur del Perú. Unpublished report, Pontificia Universidad Católica del Perú, Instituto Geofísico del Perú, Lima, 118 pp.

MONTESINOS, F., 1642 [1906]: *Anales del Perú*, 2 vol. Madrid: V. M. Maúrtua.

MORENO, G., 1804: *Almanaque Peruano y Guia de Forasteros, para el Año 1800*. Lima: Imprenta Real del Telégrafo Peruano.

MURPHY, R. C., 1923: The oceanography of the Peruvian littoral. *Geographical Review*, **13**, 64–85.

MURPHY, R. C., 1925: *Bird Islands of Peru: The Record of a Sojourn on the West Coast*. New York: Putnam, 362 pp.

MURPHY, R. C., 1926: Oceanic and climatic phenomena along the west coast of South America during 1925. *Geographical Review*, **16**, 26–54. (Spanish translation also published in 1926: Fenómenos oceánicos y climatéricos en la costa occidental de Sur-America durante el año 1925, *Boletín de la Sociedad Geográfica de Lima*, **63**(2): 87–125.)

OCAÑA, D. de and ALVAREZ, A., 1596 [1969]: *Un Viaje Fascinante por la América Hispana del Siglo XVI*. Lima: Studium.

OLIVA, A., 1631 [1895]: *Historia del Perú y Varones Insignes en Santidad de la Compañia de Jesús*. Varela, J. F. P. and Varela y Orbegozo, L. (eds.). Lima: Imprenta y Libreria de San Pedro, 216 pp.

ORTLIEB, L., 1994: Las mayores precipitaciones históricas en Chile central y la cronología de eventos "ENSO" en los siglos XVI–XIX. *Revista Chilena de Historia Natural*, **67**(3), 117–139.

ORTLIEB, L., 1995: Eventos El Niño y episodios lluviosos en el Desierto de Atacama: El registro de los últimos dos siglos. *Bulletin de l'Institut Français d'Etudes Andines*, **24**(3), 519–537.

ORTLIEB, L., HOCQUENGHEM, A. M., and MINAYA, A., 1995: Toward a revised historical chronology of El Niño events registered in western South America. XIV INQUA Congress, Berlin, 1995. Abstr. vol., *Terra Nostra*, 2/95, p. 113.

ORTLIEB, L. and MACHARÉ. J., 1993: Former El Niño events: Records from western South America. *Global and Planetary Changes*, **7**, 181–202.

PALMA, R., 1894 [1961]: *Tradiciones Peruanas*. Madrid: Aguilar, 1783 pp.

PAREDES, J. G., n. d., Calendario y guía de forasteros para 1829 (as cited by Eguiguren 1894).

PAZ SOLDÁN, D. D. M., 1862: *Geografia del Perú*. Librairie M. A. Paris: Durand.

PERALTA, H., 1985: *El Niño en el Perú*. Lima: Instituto José María Arguedas Editores, 287 pp.

PETERSEN, C., 1935: Estudios climatológicos en el noroeste Peruano. *Boletín de la Sociedad Geológica del Perú*, **7**(2), 1–142. Reedition (in two parts) in *Boletín de la Sociedad Nacional de Minería y Petróleo*, Lima, 1956, **48**, 1–112, and **49**, 1–55.

PHILANDER, G. S., 1991: *El Niño, La Niña and the Southern Oscillation*. San Diego: Academic Press.

PORTAL, I. I., 1932: Del pasado limeño, Cuatro tempestades: La de 1877, Pánico en la ciudad. *El Comercio*, 27 Junio 1932, Lima.

PORTOCARRERO, J., 1926: Contribución al estudio hidrológico del territorio Peruano. *Informes Y Memorias de la Sociedad de Ingenieros del Perú*, **28**(2), 68–93.

PRESCOTT, W. H., 1892: *History of the Conquest of Peru*, Vol. 1. Philadelphia: J. B. Lippincott, 469 pp. Spanish edition: Prescott, G. H. [1955]: *Historia de la Conquista del Perú*. Buenos Aires. Ediciones Suma, 624 pp.

PRIETO, M. del R., 1994: Reconstrucción del clima de América del Sur mediante fuentes históricas, Estado de la cuestión. *Revista del Museo de Historia Natural de San Rafeal* (Mendoza), **XII** (4), 323–342.

PRIETO, M. del R., HERRERA, R., and DUSSEL, P., 1999: Clima y disponibilidad hídrica en el sur de Bolivia y noroeste de Argentina entre 1560 y 1710. Los documentos españoles como fuente de datos ambientales. *Bamberger Geographische Schriften*, **15**(36–37), 16 pp.

PUENTE, J. A. de la, 1885: *Diccionario de la Legislacion de Aguas y Agricultura del Perúi*. Lima: Imprenta de J. Francisco Sofis, 44 pp.

PULS, C., 1895: Oberflachtemperaturen und stromungsverhaltnisse des aequatorial gurtels des Stillen Ozeans. *Archiv Deutsch Seewarte*, **18**(1), 1–37, +3 charts.

QUINN, W. H., 1992: A study of Southern Oscillation–related climatic activity for A. D. 622–1900 incorporating Nile River flood data. *In* Diaz, H. F. and Markgraf, V. (eds.), *El Niño: Historical and Paleoclimatic Aspects of the Southern Oscillation*. Cambridge: Cambridge University Press, pp. 119–149.

QUINN, W. H., 1993: The large-scale ENSO event, the El Niño, and other important features. *Bulletin de l'Institut Français d'Etudes Andines*, Lima, **22**(1), 13–34.

QUINN, W. H. and NEAL, V. T., 1983a: Long-term variations in the Southern Oscillation, El Niño, and Chilean subtropical rainfall. *Fishery Bulletin*, **81**(2), 363–374.

QUINN, W. H. and NEAL, V. T., 1983b: Southern Oscillation–related climatic changes and the 1982–82 El Niño. *In Proceedings of the International Conference on Marine Resources of the Pacific*. Arana, P. M. (ed.), Santiago, pp. 71–82.

QUINN, W. H. and NEAL, V. T., 1992: The historical record of El Niño events. *In* Bradley, R. S. and Jones P. D. (eds.), *Climate Since* A.D. *1500*. London: Routledge, pp. 623–648.

QUINN, W. H., NEAL, V. T., and ANTÚNEZ de MAYOLO, S., 1987: El Niño occurrences over the past four and a half centuries. *Journal of Geophysical Research*, **92**(C13), 14449–14461.

RAIMONDI, A., 1874 [1965]: *El Perú*, tomo 1. Lima: Imprenta del Estado, 444 pp.

RAIMONDI, A., 1876: Historia de la geografía del Perú. *In* del Campo, J. E. (ed.), *Geografía del Perú*, tomo 2. Lima: Imprenta del Estado, 475 pp.

RAIMONDI, A., 1897: Geografía física. *Boletín de la Sociedad Geográfica de Lima*, **7**(7–9), 268–278.

RAMIREZ ZENÓN, ?, 1888: Manuel Tafur (biographical note). *El Perú Ilustrado*, Año 2, no. 62 (14 Julio 1888), p. 162. (Reproduced in San Cristoval, 1938).

RAMOS SEMINARIO, I., n. d.: Cartas de Don Pablo Seminario y Echeandía a su hermana Mariana Seminario y Echeandia de Shaefer, Archivos de familias en Piura, manuscript. (Reproduced in Hocquenghem and Ortlieb 1992b.)

REMY, F. E., 1931: De la lluvia en Lima. *El Comercio*, 21 August 1931, Lima.

ROCHA, D. A., 1681 [1891]: *Tratado Único v Singular del Origen de los Indios del Perú, México, Santa Fé y Chile*, 2 vol. (222 pp; 218 pp.). Lima: Imprenta Manuel de los Olivos.

ROPELEWSKI, C. C. and HALPERT, M. S., 1987: Global and regional scale precipitation patterns associated with El Niño/Southern Oscillation. *Monthly Weather Review*, **115**, 1606–1626.

ROSTWOROWSKI, M., 1985: El diluvio de 1578. Manuscript Biblioteca Nacional BN-534. (Reproduced in Peralta 1985.)

RUBIÑOS Y ANDRADE, J. M., 1782 [1936]: Noticia previa por el Liz. D. Justo Modesto Rubiños y Andrade, Cura de Mórrope año de 1782. *In* Un manuscrito interesante: Sucesión cronológica de los curas dé Mórrope y Pacora, Romero, C. E. (ed.), *Revista Histórica*, **10**(3), 289–363.

RUIZ de ARCE, J., 1545 [1968]: Advertencias. *In El Perú a Través de los Siglos*. Biblioteca Peruana, Tomo 1, 405–437. Lima: Editores Técnicos Asociados.

RUSCHENBERGER, W. S. W., 1835: *Three Years in the Pacific, Containing Notices of Brazil, Chile, Bolivia, Peru and Colombia in 1831, 1832, 1833, 1834*, 2 vol. (403 pp.; 440 pp.). London: R. Bentley. (The reference cited by QNA and Q&N is another edition published in 1834 by Carey, Lea and Blanchard, in Philadelphia.)

RUTTLANT, J. and FUENZALIDA, H., 1991: Synoptic aspects of the central Chile rainfall variability associated with the Southern Oscillation. *International Journal of Climatology*, **111**, 63–76.

SAN CRISTOVAL, E., 1938: *Apéndice al Diccionario Histórico-Biográfico del Perú*, Tomo 4. Lima: Libreria e Imprenta Gil, 526 pp.

SCHLÜPMANN, J., 1988: Piura du XVIème au XIXème siècle: Évolution de la structure agraire et formation d'une société régionale au nord du Pérou. Mémoire de DEA d'Histoire, Université de Paris VII, Paris, 125 pp.

SCHLÜPMANN, J., 1994: La structure agraire et le développement d'une société régionale au nord du Pérou, Piura, 1588–1854, 2 tomes. Thèse Doctorat, UFR Géographie, Histoire et Sciences de la Société, Université de Paris VII, Paris.

SCHOTT, G., 1938: Klimakunde der Sudsee Inseln. *In* Köppen, W. and Geiger, R. (eds.), *Handbuch der Klimatologie*. Berlin: Verlag von Gebruden Borntraeger.

SCHWEIGGER, E. H., 1959: *Die Westkuste Sudamerikas in Berich des Peru Stroms*. Heidelberg & Munchen: Keyserche Verslagsbuchhandling, 513 pp.

SCHWEIGGER, E. H., 1964: *El Litoral Peruano*, 2nd ed. Lima: Universidad Nacional Federico Villarreal, 414 pp.

SEMINARIO OJEDA, M. A., 1994: *Piura y la Independencia*. Piura: Gobierno local de Piura, 206 pp.

SHELVOCKE, C., 1726 [1971]: *A Voyage Round the World by the Way of the Great South Sea*. New York: Da Capo Press.

SIEVERS, W., 1914: *Reise in Peru und Ecuador Ausgefurt 1900*. Munich: Verlag von Duncker und Humboldt, 411 pp.

SPRUCE, R., 1864: *Notes on the Valleys of Piura and Chira in Northern Peru and on the Cultivation of Cotton Therein*. London: Eyre and Spottswoodie, 81 pp.

STEVENSON, W. B., 1825: *A Historical and Descriptive Narrative of Twenty Years Residence in South America*, Vol. 2. London: Longman Rees, Orme, Brown and Green, 434 pp. (Edition consulted published by Hurst, Robinson and Co., London, 1825, Vol. 2; Quinn used an edition dated 1829.)

SUARDO, J. A., 1629–39 [1936]: *Diario de Lima de Juan Antonio Suardo*, Tomo II. Biblioteca Histórica Peruana, Universidad Católica del Perú, Lima, 201 pp.

TAULIS, E., 1934: De la distribution des pluies au Chili. *In Matériaux pour l'Étude des Calamités*, Part 1, 3–20. Genève: Société de Géographie de Genève.

TÁVARA, D. S., 1854: *Proyecto de Irrigación con el Rio de La Chira, en la Provincia de Piura*. Report published by Imprenta El Comercio, Lima, and partially reproduced in *El Amigo del Pueblo*, Piura, **101** (28 November 1906), 4.

TIZÓN Y BUENO, R., 1907: Descripción sintética de las condiciones hidrológicas de la quebrada del Rímac. *Informes y Memorias de la Sociedad de Ingenieros del Perú* **9**(5), 97–119.

TRUJILLO, D. de, 1571 [1968]: *Relación del Descubrimiento del Reyno del Perú*, **2**, 9–103. Biblioteca Peruana, Editores Técnicos Asociados, Lima.

UNANUE, H., 1806 [1815]: *Observaciones sobre el Clima de Lima y sus Influencias en los Seres organizados, en especial el Hombre*, Madrid: Imprenta de Sancha.

URRUTIA de HAZBÚN, R. and LANZA LAZCANO, C., 1993: *Catástrofes en Chile 1541–1992*. Santiago: Editorial la Noria, 440 pp.

VÁSQUEZ de ESPINOZA, A., 1629 [1942]: *Compendium and Description of the West Indies* (translated by C. U. Clark), Washington, D.C.: Publ. 3646, Smithsonian Institution.

VICUÑA MACKENNA, B., 1877 [1970]: *El Clima de Chile*, 2nd ed. Buenos Aires: Editorial Francisco de Aguirre, 399 pp. (Cited by Q&N as Mackenna, B. V.)

VIDAL GORMAZ, P., 1901: *Naufrajios ocurridos en las costas de Chile desde su descubrimiento hasta nuestros dias*. Santiago: Imprenta Elzevier. (Cited by Q&N as Gormaz, P. V.)

von HUMBOLDT, A., 1804: La corriente de agua fría a lo largo de la costa occidental de Sudamerica. *Revista del Instituto de Geografía*, Universidad Nacional Mayor de San Marcos, Lima, Peru, 1960, **6**, 7–22.

WEBERBAUER, A., 1914: Die Vegetationsgliederung des noerdlichen Peru, im 5° S.B. *Englers. Bot. Fahrbuecher*, **50**, 77.

WHETTON, P. H. and RUTHERFURD, I., 1994: Historical ENSO teleconnections in the Eastern Hemisphere. *Climatic Change*, **28**, 221–253.

WHETTON, P. H., ALLAN, R. J., and RUTHERFURD, I., 1996: Historical ENSO teleconnections in the Eastern Hemisphere: Comparison with latest El Niño series of Quinn. *Climatic Change*, **32**, 103–109.

WOODMAN, R. F., 1985: Recurrencia del fenómeno El Niño con intensidad comparable a la del año 1982–1983. *In Proceedings of the Seminario Regional Ciencia, Tecnologia y Agresión Ambiental, El Fenómeno El Niño*. Lima: CONCYTEC, 301–332.

XEREZ, F. de, 1534 [1968]: Verdadera relación de la conquista del Perú y provincia del Cuzco llamada la Nueva Castilla. *In El Perú a Través de los Siglos*, **1**, 191–272. Biblioteca Peruana, Editores Técnicos Asociados, Lima. (Quinn et al. 1987 referred to an 1872 edition, translated by E. R. Markham, and published by Burt Frankin, New York.)

ZÁRATE, A. de, 1545 [1968]: *Historia del Descubrimiento y Conquista del Perú*, **2**, 105–413. Biblioteca Peruana, Editores Técnicos Asociados, Lima.

ZEGARRA, J. M., 1926: Las Lluvias y avenidas extraordinarias del verano de 1925 y su influencia sobre la agricultura del Departamento de La Libertad. *Informes y Memorias de la Sociedad de Ingenieros del Perú*, **28**(1), 1–46.

8

Tree-Ring Records of Past ENSO Variability and Forcing

EDWARD R. COOK AND ROSANNE D. D'ARRIGO

Tree-Ring Laboratory, Lamont-Doherty Earth Observatory, Columbia University, Palisades, New York 10964, U.S.A.

JULIA E. COLE*

Department of Geological Sciences and Institute of Arctic and Alpine Research, University of Colorado, Boulder, Colorado 80309, U.S.A.

DAVID W. STAHLE

Department of Geography, University of Arkansas, Fayetteville, Arkansas 72701, U.S.A.

RICARDO VILLALBA

Laboratorio de Dendrocronologia, CRICYT-Mendoza, CONICET, C.C. 330, 5500 Mendoza, Argentina

Abstract

We review results of recent and ongoing research in using long, exactly dated, annual tree-ring chronologies to study and reconstruct past El Niño/Southern Oscillation (ENSO) variability and forcing. The research covered here includes (1) the development and testing of long teak chronologies from Indonesia for ENSO signals; (2) the reconstruction of the Tahiti–Darwin Southern Oscillation Index (SOI) using tree-ring chronologies from North America and Java; (3) the recent discovery of some tree species suitable for developing tree-ring chronologies from ENSO-sensitive regions of East Africa; (4) the discovery of long-term ENSO forcing of climate in the mid- to high-latitude Southern Hemisphere from a tree-ring reconstruction of the summer transpolar sea-level pressure (SLP) index; and (5) the characterization of the ENSO teleconnection with drought in the United States and its temporal stability in a network of long drought reconstructions from tree rings.

Introduction

The El Niño/Southern Oscillation (ENSO) is the most important known cause of interannual climate variability on Earth (see, e.g., Ropelewski and Halpert 1987; Kiladis and Diaz 1989). El Niño/Southern Oscillation has exhibited large changes in amplitude and

* Present Affiliation: Geoscience Department, University of Arizona, Tucson, AZ 85721.

phase of the annual cycle and in the frequency and intensity of warm and cold events during the past 100 years of instrumental observation. Dynamic models of ENSO have successfully anticipated some recent ENSO activity (Cane et al. 1986) and have also simulated dramatic amplitude modulation of warm events on decadal to century timescales (Cane et al. 1995). It appears that the ENSO system itself interacts with the global ocean–atmosphere–land system at decadal timescales, and these ill-defined long-term interactions may reduce the climate prediction skill for ENSO (Lukas 1995). The instrumental meteorological and oceanographic records available for the past century do not necessarily provide a representative sampling of the interdecadal variability of ENSO or the higher latitude ocean–atmosphere–land processes with which the ENSO system appears to interact. Fortunately, the widely distributed array of paleoclimatic proxies dated with seasonal to annual precision can extend the climate and oceanographic database centuries into the past and can be used to document the interdecadal variability of ENSO and its associated suite of climate teleconnections. This long-term high-resolution paleoclimatic perspective, most notably provided by long tropical and subtropical tree-ring chronologies and annually banded corals in tropical oceans, will be crucial to our understanding of the physical causes of ENSO variance on annual, decadal, and century timescales.

Exactly dated annual tree-ring chronologies from certain key climate regions in North America and Southeast Asia provide one of the best proxies of ENSO variability yet demonstrated (Lough and Fritts 1985; D'Arrigo and Jacoby 1992; Stahle and Cleaveland 1993). Unfortunately, most of the ENSO teleconnection zones identified with modern instrumental data are not represented by long tree-ring chronologies, particularly in the tropical latitudes where ENSO impacts are most prevalent. However, recent success with the development of annual tree-ring chronologies from a few select tropical tree species indicates real potential to expand the tropical tree-ring record of ENSO.

In this chapter, we will review some of the recent developments in using long tree-ring records of past climate to study and reconstruct ENSO variability over the past few centuries. This work builds upon previous successes that were well covered by Diaz and Markgraf (1992).

The first section describes recent successful efforts to establish long teak tree-ring chronologies for Indonesia that are sensitive to ENSO as it affects rainfall variability over that region. This work is also used to show that the often analyzed teak chronology of Berlage (1931), which covers the interval A.D. 1514–1929, is reliably dated only back to 1800. The second section describes a highly successful effort in reconstructing the Tahiti–Darwin Southern Oscillation Index (SOI) from a network of ENSO-sensitive tree-ring chronologies from northwestern Mexico, the southwestern/south-central United States, and Java. The third section describes promising results from an ongoing effort to develop exactly dated tree-ring chronologies from ENSO-sensitive regions of eastern Africa, particularly Kenya and Zimbabwe. The fourth section describes an intriguing ENSO signal in a reconstruction of the summer transpolar sea-level pressure (SLP) index based solely on mid- to high-latitude southern tree-ring chronologies. This result provides some of the strongest evidence yet for long-term forcing of the mid- to high-latitude climate system in the Southern Hemisphere by ENSO. The fifth section describes the impact of ENSO on drought and wetness in the United States by using a network of long drought reconstructions from tree rings to generate teleconnection maps. In particular, they show that the spatial extent of the ENSO teleconnection has changed through time, with an especially large change occurring in the 1920s.

ENSO-Sensitive Teak Tree-Ring Records from the Indonesian Archipelago

High-resolution paleoclimatic time series from tree rings and corals can be used to evaluate the long-term behavior of the ENSO system prior to the twentieth century. Yet very few such records exist for the ENSO tropical centers of action. One of these centers of action is over Indonesia. The Indonesian archipelago is situated under the Indonesian low-pressure cell, the western equatorial arm of ENSO (Allan et al. 1996). Long paleoclimate records from Indonesia could thus potentially yield extended information on ENSO and monsoon rainfall variability for several centuries or more prior to the period of instrumental record. Development of tropical tree-ring records has been limited, however, by the absence of pronounced climate seasonality in many low-latitude regions and by the scarcity of suitable tree species known to produce distinct annual rings and to grow old enough for tree-ring analyses (Jacoby 1989).

Teak (*Tectona grandis* L.F.) was first introduced to Indonesia many centuries ago from India and Burma (Whitten et al. 1988; Soerianegara and Lemmens 1993). It is one of very few tree species successfully shown to cross-date in Southeast Asia (Berlage 1931; Buckley et al. 1995; Pumijumnong et al. 1995; Stahle et al. 1998; Yadav and Bhattacharyya 1996). Teak is also one of the most commercially valuable tropical tree species, making it increasingly scarce because of extensive logging in many areas (D'Arrigo et al. 1994a). Below we describe three new tree-ring chronologies of teak from the islands of Java and Sulawesi, Indonesia, and discuss their link to ENSO variability (Fig. 8.1).

Fig. 8.1 Map of teak tree-ring sites in central Java (1) and on the island of Muna in Sulawesi (2).

Central Java

Only one tree-ring record has been previously published for Indonesia: a teak chronology (1514–1929) for Cepu, central Java (Berlage 1931). This chronology was reanalyzed by Jacoby and D'Arrigo (1990) and Murphy and Whetton (1989). More recently, it has been updated by using samples from natural plantation (i.e., planted but unmanaged) teak trees up to about 120 years old collected in the Cepu area (D'Arrigo et al. 1994b; and see Palmer and Murphy 1993). The updated Cepu series extends from 1514 to 1991. The Cepu record is positively correlated with rainfall and with the SOI (Jacoby and D'Arrigo 1990; D'Arrigo et al. 1994b). During warm ENSO events (negative SOI), drought (and decreased tree growth) tends to occur over much of Indonesia (Hackert and Hastenrath 1986).

The Berlage and updated series for Cepu correlate at $r = 0.57$ for the overlapping interval from 1880 to 1929, suggesting that the Berlage record is correctly dated for this recent period (D'Arrigo et al. 1994b). However, the dating of the earlier part of the Berlage record, based on only a few samples, is considered less certain (Jacoby and D'Arrigo 1990; D'Arrigo et al. 1994b; Murphy and Whetton 1989; Palmer and Murphy 1993). Until now there has been no independent long record available for comparison.

Two additional teak chronologies have also been recently developed for central Java: Bigin (1839–1995) and Sara (1689–1995) (Table 8.1; Figs. 8.1 and 8.2). As was previously found for the Cepu site (Jacoby and D'Arrigo 1990; D'Arrigo et al. 1994b), both of these records significantly correlate with the SOI for the months just prior to and during the growing or wet season (Fig. 8.3). The record from Bigin, the site closest to Cepu, correlates with the updated Berlage chronology at $r = 0.50$ for the overlapping interval (1840–1991). Comparison of the raw ring-width measurements from the Sara and Bigin sites with the original Cepu chronology, done by comparing correlations between overlapping time series segments (program COFECHA; Holmes 1983), indicates that the original Berlage series is dated correctly back to 1800. However, no match was found with the longer Sara record for the 1700s, suggesting that the Berlage series may be misdated prior to about 1800. Sara, the longest of the two recently developed Java series, has been included in multiproxy experimental reconstructions of the Tahiti–Darwin SOI (Stahle et al. 1998).

Table 8.1 *Chronology statistics for teak chronologies from Indonesia. Indicated are the length of record, number of samples (i.e., number of measured radii), intercorrelation among all samples, mean sensitivity (index of the relative change in ring width from one year to the next), and the autocorrelation coefficients of the unfiltered series (Fritts 1976).*

Site	Years	Number of samples	Inter-correlation	Mean sensitivity	Auto correlation
Bigin	1839–1995	20	0.567	0.410	0.619
Sara	1689–1995	15	0.503	0.381	0.706
Muna	1565–1995	39	0.421	0.384	0.688

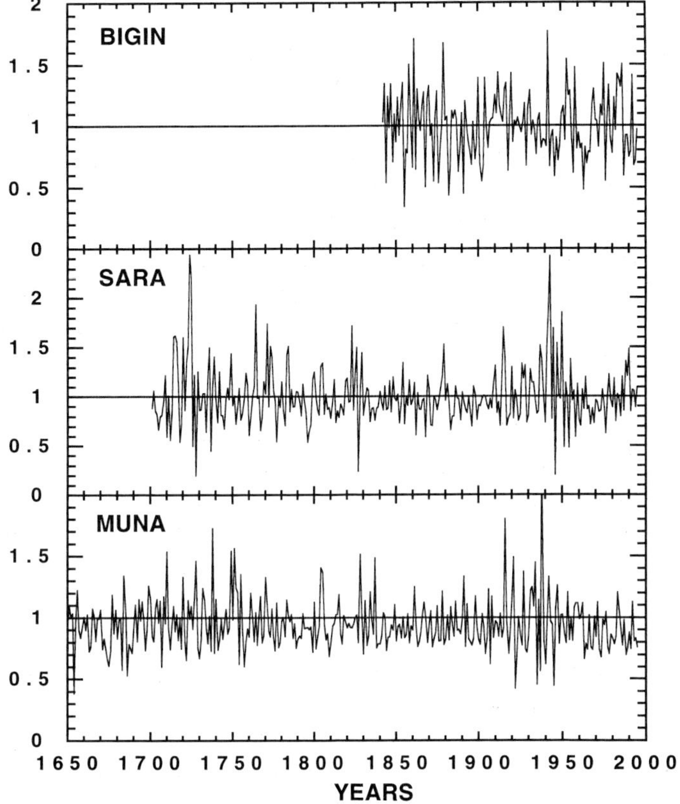

Fig. 8.2 Three new ring-width chronologies of teak for Indonesia: Bigin and Sara series for central Java and Muna chronology for southeastern Sulawesi. The chronology units are standardized tree-ring indices.

Muna, Southeastern Sulawesi

Teak trees were also sampled in a natural plantation setting on the island of Muna, southeastern Sulawesi (Fig. 8.1). This site is Sulawesi's oldest teak conservation area, a 1.9-hectare tract of land with 20 to 30 trees. This part of Sulawesi has a pronounced dry season (Whitten et al. 1988). As is the case in central Java, local woodcutters have begun exhuming ancient logs that have been preserved in stream and lake sediments to sell for commercial use. Twenty stem disks with 80 to 400 visible growth rings and unknown calendar age were obtained. They are believed to originate from a forest that once covered the northern half of the island. Of these, three sections were found to cross-date with the living trees, extending the living tree chronology by an additional 60 years. The final Muna chronology consists of 39 series and dates from 1565 to 1995, although it is truncated here at 1650 owing to low sample size in the earlier part of the record (Table 8.1; Fig. 8.2). As was found for the three Java series, the Muna chronology is positively and significantly correlated with SOI for the months just prior to and during the growing season (Fig. 8.3).

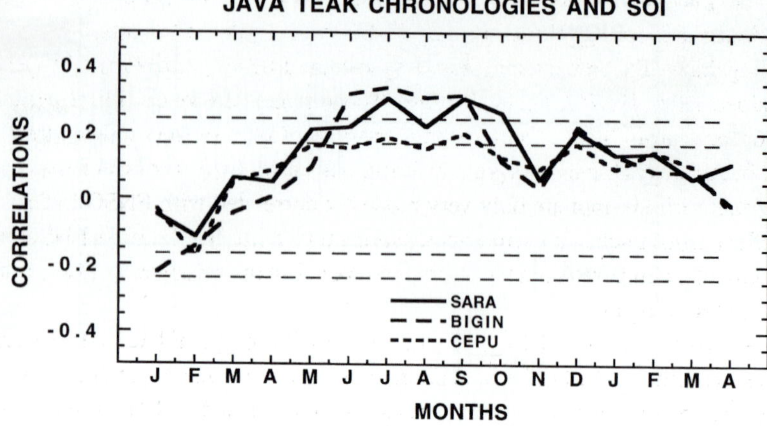

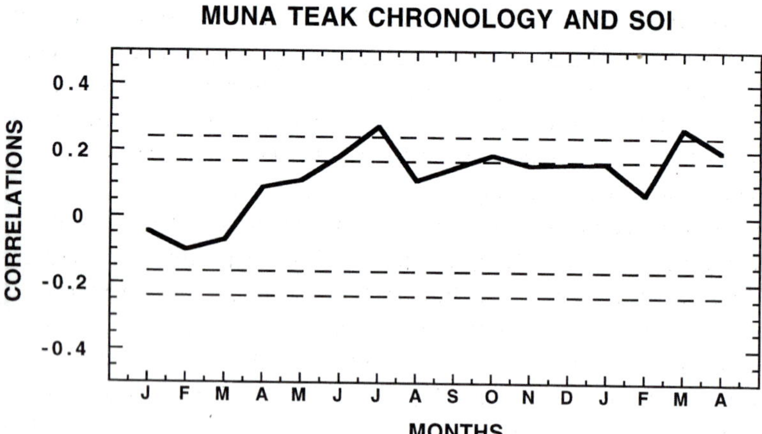

Fig. 8.3 Correlations of (top graph) central Java (Bigin, Sara, along with Cepu) and (bottom graph) Muna, Sulawesi chronologies with the Tahiti–Darwin SOI of R. Allan. Data are prewhitened for the common interval since 1870. Significance levels (5%, 1%) are also indicated.

Blackman-Tukey spectral analysis (Jenkins and Watts 1968) was used to test for the statistical significance of spectral peaks in the teak chronologies. The analyses were based on lags of the autocorrelation function equal to one-third the length of each series. All four chronologies display significant spectral peaks in the preferred ENSO bandwidth of 3–7 years. In particular, peaks at about 5.6 years are statistically significant at or above the 95% level in the Cepu, Sara, and Muna chronologies. This spectral peak is also present (although not significant) in the shorter Bigin record.

Reconstructing the Winter SOI from Long Tree-Ring Records

The strongest ENSO signal yet detected in any tree-ring data worldwide is found in the conifer chronologies of northwestern Mexico and the southwestern United States, and in the post oak chronologies of Texas and Oklahoma. This portion of subtropical North America is strongly influenced by ENSO, with cool-wet (warm-dry) conditions prevailing during the boreal winter and early spring of warm (cold) events (Ropelewski and Halpert 1986, 1987; Allan et al. 1996). This regional teleconnection zone includes much of the Gulf of Mexico and the adjacent coastal zone of the southeastern United States, but the available bald cypress and post oak chronologies from northern Florida and southern Mississippi are only very modestly correlated with ENSO indices. This may reflect a mismatch in the winter-early spring teleconnection between ENSO and regional climate, compared with the spring-summer climatic response of these particular tree-ring chronologies.

The relatively strong ENSO signal in tree growth of Mexico and New Mexico reflects both the strength of ENSO teleconnection to regional climate (as much as 30% of the variance in precipitation in northern Mexico may be linked to ENSO during winter [Douglas and Englehart 1981]) and the excellent regional climatic signal recorded by the growth rings of Douglas fir, ponderosa pine, and other southwestern conifers. Annual ring width of these arid-site conifers tends to be significantly correlated with winter precipitation, which may recharge available soil moisture before the onset of tree growth. This winter-early spring season climatic response, and the overall strength of the ENSO signal registered by these southwestern conifers, can both be enhanced by developing separate chronologies of earlywood width.

Douglas fir from northern Mexico and the southwestern United States often exhibit an abrupt transition from earlywood to latewood (i.e., springwood to summerwood). This intra-annual ring boundary can be identified optically, and separate chronologies of earlywood, latewood, and total ring width can be developed. An array of ten earlywood width chronologies of Douglas fir, extending from the Tropic of Cancer in Durango to Colorado, has recently been developed (Stahle et al. 1998). These chronologies correlate strongly with winter precipitation amounts (i.e., October through March) in the region surrounding each chronology, and with the winter SOI. In fact, the winter SOI signal in these earlywood chronologies is the strongest and most spatially coherent ENSO signal yet detected in tree-ring data (Fig. 8.4). The coniferous forests of Durango are the most commercially important in all of Mexico, and the strong ENSO signal evident in selected chronologies from this region suggests that long-range ENSO-based climate forecasts could have considerable relevance to the management of this valuable forest resource.

These earlywood width chronologies were included with additional tree-ring data from Arizona, New Mexico, Texas, Oklahoma, and Java, Indonesia, to reconstruct the winter SOI from A.D. 1706 to 1977 (Stahle et al. 1998). These tree-ring data explain 53% of the winter Southern Oscillation Index (SOI) variance, and the tree-ring estimates can be verified when compared with instrumental winter Southern Oscillation indices not included in the calibration. This winter SOI reconstruction is displayed in Figure 8.5, along with the waveforms for the three statistically significant modes of variance identified in the ENSO frequency band by using singular spectrum analysis (Vautard et al. 1992; see discussion later in the chapter). The tree-ring reconstructed waveforms at approximately 3.5, 4.1, and 5.8 years are strongly consistent with the modes of variance

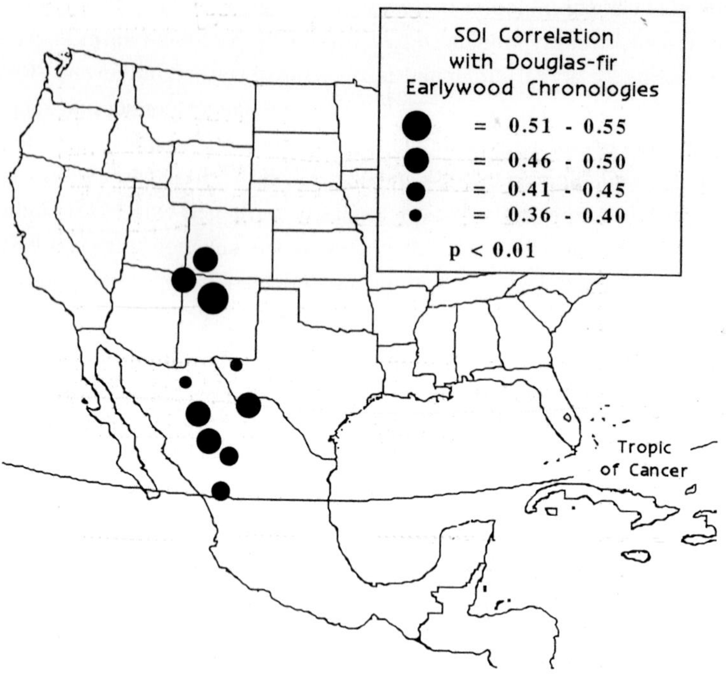

Fig. 8.4 SOI correlation with Douglas fir earlywood chronologies. The sizes of the filled circles are directly proportional to the correlations.

identified in the instrumental winter SOI (Stahle et al. 1998) and reveal interesting changes in amplitude over the past 272 years (Fig. 8.5).

There is considerable potential to expand upon the available tree-ring proxies of ENSO in Mexico and the southwestern United States. Douglas fir earlywood width chronologies up to 618 years long are already available from Durango, Mexico (the Cerro Barajas series dates from 1376 to 1993, and El Salto dates from 1481 to 1993), and a 2,129-year-long chronology of tree ring width has been developed from living and subfossil Douglas fir at El Malpais, New Mexico (Grissino-Mayer 1995). Subfossil Douglas fir samples have been recovered from the El Tabacote site in Chihuahua, which could extend that series back to A.D. 1200 or earlier. Well-preserved wood recovered from prehistoric archaeological sites in northern New Mexico have also been used to compile a 1,000-year tree-ring record (Dean and Robinson 1978), which has already been demonstrated to contain an ENSO signal (D'Arrigo and Jacoby 1992).

Efforts to extend the network of tree-ring chronologies into the tropics of northeastern and central Mexico are presently under way. The ENSO teleconnection to climate appears to change from a winter signal in northwestern Mexico to a summer signal in portions of the northeast and Yucátan Peninsula (Cavazos and Hastenrath 1990); one goal is to develop tree-ring proxies of this regional ENSO forcing. New tree-ring chronologies of Douglas fir and Montezuma pine have been developed for 150 to 200 years in Nuevo León and Tamaulipas, respectively. A 500-year chronology of Mexican

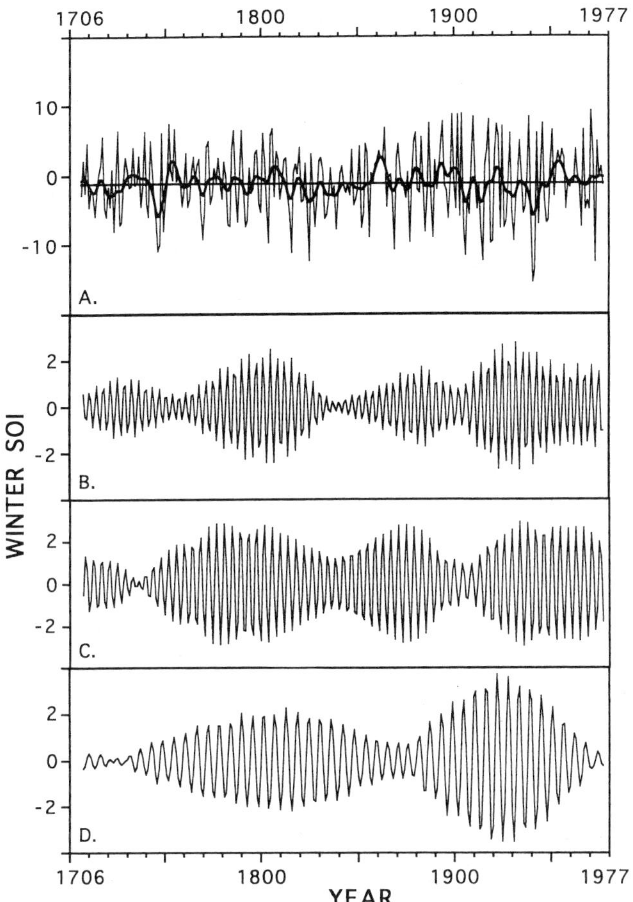

Fig. 8.5 The winter SOI reconstruction from tree rings (A) and its three significant oscillatory modes (B–D) extracted by singular spectrum analysis (SSA). The oscillatory modes have periods of 3.5, 4.1, and 5.8 years, respectively.

bald cypress has also recently been developed along the Rio Sabinas, Tamaulipas, and is highly correlated with spring–early summer rainfall.

Recent Developments in Africa

El Niño/Southern Oscillation teleconnections to regional climate have also been detected in tropical eastern and southern Africa, and a survey of the incredibly diverse indigenous forests has begun for species suitable for dendroclimatology (Stahle et al. 1995, 1997). *Vitex keniensis* and *Premna maxima* are native to the forests flanking Mt. Kenya, and both produce annual growth rings and have been used to develop exactly dated tree-ring chronologies (Stahle et al. 1995). Weak regional climate and ENSO signals have been detected in these time series, but both chronologies are unfortunately only 60 years long. Efforts to develop longer records from the few 200- to 300-year-old *Vitex* known have not yet proved successful. Poor ring definition and erratic, lobate growth of many individual growth rings necessitate much larger wood sections for analysis than can be obtained with a 5 mm diameter increment borer. However, only a very few old-growth *V. keniensis* survive, and they of course cannot be destructively sampled. The best opportunity for development of long *Vitex* chronologies may rest with the sampling of historic *Vitex* wood in old buildings or furniture.

Progress with the development of ENSO-sensitive tree-ring proxies has been somewhat more substantial in Zimbabwe, where two excellent chronologies of African bloodwood (*Pterocarpus angolensis*) have been completed. Both chronologies date back only to 1870 at present, but they are well correlated with total rainfall amounts during the wet season when the Zimbabwe climate is most influenced by ENSO. The ENSO teleconnection to Zimbabwe climate has not been entirely stationary during the twentieth century, but after 1960, warm ENSO events have been strongly coupled with extensive drought over Zimbabwe and adjacent southeastern Africa (Ropelewski and Halpert 1987). The available bloodwood chronologies were collected in western Zimbabwe near the fringe of the ENSO teleconnection zone, but they appear to contain a useful record of this episodic ENSO forcing, particularly during warm events.

The maximum longevity of *P. angolensis* is not known. Fortunately, however, African bloodwood is the most important timber species in south tropical Africa (Coates Palgrave 1983), and old bloodwood survives in buildings, furniture, canoes, and miscellaneous implements. These sources of older wood may permit the eventual development of 200- to 300-year-long chronologies of *P. angolensis* in southeastern Africa.

Trans-Polar ENSO Teleconnections in the Southern Hemisphere

Recent developments in chronology networks have occurred in mid- to relatively high latitudes in southern South America, Tasmania, and New Zealand (Boninsegna 1992; Boninsegna and Villalba 1996; Cook et al. 1992, 1996a; D'Arrigo et al. 1995, 1996; Norton and Palmer 1992; Salinger et al. 1994; Villalba et al. 1996; Villalba and Veblen 1997). With the increasing number of tree-ring collections from climatically sensitive areas in the Southern Hemisphere, the use of a spatial approach to study large-scale atmospheric variations connecting mid- to high-latitude climatic changes appears now to be feasible. A key incentive for these studies is the documented existence of teleconnections relating climatic variations between middle and high latitudes in the Southern Hemisphere (Pittock 1984; Mo and White 1985; Carleton 1992; van Loon et al. 1993; Karoly et al. 1996).

The Trans-Polar Index

The Trans-Polar Index (TPI), defined as the difference in mean sea-level pressure (MSLP) between Hobart, Tasmania (43°S 147°E) and Stanley, South Atlantic Ocean (52°S 58°W), was proposed by Pittock (1980) to measure the eccentricity of the polar

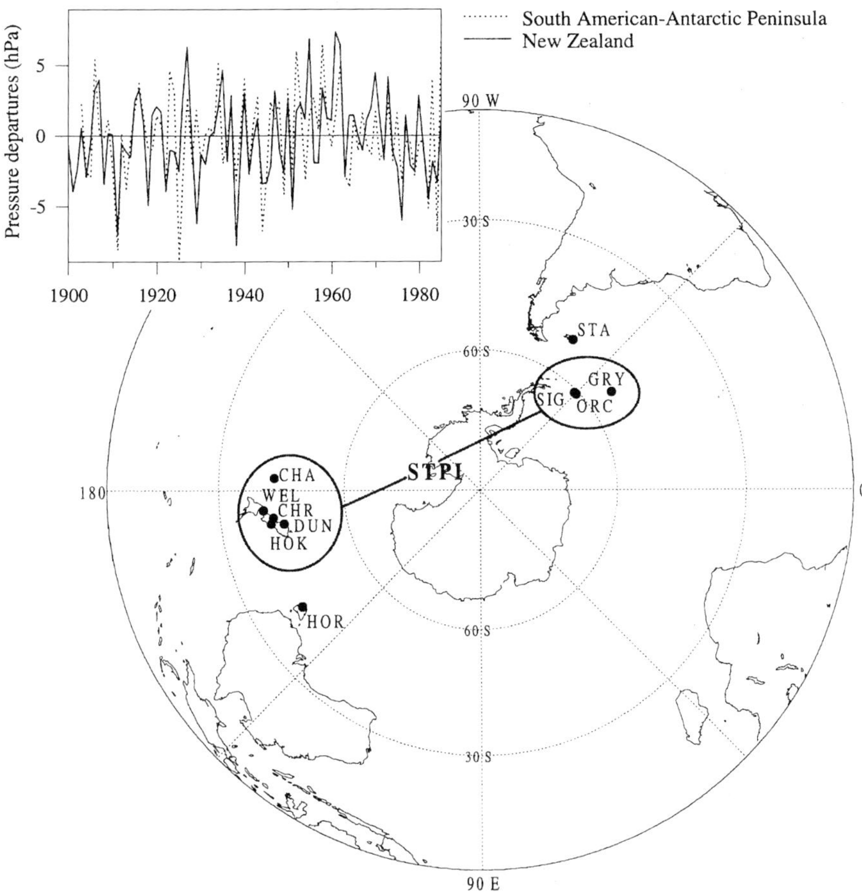

Fig. 8.6 Map of the Southern Hemisphere showing the geographical setting for the Summer Trans-Polar Index (STPI). Abbreviations for meteorological stations are ORC: Orcadas, SIG: Signy Island, GRY: Grytviken, and STA: Stanley in the South American–Antarctic Peninsula sector of the Southern Ocean and WEL: Wellington, HOK: Hokitika, CHR: Christchurch, DUN: Dunedin, CHA: Chatham Island, and HOR: Hobart in the Australia–New Zealand region. A comparison of variations in mean sea-level pressure (MSLP) anomalies between the South American–Antarctic Peninsula and New Zealand sectors around Antarctica during the summer (December and January) since 1903 is shown in the upper, left corner. The South American–Antarctic Peninsula regional record of MSLP consists of the Grytviken, Orcadas, and Signy Island stations, whereas the MSLP regional record from New Zealand includes Wellington, Hokitika, Christchurch, Dunedin, and Chatham Island records. To facilitate the comparison, the South American–Antarctic Peninsula MSLP anomalies have been multiplied by −1.

vortex around the South Pole (Fig. 8.6). Consequently, the TPI represents a measure of the extratropical wave number 1, which in the Southern Hemisphere dominates the mean pressure and height fields (Trenberth 1980) and contributes to the interannual variability of the westerlies (Rogers and van Loon 1982). When the polar vortex is displaced in relation to its average position toward the South American–South Atlantic sector, the TPI is more positive. When the TPI is positive, the subtropical anticyclones and the westerlies are displaced farther south in the Australian–New Zealand sector. In contrast, in the South American–South African sector, the subtropical anticyclone and the westerlies are shifted toward the equator. These changes in synoptic conditions result in major precipitation and temperature variations in the affected regions (Rogers and van Loon 1982; Pittock 1984).

For the austral summer months, the strongest spatial teleconnection in the MSLP field occurs between the Orcadas–Signy Island–Grytviken stations in the southern South America–Antarctic Peninsula sector and a set of New Zealand stations, including Wellington on the North Island, Hokitika, Christchurch, and Dunedin on the South Island, and Chatham Island (Fig. 8.6). For the 1903–90 interval, correlations in MSLP between these two regions are significant from November through February, peaking in December and January. The regional records of MSLP from the South America–Antarctic Peninsula and New Zealand sectors were normalized, and the difference in pressure between the two sectors was used to develop a Summer Trans-Polar Index (STPI, Fig. 8.6; Villalba et al. 1997). The correlation coefficient for December–January MSLP between the South America–Antarctic Peninsula and the New Zealand stations is $r = -0.53(n = 88, p < 0.001$, Fig. 8.6).

Tree-Ring Reconstruction of the STPI

In Tierra del Fuego (53°–55°S), the southernmost extension of forested land in the world, twenty-one tree-ring width chronologies were developed by Boninsegna et al. (1989) using two species (*Nothofagus pumilio* and *N. betuloides*) of southern beeches. Some of these chronologies extend back to A.D. 1600, but most of them are well replicated only after A.D. 1750. Three tree-ring chronologies of pink pine (*Halocarpus biformis*), dating back to the late 1600s, have been produced for Stewart Island, New Zealand (D'Arrigo et al. 1995, 1996). Two chronologies of silver pine (*Lagarostrobus colensoi*) have been developed for Mangawhero River Bridge and Ahaura on the North and South Islands of New Zealand, respectively (D'Arrigo et al. 1995, 1996). Both sites were previously sampled by LaMarche et al. (1979). Data have been merged with those from the original collections, resulting in new chronologies with improved sample sizes and lengths.

The use of a climatically sensitive network of chronologies with a consistent response to MSLP at opposite sides of Antarctica should allow a successful reconstruction of the trans-polar pressure gradient. Based on the highest correlations between tree growth and STPI variations, a subset of five chronologies from Tierra del Fuego and five from New Zealand were used as predictors of the STPI in principal components (PCs) regression analysis. In so doing, it was found that 49% of the total variance in the STPI could be explained by using the first three PCs from the tree-ring data set as predictors in the regression equation (Villalba et al. 1997). The STPI reconstruction for 1745 through 1984 (Fig. 8.7) shows a high degree of interannual variability but also long-term trends in the trans-polar pressure gradient. Tree-ring-based estimates for the STPI for the nineteenth century were substantially lower than those for the twentieth century.

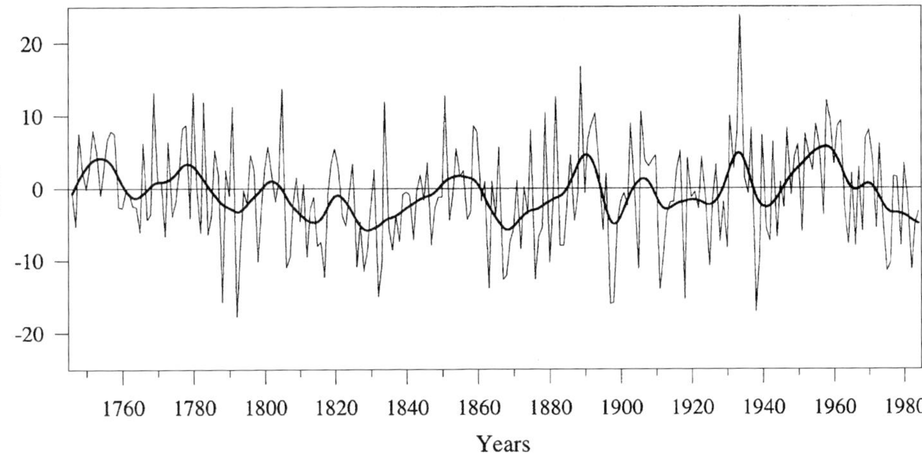

Fig. 8.7 Reconstructed variations in the STPI from A.D. 1745 to 1984. To emphasize low-frequency variations, the reconstruction is also shown in a smoothed version, based on smoothing the annual values with a cubic spline (Cook and Peters 1981) designed to reduce 50% of the variance in a sine wave with a periodicity of 15 years.

STPI Changes in Southern Latitudes and ENSO

Several authors have noted possible relationships between ENSO and MSLP at higher southern latitudes (van Loon and Shea 1985; Carleton 1988, 1989; Karoly et al. 1989, 1996; Smith and Stearns 1993; Cullather et al. 1996). Recently, White and Peterson (1996), using data obtained through a variety of observational techniques during the interval 1979–94, identified significant interannual climatic variations that propagate eastwards around the Southern Ocean with a period of 4–5 years. They postulated some association between this Antarctic circumpolar wave and ENSO activity in the equatorial Pacific.

The STPI reconstruction shows oscillatory behavior with a significant peak appearing in the power spectra at around 3.4–3.5 years (Villalba et al. 1997). This STPI oscillation is close to the 4- to 5-year period proposed by White and Peterson (1996) for the Antarctic circumpolar wave, and coincident with the 3.4-year oscillation present in the SOI (Mann and Park 1994). Cross-spectral analysis for the reconstructed STPI and the SOI during the interval 1866–1984 shows a coherent peak at 3.5 years above the 95% confidence limit (Fig. 8.8). Although more reduced cross-spectral power is associated with the oscillatory modes at 6.4–7.2 and 14.5 years, there is also significant coherence between the STPI and the SOI at these periods. An interdecadal mode in global surface temperatures in the 15- to 18-year period appears to represent long-term ENSO variability (Mann and Park 1994). With this positive result, singular spectral analysis (SSA), a data-adaptive method of extracting signals from noise in a time series (Vautard et al. 1992), was used to characterize the time domain behavior of the 3.4-year oscillation. The 3.4-year components of both STPI and SOI are compared in Figure 8.9. If STPI and SOI are related, then one might expect to see

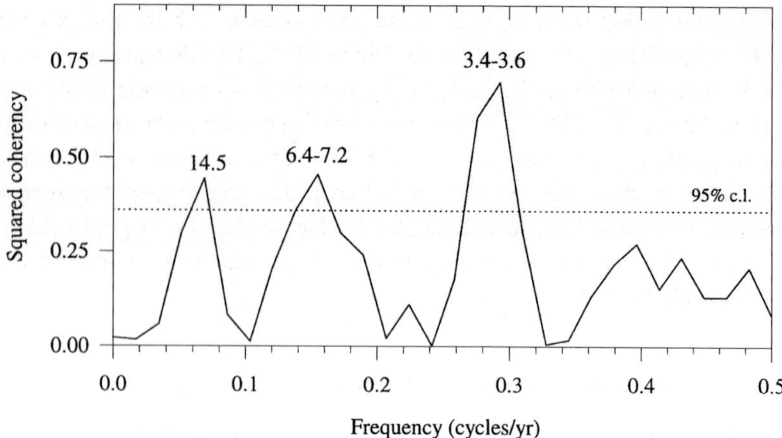

Fig. 8.8 Coherency spectrum of the reconstructed Summer Trans-Polar Index (STPI) and the Southern Oscillation Index (SOI) for the interval 1866–1984. Both indexes are highly coherent at wavelengths around 3.4–3.5 years.

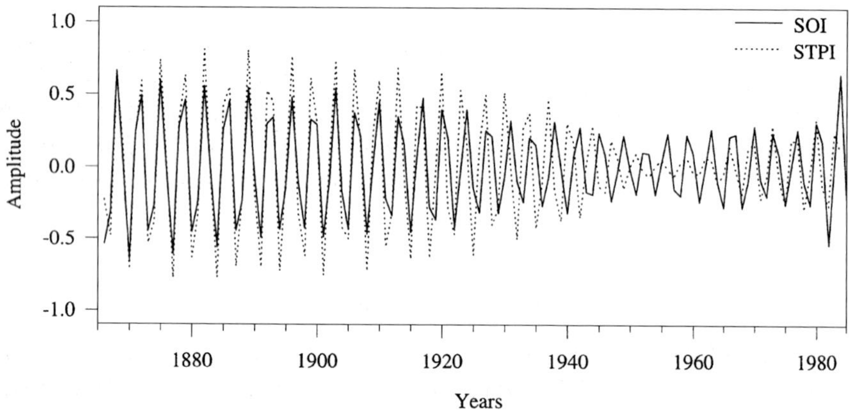

Fig. 8.9 Waveforms of the 3.4-year components of the SOI and STPI estimated by using singular spectral analysis (SSA). Units are dimensionless.

some relationship between the amplitudes of the 3.4-year cycle in the two records. The temporal evolution of STPI and ENSO components shows similar changes in amplitude over time. Both records are characterized by a trend to decreasing amplitudes from the late nineteenth century to the 1950s, followed by a gradual increase to the 1980s (Fig. 8.9). The coherence between the 3.4-year oscillatory components of the SOI and STPI is higher during those intervals in which both series show higher amplitudes.

Time series of pressure anomalies show temporal variations in the SO and reveal a number of periods when the SO was clearly evident and others when it was less active (Trenberth 1976; Allan et al. 1991; Wang and Wang 1996). Strong SO fluctuations are clearly evident for the late nineteenth century to 1920 and in recent decades, but a much quieter period occurred from about 1928 to 1950 with the exception of the strong El Niño of 1939–42 (Trenberth and Shea 1987). The changes in the structure of MSLP associated with the SO during the period 1900–40 have been noted by Allan (1993) and Allan et al. (1995) and more recently have been confirmed by Karoly et al. (1996). Regardless of the reasons behind the different correlations between the STPI and the SOI over time, the fact that the 3.4-year oscillation agrees for intervals of stronger oscillations in both records argues for the importance of producing longer proxy records of climatic forcings (such as the SO) to validate the results obtained by using temporally limited data sets.

ENSO Teleconnections with Drought in the United States

It is well known now that there is a strong teleconnection between ENSO in the tropical Pacific and precipitation and temperature anomalies in the southern United States and northern Mexico (Ropelewski and Halpert 1986; Kiladis and Diaz 1989; see also discussion earlier in this chapter). Thus, warm-phase (cold-phase) ENSO events typically result in increased (decreased) rainfall anomalies in this region. Given this well-established linkage, little effort has been made to assess the degree to which the spatial coverage of this teleconnection with North American climate has varied in time. Here we investigate this feature through its relationship with drought and wetness in the United States as measured by the Palmer Drought Severity Index (PDSI). The PDSI reflects both relative wetness (PDSI > 0) and dryness (PDSI < 0). So, through the use of the PDSI in our teleconnection analyses, we will describe the overall teleconnection with both the warm-phase and cold-phase sides of ENSO.

SOI and PDSI Data Sets

The indices of the Southern Oscillation used here to produce the ENSO teleconnection maps are the Tahiti–Darwin SOI (1866–1990, Ropelewski and Jones 1987), Darwin SLP (1869–1989, provided by R. Allan), Niño-3 sea surface temperatures (SSTs) (1856–1991, Kaplan et al. 1998), Tarawa coral oxygen isotope ratios (δ^{18}O) (1894–1992, Cole et al. 1993), and reconstructed winter SOIs from tree rings (1706–1977, Stahle et al. 1998). These series are shown in Figure 8.10 (for analysis purposes the first ten years of Niño-3 data were ultimately deleted because of large standard errors in their estimates [see Kaplan et al. 1998]). All but the tree-ring reconstruction have monthly resolution, but only the boreal winter season (December–February) will be examined here, since this is the time when the ENSO teleconnection is generally the strongest in North America (see section labeled "Reconstructing the Winter SOI from Long Tree-Ring Records"). These different SOIs are used here to demonstrate the robustness of the teleconnection patterns. The Tahiti–Darwin SOI and Darwin SLP clearly are not independent. However, there is some concern about the homogeneity of the Tahiti record (Trenberth 1984), so both are used here.

The climate data set we used here is a grid of 154 summer PDSI records reconstructed from tree rings. The grid is 2×3 and covers the entire coterminous United

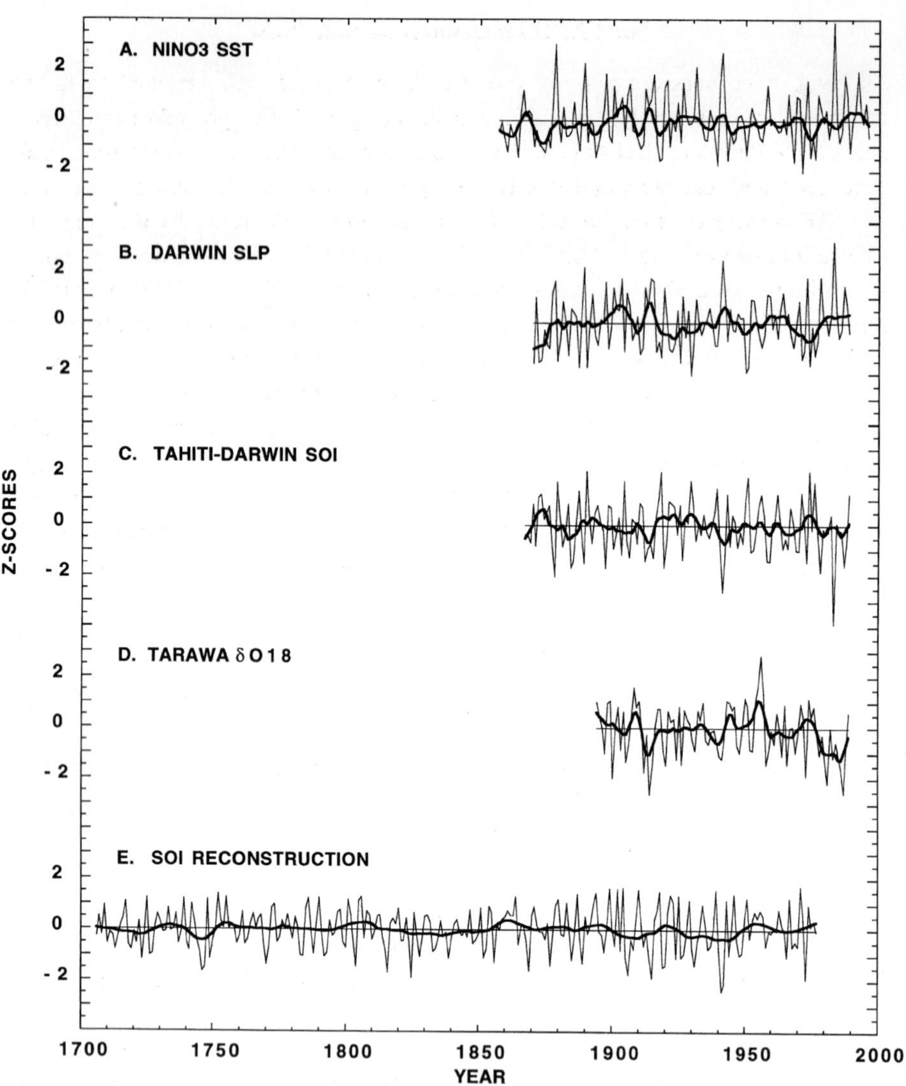

Fig. 8.10 The actual (A–D) and reconstructed (E) Southern Oscillation Indices (SOIs) used to generate the ENSO teleconnection maps for the United States. See the text for more details.

States (except for southern Florida). The time period common to the reconstructions at all grid points is 1700–1978. These reconstructions are described in more detail by Cook et al. (1996b, 1997, 1999) and are available from the National Geophysical Data Center, World Data Center-A for Paleoclimatology, in Boulder, Colorado (World Wide Web, http://www.ngdc.noaa.gov/paleo/pdsiyear.html for annual PDSI maps and http://www.ngdc.noaa.gov/paleo/usclient.html for PDSI time series). A gridded set of instrumental PDSI data are also available for comparative purposes, which cover the 1895–1995 period. These data were used to check the accuracy of the teleconnection results based on the PDSI reconstructions over that time period.

The ENSO Teleconnection with PDSI

An ENSO teleconnection map was produced by correlating each winter SOI with the 154 grid point summer PDSI series for a given time period. This procedure produced a field of 154 Pearson correlations that were then mapped. Before the correlation fields were generated, the SOIs and PDSIs were prewhitened with low-order autoregressive (AR) processes, with the order of each AR model determined by the minimum Akaike Information Criteria (AIC) procedure (Akaike 1974). This was done to remove some differences in short-lag autocorrelation between the SOIs and PDSIs that might confound the estimation of the teleconnection maps. In the process, the first two to three years of SOI data and the first two years of PDSI data were lost in each case.

Figure 8.11 shows the teleconnection maps for the four SOIs with actual and reconstructed PDSIs over the 1897–1978 time period common to all series. The color bar indicates those areas with correlations exceeding |0.30| ($p < 0.01$), with the sign being positive for Niño-3 SST and Darwin SLP and negative for Tahiti–Darwin SOI and Tarawa δ^{18}O. The winter ENSO teleconnection maps produced by the actual and reconstructed summer PDSIs are extremely similar for each species of SOI, indicating that the tree-ring estimates are capturing the regional ENSO teleconnection pattern very well. In addition, each SOI produces a similar teleconnection map, indicating that the patterns are robust. Finally, the geographic location of the highest correlation field is where it ought to be based on past analyses (e.g., Ropelewski and Halpert 1986; Kiladis and Diaz 1989; section entitled "Reconstructing the Winter SOI from Long Tree-Ring Records" in this chapter). Had the PDSI analyses been extended into northern Mexico, the regions of high correlation would have undoubtedly extended southward into that region as well.

Given the high fidelity of the teleconnection maps produced by using the reconstructed PDSIs, the analyses were repeated using this PDSI data set and the longer periods covered by the SOIs. In this case, the teleconnection maps were produced for two time periods: the beginning year of each SOI up to 1923 and 1924–78. Except for the Tarawa record, this division splits the SOI series roughly in half. Figure 8.12 shows these early- and late-period teleconnection maps. For the three longer SOIs, the early-period teleconnection maps are radically different from the late-period maps. The significant ENSO teleconnections with PDSI ($r > |0.30|$) are much more extensive in the early period and penetrate northeastward into the upper Midwest–Great Lakes region. In contrast, the late-period teleconnections are highly contracted to the southwestern region, making them remarkably similar to those based on the 1897–1978 period (cf. Fig. 8.11). An examination of the actual correlations used to produce these maps revealed that the SOI/PDSI correlations remained consistent in sign and magnitude in the southwestern region in both time periods but actually reversed sign between time periods in the Midwest–Great Lakes region. This led to a cancellation of the ENSO teleconnection in the latter region when it was estimated for the full 1897–1978 period.

The final teleconnection analyses utilized the winter SOI reconstruction of Stahle et al. (1998). The tree-ring data used in this record are not completely independent of those used in the PDSI reconstructions, but the added length of the reconstruction

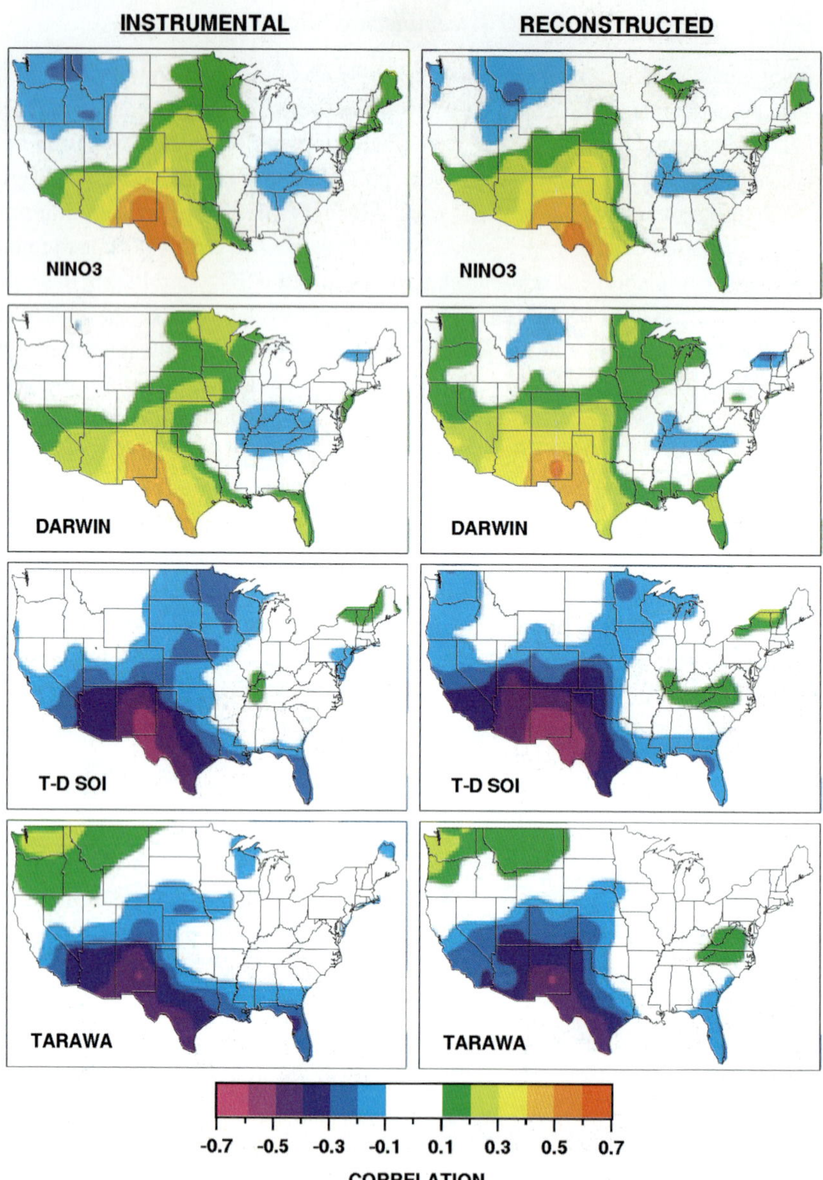

Fig. 8.11 ENSO teleconnection maps for the 1897–1978 period using the four Southern Oscillation Index (SOI) species and the actual and reconstructed summer Palmer Drought Severity Index (PDSI) records. The four maps in the left column are for the actual (ACT) PDSIs, while the right column shows the maps for the reconstructed (REC) PDSIs. The similarity between the maps for each species of SOI indicates that the reconstructions are accurately capturing the ENSO teleconnection pattern.

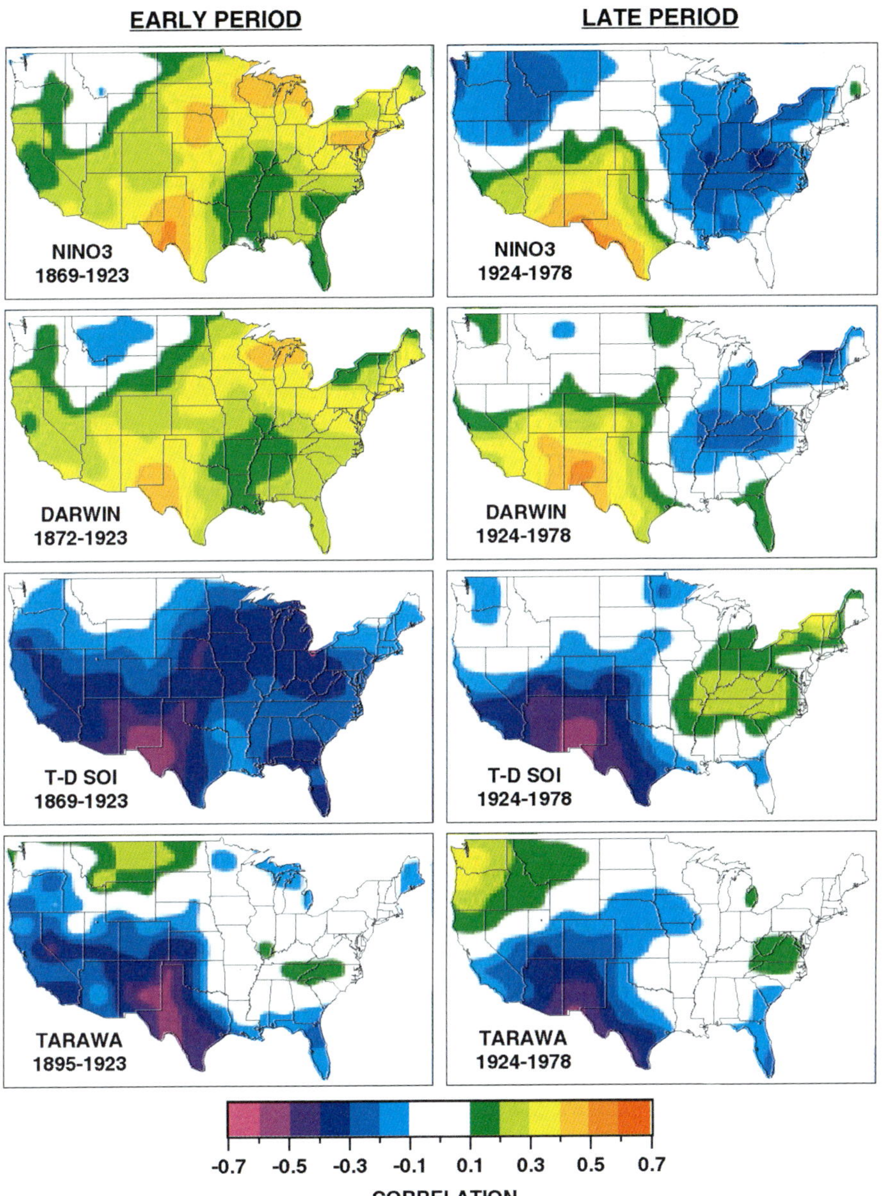

Fig. 8.12 Early- and late-period ENSO teleconnection maps based on reconstructed Palmer Drought Severity Indices (PDSIs) only. The four maps in the left column show the early-period results, which include data up to 1923. The right column maps are for the late period (1924–78). Note the great difference in extent of the ENSO teleconnection between the two time periods.

allows for a more detailed look at the ENSO teleconnection time dependence indicated above. Figure 8.13 shows these results. As a test of fidelity, the top four maps compare the results of correlating the reconstructed winter SOI with the actual and reconstructed PDSIs over the 1897–1923 and 1924–77 periods. This comparison indicates that the reconstructed SOIs capture the teleconnection differences better when the actual PDSI data are used. The early-period map based on the reconstructed PDSIs also shows the expansion into the upper Midwest, but it is more weakly defined. In contrast, the late-period maps are very similar.

The lower four maps in Figure 8.13 show the teleconnections for earlier time periods based on reconstructed SOIs and PDSIs, the periods being 1708–49, 1750–99, 1800–49, and 1850–99. Except for the earliest period, the periods are 50 years long. From these maps, two contrasting patterns are suggested: one of expanded ENSO teleconnection (1708–49 and 1850–99) and one of contracted teleconnection (1750–99 and 1800–49). Based on the previous results, it is apparent that the latter period of expansion also continued up to 1923, while the 1750–99 epoch appears to be most similar to the post-1923 epoch in spatial extent.

The teleconnection maps produced here indicate that the impact of ENSO on North American climate is spatially nonstationary. A core area in the Southwest, Texas, and Oklahoma is consistently affected by ENSO. So, this is the region where summer drought and wetness anomalies appear to be most dynamically linked to forcing from the tropical Pacific Ocean. However, during certain epochs, the ENSO teleconnection appears to expand northeastward into the upper Midwest–Great Lakes region. This expansion may last for several decades, but its ultimate transience suggests that it is less directly coupled to ENSO in a dynamical sense. Rather, it may have more to do with the interaction between ENSO and large-scale atmospheric circulation patterns that develop over the North Pacific and North America (e.g., the Pacific–North America [PNA] mode identified by Horel and Wallace [1981]). In fact, the monthly maps of correlation coefficients between a PNA index (1947–82) and divisional precipitation in the United States (Leathers et al. 1991) have spatial patterns similar to those found between the SOIs and PDSIs after 1923. That is, the correlations in the Southwest–Texas–Oklahoma and upper Midwest–Great Lakes regions are opposite in sign. This similarity suggests that the character of long-wave circulation over North America (zonal versus meridional) helps determine the degree to which the ENSO teleconnection can expand outward from its core region. It is also possible that periods of expanded ENSO teleconnection might be due to changes in the vigor of ENSO and/or its changing base state. Determining what conditions might allow expansion of the ENSO teleconnection into the upper Midwest–Great Lakes region is currently under study (Cole and Cook 1998). In any case, the results shown here indicate that the teleconnection between ENSO and climate over the United States may be difficult to predict outside the core region.

Summary and Conclusions

The contributions in this chapter show the vast potential of long tree-ring chronologies for providing information about past El Niño/Southern Oscillation (ENSO) variability

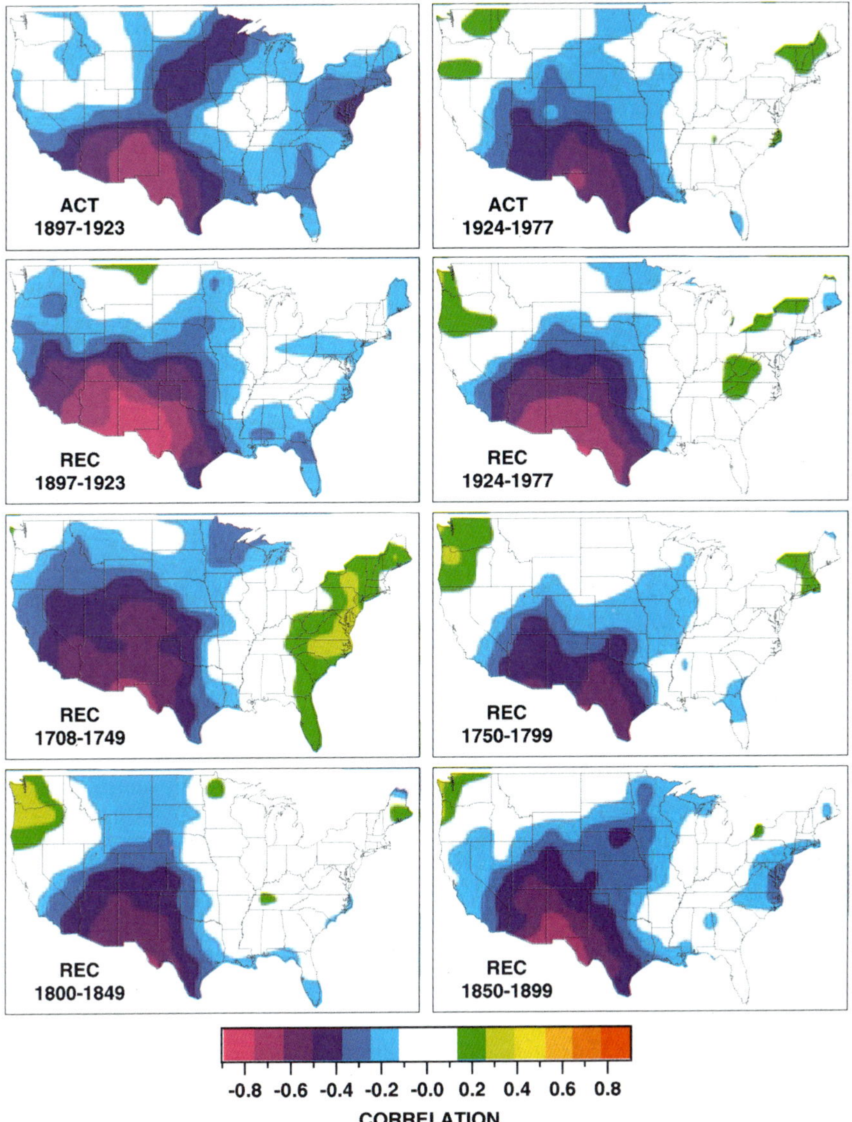

Fig. 8.13 - ENSO teleconnection maps using the winter Southern Oscillation Index (SOI) reconstruction and reconstructed Palmer Drought Severity Indices (PDSIs). The top four maps first compare the results of using actual (ACT) and reconstructed (REC) PDSI for two periods, 1897–1923 and 1924–77. This comparison indicates that the reconstructions underestimate the expanded teleconnection in the early period somewhat. The bottom four plots show maps for four time periods using the pre-1900 portion of the winter SOI reconstruction. These maps suggest that earlier periods of expansion and contraction of the ENSO teleconnection have taken place.

in a number of ways. The feasibility of directly reconstructing the Southern Oscillation Index from tropical and extratropical tree-ring series has been clearly demonstrated here. Recent and ongoing tree-ring chronology development efforts in Indonesia and East Africa have revealed the largely untapped potential for new exactly dated tree-ring chronologies from ENSO-sensitive regions of the tropics and subtropics. Tree-ring reconstructions of austral summer sea level pressure and the Trans-Polar Index have revealed strong evidence for long-term forcing of Southern Hemisphere climate in mid- to high latitudes by ENSO. And, the ENSO teleconnection with drought and wetness in the United States has been shown to be consistently strong, but spatially nonstationary, in a grid of long Palmer Drought Severity Index reconstructions from tree rings.

All of these studies strongly support the expanded development and use of long tree-ring chronologies for the study of past ENSO. Coupled with exactly dated coral records from the tropical oceans, it should be possible to develop a reasonably comprehensive history of ENSO variability and forcing over much of the world.

Acknowledgments The research reported in the chapter was funded by National Oceanic and Atmospheric Administration (NOAA) Grant NA66GPO251 (E. Cook), NOAA Grant NA56GPO217 (R. D'Arrigo), National Science Foundation (NSF) Grant ATM 95-28148 (D. Stahle), NSF Grant ATM 94-16662 (R. Villalba), and by a Lamont-Doherty post-doctoral fellowship to R. Villalba. Lamont-Doherty Earth Observatory Contribution No. 5818.

References

AKAIKE, H., 1974: A new look at statistical model identification. *IEEE Transactions on Automatic Control*, **AC-19**, 716–723.

ALLAN, R. J., 1993: Historical fluctuations in ENSO and teleconnection structure since 1879: Near-global patterns. *Quaternary Australasia*, **11**, 17–27.

ALLAN, R. J., LINDSAY, J. A., and REASON, C. J. C., 1995: Multidecadal variability in the climate system over the Indian Ocean region during the austral summer. *Journal of Climate*, **8**, 1853–1873.

ALLAN, R. J., LINDSAY, J., and PARKER, D., 1996: *El Niño–Southern Oscillation and Climatic Variability*. Collingwood, Victoria, Australia: CSIRO Publishing, 408 pp.

ALLAN, R. J., NICHOLLS, N., JONES, P. D., and BUTTERWORTH, I., 1991: A further extension of the Tahiti-Darwin SOI, early ENSO events and Darwin pressure. *Journal of Climate*, **4**, 743–749.

BERLAGE, H. P., 1931: On the relationship between thickness of tree-rings of Djati (teak) trees and rainfall on Java. *Tectona*, **24**, 939–953.

BONINSEGNA, J. A., 1992: South American dendroclimatological records. *In* Bradley, R. S. and Jones, P. D. (eds.), *Climate since A.D. 1500*. London: Routledge, pp. 446–462.

BONINSEGNA, J. A., KEEGAN, J., JACOBY, G. C., D'ARRIGO, R., and HOLMES, R. L., 1989: Dendrochronological studies in Tierra del Fuego, Argentina. *Quaternary of South America and Antarctic Peninsula*, **7**, 315–326.

BONINSEGNA, J. A. and VILLALBA, R., 1996: Dendroclimatology in the Southern Hemisphere: Review and prospects. *In* Dean, J. S., Meko, D. M., and Swetnam, T. W. (eds.), *Tree Rings, Environment, and Humanity*. Tucson: Radiocarbon, University of Arizona, pp. 127–141.

BUCKLEY, B., BARBETTI, M., WATANASAK, M., D'ARRIGO, R., BOONCHIRDCHOO, S., and SARUTANON, S., 1995: Preliminary dendrochronological investigations in Thailand. *IAWA Bulletin*, **16**, 393–409.

CANE, M. A., ZEBIAK, S. E., and DOLAN, S., 1986: Experimental forecasts of El Niño. *Nature*, **322**, 827–832.

CANE, M. A., ZEBIAK, S. E., and XUE, Y., 1995: Model studies of the long-term behavior of ENSO. *In* Martinson, D. G., Bryan, K., Ghil, M., Hall, M. M., Karl, T. R., Sarachik, E. S., Sorooshian, S., and Talley, L. D. (eds.), *Natural Climate Variability on Decade-to-Century Time Scales*. DEC-CEN Workshop, Irvine, CA. Washington, DC: National Academy Press, pp. 442–457.

CARLETON, A. M., 1988: Sea ice–atmosphere signal of the Southern Oscillation in the Weddel Sea, Antarctica. *Journal of Climate*, **1**, 379–388.

CARLETON, A. M., 1989: Antarctic sea–ice relationships with indices of the atmospheric circulation of the Southern Hemisphere. *Climate Dynamics*, **2**, 207–220.

CARLETON, A. M., 1992: Synoptic interactions between Antarctica and lower latitudes. *Australian Meteorological Magazine*, **40**, 129–147.

CAVAZOS, T. and HASTENRATH, S., 1990: Convection and rainfall over Mexico and their modulation by the Southern Oscillation. *International Journal of Climatology*, **10**, 377–386.

COATES PALGRAVE, K., 1983: *Trees of Southern Africa*. Cape Town, South Africa: Struik, 959 pp.

COLE, J. E. and COOK, E. R., 1998: The changing relationship between ENSO variability and moisture balance in the continental United States. *Geophysical Research Letters*, **25**(24), 4529–4531.

COLE, J. E., FAIRBANKS, R. G., and SHEN, G. T., 1993: Recent variability in the Southern Oscillation: Isotopic results from a Tarawa Atoll coral. *Science*, **260**, 1790–1793.

COOK, E. R., BIRD, T., PETERSON, M., BARBETTI, M., BUCKLEY, B. M., D'ARRIGO, R. D., and FRANCEY, R., 1992: Climatic change over the last millennium in Tasmania reconstructed from tree-rings. *The Holocene*, **2**, 205–217.

COOK, E. R., BUCKLEY, B. M., and D'ARRIGO, R. D., 1996a: Inter-decadal climate oscillations in the Tasmania sector of the Southern Hemisphere: Evidence from tree rings over the past three millennia. *In* Jones, P. D., Bradley, R., and Jouzel, J. (eds.), *Climatic Variations and Forcing Mechanisms of the Last 2000 Years*. NATO ASI Series, Vol. I 41, pp. 141–160.

COOK, E. R., MEKO, D. M., STAHLE, D. W., and CLEAVELAND, M. K., 1996b: Tree-ring reconstructions of past drought across the coterminous United States: Tests of a regression method and calibration/verification results. *In* Dean, J. S., Meko, D. M., and Swetnam, T. W. (eds.), *Tree Rings, Environment and Humanity*. Tucson: *Radiocarbon 1996*, pp. 155–169.

COOK, E. R. and PETERS, K., 1981: The smoothing spline: A new approach to standardizing forest interior ring-width series for dendroclimatic studies. *Tree-Ring Bulletin*, **41**, 45–53.

COOK, E. R., MEKO, D. M., and STOCKTON, C. W., 1997: A new assessment of possible solar and lunar forcing of the bi-decadal drought rhythm in the western United States. *Journal of Climate*, **10**, 1343–1356.

COOK, E. R., MEKO, D. M., STAHLE, D. W., and CLEAVELAND, M. K., 1999: Drought reconstructions for the continental United States. *Journal of Climate*, **12**(4), 1145–1162.

CULLATHER, R. I., BROMWICH, D. H., and VAN WOERT, M. L., 1996: Interannual variations in Antarctic precipitation related to El Niño–Southern Oscillation. *Journal of Geophysical Research*, **101**, 19109–19118.

D'ARRIGO, R. D., BUCKLEY, B. M., COOK, E. R., and WAGNER, W. S., 1995: Temperature-sensitive tree-ring width chronologies of Pink Pine (*Halocarpus biformis*) from Stewart Island, New Zealand. *Palaeogeography, Palaeoclimatology, Palaeoecology*, **119**, 293–300.

D'ARRIGO, R. D., COOK, E. R., JACOBY, G. C., and BUCKLEY, B. M., 1996: Tree-ring records of subantarctic climate over recent centuries to millennia. *In* Dean, J. S., Meko, D. M., and Swetnam, T. W. (eds.), *Tree Rings, Environment and Humanity*. Tucson: Radiocarbon, pp. 171–180.

D'ARRIGO, R. D. and JACOBY, G. C., 1992: A tree-ring reconstruction of New Mexico precipitation and its relation to El Niño/Southern Oscillation events. *In* Diaz, H. F. and Markgraf, V. (eds.), *El Niño: Historical and Paleoclimatic Aspects of the Southern Oscillation*. Cambridge: Cambridge University Press, pp. 243–257.

D'ARRIGO, R. D., JACOBY, G. C., and KRUSIC, P. J., 1994a: On the need to rescue teak wood from Southeast Asia for dendroclimatic studies. PAGES newsletter **2**(3), November 1994.

D'ARRIGO, R. D., JACOBY, G. C., and KRUSIC, P. J., 1994b: Progress in dendroclimatic studies in Indonesia. *Terrestrial, Atmospheric and Oceanographic Science*, **5**, 349–363.

DEAN, J. S. and ROBINSON, W. J., 1978: Expanded tree-ring chronologies for the southwestern United States. Chronology Series III. Laboratory of Tree-Ring Research. Tucson: The University of Arizona, 58 pp.

DIAZ, H. F. and MARKGRAF, V. (eds.), 1992: *El Niño: Historical and Paleoclimatic Aspects of the Southern Oscillation*. Cambridge: Cambridge University Press, 476 pp.

DOUGLAS, A. V. and ENGLEHART, P. J., 1981: On a statistical relationship between autumn rainfall in the central equatorial Pacific and subsequent winter precipitation in Florida. *Monthly Weather Review*, **109**, 2377–2382.

FRITTS, H. C., 1976: *Tree Rings* and *Climate*. New York: Academic Press.

GRISSINO-MAYER, H. D., 1995: *Tree-ring reconstructions of climate and fire history at El Malpais National Monument, New Mexico*. Ph.D. dissertation, University of Arizona, Tucson.

HACKERT, E. C. and HASTENRATH, S., 1986: Mechanisms of Java rainfall anomalies. *Monthly Weather Review*, **114**, 69–78.

HOLMES, R. L., 1983: Computer-assisted quality control in tree-ring dating and measurement. *Tree-Ring Bulletin*, **43**, 69–78.

HOREL, J. D. and WALLACE, J. M., 1981: Planetary-scale atmospheric phenomena associated with the Southern Oscillation. *Monthly Weather Review*, **109**, 813–829.

JACOBY, G. C., 1989: Overview of tree-ring analysis in tropical regions. *In* Baas, P. and Vetter, R. E. (eds.), Growth rings in tropical woods. *IAWA Bulletin*, **10**, 99–108.

JACOBY, G. C. and D'ARRIGO, R. D., 1990. Teak (*Tectona grandis*): A tropical species of large-scale dendroclimatic potential. *Dendrocronologia*, **8**, 83–98.

JENKINS, G. M. and WATTS, D. G., 1968: *Spectral Analysis and Its Applications*. San Francisco: Holden-Day.

KAPLAN, A., CANE, M. A., KUSHNIR, Y., BLUMENTHAL, B., and RAJAGOPALAN, B., 1998: Analyses of Global Sea Surface Temperature 1856–1991. *Journal of Geophysical Research*, **103**, 18567–18589.

KAROLY, D. J., HOPE, P., and JONES, P. D., 1996: Decadal variations of the Southern Hemisphere circulation. *International Journal of Climatology*, **16**, 723–738.

KAROLY, D. J., PLUMB, R. A., and TING, M., 1989: Examples of the horizontal propagation of quasi-stationary waves. *Journal of the Atmospheric Sciences*, **46**, 2802–2811.

KILADIS, G. N. and DIAZ, H. F., 1989: Global climatic anomalies associated with extremes of the Southern Oscillation. *Journal of Climate*, **2**, 1069–1090.

LaMARCHE, V. C., Jr., HOLMES, R. L., DUNWIDDIE, P. W., and DREW, L. G., 1979: Tree-ring chronologies of the Southern Hemisphere. 3. New Zealand. Chronology Series V. Laboratory of Tree-Ring Research, University of Arizona, Tucson.

LEATHERS, D. J., YARNAL, B., and PALECKI, M. A., 1991: The Pacific/North America teleconnection pattern and United States climate, Part I: Regional temperature and precipitation associations. *Journal of Climate*, **4**, 517–528.

LOUGH, J. M. and FRITTS, H. C., 1985: The Southern Oscillation and tree rings: 1600–1961. *Journal of Climate and Applied Meteorology*, **24**, 952–966.

LUKAS, R. B., 1995: Intersection of seasonal to interannual and decadal to centennial climate variability and prediction. *In A Review of the U.S. Global Change Research Program and NASA's Mission to Planet Earth/Earth Observing System*. National Research Council, Appendix A. Washington, DC: National Academy Press.

MANN, M. E. and PARK, J., 1994: Global-scale modes of surface temperature variability on interannual to century timescales. *Journal of Geophysical Research*, **99**, 25819–25833.

MO, K. C. and WHITE, G. H., 1985: Teleconnections in the Southern Hemisphere. *Monthly Weather Review*, **113**, 22–37.

MURPHY, J. O. and WHETTON, P. H., 1989: A re-analysis of a tree-ring chronology from Java. *Proc. Konink. Neder. Akad. Wetensch.* Series B, **92**, 241–257.

NORTON, D. A. and PALMER, J. G., 1992: Dendroclimatic evidence from Australasia. *In* Bradley, R. S. and Jones, P. D. (eds.), *Climate since A.D. 1500*. London: Routledge, pp. 463–482.

PALMER, J. G. and MURPHY, J. O., 1993: An extended tree ring chronology (teak) from Java. *Proc. Konink Neder Akad. Wetensch.*, **96**, 27–41.

PITTOCK, A. B., 1980: Patterns of climatic variation in Argentina and Chile. I. Precipitation, 1931–1960. *Monthly Weather Review*, **108**, 1347–1361.

PITTOCK, A. B., 1984: On the reality, stability, and usefulness of southern hemisphere teleconnections. *Australian Meteorological Magazine*, **32**, 75–82.

PUMIJUMNONG, N., ECKSTEIN, D., and SASS, U., 1995: Tree-ring research on *Tectona grandis* in northern Thailand. *IAWA Bulletin*, **16**, 385–392.

ROGERS, J. C. and van LOON, H., 1982: Spatial variability of sea level pressure and 500 mb height anomalies over the Southern Hemisphere. *Monthly Weather Review*, **110**, 1375–1392.

ROPELEWSKI, C. F. and HALPERT, M. S., 1986: North American precipitation and temperature patterns associated with the El Niño/Southern Oscillation (ENSO). *Monthly Weather Review*, **114**, 2352–2362.

ROPELEWSKI, C. F. and HALPERT, M. S., 1987: Global and regional scale precipitation patterns associated with the El Niño/Southern Oscillation. *Monthly Weather Review*, **115**, 1606–1626.

ROPELEWSKI, C.F. and JONES, P. D., 1987: An extension of the Tahiti-Darwin Southern Oscillation Index. *Monthly Weather Review*, **115**, 1261–1265.

SALINGER, M. J., PALMER, J. G., JONES, P. D., and BRIFFA, K. R., 1994: Reconstruction of New Zealand climate indices back to A.D. 1731 using dendroclimatic techniques: Some preliminary results. *International Journal of Climatology*, **14**, 1135–1149.

SMITH, S. R. and STEARNS, C. R., 1993: Antarctic climate anomalies surrounding the minimum in the Southern Oscillation index. *Antarctic Research Series*, **61**, 149–174.

SOERIANEGARA, I. and LEMMENS, R. H. M. J. (eds.), 1993: *Plant Resources of South-East Asia*. No. 5 (1). Timber trees: Major commercial timbers. Wageningen, The Netherlands: Pudoc Scientific Publishers.

STAHLE, D. W. and CLEAVELAND, M. K., 1993: Southern Oscillation extremes reconstructed from tree rings of the Sierra Madre Occidental and Southern Plains. *Journal of Climate*, **6**, 129–140.

STAHLE, D. W., CLEAVELAND, M. K., HAYNES, G. A., KLIMOWICZ, J., MUSHOVE, P., NGWENYA, P., and NICHOLSON, S. E., 1997: Development of a rainfall sensitive tree-ring chronology in Zimbabwe. *Eighth Conference on Global Change Studies*, Preprint Volume. American Meteorological Society, 77th Annual Meeting, Long Beach, CA, Feb. 1997. pp. 205–211.

STAHLE, D. W., CLEAVELAND, M. K., MAINGI, J., and MUNYAO, J., 1995: The dendroclimatology of *Vitex keniensis* in Kenya (abstract). *EOS* Supplement, Nov. 7, 1995.

STAHLE, D. W., CLEAVELAND, M. K., THERRELL, M. D., GAY, D. A., D'ARRIGO, R. D., KRUSIC, P. J., COOK, E. R., ALLAN, R. J., COLE, J. E., DUNBAR, R. B., MOORE, M. D., STOKES, M. A., BURNS, B. T., VILLANUEVA-DIAZ, J., and THOMPSON, L. G. 1998: Experimental Dendroclimatic Reconstruction of the Southern Oscillation. *Bulletin of the American Meteorological Society*, **79**(10), 2137–2152.

TRENBERTH, K. E., 1976: Fluctuations and trends in indices of the Southern Hemispheric circulation. *Quarterly Journal of the Royal Meteorological Society*, **102**, 65–75.

TRENBERTH, K. E., 1980: Planetary waves at 500 mb in the Southern Hemisphere. *Monthly Weather Review*, **108**, 1378–1389.

TRENBERTH, K. E., 1984: Signal versus noise in the Southern Oscillation. *Monthly Weather Review*, **112**, 326–332.

TRENBERTH, K. E. and SHEA, D. J., 1987: On the evolution of the Southern Oscillation. *Monthly Weather Review*, **115**, 3078–3096.

van LOON, H., KIDSON, J. W., and MULLAN, A. B., 1993: Decadal variation of the annual cycle in the Australian dataset. *Journal of Climate*, **6**, 1227–1231.

van LOON, H. and SHEA, D. J., 1985: The Southern Oscillation. Part IV: The precursors south of 15°S to the extremes of the oscillation. *Monthly Weather Review*, **113**, 2063–2074.

VAUTARD, R., YIOU, P., and GHIL, M., 1992: Singular spectrum analysis: A toolkit for short noisy chaotic time series. *Physica D*, **58**, 95–126.

VILLALBA, R., BONINSEGNA, J. A., LARA, A., VEBLEN, T. T., ROIG, F. A., ARAVENA, J. C., and RIPALTA, A., 1996: Interdecadal climatic variations in millennial temperature reconstructions from southern South America. *In* Jones, P. D., Bradley, R., and Jouzel, J. (eds.), *Climatic Variations and Forcing Mechanisms of the Last 2000 Years*. NATO ASI Series Vol. I 41, pp. 161–189.

VILLALBA, R., COOK, E. R., D'ARRIGO, R. D., JACOBY, G. C., JONES, P. D., SALINGER, J. M., and PALMER, J., 1997: Sea-level pressure variability around Antarctica since A.D. 1750 inferred from subantarctic tree-ring records. *Climate Dynamics*, **13**, 375–390.

VILLALBA, R. and VEBLEN, T. T., 1997: Spatial and temporal variations in Austrocedrus growth along the forest-steppe ecotone in northern Patagonia. *Canadian Journal of Forest Research*, **27**, 580–597.

WANG, B. and WANG, Y., 1996: Temporal structure of the Southern Oscillation as revealed by waveform and wavelets analysis. *Journal of Climate*, **9**, 1586–1598.

WHITE, W. B. and PETERSON, R. G., 1996: An Antarctic circumpolar wave in surface pressure, wind, temperature and sea-ice extent. *Nature*, **380**, 699–702.

WHITTEN, A. J., MUSTAFA, M., and HENDERSON, G. S., 1988: *The Ecology of Sulawesi*. Yogyakarta, Indonesia: Gadjah Mada University Press.

YADAV, R. R. and BHATTACHARYYA, A., 1996: Biological inferences from the growth–climate relationship in teak from India. *Proceedings of the Indian National Science Academy*, **B62**(3), 233–238.

9

The Tropical Ice Core Record of ENSO

LONNIE G. THOMPSON,[1,2] KEITH A. HENDERSON,[1,2] ELLEN MOSLEY-THOMPSON,[1,3] AND PING-NAN LIN[1]

[1]*Byrd Polar Research Center, The Ohio State University, 108 Scott Hall, 1090 Carmack Road, Columbus, Ohio 43210-1002*

[2]*Department of Geological Sciences, The Ohio State University, Columbus, Ohio*

[3]*Department of Geography, The Ohio State University, Columbus, Ohio*

Abstract

Ice core records from tropical and subtropical ice caps provide unique information about the chemical and physical character of the atmosphere. Seasonal variations in the chemical composition of the snowfall and amount of precipitation accumulating on these ice caps produce annual laminations that allow these stratigraphic sequences to be dated. The thickness of an annual lamination reflects the net accumulation, while the physical and chemical constituents (e.g., dust, $\delta^{18}O$, various ions) record atmospheric conditions during deposition. The information presented in this chapter builds upon an earlier investigation of the preservation of an El Niño/Southern Oscillation (ENSO) history in the 1,500-year record from ice cores recovered from the Quelccaya ice cap, Peru (Thompson et al. 1992).

Recent ice cores from Nevado Huascarán, Peru (90°7′S, 77°37′W, 6,048 m), which provided the first tropical ice core history extending back to the Late Glacial Stage (Thompson et al. 1995), also contain an annually resolvable record for the past 270 years. This study is based upon the most recent 68-year period from the Huascarán ice cores, from which a methodology for isolation of ENSO events is developed.

The Quelccaya ENSO study (Thompson et al. 1992) revealed that in the Peruvian Andes the ice core constituent most highly correlated with sea surface temperatures (SSTs) is $\delta^{18}O$ ($r = 0.36$, significant at the 95% level). The current investigation is based upon a 68-year monthly resolved $\delta^{18}O$ time series from the col of Huascarán. $\delta^{18}O$, as preserved on Huascarán, has been shown to reflect large-scale climatological variability over Amazonia and the western tropical Atlantic on timescales of decades to centuries (Thompson et al. 1995). Over the 25-year period (1968–93) of available climatological observations, the interannual variations of $\delta^{18}O$ are closely related to zonal wind variations over tropical South America at the 500 hPa level. Limited evidence suggests that the spatial distribution of SST anomalies in the western tropical Atlantic influences the 500 hPa circulation, which affects the isotopic fractionation of moisture advected across Amazonia and subsequently the $\delta^{18}O$ of Andean precipitation.

A composite response to ENSO developed by using superposed epoch analysis suggests that often during ENSO warming, the moisture convergence axis over the Atlantic Ocean is diverted northward, resulting in unusually warm and dry conditions over northeast Brazil and ^{18}O enrichment of snowfall on Huascarán. Roughly one year later, as ENSO declines, the Atlantic trade wind circulation strengthens and the associated cooler, moister conditions over the Amazon result in Andean moisture that is more depleted in ^{18}O. The Huascarán results illustrate how the ENSO signal is recorded in Andean snowfall and provide a methodology for application to other annually resolved ice core records. These include recently recovered tropical ice cores (Sajama, Bolivia) and subtropical ice cores (Dasuopu Glacier, Himalayas of China), which are introduced briefly in the final section of the chapter. The Dasuopu record is most promising in its potential to link ENSO histories from the Andes with monsoonal variability over southern Asia.

Introduction

Oceanographers have known for several decades that temporary redistributions of heat energy occur periodically in the surface waters of the tropical Pacific Ocean and that these events are inherently linked with other unusual phenomena in and around the Pacific basin. This system, now referred to as the El Niño/Southern Oscillation (ENSO), includes both an atmospheric component and an oceanic component that are coupled so that variations in one fluid medium are simultaneously reflected in the other (Enfield 1989). The spatial scale of ENSO influence is now recognized as nearly planet-wide (Fig. 9.1), and as a short-term climate forcing mechanism, ENSO is second only to seasonal changes in the distribution of solar irradiance. Geographical shifts in convective activity result in a variety of climatic teleconnections. Those most commonly associated with the warm phase of the ENSO include: a strong Aleutian low; stronger extratropical westerlies and enhanced rainfall (along the subtropical jets); and droughts in northern Brazil, Australia, and Zimbabwe (Bjerknes 1969). Figure 9.1 illustrates the nature of these persistent responses to ENSO in certain regions and includes for future reference the locations of six tropical and subtropical ice core records.

The term "El Niño" was initially used to identify the annual arrival of a southward warm current along coastal Peru, but now it is more commonly applied to the unusual warm events (i.e., when sea surface temperatures [SSTs] in the eastern Pacific increase by more than one standard deviation). Although the cold phase (La Niña) is observed to be the normal condition in the Pacific, unusual strengthening of the eastern subtropical high, enhanced trade wind circulation, and large negative SST anomalies along the Peruvian coast do occasionally occur.

Even though ENSO is primarily a Pacific basin phenomenon (a similar in-phase oscillation is also found in the Indian Ocean), Atlantic SSTs often reflect a lagged response (about 8 months) to the warming/cooling events in the tropical eastern Pacific.

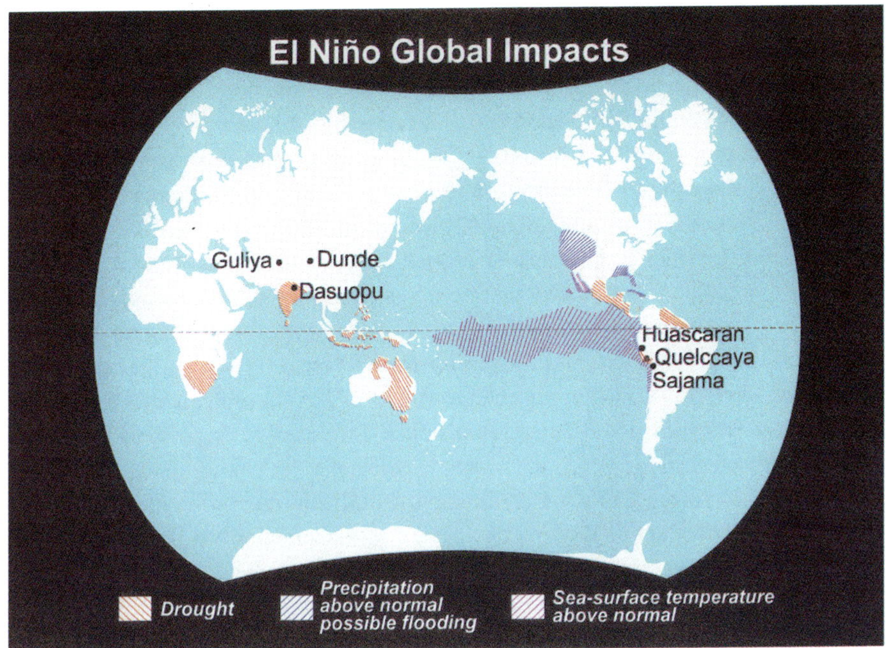

Fig. 9.1 Regional climate changes often experienced during ENSO events are shown along with the locations for tropical and subtropical ice cores discussed in the text. (Modified after NOAA map.)

The mechanism by which Pacific anomalies are transferred (or teleconnected) to the Atlantic basin was described by Covey and Hastenrath (1978). It involves a standing pressure wave between the east tropical Pacific and the west tropical Atlantic such that the nodal pressure boundary that runs generally along the Andean range defines an additional cell component of the east–west Pacific Walker Circulation (Webster 1983). This cell includes the continental surface low positioned between the subtropical anticyclones of the Pacific and Atlantic Oceans as a zone of convergence and rising air, analogous to the large-scale low over the "maritime continents" Australia and Southeast Asia. A series of studies (i.e., Weare 1977; Hastenrath 1978; Lough 1986; Servain and Legler 1986; Servain 1991) using EOF analyses of SST anomalies in the tropical Atlantic clearly showed an analogue to the ENSO variability in the Pacific. The SST anomalies in the tropical Atlantic are substantially weaker than those observed in the equatorial Pacific in association with ENSO (Wallace et al. 1998), but the pattern is quite robust, and there is good evidence based on atmospheric general circulation model simulations (Moura and Shukla 1981; Mechoso et al. 1990) that it is capable of forcing the observed rainfall anomalies. Trenberth et al. 1998 indicate that small changes in SST and SST gradients can lead to shifts in the location of the large-scale organized convection in the tropics and also to changes in the intensity of the convection and hence latent heat release in precipitation and upper tropospheric divergence.

The impact of major ENSO warm events in the tropical Andes is well demonstrated by the 1982–83 and 1997–98 events, which brought major flooding to Ecuador and northern and central Peru and drought to southern Peru and Bolivia. However, moisture feeding the Andean snowfields is derived almost solely from the east via the trade wind circulation, which implies that the tropical Atlantic is the initial moisture source. Thus, if Pacific SST variability is recorded in Andean precipitation, as the Quelccaya study suggests (Thompson et al. 1984), the signal must first be "filtered" through the Atlantic sector. In absolute terms, the magnitude of the extremes in the ENSO-type Atlantic SST oscillations is much smaller (i.e., 0.8 to 1.2°C) than the Pacific extremes, which exhibit strong coastal Pacific warming of 4 to 5°C. According to basic isotopic fractionation theory, this small range of interannual variability alone should not act as the dominant factor controlling the isotopic composition of atmospheric water vapor or condensate. However, it remains possible that some fluctuations over the Atlantic sector are enhanced via tropical atmospheric processes, such as low-level convergence within the Intertropical Convergence Zone (ITCZ) and over the Amazon basin.

The ability to forecast future ENSO activity, particularly in conjunction with observed twentieth-century warming, relies on the continued improvement of our understanding of the global climate system, as well as on lengthening the temporal perspective of climate variability through the development of high-resolution paleoclimatic records. This study is based upon the annually resolved $\delta^{18}O$ history from two ice cores recovered to bedrock from the center of the Col de Huascarán, whose south summit (6,770 m) marks the highest point in the tropics (**Photo 9.1**). The ice thickness in

Photo 9.1 Huascarán in the Cordillera Blanca of north-central Peru taken at sunset. The drill site is located on the col (6,048 m asl) between the north peak (6,658 m asl) on the left and the south peak (6,768 m above sea level) on the right.

the center of the col is 160–165 m, and it provided a continuous archive of snowfall covering the past 20,000 years (Thompson et al. 1995). Due to the natural thinning of glacier ice, layers corresponding to most of the 20,000 years are found in the bottom 3 m of ice; however, the abundant annual snow accumulation, currently 1.4 to 1.8 m H_2O water equivalent, produces annual layers sufficiently thick that a detailed record of the past 100 years may be reconstructed. With near absolute dating for much of this time period, the Huascarán ice cores constitute a high-resolution proxy record of environmental conditions and thus provide an unprecedented opportunity to explore the ENSO history preserved therein.

It has been demonstrated (Grootes et al. 1989; Rozanski et al. 1997; Thompson and Dansgaard 1975; Thompson et al. 1986) that the relationship between atmospheric temperature and the fractionation of ^{18}O and ^{16}O in moisture deposited in the tropical Andes (as well as on the Dasuopu Glacier on the southern margin of the Plateau of Tibet) is the inverse of that found in higher latitudes where the most ^{18}O-depleted snowfall occurs in the winter. Thus, on Huascarán the seasonal $\delta^{18}O$ cycle exhibits greatest ^{18}O depletion in snowfall arriving during austral summer months. The two primary factors controlling $\delta^{18}O$ here are atmospheric temperature at condensation and the precipitation history of the air mass. Over the annual cycle, precipitation appears dominant via the amount effect, but on longer temporal scales (i.e., decades to centuries), atmospheric temperature is clearly the controlling factor. This is confirmed by the presence of both major and minor large-scale climate events (e.g., Little Ice Age, Younger Dryas, Last Glacial Stage, early Holocene optimum) in Quelccaya (Thompson et al. 1986), Huascarán (Thompson et al. 1995), and Sajama (Thompson et al. 1998).

The characteristic response of $\delta^{18}O$ to ENSO (including both warm and cold events) was explored by using superposed epoch analysis. Each event was centered on the time of the peak SST anomaly along coastal Peru, and then 6-year epochs of monthly averaged $\delta^{18}O$ anomaly and other climatological indices were superposed to form composites from which the mean character of the warm vs. cold events was compared.

Special attention was focused on the most recent 25-year period (1970 to 1994), as additional observations of mid- and upper tropospheric conditions were available over limited areas of South America. These additional data made possible a more in-depth examination of the relationship between $\delta^{18}O$ and temperature, humidity, precipitation, and atmospheric circulation. This is particularly important, as the effect of air temperature on isotopic fractionation in tropical meteoric water reservoirs is opposite to that in the polar regions. Thus, the incorporation of these regional meteorological observations provides an opportunity to identify and better understand the physical mechanisms that translate Pacific events into climatic shifts in the source regions for Andean moisture.

Site Description and Analysis

In July–August 1993, two ice cores to bedrock were recovered from the col between the north and south peaks of Nevado Huascarán, Peru (9°S, 77°30′W, col elevation 6,050 m) (Fig. 9.1, **Photos 9.1 and 9.2**). Ice motion vectors determined from stake

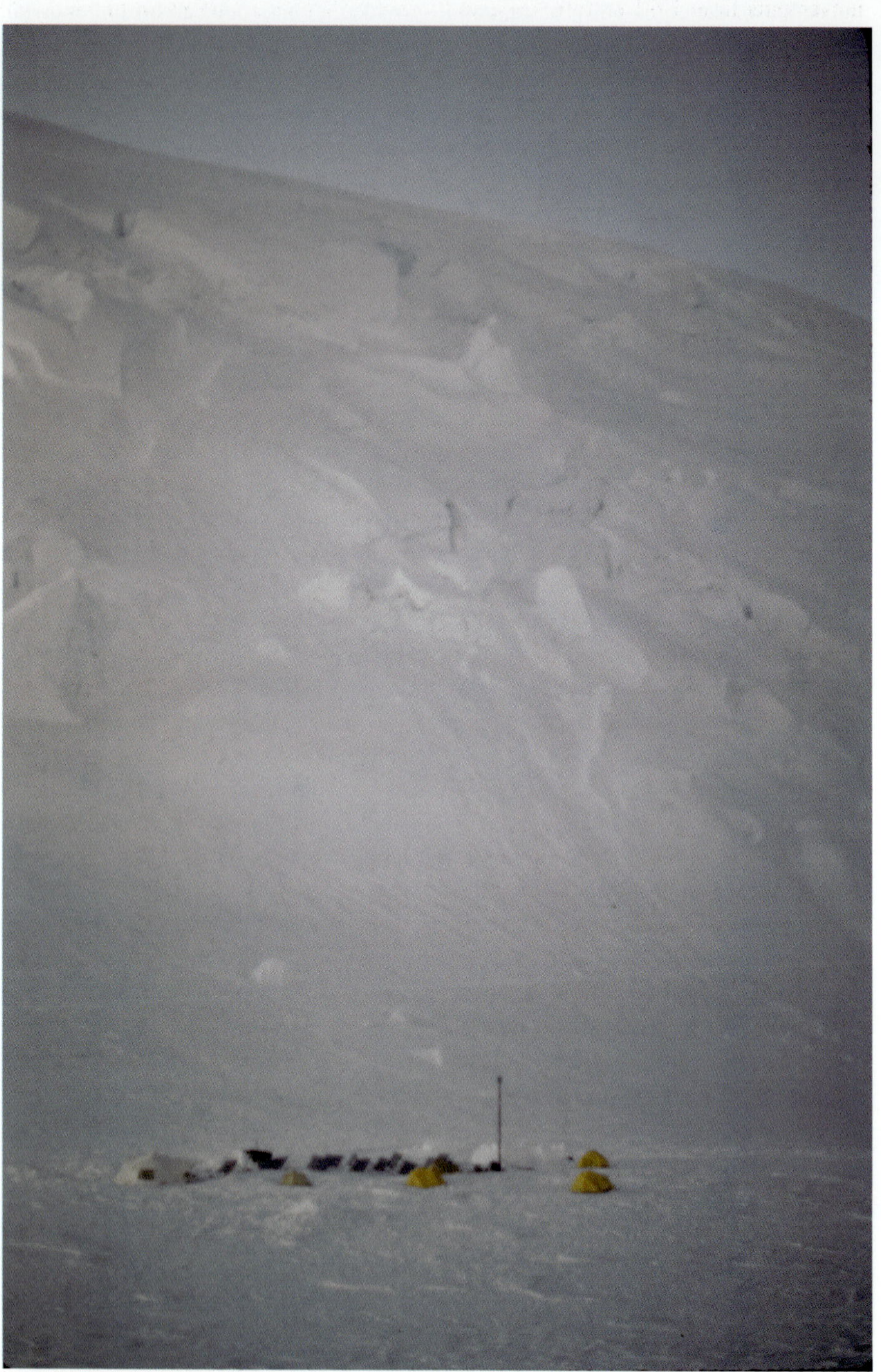

Photo 9.2 The drill camp on the col of Huascarán with the flank of the south peak in the background. The Amazon basin is to the left, and the Pacific Ocean is to the right.

movements from 1991 to 1993 indicate that the drill sites are proximal to the divide between ice flowing toward the east and west outlets of the col. Visible observations and borehole temperatures indicate that the Huascarán is a "polar" type glacier frozen to the bed (Thompson et al. 1995). Core 1, henceforth HSC1, was 160.40 m in length and was cut into 2,677 samples decreasing in thickness from 13 cm at the top to 3 cm at the base. These samples, cut in the field, were melted and poured into 2 or 4 oz. plastic (HDPE) bottles, which were sealed with wax. Core 2, (HSC2), was 166.08 m long and was drilled approximately 100 m from the HSC1 site. Core 2 was returned to the cold room facility at the Byrd Polar Research Center (BPRC) in 1 m long frozen core sections. Roughly six tons of equipment were carried up Huascarán, across crevasse fields (**Photo 9.3**) to the drill site, and over the next 53 days nine tons of equipment, including the two ice cores drilled to bedrock, were carried back down.

Each ice sample from HSC2 was prepared in a Class 100 clean room environment and analyzed for major anion concentrations (Cl^-, NO_3^-, and SO_4^{2-}) by using a Dionex 2010i ion chromatograph. $\delta^{18}O$ was measured by using a Finnigan Mat Delta E mass spectrometer (Craig 1957), and particulate (dust) concentrations and size distributions were determined with a Model TA-II Coulter Counter (Thompson 1973). A second, complete $\delta^{18}O$ profile was also produced from the bottled samples from HSC1, but potential contamination during field preparation precluded dust and anion analyses.

Development of the Time/Depth Relationship

The climate of tropical South America is marked by annual dry seasons (May–October), which are identifiable in the ice cores as elevated concentrations of anion species and dust. The nitrate (NO_3^-) concentrations (Fig. 9.2d) provided the most definitive seasonal marker, but the final timescale was constructed by using a combination of the seasonal oscillations of four ice core constituents as shown in Figure 9.2. Each annual maximum corresponds to the middle of the dry season, assumed to occur on 1 August for the purposes of this analysis. The rapid layer thinning below 120 m limited annual resolution of the record to the most recent 270 years, but the high accumulation and excellent preservation of the seasonal cycles made possible the subannual resolution of $\delta^{18}O$ for 110 annual cycles from 1884 to 1993 (Fig. 9.3).

The accuracy of the ice core timescale is critical for exploration of the relationships among ice core proxy data and tropical climate conditions. The timescale was confirmed to 1980 by comparing the $\delta^{18}O$ record in the 1993 core with that in a 10 m core drilled on Huascarán in 1980 during the original reconnaissance expedition (Thompson et al. 1984). Aside from minor variations in accumulation and modest signal attenuation, the 1993 $\delta^{18}O$ record duplicated the 1980 record for the period of overlap, thus confirming the layer counting back to 1980 as absolute. Three time stratigraphic horizons confirmed the reliability of the layer counting and support precise dating for the uppermost fifty

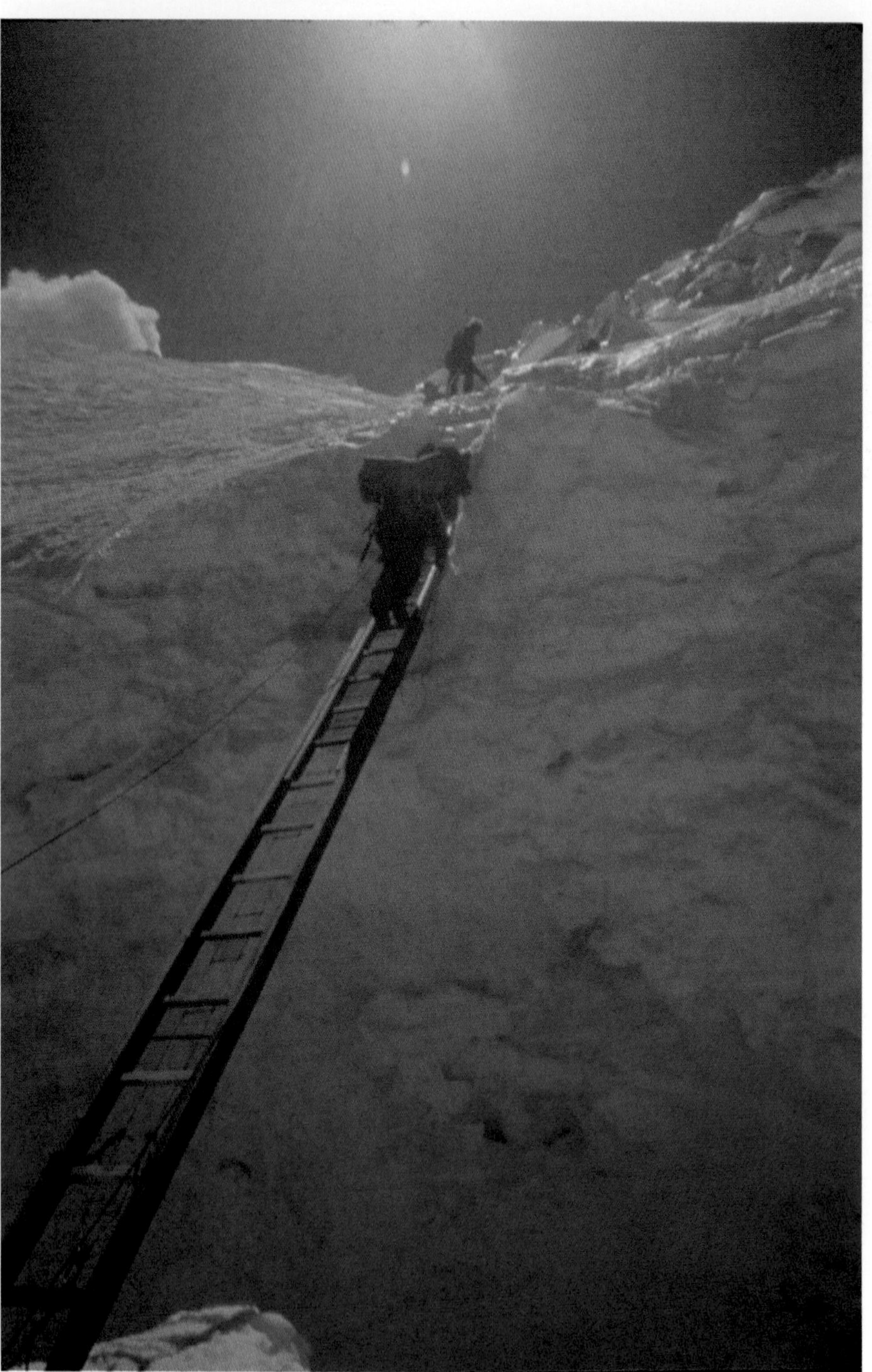

Photo 9.3 This 16 m ladder crosses a major crevasse on the trail to the col of Huascarán. Roughly six tons of equipment were transported across the crevasse to the col, and nine tons, including two ice cores drilled to bedrock, were carried back down.

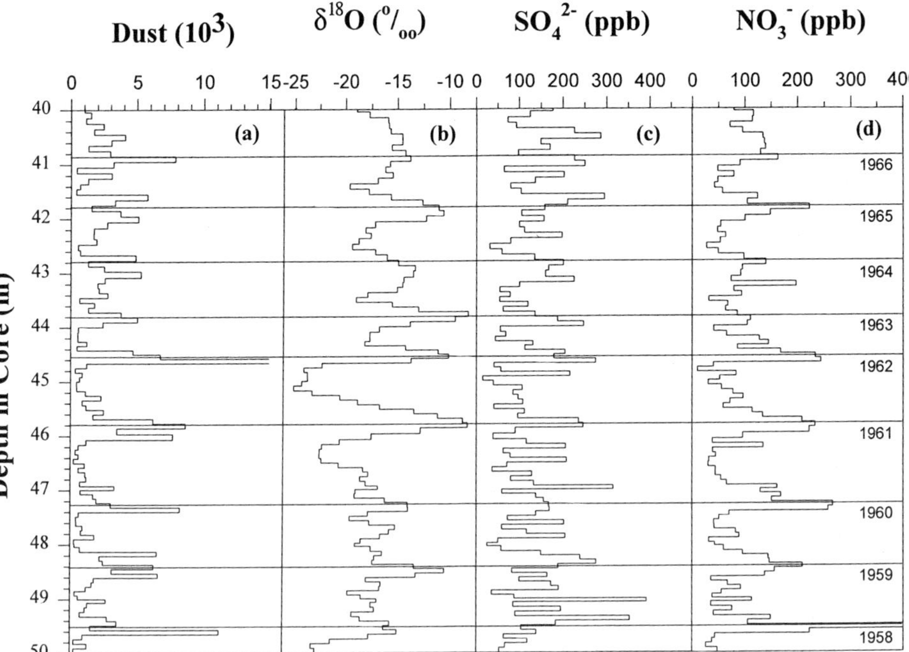

Fig. 9.2 The four seasonally varying constituents used to date the upper part of Huascarán Core 2. These are (a) dust, (b) $\delta^{18}O$, (c) sulfate, and (d) nitrate, shown here for the 40–50 m depth interval. Horizontal lines illustrate the horizons chosen as annual boundaries between successive layers and correspond to snowfall deposited during the middle of the dry season. Dust concentrations are per milliliter sample for particles with diameters of 2.0 m.

years. In May 1970 a magnitude 7.7 earthquake struck coastal Peru, generating large mud flows following the collapse of a large portion of ice from the north peak of Huascarán. The event created a locally fresh sediment source, which is evident in the ice core (Fig. 9.4a) by increased dust deposition for the following four years. Note also that the enhanced dust concentrations during the 1972 and 1973 dry seasons probably reflect the drier conditions associated with that El Niño event. Another time horizon in the HSC2 core is the abrupt increase in ^{36}Cl (Synal et al. 1997). ^{36}Cl is produced by neutron activation during explosion of atomic devices in the presence of a ^{35}Cl source, such as seawater. An abrupt rise in the ^{36}Cl concentration occurs at ~54 m depth (Fig. 9.4b) and is dated to 1952 by layer counting. This spike in ^{36}Cl originated from the 31 October 1952 U.S. "Ivy" surface test of an experimental nuclear device (Carter and Moghissi 1977) on the Eniwetok Atoll in the Pacific Ocean (11°N, 162°E). Finally, in both HSC1 and HSC2, the 1883 eruption of Krakatau, Indonesia (6°S, 105°30′E), produced an anomalous sulfate concentration of ~400 ppb at 110 m depth (Fig. 9.4c), more than twice the concentration of any other sulfate event within a 10 m depth range in the cores. A date of mid-1884 was thus considered to be an absolute time marker for both cores, within the error of the time lag (less than one year).

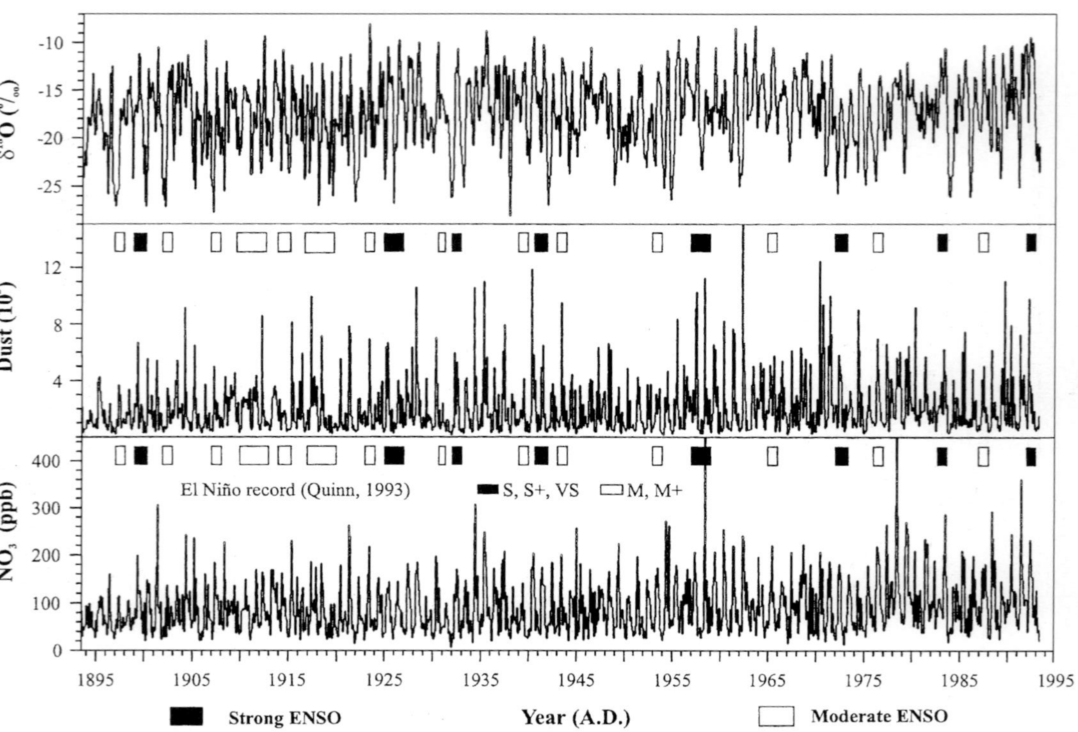

Fig. 9.3 Seasonal variations in the δ^{18}O, insoluble dust, and NO_3^- measured in the Huascarán Core 2 for the past 100 years. Included are the El Niño event identified by Quinn (1993). Note that El Niño strengths are indicated by either solid bars (very strong, strong plus, and strong events) or open bars (moderate and moderate plus events). Weaker El Niño events are not included. Dust concentrations are per milliliter sample for particles with diameters of 2.0 m.

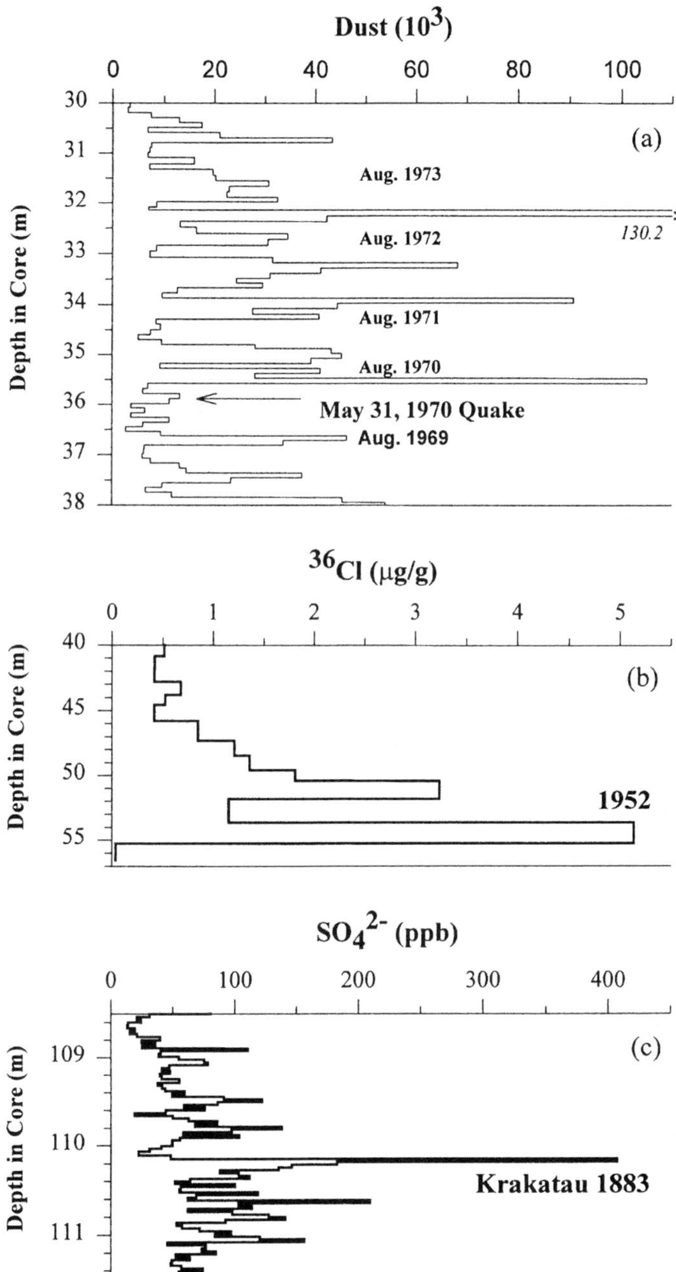

Fig. 9.4 Three different time stratigraphic horizons contribute to the calibration and confirmation of the ice core timescale determined by counting seasonal variations in four different constituents (Fig. 9.2). (a) Dust concentrations were elevated for four years after the 31 May 1970 earthquake in Peru. Dust concentrations are for particles with diameters of 0.63 m per ml sample. (b) The elevated concentration of ^{36}Cl associated with the Ivy test (thermonuclear) on 31 October 1952. (c) Increased sulfate deposition associated with the 1883 eruption of Krakatau.

Construction of Ice Core and Climatological Data Sets for Superposed Epoch Analysis

The high accumulation on Huascarán provided a sample density (number of samples cut per year of accumulation) sufficient to retain monthly resolution for the past 110 years. However, given temporal uncertainties in the earlier portion, only the most recent 68 years were analyzed. To facilitate comparisons among ice core parameters and climatological data sets, the Huascarán $\delta^{18}O$ records were transferred to a timescale with a constant time interval. To construct a continuous "quasi-monthly" time series, the layer thickness representing each "thermal" year (1 August to subsequent 31 July) was divided into 12 units, or bins, of equal thickness or depth. The information for these "artificial" months may deviate from the real months due to variations in annual accumulation rates, differing sample intervals, use of averaging techniques, and poorly defined dry season time markers. Dividing the annual oxygen isotope record into 12 "monthly" increments can be justified since despite the seasonal nature of precipitation in the tropics (measured at lower elevation meteorological stations), the annual oxygen isotopic signal preserved in the ice core is remarkably evenly proportioned between enriched (dry season) and depleted (wet season) isotopic values (Fig. 9.2). The reasons for this are at least twofold. First, although during the dry season it does not rain at lower elevations, often clouds will form at the mountaintops and snow will fall on these high elevation snowfields. Second, it has been known for years from work in the polar regions (i.e., Hammer et al. 1978) that the isotopic input signal from individual snow events is much noisier than the record preserved in the snowpack. The mass exchange by diffusion via the vapor phase in the porous snow usually obliterates the high $\delta^{18}O$ frequencies within a few years, depending on the temperature and the thickness of the individual layers. In areas of high accumulation such as Huascarán, the dominating one year $\delta^{18}O$ cycle remains in the ice. Because on Huascarán this process is largely driven by diurnal heating and cooling, the water vapor is transferred daily up and down the snow column. The net result of this process is an annual signal that represents both dry and wet season isotopic values in the snowpack. Thus, the dry season $\delta^{18}O$ signal is distributed over a greater percentage of each annual increment than might be expected from the annual precipitation patterns observed at the lower altitude meteorological stations.

Within each year, the depth of a sample determined the subannual bin into which its $\delta^{18}O$ value was placed. As many of the years did not contain exactly 12 samples (analyses), adjustments were required. Years containing less than 12 samples (e.g., a sample density $(x) < 12$) required the insertion of $12 - x$ data points, each calculated as the simple average of the two measurements on either side of the empty bin. In contrast, years with more than 12 samples contained $x - 12$ values calculated as the average of two or more measurements. Only three years were represented by more than 24 samples and thus required averaging three values to create 12 monthly averages. For HSC2, a total of 978 samples encompassed the 823 months (July 1993 to December 1925) considered in this analysis. In summary, 68.8% of the months were represented by an unmodified value, 24.3% represented an average of two values within a "monthly" bin, 0.5% represented the average of three values within a "monthly" bin, and the remaining 6.4% were created by interpolation.

The identical procedure was performed for the second core, HSC1, and confirmed that this approach captures the temporal fluctuations of $\delta^{18}O$ in Huascarán. For the 820 months (the top 3 months of accumulation were lost in the drilling) covering the top 68 years of HSC1, only 810 samples were cut. The following results were obtained: 72.7% of the months were represented by an unmodified value (a single $\delta^{18}O$ measurement), 12.7% were averages of two samples, 0.1% were averages of three samples, and 14.4% were created by interpolation. Although HSC1 was sampled more coarsely, the major subannual and interannual variations were duplicated (Fig. 9.5a,b) and correlate well, particularly over the past 40 years. Because HSC1 lacks monthly resolution before 1940, HSC2 is considered the better temporally resolved record. Figure 9.6 illustrates the conversion of the HSC2 $\delta^{18}O$ record in the upper 25 m of core (profile a) into the quasi-monthly $\delta^{18}O$ data for 1980–93. The monthly anomalies (Fig. 9.6c) are calculated with respect to their respective long-term climatological means just as monthly anomalies are calculated for meteorological time series.

For the period 1925–93, monthly averaged SST anomalies were created for both the tropical Pacific and Atlantic Oceans. Although SST anomalies are readily available for various regions of the tropical Pacific (e.g., Niño-1, Niño-2, and Niño-3), these indices were not available prior to 1950, and thus they were not used in this analysis. Furthermore, because World War II (WWII) interrupted the commercial and research ship traffic over much of this region, resulting in sparse data collection, data from several sources and various locations were combined to fill any gaps. To complete the coastal Peru record (Pacific SST data set, abbreviated PacSST), the following products were employed:

(1) Coupled Ocean–Atmosphere Data Set (COADS) Monthly Summary Trimmed Groups (MSTG) Group 3 subset (1925–79) – global 2 × 2 grid (Fletcher and Radok 1985);
(2) National Center for Atmospheric Research (NCAR) ds277.0 "recon" SST data set, filtered by empirical orthogonal function (EOF) analysis (1950–92) – COADS grid;
(3) Puerto de Chicama, Peru (74°2′S, 79°27′W) coastal data set (1925–92) (Quispe Arce 1993).

As all three data sets span the period from 1950 through 1979, all anomalies were determined with respect to this common period. Based on the COADS grid, data from five coastal grid boxes (Fig. 9.7), along with the entire Puerto de Chicama time series (continuous through WWII), were used to construct the PacSST data set. Each monthly value was a simple arithmetic mean of all anomaly values (up to six) available. The final construction of the PacSST anomaly time series involved piecing together the COADS and filtered SST series, which were linked by their common climatology (1950–79). For the period from 1950 to 1979 in the final PacSST record, the filtered anomalies series were selected because of the post-treatment. More details are available in Henderson (1996).

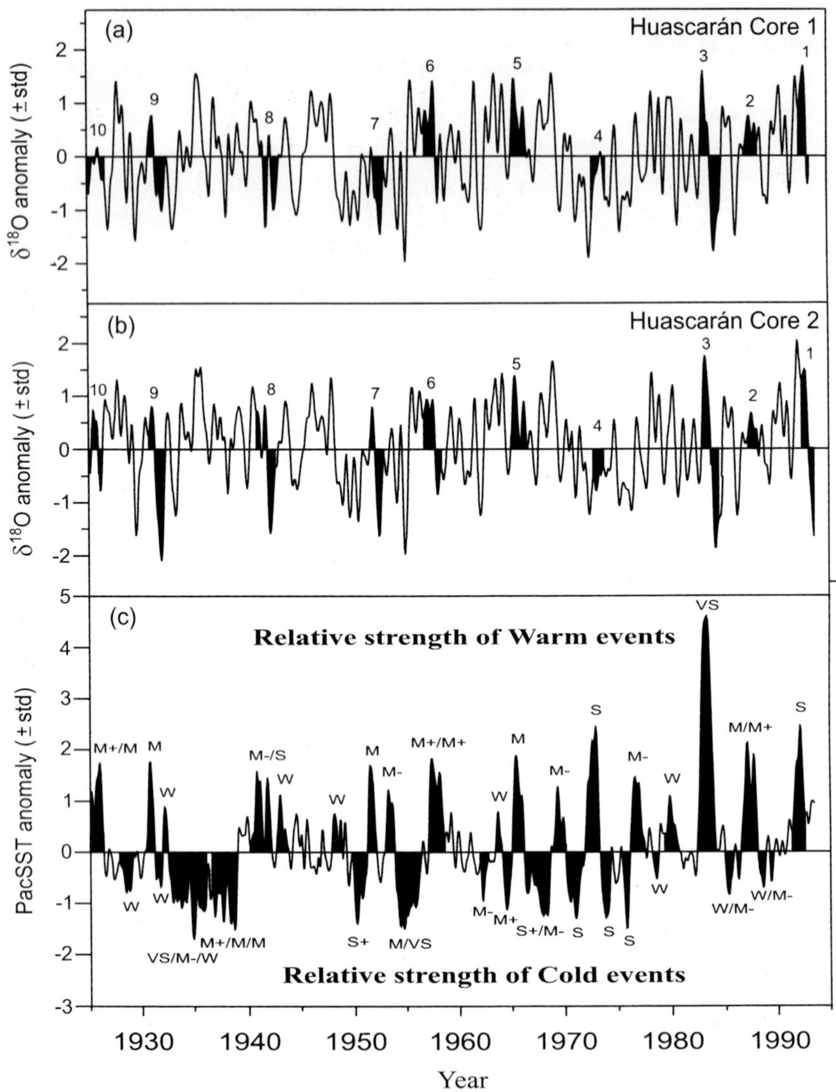

Fig. 9.5 The δ^{18}O anomaly records from (a) HSC1 and (b) HSC2, both plotted as a seven-sample running mean, with weights determined by a "cosine-bell" function, are compared with (c) the PacSST anomalies for coastal Peru derived from COADS/ds277.0("recon")/Chicama, smoothed by a weighted five-term running mean, for the period 1925–93. Numbered and shaded intervals in (a) and (b) depict the individual δ^{18}O responses to the ten largest El Niño events (Table 9.1). Letter designations in (c) indicate relative magnitudes of identified El Niño (warm) and La Niña events (see comparison with Quinn, 1993, in Henderson, 1996.)

The centered (adjusted to share a common mean) PacSST anomaly time series (smoothed) for January 1925 through July 1993 is shown in Figure 9.5c. The marked shift to warmer temperatures at about 1940 reflects in part a shift in the COADS data (Folland and Parker 1995) resulting from a sudden change in the use of canvas or

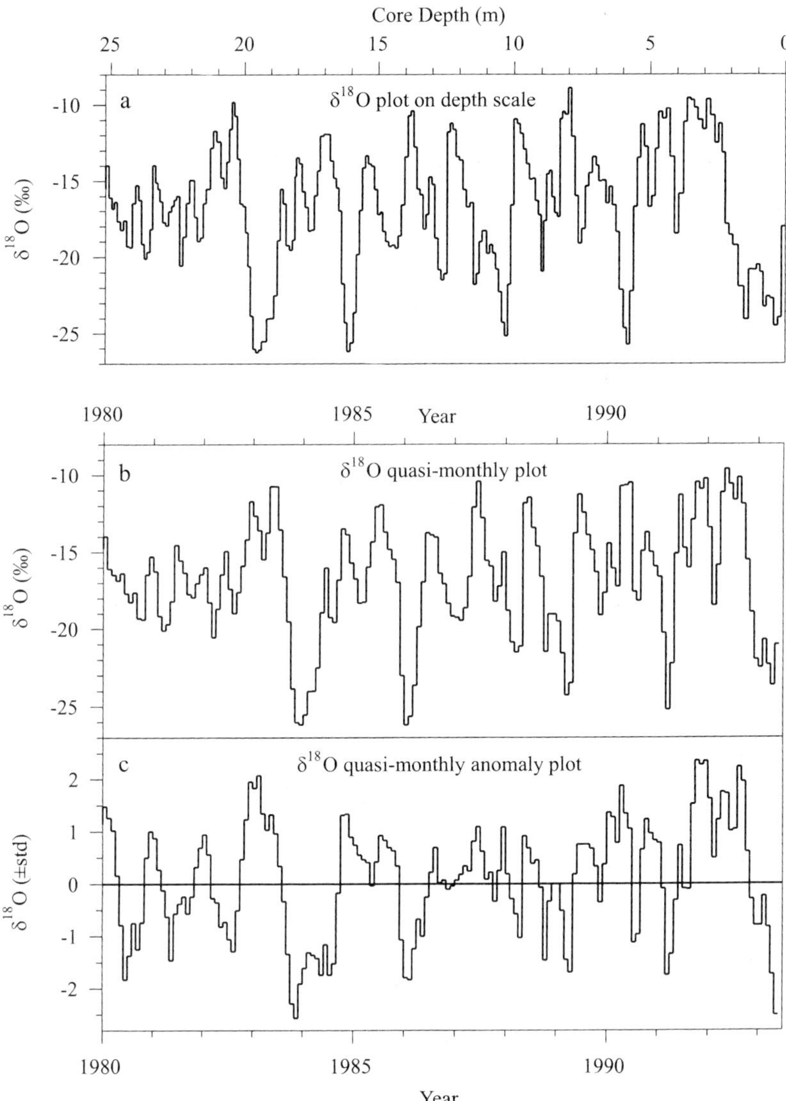

Fig. 9.6 The HSC2 δ^{18}O record for the upper 25 m of the core shown in (a) was converted to (b) a 13-year δ^{18}O history with quasi-monthly resolution. The quasi-monthly record was converted to monthly anomalies (c), each calculated with respect to its respective long-term monthly means.

wooden buckets to engine intake thermometers in about 1942. This effect has not been accounted for, as the long-term warming trend is not significant to the problem under investigation here.

The procedures for producing the SST anomaly data sets for portions of the western tropical Atlantic were similar to those for PacSST, and, of course, the Chicama data were not applicable. Because interest in the Atlantic dealt more with the recent record,

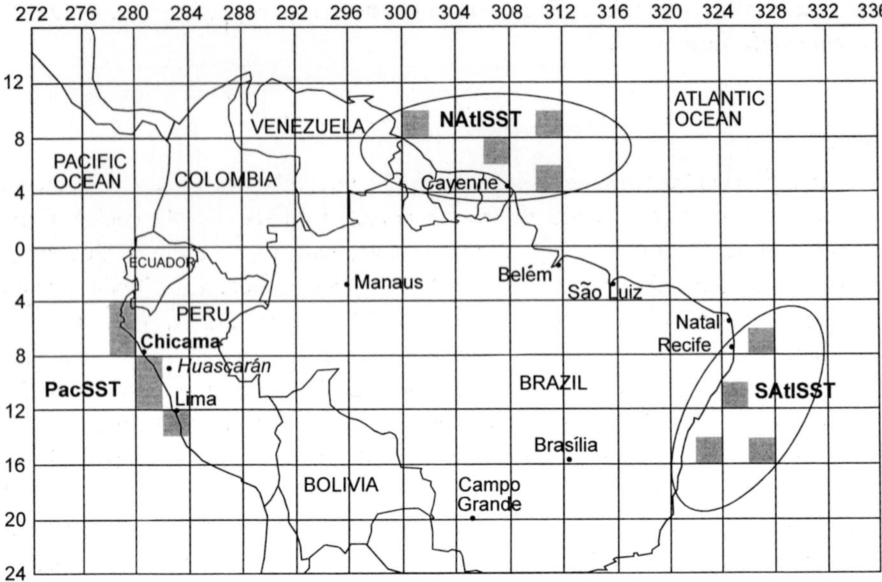

Fig. 9.7 This map shows the location of Nevado Huascarán, the COADS/ds277.0 grid boxes (shaded) used for the creation of PacSST (left), and the meridional Atlantic SST (NAtlSST – SAtlSST) time series (right). The coastal SST monitoring station (Puerto de Chicama) is indicated in boldface. Included are the locations of nine meteorological stations for which surface and upper air data were compiled to create the various tropical South America indices discussed in the text.

only the NCAR ds277.0 "recon" data were used. As the interest in the Atlantic SST data was to explore spatial patterns in the possible source areas for Huascarán moisture, a wider spread of grid points was preferable. Specific grids were selected by their record length and continuity. In earlier work, Henderson (1996) calculated the average SST anomalies for northern, central, and southern coastal regions. The current study employs the average coastal Atlantic SSTs for only two regions (Fig. 9.7): (1) the region off the coast of the Guianas (NAtlSST) and (2) that along the east-central Brazilian coast (SAtlSST). The simple difference between these two anomaly data sets provided a qualitative measure of the meridional heat flux across the moisture source region for Huascarán precipitation.

Monthly averaged surface and upper air data for nine tropical South America locations (Fig. 9.7) were extracted from the journal *Monthly Climatic Data for the World* for 1968 through 1993. Temperature, wind vectors, dew point, and geopotential height data were obtained for three standard levels of the lower and mid-troposphere (850 hPa, 700 hPa, and 500 hPa). Surface data available included temperature, pressure, dew point, and precipitation. For some years data were sporadic, and the individual records were combined into single time series encompassing the entire region surrounding the Amazon basin. Long-term monthly means were determined by using the entire length of the individual records as the basis for the climatology. Anomalies were not standardized in this case, as the standard deviations were often suspect because of the discontinuities in some of the records.

Superposed Epoch Analysis

The timing, spatial extents, and relative strengths of El Niño events are well known for the past 100 years, although the historical record differs with the choice of the specific indices (both climatological and phenological), which respond differently, some directly and some indirectly, to the ENSO phenomenon. The El Niño record developed by Quinn (1993) extends from A.D. 1525 to the present. His reconstruction was based on sea surface and air temperatures as well as recorded episodic events such as flood damage, mass mortality or migration of sea life, and travel time variations for commercial vessels. As only the physical condition of the ocean should influence the nature of high Andean snowfall, the El Niño record used in this study is based only on the PacSST time series.

The method for classifying events was necessarily subjective, as maximum warming and cooling, the duration of sustained anomalous conditions, and their peak characteristics (e.g., single versus double) were all considered. The letter designations used by Quinn (1993) – VS, S+, S, M+, M, and M– – were retained (Fig. 9.5c), and an additional classification (W) was added for borderline (weak) anomalies. These designations were applied also to cold events (La Niña) in an analogous way. Events displaying two (or three) distinct maxima or minima separated by a significant length of time were classified as 2- or 3-year-events. Although no specific time interval between successive peaks was set, most successive peaks of multiyear events were separated by 10 to 16 months. Unlike Quinn's analysis, each of the multiple peaks was given a separate magnitude rating, and the response of the ice core record was considered for each maximum or minimum.

Eighteen El Niño events (including four 2-year events) total were identified in the PacSST analysis (Table 9.1; Fig. 9.5c), two more than identified by Quinn. Only a few events failed to be identified in each (ice core and SST) record (none of them strong events), with the level of agreement generally improving through time toward the present. Fifteen cold events (six multiyear) total were identified in the PacSST analysis (Fig. 9.5c).

A simple comparison of the PacSST and Huascarán $\delta^{18}O$ anomaly records (Fig. 9.5) reveals little similarity at higher frequencies (1–10 years), with the correlation being virtually zero ($r^2 = 0.0035$). The poor correlation is not due to high-frequency noise, as the correlation of thermal year averages remains negligible ($r^2 = 0.0033$). The possibility of dating errors remains a potential complicating factor, but the absence of any easily visible 4- to 5-year periodicity in the Huascarán $\delta^{18}O$ record indicates the lack of any consistent first-order control by Pacific SSTs. This is not unexpected, as the primary moisture source for Andean snowfall is the Atlantic Ocean and not the Pacific Ocean. This raises the possibility that the $\delta^{18}O$ of Andean snowfall may contain a delayed and modulated response to ENSO via the response of the Atlantic SSTs to Pacific SST (ENSO) forcing.

Table 9.1 *El Niño and La Niña events derived from PacSST anomalies (VS = very strong, S = strong, M = medium, W = weak) for period 1925–1993; each "center month" refers to the mid-point of coastal Pacific (Peru) warming. The ten (nine) El Niño (La Niña) events listed in boldface indicate the subset of events (with M or greater magnitude, one maxima per multiyear event) used to create the $\delta^{18}O$ anomaly composites shown in Figure 9.8c. The individual responses in Huascaran $\delta^{18}O$ to the subset of ten moderate–strong El Niño events are highlighted and numbered sequentially in Figure 9.5.*

El Niño events		La Niña events	
Center month	Strength	Center month	Strength
3/1992	S	5/1989	W
10/1987	M	9/1988	M−
3/1987	M	4/1986	W
2/1983	VS	5/1985	M−
10/1979	W	8/1978	W
8/1976	M−	**11/1975**	S
10/1972	S	**11/1973**	S
5/1969	M−	**1/1971**	S
7/1965	M	**1/1968**	S+
9/1963	W	6/1966	M−
2/1958	M+	**7/1964**	M+
5/1957	M+	3/1962	M−
6/1953	M−	11/1955	M
8/1951	M+	**10/1954**	VS
9/1948	W	**4/1950**	S+
1/1943	W	8/1938	M+
11/1941	M−	10/1937	M
1/1941	S	**12/1936**	M
4/1932	W	**12/1934**	VS
11/1930	M	10/1933	M−
12/1925	M+	12/1932	W
3/1925	M	9/1931	W
		8/1928	W

Based on the conclusion that the Huascarán δ^{18}O time series does not constitute a direct ENSO paleo-record, the next logical step was to identify and define the characteristic δ^{18}O response to both warm (El Niño) and cold (La Niña) events and to explore the differences between the two. Bradley et al. (1987) used superposed epoch analysis (SEA) to explore the ENSO signal as preserved in continental temperature and precipitation records of the Northern Hemisphere. The SEA involved the identification of a "zero" point marking the transition from positive to negative Southern Oscillation Index (SOI) (or vice versa) during the onset of an El Niño (La Niña). In each case, April of Year 0 (using the chronology provided by Rasmusson and Carpenter 1982) was chosen as the "zero" month, as it corresponded to the average occurrence of the SOI sign change. Temperature anomalies were then superposed and averaged for a period beginning 36 months before and ending 36 months after the zero month. Their analysis covered the period from 1881 to 1980 and included twenty-three El Niño (warm) years and twenty La Niña (cold) years. Because the superposition of events by Bradley et al. (1987) assumed an invariant April onset, differences in the timing of the anomaly development were not investigated. In fact, El Niño events develop and peak in different times of the year. For this reason, and as this study includes only a single climate record (so that noise is a concern), the zero month identified for superposition of events was variable. For each event the selected "zero" month was the mid-point of respective coastal Pacific (PacSST) warming. With this exception, the SEA of the PacSST and Huascarán $\delta^{18}O$ anomaly time series followed the procedure described by Bradley et al. (1987).

The 73-month SEA of PacSST and HSC2 $\delta^{18}O$ anomalies for all warm and cold events is shown in Figure 9.8(a and b, respectively). In the 16 months following the peak PacSST anomaly (Fig. 9.8a) evidence appears for an ENSO signature in the Huascarán $\delta^{18}O$ composite (Fig. 9.8b). For El Niño events, $\delta^{18}O$ is enriched for approximately 6–8 months after the zero month after which it changes sign and remains

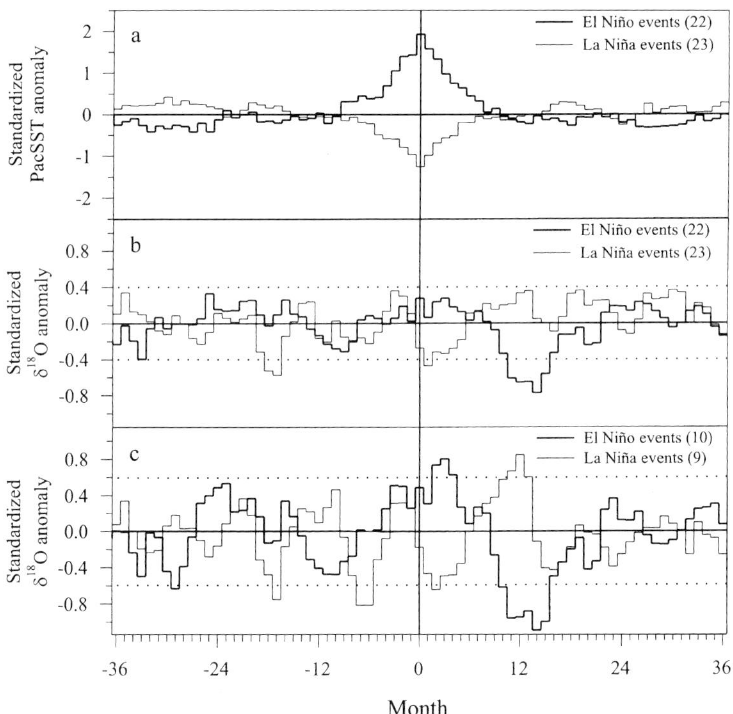

Fig. 9.8 Superposed epoch analysis results (1925–93) for (a) PacSST and (b) Huascarán $\delta^{18}O$ anomalies, both including all identified warm and cold events, and (c) Huascarán $\delta^{18}O$, but only for ENSO events with moderate (M) magnitudes or greater and including only one epoch per each multiyear event. Dotted lines (in b and c) indicate the 95% confidence interval based on 400 Monte Carlo simulations.

depleted for approximately the next 10 months. La Niña events show similar temporal sequences of $\delta^{18}O$ changes, but with opposite sign to those of the warm events. Also evident is an enrichment (depletion) of $\delta^{18}O$ during warm (cold) phases approximately 18–20 months preceding the peak anomaly, which contributes to an apparent (but unsubstantiated) 19-month periodicity. The dotted lines on Figure 9.8 represent the 95% confidence interval based on a Monte Carlo analysis involving 400 simulations of 41 randomly selected and superposed epochs (Bradley et al. 1987).

To further investigate the characteristics of the relationship between ENSO and the $\delta^{18}O$ signal, a second SEA experiment was performed. To accentuate the response to stronger ENSO, (i.e., those most likely to influence Andean precipitation), the weaker events were removed. Also, the secondary maxima (or minima) for multiyear events were eliminated from the composite, except for two cold events for which the second peak was retained, as both were at least two grade levels above those of the initial warm phase. Obviously, the close spacing of maxima or minima (10–16 months) in a 2-year event prohibits back-to-back oscillations (whether high-to-low or low-to-high), as they would overlap and cancel each other. The ten ENSO events of this subset are highlighted in Table 9.1 and numbered sequentially in Figure 9.5 (a, b), and the resulting SEA profiles are shown in Figure 9.8c. The basic characteristics of the responses to warm and cold events remain generally unchanged (comparing Figs. 9.5b and 9.5c); however, in both cases the oscillations are much clearer with the removal of minor events and complicating factors. With the exception of an intriguing 19-month periodicity, the distribution of anomalies in other parts of the 73-month SEA window remains nearly random, indicating that the HSC2 response to PacSST anomalies begin only after the ENSO event has reached peak level.

Relationship between Huascarán $\delta^{18}O$ and Tropical Meteorology

In the previous section, the characteristic ENSO signal in the isotopic composition of archived snowfall in the col of Nevado Huascarán was identified and illustrated. For the purpose of identifying a potential mechanism by which anomalous Pacific SSTs might influence the Andean ice core $\delta^{18}O$ records, it is useful to examine the relationship between the $\delta^{18}O$ interannual variability and conditions of the tropical atmosphere, particularly over the Andean moisture source region. Thus, the time series of the various climatological variables available at the surface and at the 850, 700, and 500 hPa levels were used to explore the relationship between HSC2 $\delta^{18}O$ anomalies and the concomitant tropical climatology. The only statistically significant relationship between $\delta^{18}O$ (Fig. 9.9c) and the meteorological data ($r^2 = 0.30$ for the period 1973–93, averaged quarterly) is with the 500 hPa zonal wind speed anomalies (Fig. 9.9b). Interestingly, the lower-level zonal winds (Fig. 9.9a) display different behavior over this 25-year period and show no clear relationship with either midtropospheric winds (Fig. 9.9b) or the Huascarán $\delta^{18}O$ record (Fig. 9.9c). The lack of any significant relationship between the ice core $\delta^{18}O$ and meteorological conditions (pressure, temperature [Fig. 9.9d], atmospheric layer thickness [Fig. 9.9d], humidity, and rainfall [Fig. 9.9e] anomalies) suggests that "short-term" variations in the $\delta^{18}O$ composition

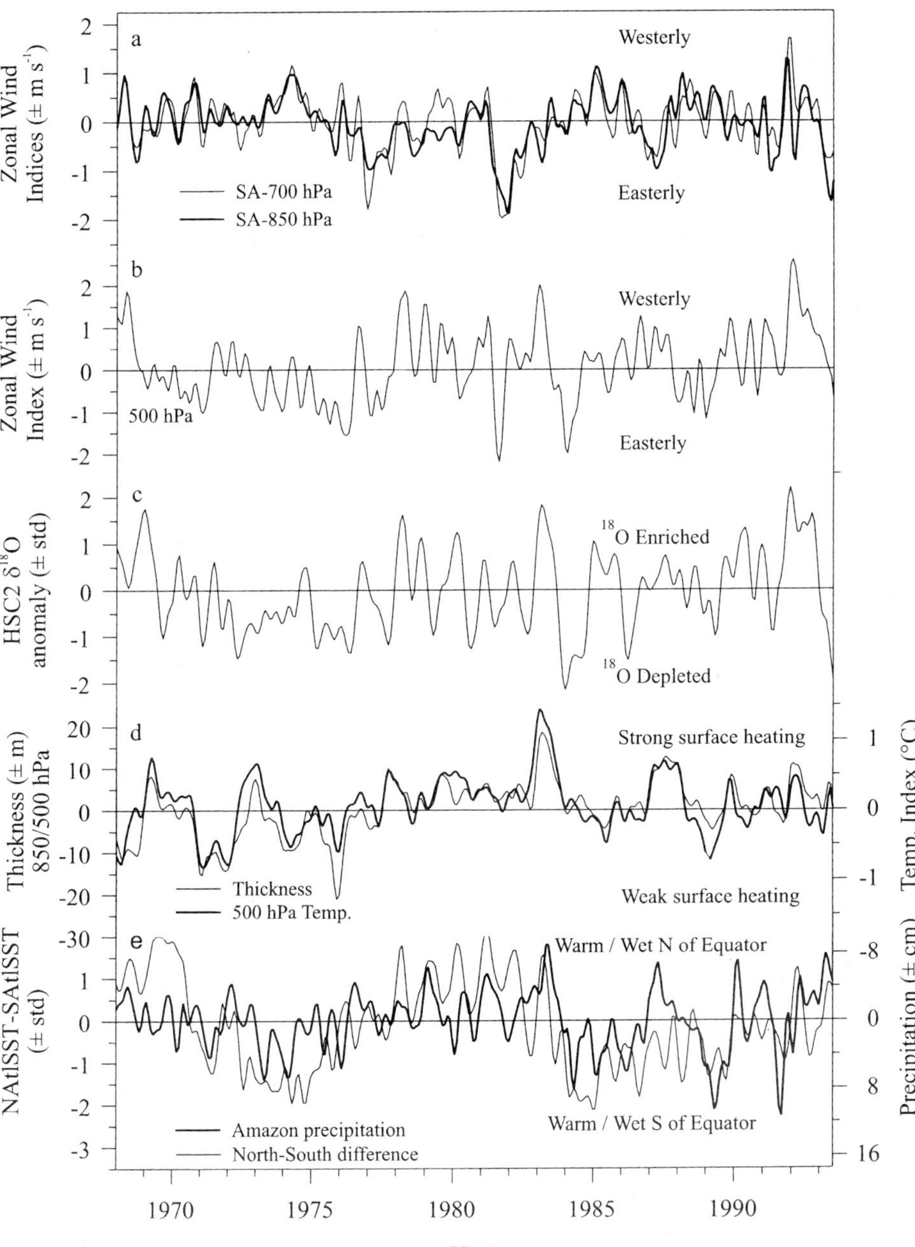

Fig. 9.9 Comparison of (a) zonal wind indices at the 850 and 700 hPa levels, (b) zonal wind index at the 500 hPa level, (c) HSC2 $\delta^{18}O$ anomalies, (d) atmospheric thickness between the 850 and 500 hPa pressure surfaces (light line) and temperature index at the 500 hPa level (dark line), and (e) the NAtlSST−SAtlSST difference in the western tropical Atlantic (light line) and Amazon precipitation anomalies (dark line). For (d) and the rainfall anomalies in (e), only the six northeastern sites (Fig. 9.1) were used. All data were smoothed by a weighted five-term running mean.

of Andean precipitation are not recording regional climate in a recognizable fashion. This is not inconsistent with the earlier observation (see Fig. 9.8) that the relationship between the δ^{18}O seasonal cycle and atmospheric temperature is reversed in the tropics.

Rayleigh fractionation theory (Jouzel 1986) has been useful in providing a physical link between the water vapor budgets necessary for isotopic depletion and the near-surface atmospheric temperatures over high-latitude ice caps (Dansgaard et al. 1973). While the exact transfer functions have been questioned recently (Cuffey et al. 1995), particularly for different past climate regimes, the temperature – δ^{18}O relationship in high-latitude environments has been shown empirically to have a consistent (positive) linear relationship (Johnsen et al. 1989). As was noted earlier, for tropical ice fields such as Huascarán, the seasonal δ^{18}O–T relationship is opposite to that in high-latitude areas (i.e., the most depleted values occur in the warmest season). A modeling study by Grootes et al. (1989) invoked Amazon basin hydrology as a controlling mechanism on the Rayleigh "f" parameter ("water vapor fraction remaining") to explain this reverse relationship.

While these attempts at connecting physical mechanisms to the seasonal cycle in tropical meteorology are pertinent to the application of geochemical measurements as climatological proxies, the actual dynamical processes responsible for the interannual isotopic variability (i.e., the δ^{18}O anomalies) are not sufficiently well established. It is readily apparent that longer (century- to millennial-scale) δ^{18}O fluctuations in Andean ice cores are primarily temperature dependent as in the higher latitudes. Well-documented recent events such as the Little Ice Age and twentieth-century warming (Thompson et al. 1984) impart strong evidence for this conclusion, but even stronger validation lies in the 6‰ shift across the glacial–interglacial (Holocene/Wisconsinan) transition and the identification of a Younger-Dryas signature in both the Huascarán (Thompson et al. 1995) and Sajama (Thompson et al. 1998) records. This raises a key question: At what temporal window length does the tropical δ^{18}O–T relationship reverse and become positive?

Although the HSC2 δ^{18}O record appears to lack direct temporal correlation with zonal wind changes in the lower troposphere, the fairly strong correlation ($r^2 = 0.30$) with the midtropospheric (500 hPa) zonal wind deserves further exploration. Wind velocities at 500 hPa are generally dependent upon the meridional temperature gradient. To approximate the meridional temperature gradient in the source region for Andean moisture, the NAtlSST−SAtlSST difference was calculated (Fig. 9.9e; lighter line). On the multi-annual scale, when SAtlSST is warmer than NAtlSST, the precipitation over the Amazon (south of the equator) is enhanced, leading to enhanced ^{18}O depletion (via the "amount effect") as water vapor traverses the Amazon to the Andes.

Discussion

The characteristic ENSO signal (from the SEA composite) in the Huascarán ice core δ^{18}O record is represented by a more or less symmetrical wave in the anomaly profiles, with a peak in enrichment (Fig. 9.8c) nearly coincident with the maximum PacSST (Fig. 9.8a) and a peak in ^{18}O depletion about 12 months later. The SEA composites of HSC1 (not shown) are very similar, indicating reproducibility in spite of the lower temporal resolution of the record. Using the HSC2 SEA δ^{18}O pattern (Fig. 9.8c) to estimate the influence of individual ENSO events on the Atlantic/Amazon climate regime leads to the conclusion that the influence of the El Niño events of 1992, 1983, 1952, 1941, and 1930 was strong but that ENSO events of 1987, 1972, and 1965 were either weak or had no influence on the Atlantic (Fig. 9.5b).

The lack of a relationship between the composite warm and cold δ^{18}O signals in the 8 months preceding the SST maximum (i.e., during anomaly growth) indicates that the Huascarán δ^{18}O ENSO signal is not forced directly by Pacific SSTs. The strong δ^{18}O response pattern for the 18 months following an ENSO maximum suggests that δ^{18}O is linked to an ENSO response related to SST in the Atlantic Ocean. This Atlantic sector response must be linked to the Pacific, which accounts for the observed lag. Huascarán remains in a strong trade wind regime throughout the year. However, there is only a moderate response in δ^{18}O in the year before the zero point, which is dwarfed by the much more significant trailing effect about 12 months later.

The multi-annual scale enrichment of δ^{18}O in Huascarán precipitation appears prevalent in conjunction with warmer continental temperatures (Fig. 9.9d) and reduced precipitation in Amazonia (Fig. 9.9e). Temperatures at 500 hPa over tropical South America (Fig. 9.9d, dark line) are concomitant with surface temperatures and atmospheric thicknesses (Fig. 9.9d, light line), as the amount of surface heating regulates the height of the tropopause and the height of constant pressure levels (isobaric surfaces). Over the 25-year period investigated (Fig. 9.9), the largest temperature (thickness) changes appear ENSO related, apart from the shift to prolonged warmer conditions occurring from 1975 to 1978. Several of these warm episodes are coincident with enriched δ^{18}O in HSC2 (particularly in 1983). In the late 1970s the δ^{18}O record shifts to a more enriched baseline, possibly reflecting the large-scale transition to warmer conditions, which have continued throughout the 1980s.

Many other features of the temperature record are not duplicated in the δ^{18}O record, suggesting that other regulating factors are at work and need to be identified. For instance, Amazon rainfall appears to have remained above normal for a few years following several of the El Niño events (e.g., 1984 and 1988–89). In the case of 1984, this is reflected in a strong depletion of ^{18}O (a +12-month lagged response), while near-normal temperatures were being recorded over Amazonia.

In comparing the temperature and precipitation anomalies, it is apparent that in many cases, the warm periods (normally during El Niño) are also dry in and around Amazonia, and vice versa (e.g., Kiladis and Diaz 1989). A number of studies have documented the larger scale regional fluctuations during the Pacific ENSO warm and cold phases that involve spatial redistributions of surface temperature anomalies across the entire tropical Atlantic basin (see Anderson et al. 1998). One early such study was that of Hastenrath (1978) who showed that a typical pattern of tropical Atlantic SST anomalies exists (warm off the coast of Angola, and cold near the Caribbean) during times when the low-level convergence axis (ITCZ) is positioned anomalously far south (a condition often coincident with the cold phase of ENSO). This phenomenon has been identified as a cause of high rainfall in the Ceará region of Brazil, the antithesis being the major droughts (secas) that occur there coincident with the stronger Pacific warm events. In Figure 9.9e, this is illustrated by the generally positive relationship between rainfall anomalies and the NAtlSST–SATlSST difference anomalies. As anecdotal evidence, the abrupt warming of 1975–78 (mentioned earlier) is larger in the Caribbean sector than along the south tropical margin of Brazil (note the positive NAtlSST–SATlSST difference anomalies) and is coincident with a reduction in precipitation over coastal and inland Amazon sites. In 1984 this relationship reversed, with warmer temperatures along the south tropical margin of Brazil coincident with increased rainfall over the Amazon and increased ^{18}O depletion in Huascarán precipitation (Fig. 9.9c).

Further examination of the data suggests that a direct link between the behavior of the zonal winds and SST distribution. Hastenrath (1984) demonstrated that a southward shift of the ITCZ enhances the zonal surface trade winds at 10°S. The expected implication for Andean precipitation is enhanced precipitation and ^{18}O depletion in Amazonian water vapor (0°–10°S) via the "amount effect" when the NAtlSST–SAtlSST index is negative, and vice versa. Support for the trade wind concept also comes from an analysis by Rao et al. (1993), who showed that the strength of the surface wind vector oriented perpendicular to the coast (i.e., southeasterly) determines the amount of rainfall received along the Atlantic margin of Brazil between 5° and 15°S. The suggestion of (unidirectional) velocity convergence as a mechanism controlling rainfall along the water vapor transport path to Huascarán could help to explain the apparent strong linkage between zonal wind strength and ^{18}O depletion in Andean snowfall (Fig. 9.10a). However, this localized "land breeze," coupled with minor topography in eastern Brazil, must be integrated with the hydrologic processes of the remainder of the Amazon basin before any further generalizations are possible. Finally, a recent study of meteorological conditions over the Bolivian altiplano (P. Aceituno, unpublished data) noted that a "plume" of water vapor exists within the midtroposphere over that region. It is worth noting that Newell and Zhu (1994) report the existence of "tropospheric rivers of water vapor" from their analysis of atmospheric specific humidity and wind velocity data. If such a layered structure in atmospheric water vapor were a common phenomenon over large areas of the Andean front, a more quantitative relationship might be derived among the upper air circulation, barrier-generated velocity convergence, and high Andean precipitation.

Future ENSO Research in Light of New Tropical Ice Core Records

The 68-year time series of $\delta^{18}O$ from the Huascarán ice cores is a well-preserved, subannually resolved archive of the environmental conditions over Amazonia and the tropical Atlantic. However, the interannual variations reflect a modified signal of either oceanic or atmospheric temperatures that appear closely related only to zonal wind variations over tropical South America at the 500 hPa level (Fig. 9.9a). There is some evidence that the spatial distribution of sea surface temperature anomalies in the western tropical Atlantic affects the 500 hPa circulation, which in turn influences precipitation over the Amazon basin and ^{18}O fractionation. Also, it is apparent that enrichment (depletion) of ^{18}O in Andean snow is related to a combination of warmer (cooler) atmospheric temperatures and less (more) pluvial conditions. On longer timescales, the control of temperature on $\delta^{18}O$ ratios begins to dominate the effect of precipitation (the amount effect).

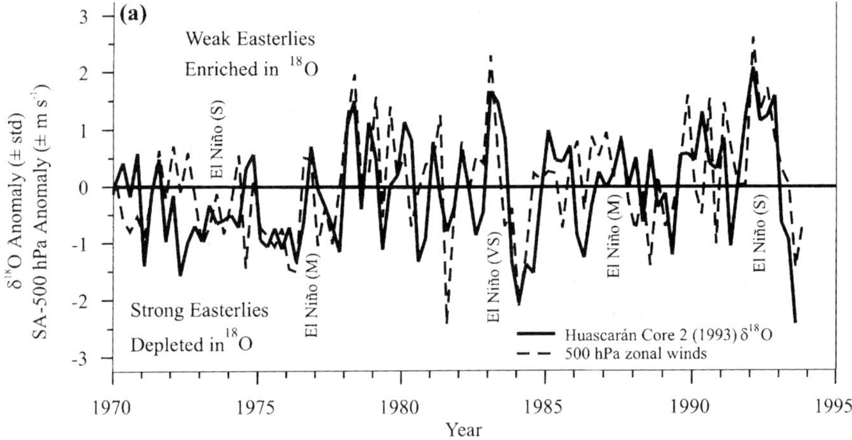

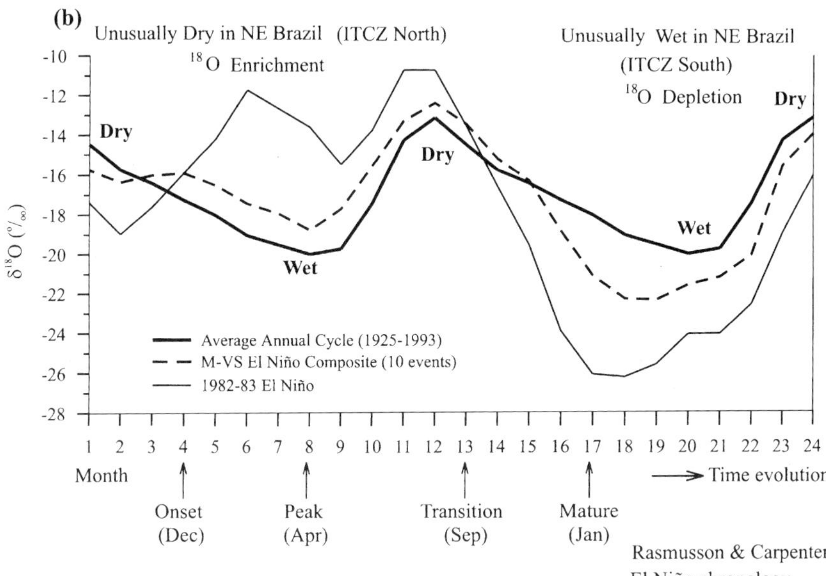

Fig. 9.10 (a) Comparison of δ^{18}O anomalies (standardized departures from long-term monthly means) from Huascarán Core 2 and a composite index of zonal wind departures at the 500 hPa level derived from radiosonde measurements over nine stations in Brazil and Peru (averaged quarterly from 1970 to 1994). (b) The composite δ^{18}O response to El Niño events is compared with the δ^{18}O response of the 1982–83 event and the normal annual variation in δ^{18}O due to the wet–dry seasonality over the Amazon.

The event-based SEA analysis, involving the superposition of individual warm (El Niño) and cold (La Niña) events, depicts the typical reaction of the Atlantic-driven tropical engine over Amazonia and identifies a characteristic 10- to 14-month half-period δ^{18}O wave in the composites (Fig. 9.10b). During El Niño, the onset of the disturbance occurs precisely at the peak positive anomaly in Pacific SST, identifiable by the development of strong enrichment of ^{18}O (significant at the 5% level) soon afterward and then a full reversal to strongly ^{18}O-depleted snowfall 10–14 months afterward.

This scenario is illustrated by the schematic in Figure 9.10b, which includes the composite ice core $\delta^{18}O$ records for the ten strong to moderate ENSO events (Fig. 9.8c, dark line), the $\delta^{18}O$ record for the 1982–83 ENSO event (Fig. 9.10b, light solid line), and average annual $\delta^{18}O$ cycle (1925–93). The characteristics of the cold event composites are very similar but opposite, resulting in a strong negative relationship between warm and cold event composites in the year following peak anomaly development (Fig. 9.8c).

In the last year, new ice cores were recovered from two tropical sites: Sajama, Bolivia (18°06′S, 68°53′W; 6,542 m; Fig. 9.1) and Dasuopu Glacier in the Himalayas of China. The records from these cores will complement earlier records from the Quelccaya ice cap (Thompson et al. 1992) and Huascarán (Thompson et al. 1995; Henderson et al. 1999) by increasing the spatial coverage of ice core records in the tropics. Figure 9.11 illustrates the seasonal variations in $\delta^{18}O$, insoluble dust, and NO_3^- concentrations in

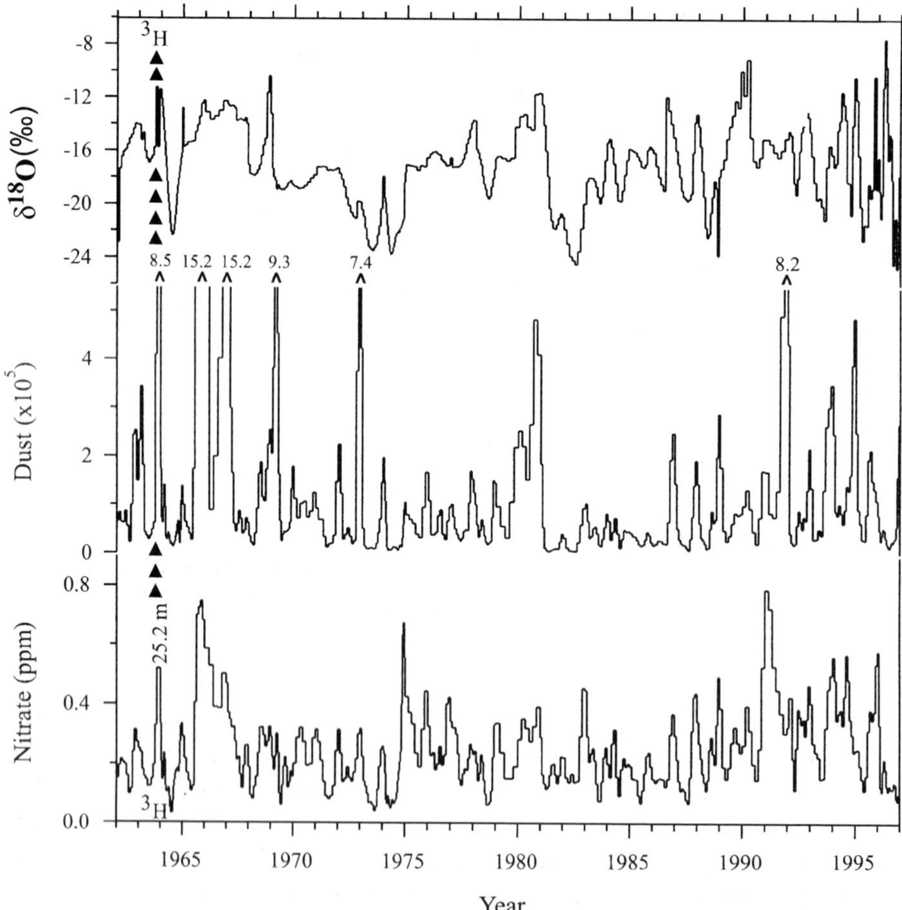

Fig. 9.11 Seasonal variations in $\delta^{18}O$, insoluble dust, and NO_3^- concentrations preserved in the Sajama, Bolivia, ice cores allow dating by layer counting in the upper 50 m of core. Dating is confirmed at 1964 by identification of the 1964 tritium (^3H) peak. The sample values in each plot are smoothed by using a three-point weighted (1,2,1) running mean. Dust concentrations are for particles with diameters of 0.63 m per ml sample.

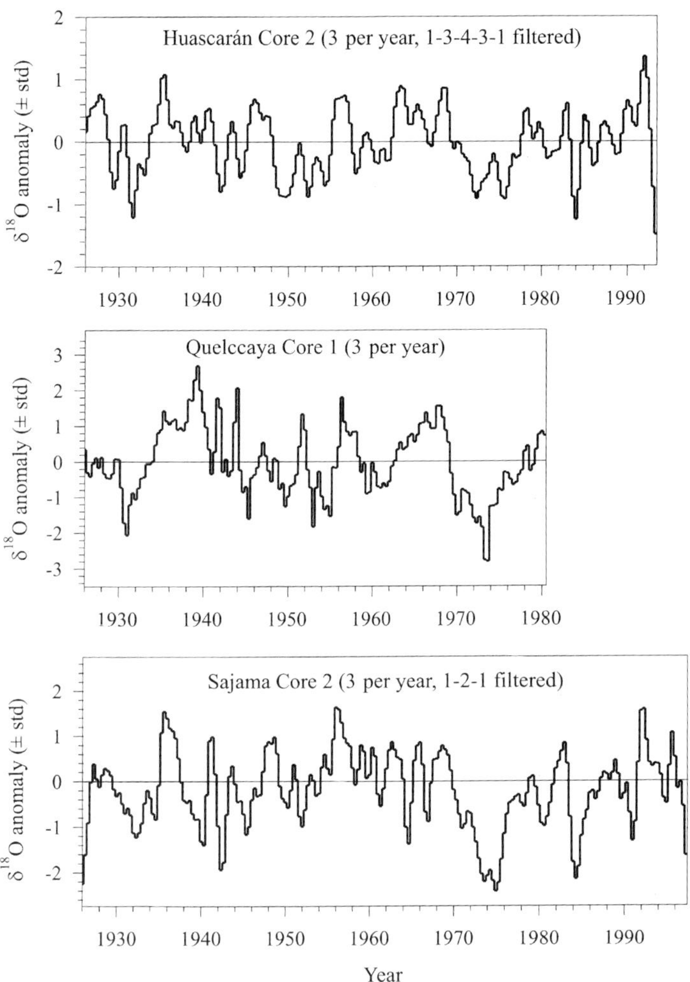

Fig. 9.12 Comparison of $\delta^{18}O$ anomalies along a north to south transect in the tropical Andes: (a) Huascarán, (b) Quelccaya, and (c) Sajama. These plots illustrate both local and regional variability among these sites ranging along the altiplano between 9°S and 18°S.

one of the Sajama cores, which will allow annual layer counting (precise dating) in the upper 50 m. The dating shown has been calibrated by identification of the 1964 tritium (^{3}H) peak. The sample values in each plot are smoothed by using a three-point (1,2,1) filter (Thompson et al. 1998). The annual resolution in the Sajama records will facilitate detailed comparison with records from Huascarán at 9°S and Quelccaya at 14°S. For example, the $\delta^{18}O$ anomalies from 1928 up to the time the ice cores were recovered (Fig. 9.12) illustrate their similarities and differences north to south in the Andes. Intervals from 1970 to 1978 and from 1930 to 1932 stand out as cold periods (^{18}O depleted) at all three sites. Similarly, warm periods from 1932 to 1970 and 1978 to the present (^{18}O enriched) are also evident at each site. As was discussed previously, in the

Huascarán core the 1982–83 El Niño is characterized by increased seasonal extremes in $\delta^{18}O$ anomalies, a feature also evident in the recently collected Sajama ice core $\delta^{18}O$ record.

There is increasing interest in the nature of the connection between ENSO and monsoonal variability and in particular in how these are linked to rice production in countries such as India (Webster et al. 1998). Several linkages between monsoonal intensity and Tibetan snow cover have been discovered (e.g., Barnett et al. 1989; Yasunari et al. 1991; Vernekar et al. 1995). The changing albedo effect due to snow-cover variations on the high Plateau of Tibet may affect the strength of the Asian monsoon (Sirocko et al. 1993). Barnett et al. (1989) suggested that monsoonal intensity may be affected by the extent and duration of Tibetan snow cover. More snow cover or snow cover extending later into the summer season reduces the sensible heat available to heat the Tibetan plateau surface, potentially resulting in a less intense summer low-pressure system and thus reduced monsoon intensity. This in turn may also affect intensities of ENSO events. On longer timescales, general circulation model simulations indicate that increases in snow and ice cover on the Tibetan plateau during the last glacial maximum, resulting in increases in albedo, may have also caused weakening of the monsoonal circulation (Kutzbach et al. 1998).

Recent ice core records from the Tibetan plateau (Dunde Ice Cap [Thompson et al. 1989] and Guliya Ice Cap [Thompson et al. 1997]) may allow these relationships to be examined on annual to millennial timescales. A new record is being reconstructed from ice cores recovered in 1997 from the Dasuopu Glacier (28°23′N, 85°43′E; 7,200 m asl). Dasuopu, located in an area likely to be influenced by ENSO events (Fig. 9.1), is of particular interest, as it offers an excellent opportunity to compare interannual climate variability in the Himalayas with that in the South American tropical ice cores. In the Himalayas and indeed across most of the Tibetan plateau, 80–90% of the precipitation comes in the wet season (monsoon season) from May to October, which produces distinct annual layers in all relevant ice core constituents. The annual records of $\delta^{18}O$, insoluble dust, and nitrates are shown in Figure 9.13 for the upper part (1973 to 1997) of the Dasuopu core. The black squares indicate major ENSO events, while the open squares indicate moderate or minor ones, over this period. Of particular interest is that the summer monsoon isotopic depletions during many of the ENSO years are not as low as during "normal" years over this period. These relationships will be explored after the analyses of the three new Dasuopu ice cores (149.2 m, 159.6 m, 167.6 m) are completed. Initial results suggest that the record will extend back into the Last Glacial Stage.

Acknowledgments We wish to thank Mary E. Davis, Paul Kinder, Bruce Koci, Vladimir Mikhalenko, Cesar Portocarrero, Willie Tamayo Alegre, Selio Villón, and the team of mountain guides, porters, cooks, and arrieros whose assistance in the field logistics was essential. We also thank Chung-Chieh Wang of The Ohio State University Department of Geography for assistance in computer programming and Steve Worley of NCAR for access to the COADS data. We also gratefully acknowledge Juerg Beer of the Swiss Federal Institute of Environmental Science and Technology, who performed the accelerator mass spectrometry measurements of ^{36}Cl on the ice core.

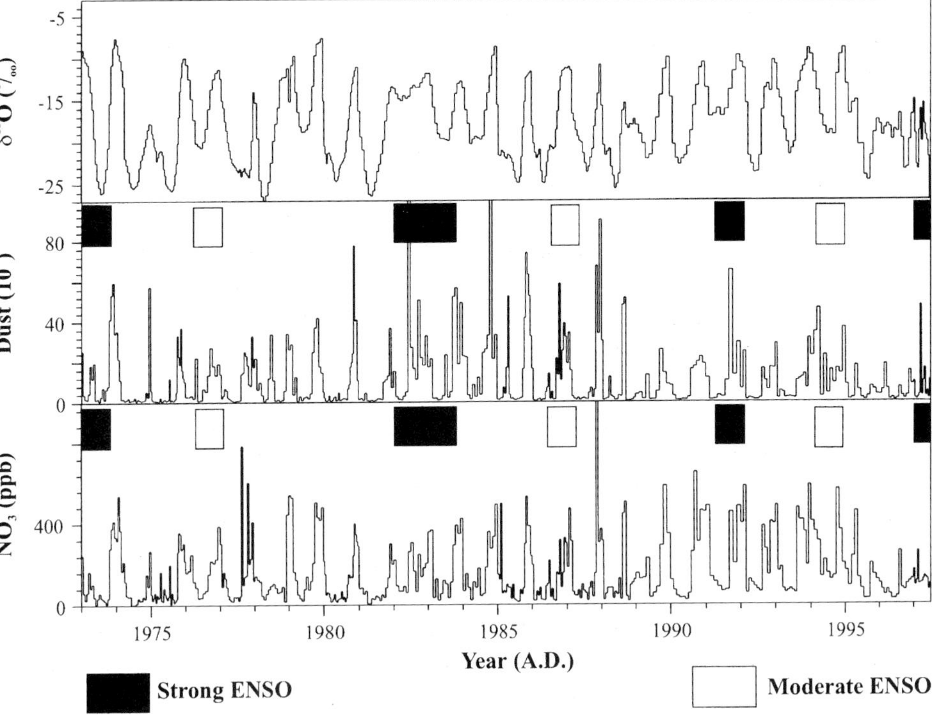

Fig. 9.13 Seasonal variations in $\delta^{18}O$, insoluble dust, and NO_3^- concentrations preserved within Core 1 from the Dasuopu Glacier (Himalayas) allow dating by annual layer counting in the upper 150 m. Dust concentrations are for particles with diameters of 0.63 m per ml sample.

This research was supported under National Oceanic and Atmospheric Administration (NOAA) awards (NA16RC0525 and NA76GP0025). Special thanks to David Goodrich and Mark Eakin of the NOAA Office of Global Programs for their extraordinary efforts to support ice core research in the Andes. This is Contribution Number 1125 of the Byrd Polar Research Center, The Ohio State University.

References

ANDERSON, D. L. T., SARACHIK, E. S., WEBSTER, P. J., and ROTHSTEIN, L. M. (eds.), 1998: *The TOGA Decade, Reviewing the Progress of El Niño Research and Prediction. Journal of Geophysical Research*, **103**, 14,167–14,510.

BARNETT, T., DÜMENIL, L., SCHLESE, U., ROECKNER, E., and LATIF, M., 1989: The effect of Eurasian snow cover on regional and global climate variations. *Journal of the Atmospheric Sciences*, **46**, 661–685.

BJERKNES, J., 1969: Atmospheric teleconnections from the equatorial Pacific. *Monthly Weather Review*, **97**, 163–172.

BRADLEY, R. S., DIAZ, H. F., KILADIS, G. N., and EISCHEID, J. K., 1987: ENSO signal in continental temperature and precipitation records. *Nature*, **327**, 497–501.

CARTER, M. W. and MOGHISSI, A. A., 1977: Three decades of nuclear testing. *Health Physics*, **33**, 55–71.

COVEY, D. L. and HASTENRATH, S., 1978: The Pacific El Niño phenomenon and the Atlantic circulation. *Monthly Weather Review*, **106**, 1280–1287.

CRAIG, H., 1957: Isotopic standards for carbon and oxygen and correction factors for mass spectrometric analysis of carbon dioxide. *Geochimica et Cosmochimica Acta*, **12**, 133–149.

CUFFEY, K. M., CLOW, G. D., ALLEY, R. B., STUVIER, M., WADDINGTON, E. D., and SALTUS, R. W., 1995: Large Arctic temperature change at the Wisconsin-Holocene glacial transition. *Science*, **270**, 455–458.

DANSGAARD, W., JOHNSEN, S., CLAUSEN, H. B., and GUNDESTRUP, N., 1973: Stable isotope glaciology. *Meddelelser Om Grønland*, **192**, 1–53.

ENFIELD, D. B., 1989: El Niño, past and present. *Reviews of Geophysics*, **27**, 159–187.

FLETCHER, J. and RADOK, U., 1985: COADS (Comprehensive Ocean-Atmosphere Data Set) Release 1 Handbook, a joint program by CIRES, NOAA-ERL, NCAR, and NCDC, Boulder, Colorado, 39 pp. plus supplements.

FOLLAND, C. K. and PARKER, D. E., 1995: Correction of instrumental biases in historical sea surface temperature data. *Quarterly Journal of the Royal Meteorological Society*, **121**, 319–367.

GROOTES, P. M., STUIVER, M., THOMPSON, L. G., and MOSLEY-THOMPSON, E., 1989: Oxygen isotope changes in tropical ice, Quelccaya, Peru. *Journal of Geophysical Research*, **94**, 1187–1194.

HAMMER, C. U., CLAUSEN, H. B, DANSGAARD, W., GUNDESTRUP, N., JOHNSEN, S. J., and REEH, N., 1978: Dating of Greenland ice cores by flow models, isotopes, volcanic debris and continental dust. *Journal of Glaciology*, **20**, 3–26.

HASTENRATH, S., 1978: On modes of tropical circulation and climate anomalies. *Journal of the Atmospheric Sciences*, **35**, 2222–2223.

HASTENRATH, S., 1984: Interannual variability and annual cycle: Mechanisms of circulation and climate in the tropical Atlantic sector. *Monthly Weather Review*, **112**, 1097–1107.

HENDERSON, K. A., 1996: *The El Niño–Southern Oscillation and other modes of interannual tropical climate variability as recorded in ice cores from the Nevado Huascarán Col, Peru*. M.S. Thesis, Dept. of Geological Sciences, The Ohio State University, Columbus, Ohio, 194 pp.

HENDERSON, K. A., THOMPSON, L. G., and LIN, P.-N., 1999. Recording of El Niño in ice core δ^{18}O records from Nevado Huascarán, Peru. *Journal of Geophysical Research* (in press).

JOHNSEN, S. J., DANSGAARD, W., and WHITE, J. C., 1989: The origin of Arctic precipitation under present and glacial conditions. *Tellus*, **41B**, 452–468.

JOUZEL, J., 1986: Isotopes in cloud physics: Multiphase and multistage condensation processes. *In* Fritz, P. and Fontes, J. Ch. (eds.), *Handbook of Environmental Isotope Geochemistry*, Vol. 2 (*The Terrestrial Environment B*). Amsterdam: Elsevier Press, 61–112.

KILADIS, G. N. and DIAZ, H. F., 1989: Global climatic anomalies associated with extremes in the Southern Oscillation, J. *Climate*, **2**, 1069–1090.

KUTZBACH, J., GALLIMORE, R., HARRISON, S., BEHLING, P., SELIN, R., and LAARIF, F., 1998: Climate and biome simulations for the past 21,000 years. *Quaternary Science Reviews*, **17**, 473–506.

LOUGH, J. M., 1986: Tropical Atlantic sea surface temperatures and rainfall variations in Subsaharan Africa, *Monthly Weather Review*, **114**, 561–570.

MECHOSO, C. R., LYONS, S. W., and SPAHR, J. A., 1990: The impact of sea surface temperature anomalies on the rainfall over northeast Brazil. *Journal of Climate*, **3**, 812–826.

MOURA, A. D. and SHUKLA, J., 1981: On the dynamics of droughts in northeast Brazil: Observations, theory and numerical experiments with a general circulation model. *Journal of Atmospheric Sciences*, **38**, 2653–2675.

NEWELL, R. E. and ZHU, Y., 1994: Tropospheric rivers: A one-year record and possible application to ice core data. *Geophysical Research Letters*, **21**, 113–116.

QUINN, W. H., 1993: The large-scale ENSO event, the El Niño and other important regional patterns. *Bulletin de l'Institut Français d'Études Andines*, **22**, 13–34.

QUISPE ARCE, J., 1993: Variaciones de la temperatura superficial del mar en Puerto Chicama y del Índice de Oscilación del Sur: 1925–1992, *Bulletin de l'Institut Français d'Études Andines*, **22**, 111–124.

RAO, V. B., de LIMA, M. C., and FRANCHITO, S. H., 1993: Seasonal and interannual variations of rainfall over eastern northeast Brazil. *Journal of Climate*, **6**, 1754–1763.

RASMUSSON, E. M. and CARPENTER, T. H., 1982: Variations in tropical sea surface temperature and surface wind fields associated with the Southern Oscillation/El Niño. *Monthly Weather Review*, **110**, 354–384.

ROZANSKI, K. S., JOHNSEN, S., SCHOTTERER, U., and THOMPSON, L. G., 1997: Reconstruction of past climates from stable isotope records of palaeo-precipitation preserved in continental archives. *Hydrological Sciences–Journal des Sciences Hydrologiques*, **42**, 725–745.

SERVAIN, J., 1991: Simple climatic indices for the tropical Atlantic Ocean and some applications. *Journal of Geophysical Research*, **96**, 15,137–15,146.

SERVAIN, J. and LEGLER, D. M., 1986: Empirical orthogonal function analysis of tropical Atlantic sea surface temperature and wind stress: 1964–1979. *Journal of Geophysical Research*, **91**, 14,181–14,191.

SIROCKO, F., SARNTHEIN, M., ERLENKEUSER, H., LANGE, H., ARNOLD, M., and DUPLESSY, J. C., 1993: Century-scale events in monsoonal climate over the past 24,000 years. *Nature*, **364**, 322–324.

SYNAL, H.-A., BEER, J., SCHOTTERER, U., SUTER, M., and THOMPSON, L. G., 1997: Bomb-produced ^{36}Cl in ice core samples from mountain glaciers. *Glaciers from the Alps: Climate and Environmental Archives, Proceedings*, Paul Scherrer Institut, Villigen, Switzerland, 99–102.

THOMPSON, L. G., 1973: Analysis of the concentration of microparticles in an ice core from Byrd Station, Antarctica. *Ohio State University, Institute of Polar Studies, Report No. 46*, 43 pp.

THOMPSON, L. G. and DANSGAARD, W., 1975. Oxygen isotope and microparticle studies of snow samples from Quelccaya Ice Cap, Peru. *Antarctic Journal of the U.S.*, **10**, 24–26.

THOMPSON, L. G., DAVIS, M. E., MOSLEY-THOMPSON, E., SOWERS, T. A., HENDERSON, K. A., ZAGORODNOV, V. S., LIN, P.-N., MIKHALENKO, V. N., CAMPERN, R. K., BOLZAN, J. F., and COLE-DAI, J., 1998: A 25,000 year tropical climate history from Bolivian ice cores. *Science*, **282**, 1858–1864.

THOMPSON, L. G., MOSLEY-THOMPSON, E., DANSGAARD, W., and GROOTES, P. M., 1986: The "Little Ice Age" as recorded in the stratigraphy of the tropical Quelccaya ice cap. *Science*, **234**, 361–364.

THOMPSON, L. G., MOSLEY-THOMPSON, E., DAVIS, M. E., BOLZAN, J., DAI, J., GUNDESTRUP, N., YAO, T., WU, X., and XIE, Z., 1989: Holocene–Late Wisconsin Pleistocene climatic ice core records from Qinghai-Tibetan Plateau. *Science*, **246**, 474–477.

THOMPSON, L. G., MOSLEY-THOMPSON, E., DAVIS, M. E., LIN, P.-N, HENDERSON, K. A., COLE-DAI, J., BOLZAN, J. F., and LIU, K.-B., 1995: Late glacial stage and Holocene tropical ice core records from Huascarán, Peru. *Science*, **269**, 46–50.

THOMPSON, L. G., MOSLEY-THOMPSON, E., and THOMPSON, P. A., 1992: Reconstructing interannual climate variability from tropical and subtropical ice-core records: *In* Diaz, H. F. and Markgraf, V. (eds.), *El Niño: Paleoclimatic Aspects of the Southern Oscillation*. Cambridge: Cambridge University Press, 295–322.

THOMPSON, L. G., MOSLEY-THOMPSON, E., and MORALES ARNAO, B., 1984: Major El Niño/Southern Oscillation events recorded in stratigraphy of the tropical Quelccaya Ice Cap. *Science*, **226**, 50–52.

THOMPSON, L. G., YAO, T., DAVIS, M. E., HENDERSON, K. A., MOSLEY-THOMPSON, E., LIN, P.-N., BEER, J., SYNAL, H.-A., COLE-DAI, J., and BOLZAN, J. F., 1997: Tropical climate instability: The last glacial cycle from a Qinghai-Tibetan ice core. *Science*, **276**, 1821–1825.

TRENBERTH, K. E., BRANSTATOR, G. W., KAROLY, D., DUMAR, A., LAU, N.-C., and ROPELEWSKI, C., 1998: Progress during TOGA in understanding and modeling global teleconnections associated with tropical sea surface temperatures. *Journal of Geophysical Research*, **103**, 14,291–14,324.

VERNEKAR, A. D., ZHOU, J., and SHUKLA, J., 1995: The effect of the Eurasian snow cover on the Indian monsoon. *Journal of Climate*, **8**, 248–266.

WALLACE. J. M, RASMUSSON, E. M., MITCHELL, T. P., KOUSKY, V. E., SARACHIK, E. S., and van STROCH, H., 1998: On the structure and evolution of ENSO-related climate variablility in the tropical Pacific: Lessons from TOGA. *Journal of Geophysical Research*, **103**, 14,242–24,259.

WEARE, B. C., 1977: Empirical orthogonal function analysis of Atlantic surface temperatures. *Quarterly Journal of the Royal Meterorological Society*, **193**, 467–478.

WEBSTER, P. J., 1983: The large-scale structure of the tropical atmosphere. *In* Hoskins, B. J. and Pierce, R. (eds.), *Large-Scale Dynamic Processes in the Atmosphere*. New York: Academic Press, 235–276.

WEBSTER, P. J., MAGAÑA, V. O., PALMER, T. N., SHUKLA, J., TOMAS, R. A., YANAI, M., and YASUNARI, T., 1998: Monsoons: Process, predictability, and the prospects for prediction. *Journal of Geophysical Research*, **103**, 14,451–14,510.

YASUNARI, T., KITOH, A., and TOKIOKA, T., 1991: Local and remote responses to excessive snow mass over Eurasia appearing in the northern spring and summer climate – A study with the MRI-GCM. *Journal of the Meteorological Society of Japan*, **69**, 473–487.

10

Long-Term Variability in the El Niño/Southern Oscillation and Associated Teleconnections

MICHAEL E. MANN AND RAYMOND S. BRADLEY

Department of Geosciences, University of Massachusetts, Amherst, Massachusetts 01003, U.S.A.

MALCOLM K. HUGHES

Laboratory of Tree Ring Research, University of Arizona, Tucson, Arizona 85721, U.S.A.

Abstract

We analyze global patterns of reconstructed surface temperature for insights into the behavior of the El Niño/Southern Oscillation (ENSO) and related climatic variability during the past three centuries. The global temperature reconstructions are based on calibrations of a large set of globally distributed proxy records, or "multiproxy" data, against the dominant patterns of surface temperature during the past century. These calibrations allow us to estimate large-scale surface temperature patterns back in time. The reconstructed eastern equatorial Pacific "Niño-3" areal-mean sea surface temperature (SST) index is used as a direct diagnostic of El Niño itself, while the global ENSO phenomenon is analyzed based on the full global temperature fields. We document low-frequency changes in the base state, amplitude of interannual variability, and extremes in El Niño, as well as in the global pattern of ENSO variability. Recent anomalous behavior in both El Niño and the global ENSO is interpreted in the context of the long-term reconstructed history and possible forcing mechanisms. The mean state of ENSO and its global patterns of influence, amplitude of interannual variability, and frequency of extreme events show considerable multidecadal and century-scale variability over the past several centuries. Many of these changes appear to be related to changes in global climate, and the histories of external forcing agents, including recent anthropogenic forcing.

Introduction

The instrumental record of roughly the past century provides many key insights into the nature of the global El Niño/Southern Oscillation (ENSO) phenomenon. For example, classic instrumental indices of ENSO, such as the Southern Oscillation Index (SOI) or Niño-3 eastern tropical Pacific sea surface temperature (SST) index, (1) demonstrate concentration of interannual variability in the interannual 3–7 year period band, (2) exhibit interdecadal modulation in the frequency and intensity of ENSO episodes, and (3) show demonstrable (though variable) influences of ENSO on extratropical storm tracks influencing the climates of North America and other regions remote from the tropical Pacific (see, e.g., the reviews of Philander 1990; Diaz and Markgraf 1993; Allan et al. 1996; and references within). There is little instrumental evidence, however, to describe the nature of variability prior to the twentieth century. A variety of theoretical, dynamical models support most of the features of ENSO evident in the instrumental record, including the interannual band-limited characteristics of ENSO (e.g., Jin et al. 1994; Tziperman et al. 1994; Lau et al. 1992), the tendency for amplitude and frequency modulation on longer timescales (either naturally – e.g., Zebiak and Cane 1991; Knutson et al. 1997 – or due to "external" forcing of the tropical coupled ocean–atmosphere system – e.g., Zebiak and Cane 1987), and the characteristic influences on extratropical atmospheric circulation and associated climate variability (e.g., Lau and Nath 1994). Such theoretical models, however, invoke a considerable number of assumptions (i.e., parameterizations) of the modern climate built into them. There thus remains considerable uncertainty as to how ENSO has varied in past centuries, and, just as important, how ENSO *ought to* vary on long timescales (due either to external climate forcing or to low-frequency internal variability) given the limitations of our current theoretical understanding of the governing dynamics.

Trends in ENSO, and the dynamics or forcings governing them, are in fact uncertain even for the past century. The instrumental record of tropical Pacific SST is sparse during the early half of the twentieth century, and depending on how one assesses the instrumental record, quite remarkably, both negative (i.e., La Niña–like) and positive (i.e., El Niño–like) significant trends can be argued for (see, e.g., Cane et al. 1997 and Meehl and Washington 1996, respectively). Furthermore, theoretical coupled models variously support both a negative (Cane et al. 1997) and a positive (e.g., Meehl and Washington 1996; Knutson et al. 1997) response of eastern tropical Pacific SST to anthropogenic greenhouse forcing. Conflicting interpretations of the observational record can be understood in terms of the competing influences of two prominent patterns of global surface temperature variability (Mann et al. 1998a), which include a globally synchronous large-scale warming trend and a regionally heterogeneous trend describing offsetting cooling in the tropical Pacific. The competition between these two large-scale trends complicates the interpretation of ENSO in terms of simple (e.g., the Niño-3) standard indices and underscores the importance of understanding possible anthropogenic-related trends in the context of low-frequency natural variability. Aside from being too short to understand possible twentieth-century trends in an appropriate context, the observational record is also too brief to settle fundamental questions regarding the proper paradigm (e.g, stochastically forced linear oscillator versus self-sustained nonlinear oscillator) for the dynamics that govern ENSO variability (Jin et al. 1994; Tziperman et al. 1994; Chang et al. 1996). Furthermore, questions regarding how unusual recent anomalous ENSO episodes may be (e.g., Trenberth and Hoar 1996, 1997;

Rajagopalan et al. 1997) remain moot in the context of the insufficient length of the instrumental record. Limited evidence for nonstationary characteristics in extratropical teleconnection patterns of ENSO during the twentieth century (Cole and Cook 1997), too, is difficult to evaluate without a longer-term perspective on interdecadal and longer-term variability in the global ENSO phenomenon.

Proxy climate records stand as our only means of assessing the long-term variability associated with ENSO and its global influence. Several recent studies have demonstrated the potential for reconstructing ENSO-scale variability before the past century. The historical chronology of Quinn and Neal (1992, henceforth "QN92") provides some documentary evidence for the timing of warm episodes during the past few centuries, but the rating is qualitative and one-sided (only warm, and not cold, events are indicated), and the reliability of the record is questionable more than a century or two back (an updated version of this chronology is provided by Ortlieb, this volume). Tropical and subtropical ice core records (Thompson 1992) can provide some inferences regarding ENSO variability, but the amount of variance explained is quite modest. Coral-based isotopic indicators of eastern tropical Pacific SST (e.g., Dunbar et al. 1994) or central tropical Pacific variability in precipitation (Cole et al. 1993) have provided site-based reconstructions of ENSO during past centuries, but these records are either too short (e.g., Cole et al. 1993) or too imperfectly dated (e.g., Dunbar et al. 1994) to provide a true year-by-year assessment of ENSO events in past centuries. A number of dendroclimatic (tree-ring-based) reconstructions of the SOI have been performed, based on dendroclimatic series in ENSO-sensitive regions such as western North America (Lough 1991; Michaelsen 1989; Michaelsen and Thompson 1992; Meko 1992; Cleaveland et al. 1992), combined Mexican/southwestern U.S. latewood density (Stahle and Cleaveland 1993), and combinations of tropical Pacific and North American chronologies (Stahle et al. 1998). While such reconstructions can provide long-term descriptions of interannual ENSO-related climate variability, variations on longer timescales (i.e., greater than several decades) are often limited by removal of nonclimatic tree growth trends, and variations at shorter timescales (i.e., year-to-year variations) are hampered by the nontrivial biological persistence implicit to tree growth. Reliance on purely extratropical dendroclimatic networks, moreover, assumes a "stationarity" in the extratropical response to tropical ENSO events that may not be warranted (e.g., Meehl and Branstator 1992).

An alternative approach to ENSO reconstruction employs a network of diverse, globally distributed, high-resolution (i.e., seasonal or annual) proxy indicators or a "multiproxy" network (Bradley and Jones 1993; Hughes and Diaz 1994; Diaz and Pulwarty 1994; Mann et al. 1995; Barnett et al. 1996). Through exploiting the complementary information shared by a wide network of different types of proxy climate indicators, the multiproxy approach to climatic reconstruction diminishes the impact of weaknesses in any individual type or location of indicator and makes use of the mutual strength of diverse data. We make use of recent global temperature reconstructions that are based on the calibration of multiproxy data networks against the twentieth-century instrumental record of global surface temperatures. Details regarding the global pattern reconstructions and the underlying multiproxy data network (Mann et al. 1998a,

Table 10.1 *Proxy, historical, and long instrumental data records used. Entries give description/type of record, location (latitude, longitude), meteorological variable indicated (or most likely indicated), beginning year (A.D.) of record, whether record is used in "tropical" ("trop") sub network, and reference.*

Record	Variable	Latitude	Longitude	Beginning	Trop.	Reference
1. Gridded temperature	Temp	42.5 N	92.5 W	1820		Jones and Bradley 1992
2. "	Temp	47.5 N	2.5 E	1757		
3. "	Temp	47.5 N	7.5 E	1753		"
4. "	Temp	47.5 N	12.5 E	1767		"
5. "	Temp	47.5 N	12.5 E	1775		"
6. "	Temp	52.5 N	17.5 E	1792		"
7. "	Temp	57.5 N	17.5 E	1756		"
8. "	Temp	57.5 N	17.5 E	1752		"
9. "	Temp	62.5 N	7.5 E	1816		"
10. "	Temp	62.5 N	12.5 E	1761		"
11. "	Temp	62.5 N	42.5 E	1814		"
12. Gridded precip	Precip	12.5 N	82.5 E	1813	×	"
13. "	Precip	17.5 N	72.5 W	1817	×	"
14. "	Precip	37.5 N	77.5 W	1809		"
15. "	Precip	42.5 N	2.5 E	1749		"
16. "	Precip	42.5 N	7.5 E	1804		"
17. "	Precip	42.5 N	7.5 E	1770		"
18. "	Precip	42.5 N	7.5 E	1813		"
19. "	Precip	42.5 N	7.5 E	1805		"
20. "	Precip	42.5 N	7.5 E	1697		"
21. "	Precip	42.5 N	7.5 E	1809		"
22. "	Precip	42.5 N	7.5 E	1785		"

			Historical			
23. Cent England temp	Temp	52 N	0 E	1730		Manley 1959
24. Cent Europe temp	Temp	45 N	10 E	1550		Pfister 1992
			Coral			
25. Burdekin River (fluoresc)	Precip/runoff	20 S	147 E	1746	×	Lough 1991
26. Galapagos Isabel Island ($\delta^{18}O$)	SST	1 S	91 W	1607	×	Dunbar et al. 1994
27. Gulf of Chiriqui, Panama ($\delta^{18}O$)	Precip	7.5 N	81 W	1708	×	Linsley et al. 1994
28. Gulf of Chiriqui, Panama ($\delta^{18}C$)	Ocean circ				×	
29. Espiritu Santu ($\delta^{18}O$)	SST	15 S	167 E	1806	×	Quinn et al. 1993
30. New Caledonia ($\delta^{18}O$)	SST	22 S	166 E	1658	×	Quinn et al. 1996
31. Great Barrier Reef (band thickness)	SST	19 S	148 E	1615	×	J. M. Lough, pers. comm.
32. Red Sea ($\delta^{18}O$)	SST/precip	29.5 N	35 E	1788	×	Heiss 1994
33. Red Sea ($\delta^{18}C$)	Ocean circ				×	
			Ice Core			
34. Quelccaya summit ice core ($\delta^{18}O$)	(Air temp)	14 S	71 W	470	×	Thompson 1992
35. Quelccaya summit ice core (accum)	Precip			488	×	"
36. Quelccaya ice core 2 ($\delta^{18}O$)	(Air temp)			744	×	"
37. Quelccaya ice core 2 (accum)	Precip				×	"
38. Dunde ice core ($\delta^{18}O$)	(Air temp)	38 N	96 E	1606	×	Thompson 1992
39. Southern Greenland ice core (melt)	Summer temp	66 N	45 W	1545		Kameda et al. 1996
40. Svalbard ice core (melt)	Summer temp	79 N	17 W	1400		Tarussov 1992
41. Penny ice core ($\delta^{18}O$)	(Air temp)	70 N	70 W	1718		Fisher et al. 1998
42. Central Greenland stack ($\delta^{18}O$)	(Air temp)	77 N	60 W	553		Fisher et al. 1996

Table 10.1 (*cont.*)

Record	Variable	Latitude	Longitude	Beginning	Trop.	Reference
		Dendroclimatic				
43. Upper Kolyma River (ring widths)	Temp	68 N	155 E	1550		F. H. Schweingruber, pers. comm.
44. Java teak (ring widths)	Precip	8 S	113 E	1746	×	Jacoby and D'Arrigo 1990
45. Tasmanian Huon Pine (ring widths)	Temp	43 S	148 E	900	×	Cook et al. 1991
46. New Zealand S. Island (ring widths)	Temp	44 S	170 E	1730	×	Norton and Palmer 1992
47. Central Patagonia (ring widths)	Temp	41 S	68 W	1500	×	Boninsegna 1992
48. Fernoscandian Scots Pine (density)	Temp	68 N	23 E	500		Briffa et al. 1992a
49. Northern Urals (density)	Temp	67 N	65 E	914		Briffa et al. 1995
50. Western North America (ring widths)	Temp	39 N	111 W	1602		Fritts and Shao 1992
51. North American tree line (ring widths)	Temp	69 N	163 W	1515		Jacoby and D'Arrigo 1989
52. "	Temp	66 N	157 W	1586		"
53. "	Temp	68 N	142 W	1580		"
54. "	Temp	64 N	137 W	1459		"
55. "	Temp	66 N	132 W	1626		"
56. "	Temp	68 N	115 W	1428		"
57. "	Temp	64 N	102 W	1491		"
58. "	Temp	58 N	93 W	1650		"
59. "	Temp	57 N	77 W	1663		"
60. "	Temp	59 N	71 W	1641		"
61. "	Temp	48 N	66 W	1400		"
62. S.E. U.S., N. Carolina (ring widths)	Precip	36 N	80 W	1005	×	Stahle et al. 1988
63. S.E. U.S., S.Carolina (ring widths)	Precip	34 N	81 W	1005	×	"

64. S.E. U.S.,Georgia (ring widths)	Precip	33 N	83 W	1005	×	"
65. Mongolian Siberian Pine (ring widths)	Precip	48 N	100 E	1550	×	Jacoby et al. 1996
66. Yakutia (ring widths)	Temp	62 N	130 E	1400		Hughes et al. 1999
67. OK/TX U.S. ring widths (PC #1)	Precip	29–37 N	94–99 W	1600	×	Stable and Cleaveland 1993
68–69. OK/TX U.S. ring widths (PC #2,3)	Precip	29–37 N	94–99 W	1700	×	"
70. S.W. U.S./Mex widths (PC #l)	Precip	24–37 N	103–109 W	1400	×	"
71. S.W. U.S./Mex widths (PC #2)	Precip	24–37 N	103–109 W	1500	×	"
72–73. S.W. U.S./Mex widths (PC #3,4)	Precip	24–37 N	103–109 W	1600	×	"
74–76. S.W. U.S./Mex widths (PC #5–7)	Precip	24–37 N	103–109 W	1700	×	"
77–78. S.W. U.S./Mex widths (PC #8,9)	Precip	24–37 N	103–109 W	1700	×	"
79. Eurasian tree line ring widths (PC #1)	Temp	60–70 N	100–180 E	1450		Vaganov et al. 1996
80. Eurasian tree line ring widths (PC #2)	Temp	60–70 N	100–180 E	1600		"
81. Eurasian tree line ring widths (PC #3)	Temp	60–70 N	100–180 E	1750		"
82–84. Aust/NZ chronologies (PC #1–3)	Mixed	37–42 S	145–175 E	1600	×	1TRDB[1]
85. Aust/NZ chronologies (PC #4)	Mixed	37–44 S	145–175 E	1750	×	"
86–88. S. American chronologies (PC #1–3)	Mixed	37–49 S	68–72 W	1600	×	"
89–94. N. American chronologies (PC #1–6)	Mixed	24–66 N	70–160 W	1400		"
95. N. American chronologies (PC #7)	Mixed	24–66 N	70–160 W	1600		"
96–97. N. American chronologies (PC #8,9)	Mixed	24–66 N	70–160 W	1750		"
98–115. Miscellaneous chronologies	Mixed	Global	Global	<1820		"

[1] Records taken from the International Tree Ring Data Bank (ITRDB). Full references are available through the web site sponsored by the National Geophysical Data Center (NGDC): *http://julius.ngdc.noaa.gov/paleo.treering.html*

henceforth "MBH98"; Mann et al. 1998b, 1999; Jones 1998), the evidence for external climate forcings in the reconstructed climate record (MBH98), and the patterns of variability in the North Atlantic and their relationships to coupled ocean–atmosphere model simulations (Delworth and Mann 1999) are provided elsewhere. Here we report findings from these reconstructions regarding long-term variability in ENSO and its large-scale patterns of influence. An attempt is made to separately assess the true, tropical ENSO phenomenon and its potentially more tenuous extratropical influences through the use of distinct tropical (or predominantly tropical) and global multiproxy networks. Though not without certain limitations, the reliability of the climate reconstructions are established through independent cross-validation and other internal consistency checks. With these reconstructions, we are able to take a global view of ENSO-related climate variability several centuries back in time. We examine questions of how the baseline state of ENSO has changed on long (multidecadal and century) timescales, how interannual variability associated with ENSO has been modulated on long timescales, and how global patterns of influence of El Niño may have changed over time.

In the second section, we describe the data used in this study, both proxy and instrumental, in more detail. In the third section, we describe the multiproxy pattern reconstruction approach that forms the basis of our climate reconstructions. We describe separate applications of the approach to annual-mean reconstruction of both global surface temperature patterns and the more regional Niño-3 index of ENSO-related tropical surface temperature variations. In the fourth section, we discuss the global temperature and Niño-3 reconstructions, and we analyze these reconstructions for low-frequency variability in ENSO and its global patterns of influence. We present our conclusions in the fifth section.

Data

We employ a multiproxy network consisting of diverse high-quality annual resolution proxy climate indicators and long historical or instrumental records collected and analyzed by many different paleoclimatologists (details provided in Table 10.1). All data have annual resolution (or represent annual means in the case of instrumental data available at monthly resolution). Small gaps were interpolated and any records that terminate slightly before the end of the 1902–80 calibration interval (see next section) are extended by using persistence to 1980. Certain subnetworks of the full multiproxy data network (e.g., regional dendroclimatic networks) have been represented by a smaller number of their leading principal components (PCs), the maximum number of which (available back to 1820) is indicated in parentheses, to ensure a more regionally globally homogenous network. Two different sets of calibration experiments are performed. The first set uses the entire global multiproxy network (Fig. 10.1; this is essentially the same network used by MBH98) to reconstruct the "global ENSO" signal, while a more restricted "tropical" network (Fig. 10.2) of indicators in tropical regions or subtropical regions most consistently impacted by ENSO is used to reconstruct the specific tropical Pacific El Niño signal. In using largely independent networks to describe both the

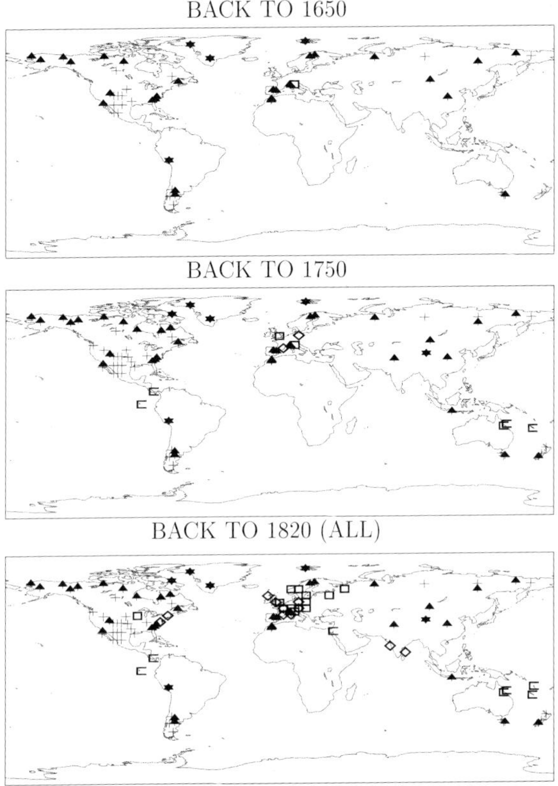

Fig. 10.1 Distribution of multiproxy network back to (a) 1650, (b) 1750, and (c) 1820 (all proxy indicators). Squares denote historical and instrumental temperature records; diamonds denote instrumental precipitation records; umbrella, or "tree," symbols denote dendroclimatic indicators; and "C" symbols indicate corals. Groups of "+" symbols indicate PCs of dense tree-ring subnetworks, with the number of such symbols indicating the number of PCs that have been used to represent a dense regional dendroclimatic network. Certain sites (e.g., the Quelccaya ice core; see Table 10.1) consist of multiple proxy indicators (e.g., multiple cores and both $\delta^{18}O$ isotope and accumulation measurements).

global ENSO and the tropical El Niño signals, we are able to independently assess both the tropical and global ENSO histories. We thereby in large part avoid the potentially flawed assumption of a stationary extratropical response to tropical ENSO forcing.

While a diverse and widely distributed set of independent climatic indicators can more faithfully capture a consistent climate, reducing the compromising effects of biases and weaknesses in the individual indicators, certain limitations must be carefully kept in mind in building an appropriate network. Dating errors in a given record (e.g., miscounted annual layers or rings) predating the calibration period can be problematic. It is well known that standardization by estimated nonclimatic tree growth trends, and the limits of maximum segment lengths in dendroclimatic reconstructions, can strongly limit the maximum timescale climatic trends that are resolvable (Cook et al. 1995). We have, however, employed a careful screening of the dendroclimatic network to ensure conservative standardization procedures and long constituent segment lengths, and

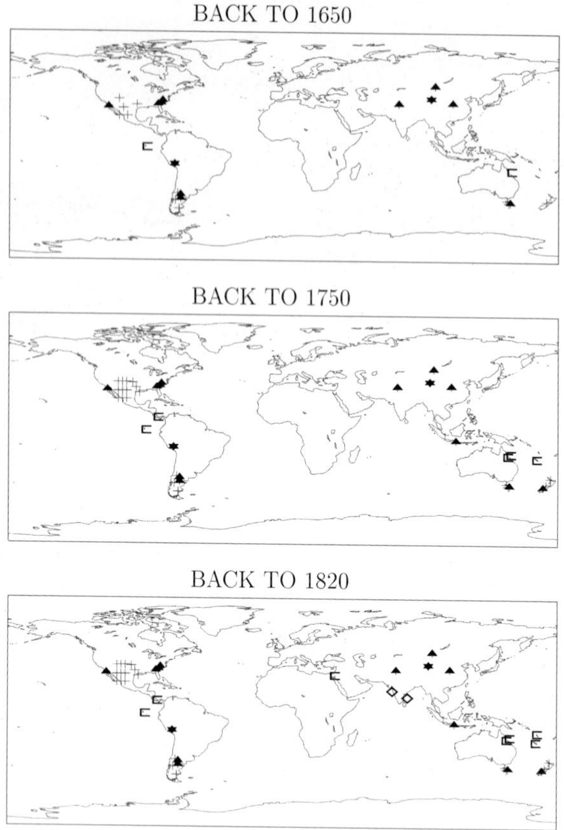

Fig. 10.2 Distribution of restricted multiproxy network used for Niño-3 reconstruction back to (a) 1650, (b) 1750, (c) 1820. Symbols as in Figure 10.1.

variations on century and even multiple-century timescales are resolved typically. An analysis of the resolvability of multicentury timescale trends in these data and implications for long-term climate reconstruction are provided elsewhere (Mann et al. 1999).

To calibrate the multiproxy networks, we make use of all available grid point (5° longitude by 5° latitude) land, air, and sea surface temperature anomaly data (Jones and Briffa 1992) with nearly continuous monthly sampling back to at least 1902. Short gaps in the monthly instrumental data are filled by linear interpolation. For the global temperature pattern reconstructions, we use all ($M = 1{,}082$) available nearly continuous grid point data available back to 1902. For tropical Pacific SST/Niño reconstructions, we use the available $M = 121$ SST grid point data in the tropical Pacific region bounded by the two tropics and the longitudes 115°E to 80°W. These data are shown in Figure 10.3. Although there are notable gaps in the spatial sampling, significant enough portions of the globe are sampled to estimate (in the case of the global field) the Northern Hemisphere (NH) mean temperature series and, in both cases, the Niño-3 index of the El Niño phenomenon. Northern Hemisphere temperature is estimated as areally weighted (i.e., cosine latitude) averages. The Niño-3 eastern equatorial Pacific SST index is constructed from the ten grid points available over the conventional Niño-3

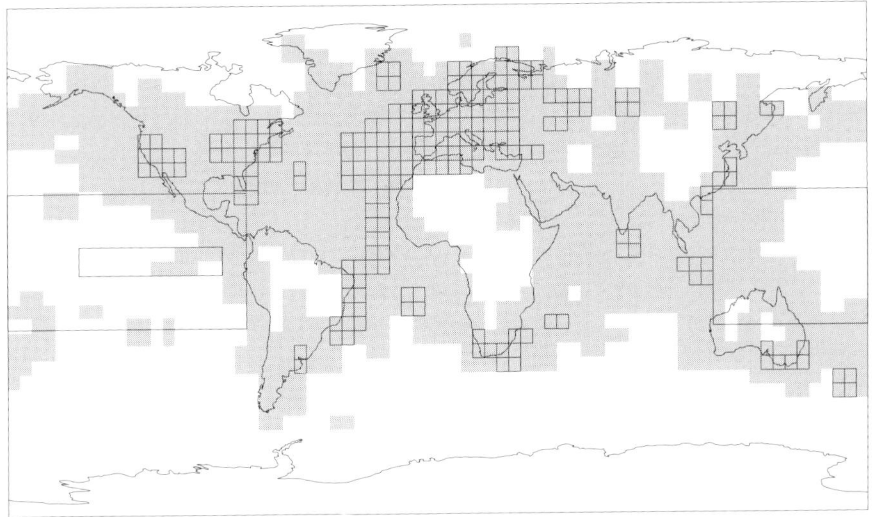

Fig. 10.3 Distribution of the (1,082) nearly continuous available land air/sea surface temperature grid point data available from 1902 onward, indicated by shading. The squares indicate the subset of 219 grid points, with nearly continuous records extending back to 1854 that are used for verification. Northern Hemisphere (NH) and global (GLB) mean temperature are estimated as areally weighted (i.e., cosine latitude) averages over the Northern Hemisphere and global domains, respectively. The large rectangle indicates the tropical Pacific subdomain discussed in the text. The small rectangle in the eastern tropical Pacific shows the traditional Niño-3 region. An arithmetic mean of grid point anomalies in our sampling of the Niño-3 region provides a reliable estimate of the fully sampled Niño-3 index (see text for discussion).

box (see Fig. 10.3) for the region 5°S to 5°N, 90°W to 150°W. While this spatial sampling covers only about one third of the total Niño-3 region, the associated time series is nearly identical to the fully sampled Niño-3 region time series available (both series share 90% of their variance in common during their overlap back to 1953). The (negative) correlation of our Niño-3 average with the SOI – an alternative ENSO indicator which tends to covary oppositely with El Niño – is consistently high ($r^2 \approx 0.5$) and robust throughout the twentieth century for the intervals during which both are available ($0.44 \leq r^2 \leq 0.51$ for each of the 1902–49, 1950–80, 1950–93, and 1902–93 intervals). We thus assert that the sparser early grid point SST data in the eastern tropical Pacific are sufficient for a reliable Niño-3 index back to at least 1902.

Method

The advantage of using multiproxy data networks in describing and reconstructing patterns of long-term climate variability has been discussed elsewhere (Bradley and Jones 1993; Hughes and Diaz 1994; Mann et al. 1995; Mann et al. 1998a,b, 1999) and relies upon the complementary characteristics and spatial sampling provided by a combined network of long historical and instrumental records and dendroclimatic, coral, and ice core–based climate proxy indicators of climate change. In our approach to climate field reconstruction, we seek to "train" the multiproxy network by calibrating it against the dominant patterns of variability in the modern instrumental record. In the case of the global temperature field reconstructions, we use the entire multiproxy network shown earlier in Figure 10.1, and we "train" this network by calibrating it, series-by-series, against the dominant global patterns of temperature variability using the full sampling of grid point data shown in Figure 10.3. In the case of Niño-3 index reconstruction, we use the restricted "tropical" multiproxy network (Fig. 10.2) and train it against the dominant patterns of SST variations confined to the tropical Pacific sub-domain (see region indicated in Fig. 10.3).

The calibration of the multiproxy network is accomplished by a multivariate regression against instrumental temperature data. Conventional multivariate regression approaches to paleoclimate reconstruction, such as canonical correlation analysis (CCA; see Preisendorfer 1988), which simultaneously decompose both the proxy "predictor" and instrumental "predictand" data in the regression process, work well when applied to relatively homogeneous networks of proxy data (e.g., regional dendroclimatic networks; see Cook et al. [1994] for an excellent review). However, we found that such approaches were not effective with the far more inhomogeneous multiproxy data networks. Different types of proxies (e.g., tree rings, ice cores, and historical records) differ considerably in their statistical properties, reflecting, for example, different seasonal windows of variability, and exhibiting different spectral attributes in their resolvability of climatic variations. Thus, rather than seeking to combine the disparate information from different proxy indicators initially through a multivariate decomposition of the proxy network itself, we instead choose to use the information contained in the network in a later stage of the regression process. Our approach is similar, in principle, to methods applied recently to the reconstruction of early instrumental temperature fields from sparse data, based on eigenvector interpolation techniques (Smith et al. 1996; Kaplan et al. 1998).

Our calibration approach, which was described in MBH98, is further detailed here. It involves first decomposing the twentieth-century instrumental surface temperature data into their dominant patterns of variability or "eigenvectors." Second, the individual climate proxy indicators are regressed against the time histories of these distinct patterns during their shared twentieth-century interval of overlap. One can think of the instrumental patterns as templates against which we "train" the much longer proxy "predictor." This calibration process allows us, finally, to solve an "inverse problem" whereby the closest-match estimates of surface temperature patterns are deduced back in time before the calibration period, on a year-by-year basis, based on information in the multiproxy network.

At least three fundamental assumptions are implicit in this approach. We assume (1) that the indicators in our multiproxy trainee network are statistically related to some linear combination of the eigenvectors of the instrumental global annual-mean temperature data. This assumption contrasts with that of typical paleoclimate calibration approaches. Rather than seeking, as in typical studies, to relate a particular proxy record locally to an at-site instrumental record of some a priori selected meteorological variable (e.g., temperature, precipitation, or atmospheric wind direction), during some a priori selected seasonal (e.g., summer) window, we make instead a less restrictive assumption that whatever combination of local meteorological variables influence the proxy record, they find expression in one or more of the largest-scale patterns of annual climate variability. Ice core and coral proxy indicators reflect, in general, a variety of seasonal influences. Many extratropical tree-ring (ring widths and density) series primarily reflect warm-season temperature influences (see, e.g., Schweingruber et al. 1991; Bradley and Jones 1993). Tree-ring-width series in subtropical semiarid regions, however, are reflective in large part of cold-season precipitation influences. These influences are, in turn, tied to larger-scale atmospheric circulation variations (e.g., the PNA or NAO atmospheric patterns; see, e.g., D'Arrigo et al. 1993) that have important influences on large-scale temperature. When a proxy indicator represents some complex combination of local meteorological and seasonal influences, we make more complete use of the available information in the calibration process than do conventional approaches. While our method seeks only to recognize that variability in the proxy data tied to large-scale climatic patterns, it appears that temperature variability at interannual and longer timescales, for which the effective number of spatial degrees of freedom are greatly reduced (see, e.g., Jones and Briffa 1992), is inherently large scale in nature. This latter observation underlies our assertion (2) that a relatively sparse but widely distributed sampling of long proxy and instrumental records may sample much of the large-scale variability that is important at interannual and longer timescales (see also Bradley 1996). Regions not directly represented in the trainee network may nonetheless be indirectly represented through teleconnections with regions that are. The El Niño/Southern Oscillation is itself an example of a pattern of climatic variability that exhibits global-scale influence (e.g., Halpert and Ropelewski 1992). Finally, we assume (3) that the patterns of variability captured by the multiproxy network have analogs in the patterns we resolve in the shorter instrumental data. This latter assumption represents a fairly weak stationarity requirement – we do not require that the climate itself be stationary. In fact, we expect that some sizable trends in the climate may be resolved by our reconstructions. We do, however, assume that the fundamental *spatial patterns* of variation which the climate has exhibited during the past century are similar to those by which it has varied during past recent centuries. Studies of instrumental surface temperature patterns suggest that such a form of stationarity holds up at least on multidecadal timescales during the past century (Kaplan et al. 1998). Independent statistical cross-validation exercises and careful examination of the properties of the calibration residuals, as described later, provide the best evidence that these key assumptions hold, at least for the purposes of our reconstructions of temperature patterns during the past few centuries.

The first step in the pattern reconstruction process, as discussed above, is the decomposition of the modern instrumental data into its dominant spatiotemporal eigenvectors. The $N = 1{,}104$ monthly samples available between 1902 and 1993 are sufficient for both the global temperature field (the $M = 1{,}082$ grid point series described in the "Data" section; see Fig. 10.3) and the $M' = 121$ grid point series in the tropical Pacific region used in Niño-3 reconstruction, in providing a unique, well-posed eigendecomposition of the instrumental surface temperature data into its leading empirical eigenvectors. The method as described henceforth is understood to apply both to the global pattern and tropical Pacific SST/Niño-3 reconstructions, which are performed independently.

For each grid point, the mean was removed, and the series was normalized by its standard deviation. The standardized data matrix can be written

$$\underline{\mathbf{T}} = \begin{bmatrix} w_1 T_{t1}^{(1)} & w_2 T_{t1}^{(2)} \ldots w_M T_{t1}^{(M)} \\ w_1 T_{t2}^{(1)} & w_2 T_{t2}^{(2)} \ldots w_M T_{t2}^{(M)} \\ \vdots & \vdots \qquad \vdots \\ w_1 T_{tN}^{(1)} & w_2 T_{tN}^{(2)} \ldots w_M T_{tN}^{(M)} \end{bmatrix},$$

where $t_1, t_2, \ldots, t_N$ span over the $N = 1{,}104$ months, and $m = 1, 2, \ldots, M$ spans the $M = 1{,}082$ grid points. The coefficients w_m indicate weighting by the cosine of the central latitude of each grid point to ensure that grid points contribute in proportion to the area represented.

A standard principal component analysis (PCA) is then applied to the standardized data matrix $\underline{\mathbf{T}}$:

$$\underline{\mathbf{T}} = \sum_{k=1}^{K} \lambda_k u_k v_k,$$

decomposing the data set into its dominant spatiotemporal eigenvectors. The M-vector or empirical orthogonal function (EOF)v_k describes the relative spatial pattern of the kth eigenvector, the N-vector u_k or PC describes its variation over time, and the scalar l_k describes the associated fraction of resolved (standardized and weighted) data variance.

The first five eigenvectors of the global temperature data are particularly important in the global temperature pattern reconstructions, as is described later. The EOFs (Fig. 10.4) and PCs (Fig. 10.5) are shown for these five eigenvectors, in decreasing order of the variance they explain in the global instrumental temperature data shown in Figure 10.3 (1 = 12%, 2 = 6.5%, 3 = 5%, 4 = 4%, 5 = 3.5%). The first eigenvector describes much of the variability in the global (GLB = 88%) and hemispheric (NH = 73%) means and is associated with the significant global warming trend of the past century (see Fig. 10.5). Subsequent eigenvectors, in contrast, describe much of the spatial variability relative to the large-scale means (i.e., much of the remaining multivariate variance "MULT"). The second eigenvector is the dominant ENSO-related component of the global temperature data, describing 41% of the variance in the Niño-3 index and exhibiting certain classic extratropical ENSO signatures (e.g., the "horseshoe" pattern of warm and cold SST anomalies in the North Pacific and an anomaly pattern over North America reminiscent of the Pacific–North American (PNA) or related Tropical/Northern Hemisphere (TNH) atmospheric teleconnections of ENSO; see Barnston and Livezey [1987]). This eigenvector exhibits a modest long-term negative trend, which, in the eastern tropical Pacific, describes a La Niña–like cooling trend (Cane et al. 1997) that opposes warming in the same region associated with the global warming pattern of the first eigenvector. This cooling trend has been punctuated recently, however, by a few large positive excursions (see Fig. 10.5) associated with the large El Niños of the 1980s and 1990s. The third eigenvector is associated

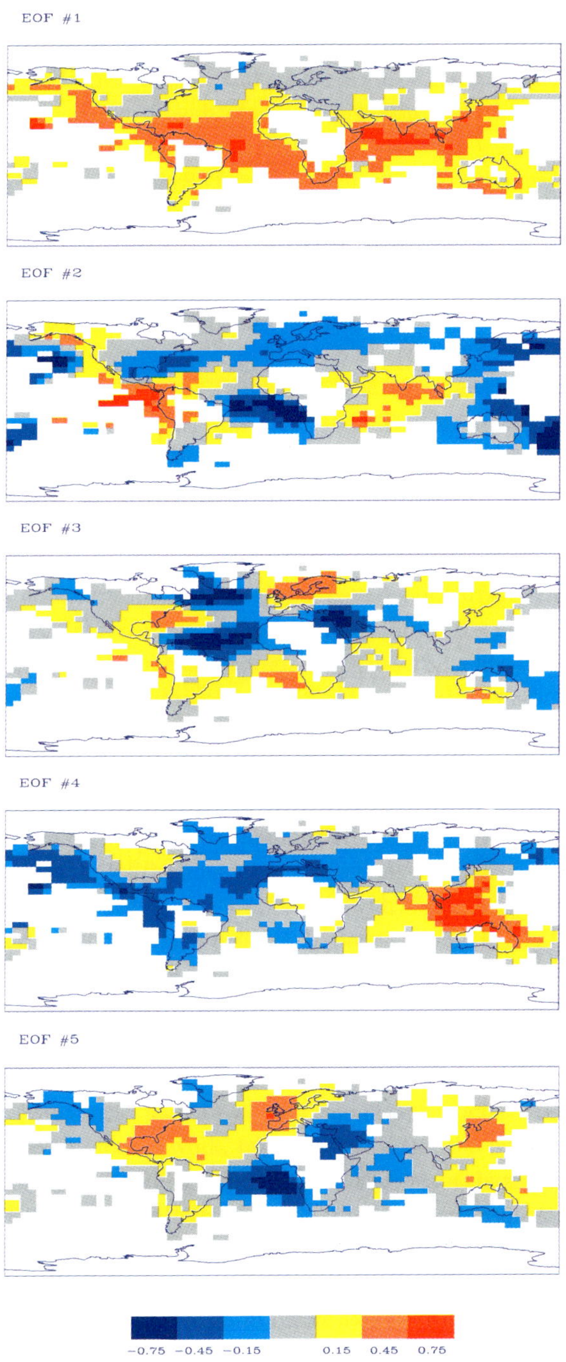

Fig. 10.4 Empirical orthogonal functions of the five leading eigenvectors of the 1902–93 global temperature data. The areal weighting factor used in the PCA (see text) has been removed, and the EOFs have been normalized by their maximum regional amplitude, so that relative spatial temperature variations can be inferred from the patterns shown.

Table 10.2 *Calibration and verification statistics for global reconstructions. Resolved variance statistics (β) are provided for the raw (1902–80) data (upper left corner), along with the squared correlation (r^2) of the raw Niño-3 series with the SOI series and its squared congruence statistic (g^2) with the historical QN92 El Niño chronology for reference. β statistics are also provided for the raw data after being filtered with the sets of eigenvectors used in the actual calibration experiments (upper right corner). For the calibration and verification experiments (lower table), the beginning year of the network employed ("NET") is given, along with the total number of (proxy and historical/instrumental) indicators ("$\#^{T}$"), number of historical/instrumental only indicators ("$\#^{I}$"), the total number and specific set of eigenvectors retained in calibration, calibration resolved variance β for global mean temperature (GLB), Northern Hemisphere mean temperature (NH), detrended Northern Hemisphere mean temperature (DET), Niño-3 index (NIN), total spatial temperature field (MT), calibration squared correlation statistic of Niño-3 (NIN) with SOI series (r^2), and congruence statistic with QN92 chronology g^2. For verification, β statistics are given for GLB, NH, and MLT (the latter based on both [A] the 1854–1901 grid point data and the [B] eleven long instrumental temperature grid point records available; see text). Any positive value of β is statistically significant (see text) at greater than 95% (or for MULTb, 99%), while the statistical significance of the verification r^2 of Niño-3 with Southern Oscillation Index (SOI, 1865–1901) and squared congruence g^2 with the QN92 chronology (back through 1750) are explicitly denoted by the indicated symbols.*

Raw data (1902–08)						
		β			r^{2a}	g^{2b}
GLB	NH	DET	NIN	MT	NIN	NIN
1.00	1.00	1.00	1.00	1.00	0.51	0.50

Eigenvector filtering (1902–80)					
EVs			β		
	GLB	NH	DET	NIN	MT
1st 40	1.00	0.99	0.98	0.91	0.73
1st 20	0.99	0.99	0.97	0.74	0.58
1–5, 9, 11, 14–16	0.95	0.93	0.87	0.72	0.40
1–5, 9, 11, 15	0.95	0.93	0.78	0.70	0.38

1–3, 6, 8, 11, 15	0.92	0.83	0.69	0.65	0.31
1st 5	0.93	0.85	0.76	0.67	0.27
1, 2, 5, 11, 15	0.92	0.83	0.70	0.55	0.23
1, 2, 11, 15	0.90	0.78	0.61	0.53	0.21
1, 2	0.91	0.76	0.57	0.50	0.18
1	0.88	0.73	0.53	0.09	0.12

Net	#T	#T		Calibration (1902–80)							Verification (pre-1902)					
				β					r^2	g^2	β				r^2	g^2
				GLB	NH	DET	NIN	MT	NIN	NIN	GLB	NH	MT[a]	MT[b]	NIN	NIN
1820	112	24	11 (1–5, 7, 9, 11, 14–16)	0.77	0.76	0.56	0.48	0.30	0.51	0.29	0.76	0.69	0.22	0.55	0.14**	0.28**
1800	102	15	”	0.76	0.75	0.54	0.50	0.27	0.52	0.30	0.75	0.68	0.19	0.45	0.10**	0.22**
1780	97	12	”	0.76	0.74	0.54	0.51	0.27	0.53	0.29	0.76	0.69	0.17	0.40	0.11**	0.20**
1760	93	8	9 (1–5, 7, 9, 11, 15)	0.76	0.74	0.52	0.49	0.26	0.52	0.29	0.75	0.70	0.17	0.33	0.10**	0.17*
1750	89	5	8 (1–3, 5, 6, 8, 11, 15)	0.76	0.74	0.53	0.34	0.18	0.39	0.33	0.64	0.57	0.11	0.13	0.10**	0.19*
1730	79	3	5 (1, 2, 5, 11, 15)	0.74	0.71	0.47	0.23	0.15	0.30	0.29	0.65	0.61	0.11	0.13	0.05*	0.17*
1700	74	2	”	0.74	0.71	0.47	0.22	0.14	0.29	0.29	0.63	0.57	0.10	0.12	0.05*	0.15+
1650	57	1	4 (1, 2, 11, 15)	0.72	0.67	0.42	0.05	0.14	0.19	0.23	0.61	0.53	0.12	0.10	0.02+	0.11*
1820[1]	88	0	8 (1–3, 5, 6, 8, 11, 15)	0.76	0.73	0.51	0.31	0.19	0.37	0.31	0.65	0.56	0.11	0.19	0.12**	0.29*
1820[2]	24	24	2 (1, 2)	0.32	0.28	−0.28	−0.27	0.10	0.07	0.20	0.30	0.37	0.11	0.26	0.0	0.16
1820[3]	42	24	7 (1–3, 5, 11, 15, 16)	0.50	0.50	0.13	0.05	0.17	0.22	0.18	0.56	0.53	0.17	0.47	0.09*	0.10
1750[3]	19	2	2 (1, 2)	0.46	0.47	0.05	0.2	0.09	0.30	0.27	0.28	0.27	0.06	0.10	0.03+	0.15

+85% significant; *90% significant; **95% significant; ***99% significant; × unphysical positive correlation obtained.

[a] r^2 with SOI. [b] g^2 w/QN92 chron.

[1] No instrumental/historical data – proxy only. [2] Instrumental/historical data only. [3] Dendroclimatic indicators excluded.

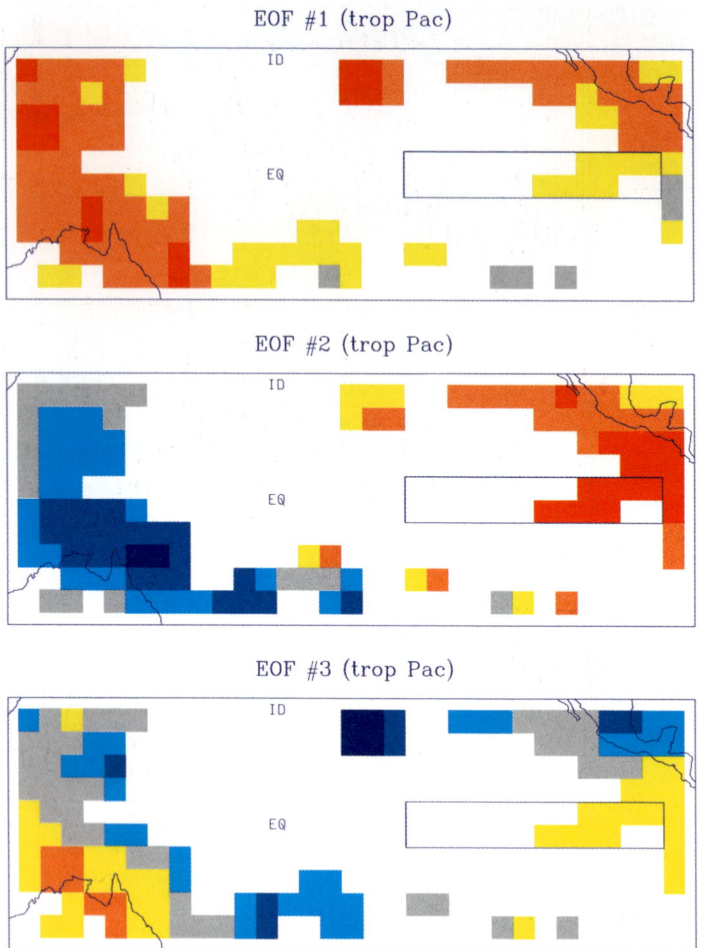

Fig. 10.6 Empirical orthogonal functions of the three leading eigenvectors of the 1902–93 restricted tropical Pacific region temperature data. Color scale and other conventions are as in Figure 10.4.

optimized, or nearly optimized, the verification explained verification statistics (discussed below).

These N_{eof} eigenvectors were trained against the N_{proxy} indicators, by finding the least-squares optimal combination of the N_{eof} PCs represented by each individual proxy indicator during the $\bar{N} = 79$ year training interval from 1902 to 1980 (the training interval is terminated at 1980 because many of the proxy series terminate at or shortly after 1980). The proxy series and PCs were formed into anomalies relative to the same 1902–08 reference period mean, and the proxy series were also normalized by their standard deviations during that period. This proxy-by-proxy calibration is well posed (i.e., a unique optimal solution exists) as long as $\bar{N} > N_{eof}$ (a limit never approached in this study) and can be expressed as the least-squares solution to the overdetermined

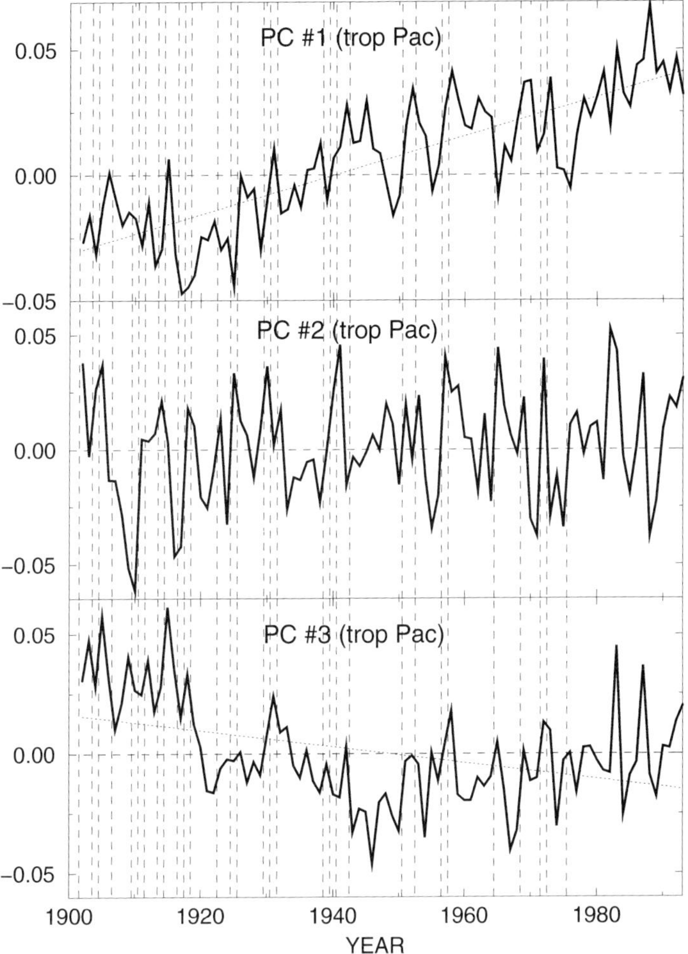

Fig. 10.7 Annual-mean PCs of the three leading eigenvectors of the 1902–93 restricted tropical Pacific region temperature data. Conventions are as in Figure 10.5. Vertical dashed lines indicate the timing of El Niño events as recorded in the QN92 chronology.

matrix equation

$$\underline{\mathbf{U}}\mathbf{x} = \mathbf{y}^{(\mathbf{p})},$$

where

$$\underline{\mathbf{U}} = \begin{bmatrix} \bar{u}_1^{(1)} & \bar{u}_2^{(1)} \ldots u_1^{(N_{\text{eof}})} \\ \bar{u}_2^{(1)} & \bar{u}_2^{(2)} \ldots u_2^{(N_{\text{eof}})} \\ \vdots & \vdots \quad\quad \vdots \\ \bar{u}_{\bar{N}}^{(1)} & \bar{u}_{\bar{N}}^{(2)} \ldots u_{\bar{N}}^{(N_{\text{eof}})} \end{bmatrix}$$

is the matrix of annual PCs and

$$\mathbf{y^{(p)}} = \begin{bmatrix} y_1^{(p)} \\ y_2^{(p)} \\ \vdots \\ y_{\bar{N}}^{(p)} \end{bmatrix}$$

is the time series $\bar{N}$-vector for proxy record p.

The N_{eof}-length solution vector $\mathbf{x} = \mathbf{G^{(p)}}$ is obtained by solving the above overdetermined optimization problem by singular value decomposition (SVD) for each proxy record $p = 1, \ldots, P$. This yields a matrix of coefficients relating the different proxies to their closest linear combination of the N_{eof} PCs,

$$\underline{\mathbf{G}} = \begin{bmatrix} G_1^{(1)} & G_2^{(1)} \ldots G_{N_{\mathrm{eof}}}^{(1)} \\ G_1^{(2)} & G_2^{(2)} \ldots G_{N_{\mathrm{eof}}}^{(2)} \\ \vdots & \vdots \quad\quad \vdots \\ G_1^{(P)} & G_2^{(P)} \ldots G_{N_{\mathrm{eof}}}^{(P)} \end{bmatrix}.$$

This matrix of coefficients cannot simultaneously be satisfied in the regression but rather represents a highly overdetermined relationship between the optimal weights on each of the N_{eof} PCs and the information in the multiproxy network at any given time.

Proxy-reconstructed patterns for the global temperature and tropical Pacific SST fields are finally obtained based on the year-by-year solution of the overdetermined matrix equation

$$\underline{\mathbf{G}}\mathbf{z} = \mathbf{y}_{(j)},$$

where $\mathbf{y}_{(j)}$ is the predictor vector of values of each of the P proxy indicators during year j. The predictand solution vector $\mathbf{z} = \hat{\mathbf{U}}$ contains the least-squares optimal values of each of the N_{eof} PCs for a given year.

This yearly reconstruction process leads to annual sequences of the optimal reconstructions of the retained PCs, which we term the reconstructed principal components, or RPCs, and denote by $\mathbf{u^k}$. Once the RPCs are determined, the associated temperature patterns (of either the global field or tropical Pacific SST) are readily obtained through the appropriate eigenvector expansion

$$\underline{\hat{\mathbf{T}}} = \sum_{k=1}^{N_{\mathrm{eof}}} \lambda_k \hat{u}_k v_k,$$

while quantities of interest (e.g., NH, Niño-3) are calculated from the appropriate spatial averages.

The optimization procedure described above to yield the RPCs is overdetermined (and thus well constrained) as long as $P > N_{\mathrm{eof}}$, which is always realized in this study. An important feature of our approach is that, unlike conventional transfer function approaches to paleoclimate reconstruction (see, e.g., Cook et al. 1994), there is no fixed relationship between a given proxy indicator (e.g., a particular coral or tree-ring series) and a given predictand (i.e., RPC) over the calibration interval. Instead, the best common choice of values for the small number of N_{eof} RPCs is determined from the mutual statistical information present in the network of multiproxy data on a year-by-year basis. The reconstruction approach is thus relatively resistant to errors or biases (e.g., dating errors) specific to any small number of indicators, in contrast to most conventional approaches.

To ensure the validity of the linear calibration procedure, we examined the calibration residuals, defined as the difference between the reconstructed and raw instrumental NH and Niño-3 series during the calibration interval (Fig. 10.8), for any evidence of bias. The distributions of the calibration residuals were found to be consistent with parent distributions of normal random deviates based on χ^2 tests applied to their respective histograms (NH significant at the 95% level; Niño-3 at the 99% level). These tests are shown graphically in Figure 10.9. The spectrum of the calibration residuals for these quantities were, furthermore, found to be consistent with a "white" distribution showing little evidence for preferred or deficiently resolved timescales in the calibration process (see Mann et al. [1999] for a more detailed analysis). Having established unbiased calibration residuals, we were thus able to calculate uncertainties in the reconstructions from the uncalibrated calibration period variance. This uncalibrated variance increases back in time, as increasingly sparse multiproxy networks calibrate fewer numbers of eigenvectors back in time and smaller fractions of instrumental variance (see Tables 10.2, and 10.3). This decreased calibrated variance back in time is thus expressed in terms of uncertainties in the reconstructions that expand back in time. For the global temperature reconstructions back to 1650 of interest in the present study, the decreases in calibrated variance, and associated increases in uncertainties back in time, are quite modest (Table 10.2). For the Niño-3 reconstructions, they are negligible (Table 10.3).

Figure 10.10 compares the spectrum of the raw and reconstructed Niño-3 series during the calibration period (separately scaled so that their median power levels align). The calibrated series clearly captures the interannual band-limited features of the raw Niño-3 series. The calibrated series exhibits relatively more interdecadal-to-secular power and relatively less quasi-biennial power than does the raw series. As was noted earlier, however, the spectrum of the calibration residuals (not shown) exhibits no statistically significant departure from the white spectrum expected for an unbiased reconstruction. It is particularly important that in the context of reconstructing low-frequency trends, there is no evidence of any loss of secular variance in calibration and no reason to presuppose that multidecadal- to century-scale variations in ENSO are not adequately described in the reconstructions. Figure 10.11 compares the actual Niño-3 annual global temperature correlation pattern (that is, the pattern of the correlation coefficient of the annual Niño-3 index with the annual global temperature field reconstructions) with the correlation pattern based on the multiproxy-reconstructed counterparts, for the 1902–80 calibration period. The multiproxy-based correlation pattern clearly captures the details of the observed correlation pattern during the calibration period, giving some justification to analyzing changing correlation patterns back in time from the reconstructions.

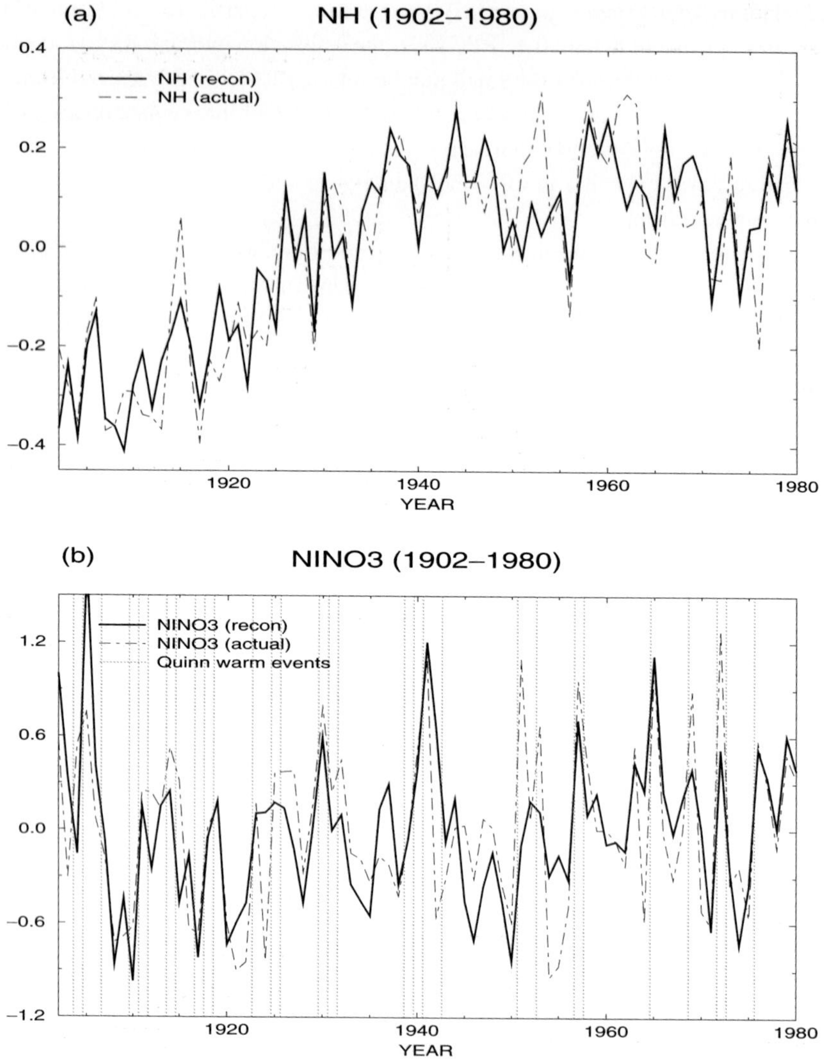

Fig. 10.8 Comparison of raw and reconstructed/calibrated (a) NH and (b) Niño-3 surface temperature series during the 1902–80 calibration interval.

The skill of the temperature reconstructions (i.e., their statistical validity) is independently established through a variety of complementary independent cross-validation or "verification" exercises. The numerical results of the calibration and verification experiments are summarized in Table 10.2 for the global pattern reconstructions and in Table 10.3 for the Pacific SST reconstructions upon which the Niño-3 index is ultimately based. Four distinct sources of verification were used for the global reconstructions: (1) widespread instrumental grid point data ($M' = 219$ grid points; see Fig. 10.3) available for the period 1854–1901, (2) a small subset of eleven very long estimated instrumental temperature grid point averages (ten in Eurasia, one in North America; see Fig. 10.1)

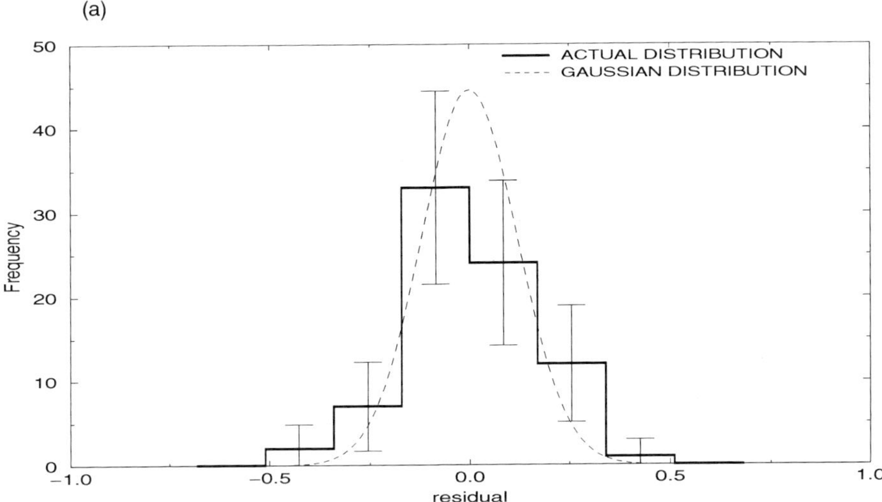

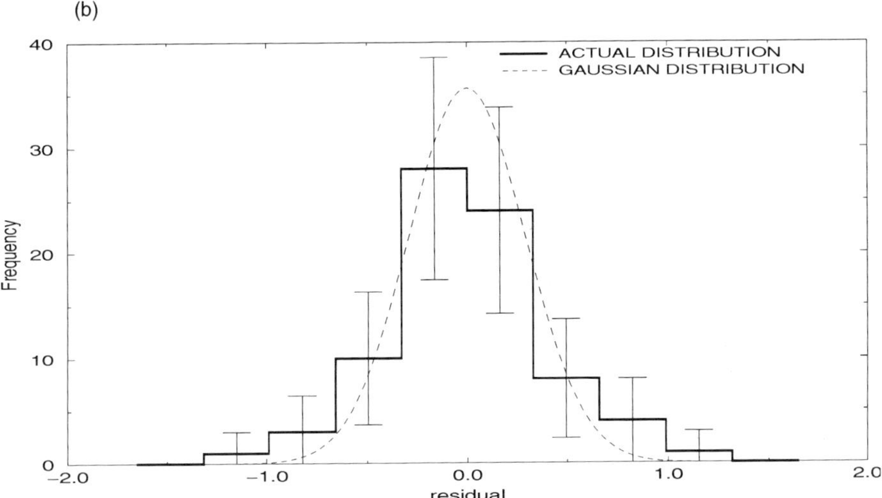

Fig. 10.9 Histograms of calibration residuals for (a) NH and (b) Niño-3 surface temperature series. A Gaussian (normal) parent distribution is shown for comparison, along with two standard error bars for the frequencies of each bin. Both distributions are consistent with a Gaussian distribution at high levels of confidence (95% for NH, 99% for Niño-3).

constructed from the longest available station measurement (each of these "grid point" series shared at least 70% of their variance with the corresponding temperature grid point available during 1854–1980, providing verification back to at least 1820 in all cases [and back through the mid- and early eighteenth century in many cases]), (3) the SOI available for corroboration of ENSO-scale variability during the late nineteenth century, and (4) the QN92 historical chronology of El Niño events dating back into the sixteenth century. Exercise (1) provides the only means of widespread instrumental

Table 10.3 *Calibration and verification statistics for tropical Pacific SST reconstructions, focusing on the diagnosed Niño-3 SST index. Resolved variance (β) statistics are provided for the raw (1902–80) Niño-3 index after being filtered with the sets of eigenvectors used in the actual calibration experiments and the reconstructed/calibrated Niño-3 index. Other details are as described for Table 10.2.*

				Calibration		Verification	
Net	$\#^T$	$\#^T$	EVs	β (1902–80)		r^2(SOI, 1865–1901)	g^2(QN, 1750–1901)
1820	49	2	3(1–3)	0.80	0.38	0.27***	0.39***
1800	45	0	”	”	0.42	0.22***	0.40***
1780	44	0	”	”	0.42	0.23***	0.38***
1750	43	0	”	”	0.42	0.23***	0.38***
1700	37	0	”	”	0.38	0.22***	0.38***
1650	29	0	”	”	0.34	0.20**	0.32*

*95% significant.
**99% significant.
***99.5% significant.

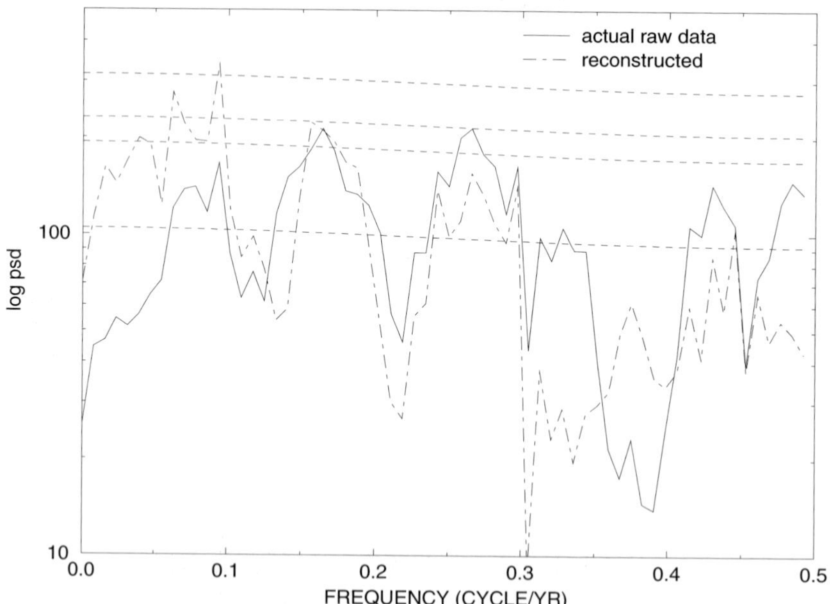

Fig. 10.10 Spectrum of raw instrumental (solid) and reconstructed/calibrated (dashed) Niño-3 series during the 1902–80 calibration period. The spectrum is calculated by using the multitaper method (Thomson 1982; Park et al. 1987; Mann and Lees 1996) based on three tapers and a time-frequency bandwidth product of $W = 2N$.

NINO3 CORRELATION MAP (1902–1980, ACTUAL DATA)

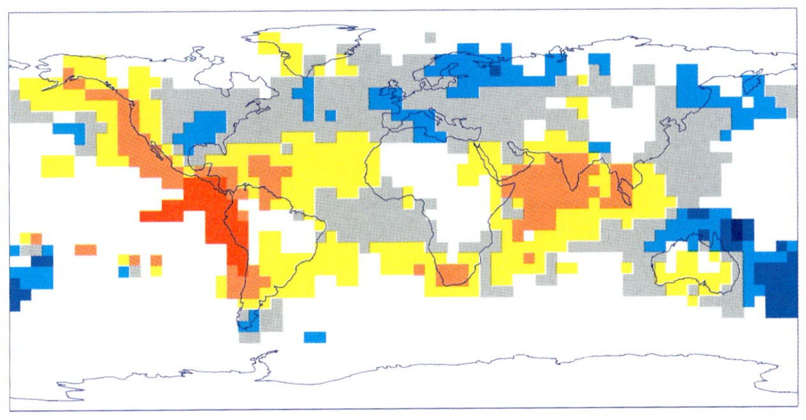

NINO3 CORRELATION MAP (1902–1980, RECONSTRUCTED)

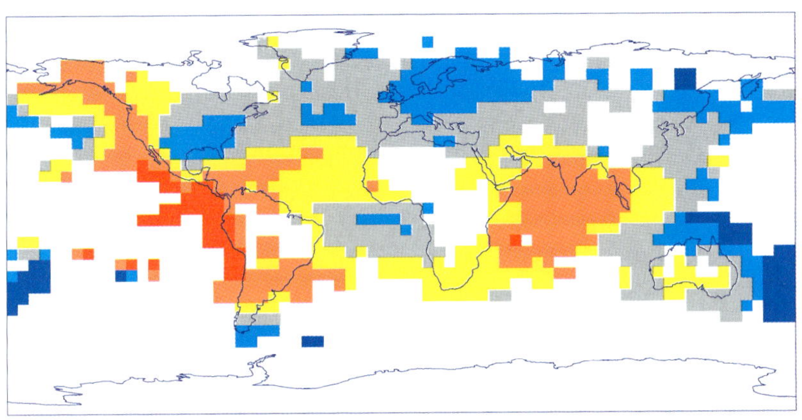

Fig. 10.11 Niño-3 correlation map during 1902–80 calibration period based on (top) raw data and (bottom) reconstructed/calibrated Niño-3 and global temperature patterns. The color scale indicates the specific level of correlation with a given grid point, with any color value outside of gray significant at greater than roughly the 80% level based on a two-sided test. Red and blue indicate significance well above the ~99% confidence level.

spatial verification, but only experiment (3) provides instrumental verification of the Niño-3 region, while exercises (2) and (4) address the fidelity of long-term variability. Only verification experiments (3) and (4) were used to diagnose (in terms of the Niño-3 index) the quality of the tropical Pacific pattern reconstructions. Note (compare Tables 10.2 and 10.3) that the calibration and verification experiments indicate a more skillful Niño-3 index from the tropical Pacific SST-based reconstructions than from the global reconstructions.

The spatial patterns of described variance (β) for both calibration (based on 1902–80 data) and verification (based on verification experiments [1] consisting of the withheld instrumental grid point data for 1854–1901) are shown for the global reconstructions in Figure 10.12. Experiment [2] provides a longer-term, albeit an even less spatially representative, multivariate verification statistic ("MULTb"). In this case, the spatial sampling does not permit meaningful estimates of NH or GLB mean quantities. In any of these diagnostics, a positive value of β is statistically significant at greater than 99% confidence as established from Monte Carlo simulations. Verification skills for the Niño-3 reconstructions were estimated indirectly by experiments [3] and [4], since a reliable instrumental Niño-3 index was not available far beyond the beginning of the calibration period. The (negative) correlation r of Niño-3 with the SOI annual mean for 1865–1901 (obtained from P. D. Jones, personal communication) and a squared congruence statistic g^2 measuring the categorical match between the distribution of warm Niño-3 events and the distribution of warm episodes according to the QN92 historical chronology (available back through 1525) were used for statistical cross-validation of the Niño-3 index, with significance levels estimated from standard one-sided tables and Monte Carlo simulations, respectively. The verification period comparisons are shown in Figure 10.13. The r^2 of the Niño-3 index with this instrumental SOI is a lower bound on a true Niño-3 verification b, since the shared variance between the instrumental annual-mean Niño-3 and SOI indices, as commented upon earlier, is itself only $\approx$50% (see Table 10.2). Thus, it is possible that a $r^2 = 0.25$ might be tantamount to a true verification described variance score of $\beta \approx 0.5$, and the verification r^2 scores should be interpreted in this context. The calibration and verification experiments thus suggest that we skillfully describe about 70–80% of the true NH variance back to 1820 (see Table 10.2) and about 38–42% of the true Niño-3 variance in the optimal index back to 1700 (see Table 10.3). Scores for earlier reconstructions are slightly lower in each case (see Tables 10.2 and 10.3). We note that the verifiable statistical skill in our annual-mean Niño-3 index is similar to that of the best independent recent ENSO reconstructions. For example, Stahle et al. (1998) reconstruct a winter SOI back to 1706, indicating 53% resolved instrumental variance for a more restricted December–February season. Multidecadal and longer-term trends of interest here, however, were not resolved in that study. In addition to the above means of cross-validation, we also tested the network for sensitivity to the inclusion or elimination of particular trainee data (e.g., instrumental records, the northern tree line dendroclimatic indicators of North America, and tropical coral records). These sensitivity experiments are also described (for the global pattern reconstructions) in Table 10.2.

CALIBRATION BETA

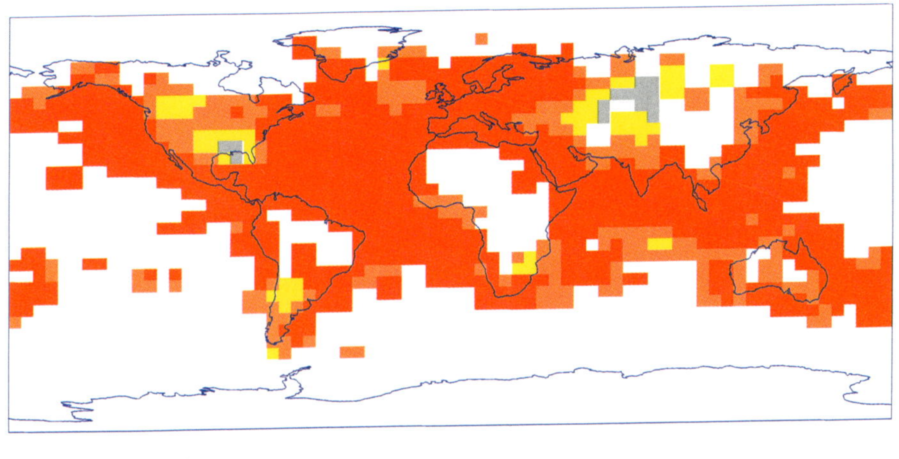

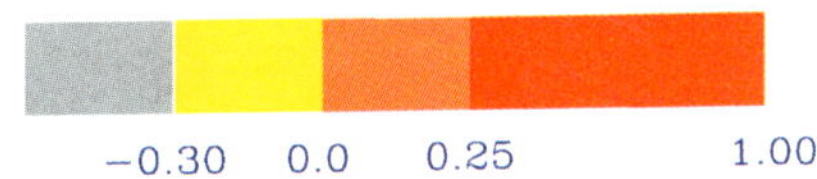

VERIFICATION BETA

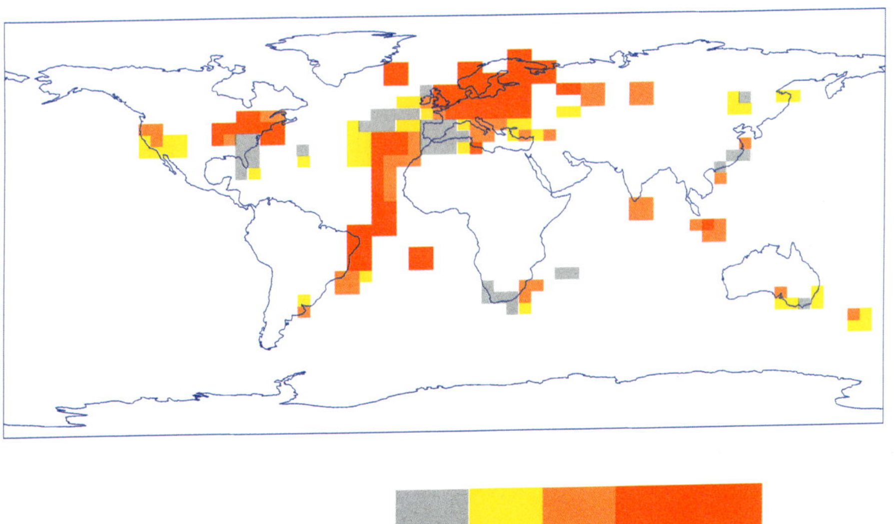

−0.30 0.0 0.25 1.00

Fig. 10.12 Pattern of calibration (top, based on 1902–80 data) and verification (bottom, based on 1854–1901 "withheld" data) resolved variance statistic β for the global reconstructions. Values that are insignificant at the 99% level are shown in gray, while negative but 99% significant values are shown in yellow, and significant positive values are shown in two shades of red. The significance estimation procedure is described in the text.

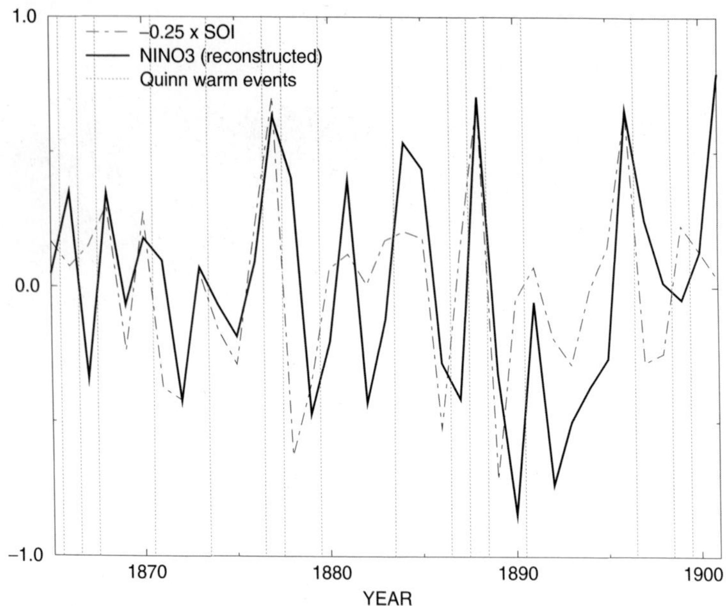

Fig. 10.13 Reconstructed Niño-3 series along with the (negative and rescaled) SOI series (dot-dashed) during the 1865–1901 subinterval of the precalibration "verification" period during which the SOI is available. The timing of El Niño events as recorded in the QN92 chronology is shown for comparison.

To illustrate the effectiveness of the proxy pattern reconstruction procedure, we show as an example (Fig.10.14) the raw, EOF-filtered, and reconstructed temperature patterns for a year during the calibration interval (1941) for which the true large-scale surface temperature pattern is known. The reconstructed pattern is a good approximation to the smoothed anomaly pattern obtained by filtering the raw data by the same eleven eigenvectors used in the calibration process (retaining about 40% of the total spatial variance). The historically documented El Niño and associated Pacific/North American temperature anomalies are clearly captured in the reconstruction, as are the prominent cold anomalies in Eurasia. This example serves to demonstrate the level of resolved spatial variance that can typically be expected in the pre-instrumental reconstructed global temperature patterns discussed below.

Temperature Reconstructions

We analyze the reconstructions of global temperature patterns and the Niño-3 index back to 1650. Temperature pattern reconstructions back to 1400 are discussed elsewhere (MBH98; Mann et al. 1998b); here, we choose to focus on the past four centuries, for which the global ENSO phenomenon can be most skillfully reconstructed. The reconstructed global temperature patterns are derived by using the optimal eigenvector subsets determined in the global calibration experiments (see Table 10.2, eleven

1941 TEMPERATURE ANOMALY (RAW)

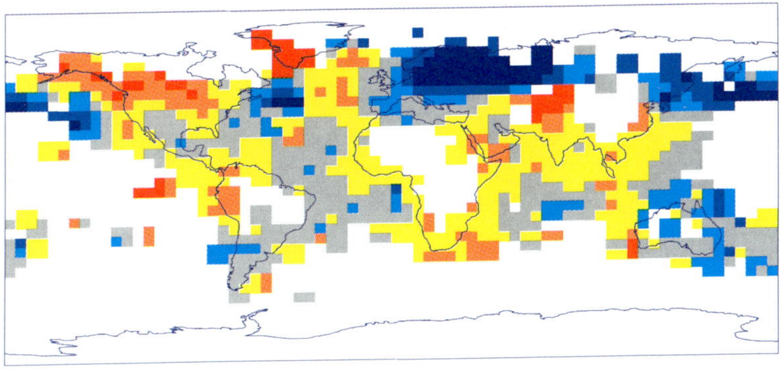

1941 TEMPERATURE ANOMALY (FILTERED)

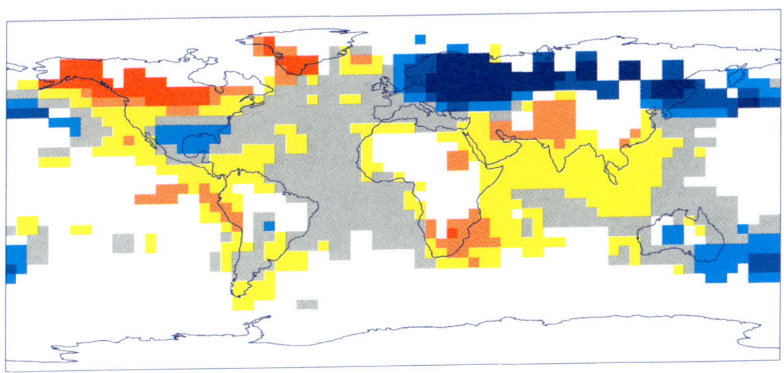

1941 TEMPERATURE ANOMALY (RECONSTRUCTED)

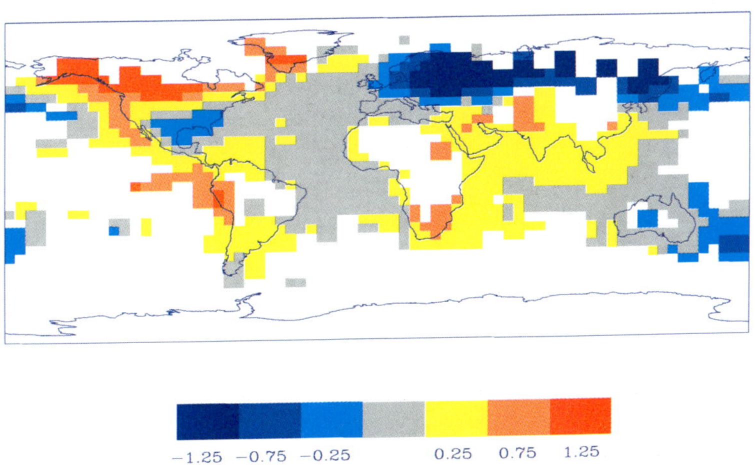

Fig. 10.14 Comparison of the proxy-reconstructed temperature anomaly pattern for 1941 with raw data, showing (top) actual, middle (EOF-filtered), and (bottom) reconstructed/calibrated pattern. Anomalies are indicated by the color scale shown, relative to the 1902–80 climatology.

eigenvectors for 1780–1980, nine eigenvectors for 1760–79, eight eigenvectors for 1750–59, five eigenvectors for 1700–49, four eigenvectors for 1650–99), while the Niño-3 index is based on the tropical Pacific SST reconstructions for which the optimal eigenvector subset includes the first three eigenvectors for the period 1650–1980 (see Table 10.3).

Large-Scale Trends

It is interesting to consider the temporal variations in the first five global RPCs (Fig. 10.15) associated with the estimated variations in the different spatial patterns (EOFs) shown in Figure 10.4 prior to the twentieth century (see MBH98). The positive trend in RPC #1 during the twentieth century is clearly exceptional in the context of the long-term variability in the associated eigenvector, and indeed it describes much of the unprecedented warming of the twentieth century as was discussed earlier. Reconstructed principal components #2 and #4 are associated with the primary ENSO-related eigenvectors in the global data set and serve to describe much of the long-term variation in the global ENSO phenomenon expressed in the global pattern

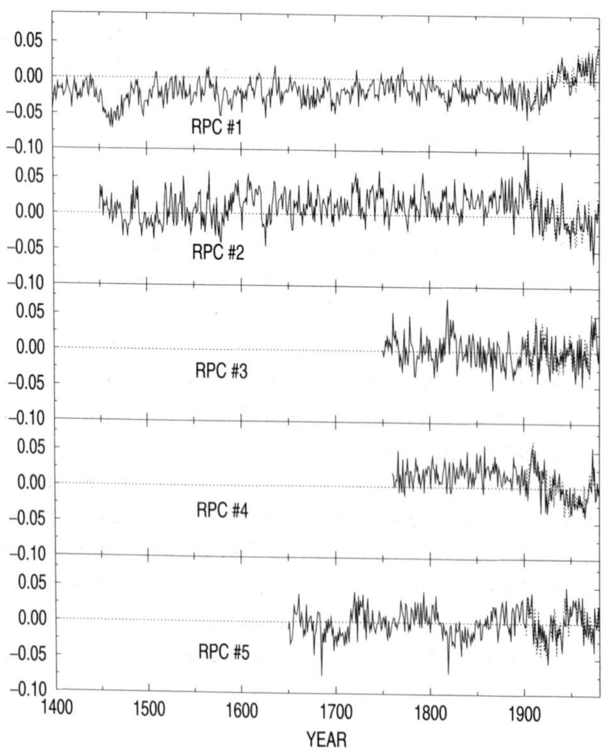

Fig. 10.15 Time reconstructions (RPCs – solid) and 1902–93 raw data (PCs – dotted) for the first five global eigenvectors. 1902–80 calibration period means are indicated by the horizontal dotted lines.

reconstructions. The negative trend in RPC #2 during the past century (see earlier discussion), which is anomalous in the context of the longer term evolution of the associated eigenvector, is particularly interesting in the context of ENSO-scale variations. This recent negative trend, describing a cooling in the eastern tropical Pacific superimposed on the warming trend in the same region associated with the pattern of eigenvector #1, could plausibly be a modulating negative feedback on global warming (Cane et al. 1997). Reconstructed PC #4, associated with a more modest share of ENSO-related variability, shows a pronounced multidecadal variation, though no obvious linear trend, during the twentieth century, but more muted variability prior to the twentieth century. Reconstructed PC #5 exhibits notable multidecadal variability throughout the modern and pre-calibration periods, associated with the wavelike trend of warming and subsequent cooling of the North Atlantic this century discussed earlier and the robust multidecadal oscillations in that region detected in a previous analysis of multiproxy data networks (Mann et al. 1995). This variability may be associated with ocean–atmosphere processes related to the North Atlantic thermohaline circulation (Delworth et al. 1993, 1997) and is further discussed by Delworth and Mann (1999).

Tropical Pacific and Niño-3 Variations

In Figure 10.16, we show the three RPCs corresponding to the separate tropical Pacific reconstructions. As in the calibration interval (see Fig. 10.7), the interannual ENSO-related variability remains dominated by the RPC #2, while RPCs #1 and #3 describe in large part the anomalous trends during the twentieth century in the region discussed earlier. The reconstructed Niño-3 index (Fig. 10.17) places recent large ENSO events in the context of a several centuries–long record, allowing us to better gauge how unusual recent behavior might be. Indeed, the anomalous 1982/1983 El Niño event represents the largest positive event in our reconstructed chronology back to 1650. A similar statement holds for the current 1997/1998 event (not shown in Fig. 10.17), which is of similar magnitude. Taking into account the uncertainties in the pre–twentieth-century reconstructions, however, a more guarded statement is warranted. The 1982/1983 and 1997/1998 events are both slightly greater than 2 standard errors (2σ) warmer than any reconstructed, pre-1902 event. However, there are nine reconstructed events that are within 1.5–2.0 standard errors of either of these two events. Roughly speaking, this lowers the probability to about 80% that either of the two events in question is warmer than *any other single event* back to 1650, taking into account the current uncertainties in the reconstruction. The significance of the occurrence of two such *reasonably* unlikely events in the past sixteen years, however, should clearly not be dismissed. Possible indications of a recent change in the state of El Niño are discussed further in later sections.

ENSO and Global Influences

Details of the NH mean-annual temperature series and associated implications for natural and anthropogenic forcing of global climate during past centuries are discussed

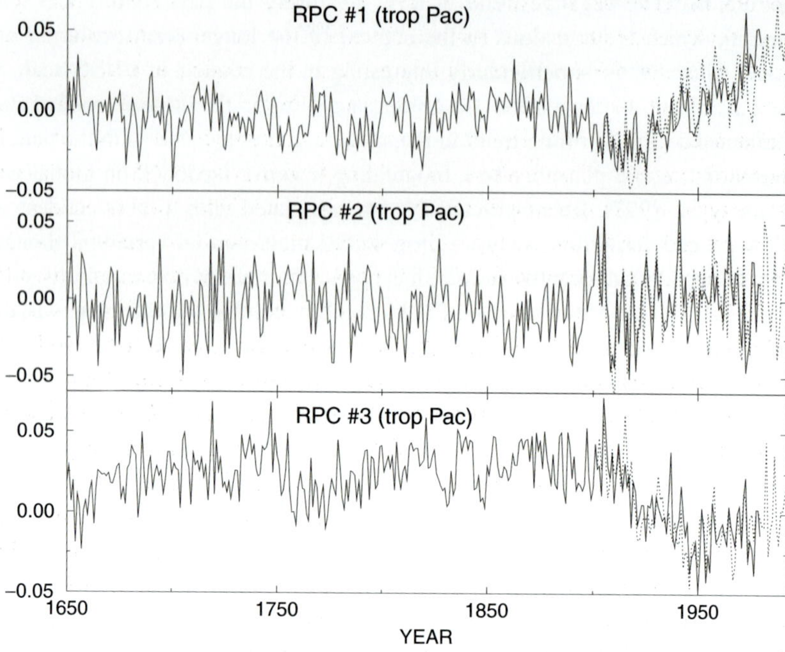

Fig. 10.16 Time reconstructions (RPCs – solid) and 1902–93 raw data (PCs – dotted) for the three eigenvectors of the restricted tropical Pacific temperature data.

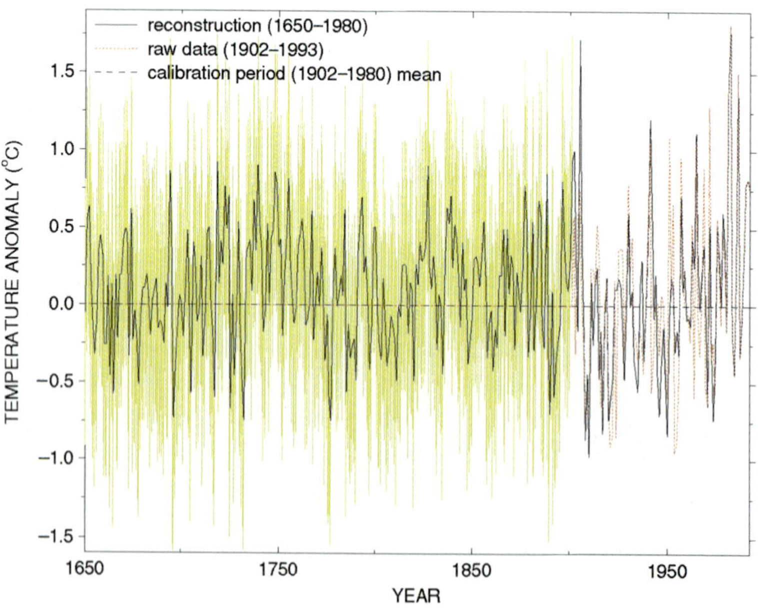

Fig. 10.17 Reconstructed Niño-3 index back to 1650, along with instrumental series from 1902–93. The reconstructed segment prior to 1902 is shown with confidence bands (in yellow) corresponding to ±2 standard errors of estimate. The calibration period (1902–80) mean is shown by the horizontal dashed line.

by MBH98. Here, we use the NH series to focus on relationships between hemispheric to global-scale temperature variation and the ENSO-scale variations captured in the separately determined Niño-3 reconstruction. We seek in particular to understand how changes in ENSO, themselves either internally or externally forced in nature, may have influenced global or hemispheric temperature trends over time. Due primarily to the poleward transport of anomalous tropical Pacific heat, warm ENSO events (i.e., El Niños) are typically associated with warm hemispheric or global conditions, while cold ENSO events (i.e., La Niñas) are typically associated with cold conditions (typically on the order of positive and negative 0.1°C for the globe, respectively; see, e.g., Jones 1989; Angell 1990). Thus, it is not unreasonable to suppose that changes in ENSO could lead to modest changes in global and hemispheric temperatures. The relationship between the reconstructed NH and Niño-3 series is shown in Figure 10.18 (the Niño-3 index is scaled by a factor of 0.25 so that a typical ENSO event [e.g., a 0.5°C Niño-3 anomaly] is roughly in scale with an expected ~0.1°C NH mean temperature anomaly). Overall, the two series are modestly correlated ($r = 0.14$ for the 244-year period 1650–1993), which is significant at well above the 99% level whether or not a one-sided or two-sided significance test is used (one could argue that only a positive correlation should be sought on physical grounds). However, the implied shared variance between the two series is less than 2%, which suggests that ENSO is not a dominant influence on changes in hemispheric or global mean temperatures, at least in comparison with external forcing agents (see MBH98). Both Niño-3 and NH do show similar low-frequency variations

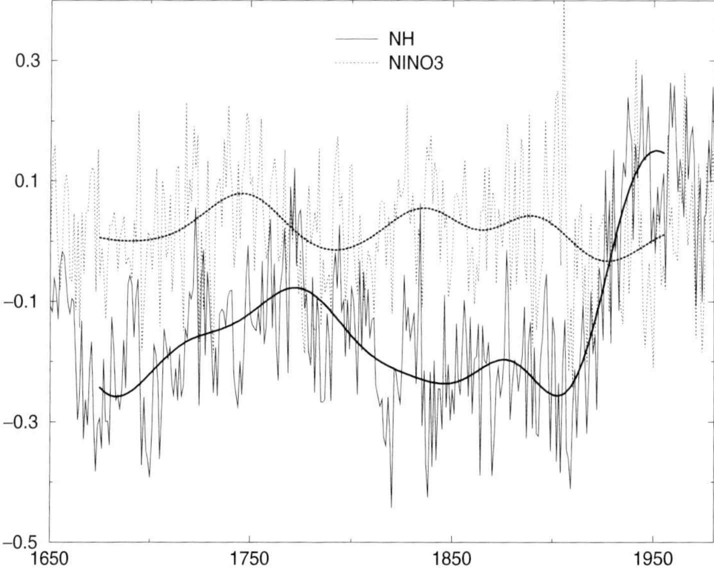

Fig. 10.18 Comparison of reconstructed Northern Hemisphere (NH) mean annual temperature series and Niño-3 index (rescaled by a factor of 0.25) from 1650–1980, along with 50-year smoothed versions of each series.

at times, however (compare the 50-year smoothed curves in Fig. 10.18), with warming occurring in both cases during the seventeenth/early eighteenth and twentieth centuries. The low-frequency variations in Niño-3 are nonetheless relatively muted compared to those in NH, particularly during the twentieth century, when NH warming appears to become dominated by greenhouse forcing (IPCC 1995; MBH98). In contrast to the NH series, there is no evidence of any significant long-term departures from the twentieth-century climatology (see Fig. 10.17). There are in fact notable periods during which the low-frequency trends are uncorrelated or even opposite (e.g., late eighteenth century through mid-nineteenth century). It is interesting, though perhaps coincidental, that this latter period corresponds to an anomalously cool period for the Northern Hemisphere on the whole, associated with low solar irradiance, and preceding any greenhouse warming (see MBH98, Fig. 10.7). As is discussed in later sections, this is also a period of anomalously weak amplitude interannual variability in ENSO and unusual global patterns of ENSO temperature influence.

The muted positive trend in Niño-3 relative to that in NH during the twentieth century can be attributed to the negative twentieth-century trend in eigenvector #2 noted earlier. As was also discussed earlier, this offsetting trend might in turn be related to the negative feedback to greenhouse warming discussed by Cane et al. (1997). However, there is a strong apparent relationship between global temperature increases and warm ENSO conditions during the more brief period of the past couple decades (e.g., Trenberth 1990), and this period coincides with what appear to be the warmest hemispheric conditions of the past six centuries (MBH98). There is, furthermore, evidence of distinct and unusual dynamical behavior of ENSO during this most recent period. For example, Goddard and Graham (1998) argue for a breakdown of the classical delayed oscillator mechanism in describing the vertical oceanographic evolution of ENSO-like variations during the 1990s. Interdecadal ENSO-like variability (as discussed in the next section below) complicates any simple interpretation of trends in ENSO during the most recent decades, however. Thus, while the long-term relationships between ENSO and global temperature changes appear modest based on our reconstructions, the possible relationships between recent global warming and changes in ENSO are not yet clearly resolved.

Stationarity of ENSO Teleconnections

It is also of interest to estimate the changes over time in the global pattern of influence of ENSO (i.e., the "teleconnections" of ENSO). Figure 10.11 showed that the modern global ENSO pattern – as described through the global correlation map of the Niño-3 index – is well captured by our multiproxy calibrations. Though the global patterns of ENSO in our reconstructions are described by a limited number of distinct global patterns or degrees of freedom, not all features are expected to be well described given the variable spatial reconstructive skill of the global reconstructions (e.g., parts of the southeastern United States; see Fig. 10.12). However, while the details of different ENSO events may be described only approximately, it is certainly true that different "flavors" of the global ENSO phenomenon are reasonably well described. This is evident in the reconstructions of global temperature patterns for specific ENSO years shown in section on "trends in extremes."

We examine how the global ENSO pattern changes over time by evaluating the correlation map of the reconstructed Niño-3 index against the global temperature pattern reconstructions in successive 50-year periods back to 1650 (Fig. 10.19). Certain features of the pattern appear remarkably robust, while other features are more variable. In all cases, the classic "horseshoe" pattern of a warm eastern tropical Pacific and west coast of North America, and a cold central North Pacific, is observed. This pattern might be regarded as an extremely robust signature of ENSO, at least over multidecadal and longer periods of time, whereas teleconnections into North America, and especially the Atlantic, and Eurasia are clearly more variable (the reader should keep in mind that the weaker features of these correlation maps are only marginally significant). The considerable variability between different 50-year periods in the North American teleconnection is consistent with the observations of Cole and Cook (1997) of multidecadal changes in ENSO-related patterns of drought in North America. The most striking evidence of a nonstationarity in ENSO teleconnections is the departure of the pattern of 1801–50 from the pattern dominated by all other periods. In particular, the pattern for this period shows a distinct absence of the typical pattern of warming in the far western tropical Pacific, an expanded region of cooling over eastern and southern North America, and cooling, rather than the typical warming, in the western equatorial Atlantic, coinciding with warm-event conditions. We believe that this pattern arises from a breakdown of interannual band ENSO variability during this period that may be related to external forcing of the global climate (see the following section). In the absence of the more typical interannual pattern, a somewhat different decadal-scale ENSO-like pattern of variability, which appears more robust over time, dominates during this period. It is through this suppression of the typical interannual pattern of variability that external forcing may lead to changes in extratropical ENSO teleconnections (see, e.g., the modeling study of Meehl and Branstator 1992 of the response to doubled CO_2; unfortunately, changes in the frequency domain characteristics of ENSO were not addressed in that study). The distinction between patterns of interannual and decadal ENSO variability has been discussed elsewhere (e.g., Trenberth and Hurrell 1994; Graham 1994; Mann and Park 1994, 1996; Knutson et al. 1997; Goddard and Graham 1998).

Frequency Domain Behavior

In this section, we investigate the frequency domain properties of the reconstructed history of ENSO and its large-scale teleconnections. A detailed study of the instrumental global temperature record for changes in the frequency and amplitude characteristics of ENSO and its large-scale teleconnections has been recently undertaken by Park and Mann (1999). Possible amplitude and frequency modulation of ENSO on longer timescales has been investigated in previous studies based on univariate (Michaelsen

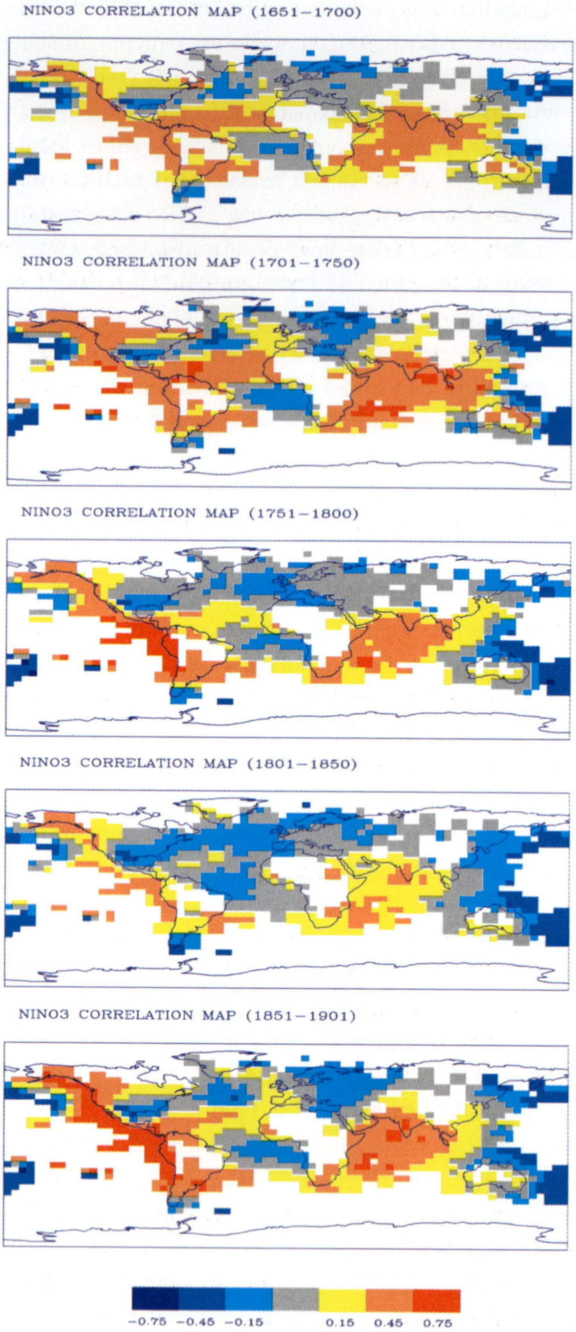

Fig. 10.19 Niño-3 temperature correlation map in successive 50-year periods between 1650 and present. The color scale shown indicates the specific level of correlation with a given grid point, with any color value outside of the gray category indicating significance at greater than a 70% level (based on a two-sided test), and with purple and red indicating significance at well above the 99% level.

1989; Meko 1992; Dunbar et al. 1994) and multivariate (Bradley et al. 1994) spectral analyses of ENSO-sensitive proxy data. Here, we analyze the reconstructions described in earlier sections to investigate low-frequency amplitude and frequency modulation of interannual variability and the characteristics of decadal and longer timescale changes in the base state of ENSO. We analyze long-term changes in the frequency domain aspects of El Niño and the global ENSO signal based on the evolutive spectrum of the reconstructed Niño-3 index (Fig. 10.20, top; calculated based on the "multitaper method"; see Thomson 1982; Park et al. 1987; Mann and Lees 1996) and the evolutive multivariate spectrum of the global temperature field (Fig. 10.20, bottom; based on the "MTM-SVD" method of Mann and Park 1996, 1999; see also Rajagopalan et al. 1998; Tourre et al. 1999), respectively. In both cases, we employ an 80-year moving window in time to investigate multidecadal amplitude modulation in ENSO and its global teleconnections. Similar results are obtained for 50- or 100-year window choices.

A general tendency toward relatively higher-frequency interannual variability during the past few centuries (e.g., from a 5- to 7-year period to a 4- to 5-year period) has been argued for in previous proxy ENSO studies (e.g., Dunbar et al. 1994; Michaelsen 1989; Diaz and Pulwarty 1994; Knutson et al. 1997). Such a trend is, arguably, evident in the evolutive spectral analysis of the Niño-3 index (Fig. 10.20, top). Frequency modulation of ENSO is not necessarily a fingerprint of anthropogenic climate change (see Knutson et al. 1997) and is consistent with natural variability in models of ENSO (Zebiak and Cane 1991; Knutson et al. 1997). What is more striking in our analysis, however, is not the frequency modulation, but rather the strong amplitude modulation, of the variability. There is evidence for relatively enhanced interdecadal variance in the reconstructed Niño-3 series during the twentieth century and relatively enhanced secular variance during the late eighteenth century (Fig. 10.20). The most pronounced amplitude modulation, however, is the clear breakdown in interannual variability during the early and mid-nineteenth century in both the Niño-3 index (Fig. 10.20, top) and the global temperature field (Fig. 10.20, bottom). This latter behavior in the Niño-3 index is consistent with the evolutive spectral analysis of the Galapagos coral δ^{18}O record (Dunbar et al. 1994) and time series waveform decompositions of SOI reconstructions (Stahle et al. 1998), both of which show weakened interannual variability during this period. This amplitude modulation is intriguing from the standpoint of possible external forcing. This period is also the coldest during the interval examined here in terms of NH mean temperature, associated with the "Maunder Minimum" period of relatively low apparent levels of solar irradiance, and preceding the marked twentieth-century greenhouse gas increases (see MBH98, Fig. 7). This coincidence suggests the possibility that interannual ENSO variability is a feature of a relatively warm climate. If this is the case, we might expect accelerated greenhouse forcing to increase the amplitude of interannual ENSO variability. However, there may also be an upper limit on the global temperature regime within which interannual ENSO variability can reside. The latter scenario is hinted at by the experiments of Knutson et al. (1997), which explored the response of a low-resolution version of the Geophysical Fluid Dynamics Laboratory (GFDL) coupled model to CO_2 forcing and showed a decrease in the amplitude of

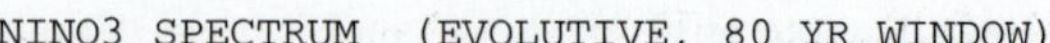

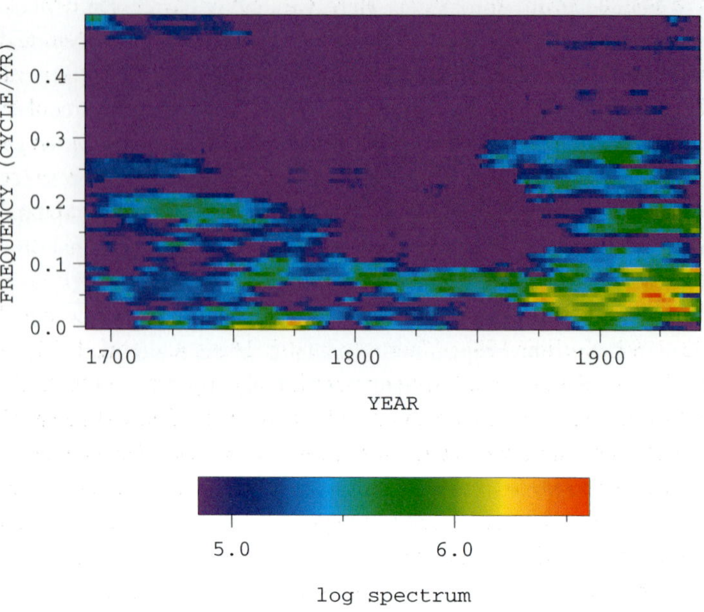

MTM-SVD LFV SPECTRUM (EVOLUTIVE, 80 YR WINDOW)

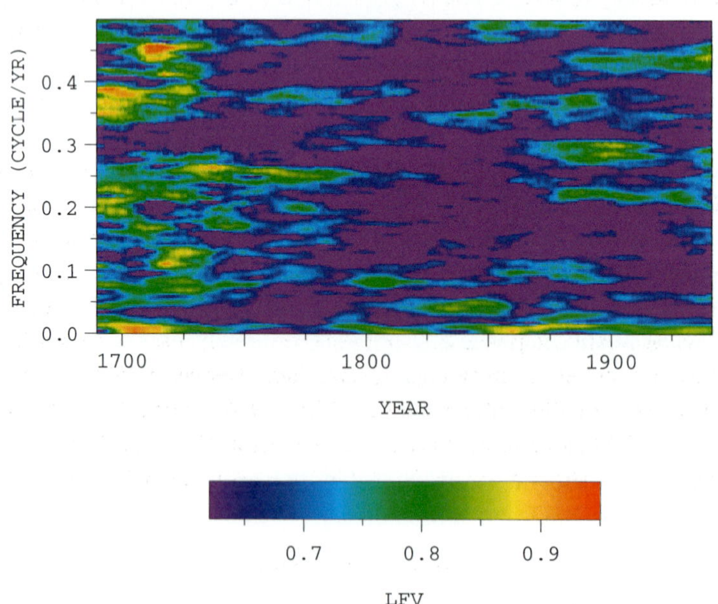

Fig. 10.20 Evolutive spectrum of Niño-3 index (top) and evolutive multivariate spectrum of global temperature fields (see text for methodological details) for the period 1650–1980, employing an 80-year moving window through time.

interannual ENSO-like variability in an enhanced greenhouse climate (Knutson et al. 1997). (We note, however, that other models [e.g., Zebiak and Cane 1991; Meehl and Branstator 1992] do not show such a response.) Furthermore, Goddard and Graham (1998) point out that the interannual delayed oscillatory mechanism appears to have broken down in describing the more trendlike ENSO-related variability of the 1990s.

The primary differences between the evolutive Niño-3 and multivariate global temperature spectra (Fig. 10.20) are at the quasi-biennial and decadal-to-secular frequencies. The quasi-biennial signal appears to be associated with extratropical features largely unrelated to ENSO (see, e.g., Mann and Park 1994, 1996), while the multidecadal (secular within the context of an 80-year window) global variations are associated with the North Atlantic multidecadal pattern discussed earlier (see, e.g., Delworth et al. 1993, 1997; Kushnir 1994; Mann et al. 1995; Mann and Park 1994, 1996). The interdecadal (15–30) year signal, which intermittently appears in both the Niño-3 and global multivariate spectra, is associated with interdecadal ENSO-like variability that has been established in previous analyses of multiproxy data (Mann et al. 1995), and which is discussed in more detail in the following section. As was noted earlier, the reconstructed Niño-3 index has slightly exaggerated power at decadal and lower frequencies in the twentieth century, relative to the instrumental Niño-3 index (see Fig. 10.10). The amplitude of decadal, relative to interannual, power may thus be somewhat exaggerated in the reconstructed Niño-3 index.

To illustrate the fundamental differences referred to earlier in the spatial pattern of the interannual ("true") ENSO signal, and the decadal ENSO-like signal, we reconstruct a typical oscillatory episode of the roughly 4- to 5-year (Fig. 10.21) and roughly 20- to 25-year (Fig. 10.22) global temperature signals based on the MTM-SVD method described by Mann and Park (1994, 1996, 1999). The primary departure of the interdecadal pattern from the ENSO pattern is the relatively subdued amplitude in the tropical Pacific and a cooling (rather than warming) influence over both the North American and tropical Atlantic sectors coincident with warm eastern tropical Pacific (i.e., El Niño) conditions. The relative prominence of the decadal-scale ENSO pattern during the early nineteenth century thus appears to explain the unusual Pacific North American and Atlantic teleconnection pattern commented on earlier (Fig. 10.19). Possible origins and extratropical–tropical connections that may describe the decadal signal (and its possible relationship with ENSO) have been widely discussed elsewhere based on analyses of twentieth-century instrumental data (Trenberth 1990; Graham 1994; Trenberth and Hurrell 1994; Mann and Park 1994, 1996; Goddard and Graham 1998; Park and Mann 1999), long-term proxy climate indicators (Mann et al. 1995), and modeling experiments (Latif and Barnett 1994; Graham 1994; Gu and Philander 1997).

Trends in Extremes

We can further quantify variations in the rates of extreme events (e.g., both large El Niños and La Niñas) commented upon earlier, by looking at how frequencies of occurrences of anomalous events have changed over time in our Niño-3 reconstructions. Because the variance properties appear to be homogeneous back in time (the resolved

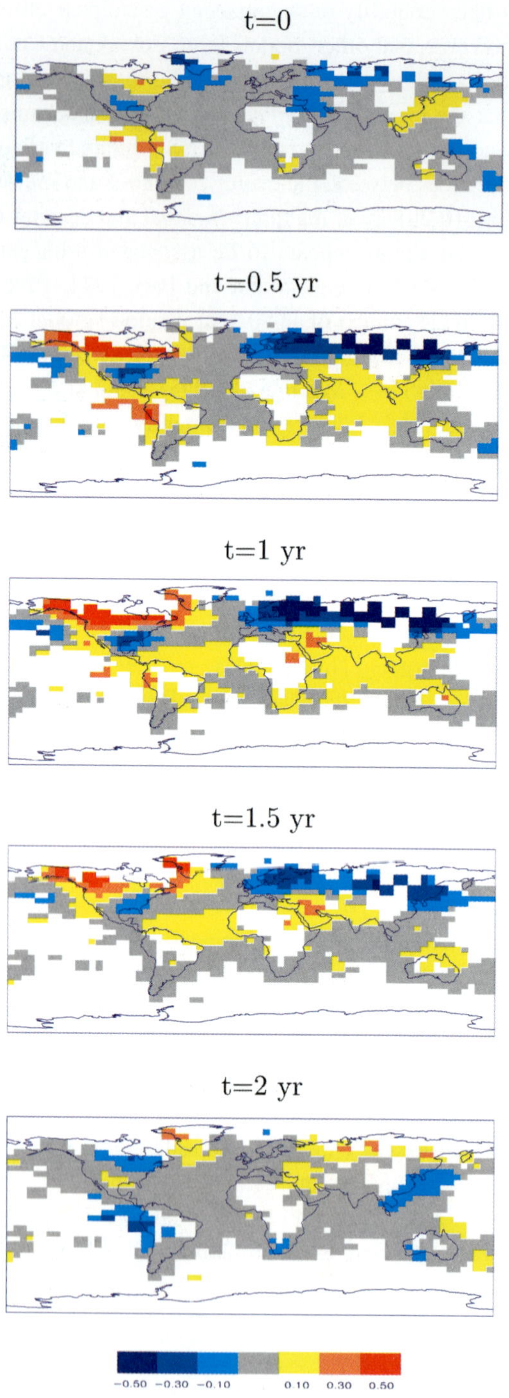

Fig. 10.21 Canonical pattern of the interannual ENSO signal in the temperature reconstructions as it evolves over half of a typical (roughly 4-year period) cycle.

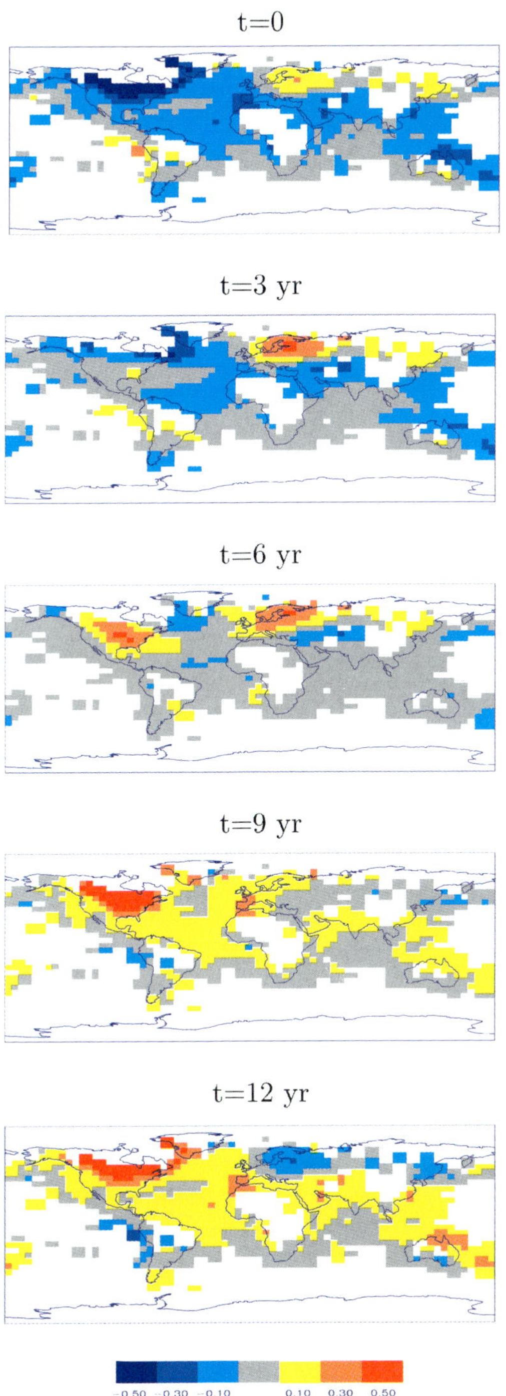

Fig. 10.22 Canonical pattern of the interdecadal ENSO-like signal in the temperature reconstructions as it evolves over half of a typical cycle. The cycle length is variable from about a 16- to 25-year period, with the case shown corresponding to the dominant roughly 25-year period evident during the early nineteenth century.

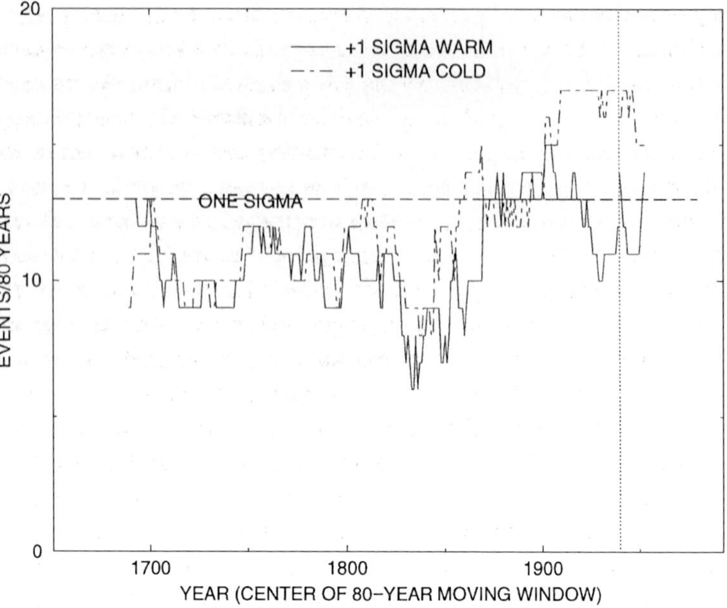

Fig. 10.23 Frequency of extreme warm or cold ENSO events per 80-year period, based on the reconstructed Niño-3 index from 1650–1980. The number of 1 sigma "outliers" expected for a Gaussian distribution with the same variance as the observed distribution during the 1902–80 calibration period is shown (horizontal dashed line) for comparison. Recent results are shown (boundary marked by the vertical line) based on the instrumental Niño-3 index available from 1902 to 1993.

variance in our Niño-3 reconstructions is a roughly constant ~40% back in time; see Table 10.3, and also the error bars in Fig. 10.17), we consider it reasonable to examine the changes in the distribution – at least the second-order statistics – back in time. Figure 10.23 shows the variations over time (in an 80-year moving window) in the frequency of anomalous cold and warm events (defined by an event amplitude greater than one standard deviation, based on the calibration period [1902–80] mean and variance of the instrumental Niño-3 index). The low-frequency variations in the frequency of both anomalous warm and cold events are similar to estimates of the low-frequency variations in the probability of all (weak and strong) warm events based on combined documentary, tropical ice core δ^{18}O, and extratropical dendroclimatic-based indices of ENSO-related variability (Michaelsen and Thompson 1992). In each case, a pronounced decreasing trend in occurrences is observed from the eighteenth through mid-nineteenth century, and a pronounced increasing trend from the mid-nineteenth through early twentieth century. The unusually low incidence of anomalous warm events during the early nineteenth century coincides closely with the breakdown of interannual variability (Fig. 10.20) and altered global teleconnection patterns (Fig. 10.19) during that period of time. Aside from the overall trend of increases in both cold and warm extremes since the mid-nineteenth century, our analysis indicates a trend toward enhanced skewness of the Niño-3 distribution during the twentieth century. The frequency of anomalous cold events has increased dramatically during the twentieth century, similar to the warm event occurrence rates of Michaelsen and Thompson (1992), while the frequency of anomalous warm events has actually decreased during the early twentieth century. Cold and warm event occurrences, otherwise, largely track each at multidecadal and longer timescales. It is interesting that this skewness of the Niño-3 distribution appears to be reversing in recent decades, owing to the recent run of anomalous warm events (see, e.g., Trenberth 1990).

The possibility that El Niño episode durations or "spell lengths" have increased in association with global warming has been highlighted by the recent study of Trenberth and Hoar (1996; see also the follow-ups by Rajagopalan et al. 1997; Trenberth and Hoar 1997), which emphasized the unusually long nature of the 1990–95 warm episode. We investigate how unusual such a warm spell is within the context of our several centuries–long reconstructions. Figure 10.24 shows the frequency of 5-year-long "warm spells," defined as the number of distinct, uninterrupted stretches of positive Niño-3 values per 80-year period, as a function of time. The most dramatic feature is the trend toward far fewer long "warm spells" during the twentieth century. In this context, the 1990–95 event is particularly striking, occurring during the period of recent increases in warm event frequencies (Fig. 10.23) and anomalous warm events (Fig. 10.17). However, this spell does not appear especially unusual when viewed in the context of certain previous periods, such as the early nineteenth century, during which there were between three

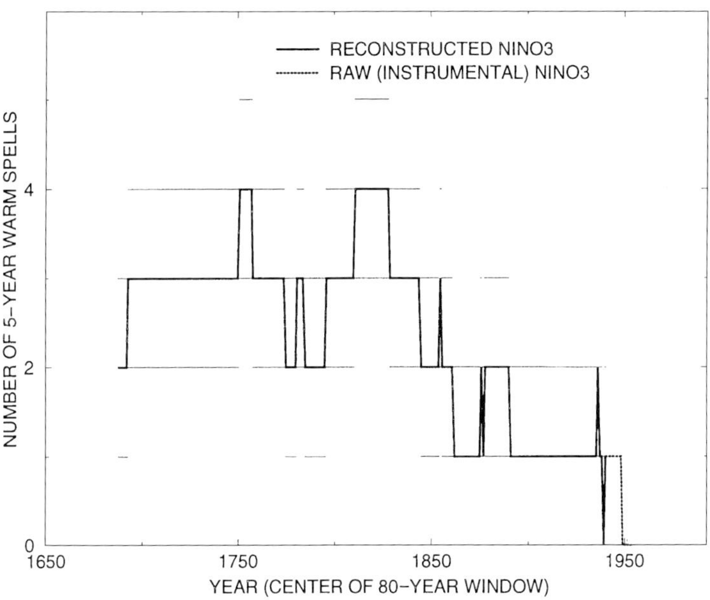

Fig. 10.24 Frequency of unusually (5-year) long warm ENSO "spells" per 80-year period, for the reconstructed Niño-3 index for 1650–1980 (solid curve). Recent results are shown (dotted curve) based on the instrumental Niño-3 index available for 1902–93; 1 standard error limits are indicated (dashed) for the estimated number of warm spells in a given 80-year period. These limits were determined from Monte Carlo simulations taking into account the uncertainties in the Niño-3 reconstruction itself (see Fig. 10.17).

and five 5-year-long warm events within the confines of our estimated uncertainties. This finding is supported by an independent proxy-based study by Allan and D'Arrigo (1999), which provides evidence for numerous "persistent" warm episodes in past centuries. It is worth emphasizing that the early nineteenth-century period associated with the greatest probability of warm spells is associated, as was commented upon earlier, with the coldest global temperatures back to 1650 and an apparent breakdown of interannual ENSO variability. The unusually protracted 1990–95 event may, in turn, be signaling a breakdown of interannual ENSO variability at the "warm global temperature" boundary of its dynamical regime.

Selected ENSO Events

Here, we look at global patterns for eight selected warm and cold ENSO years. Figure 10.25 shows the patterns for various strong (S) and very strong (VS) category El Niño years indicated by the QN92 chronology and associated with warm values of our reconstructed Niño-3 index. The correspondence between our reconstructed Niño-3 index and the QN92 chronology is far from perfect: Many QN92 warm events do not correspond to warm reconstructed Niño-3 values, and many warm reconstructed Niño-3 years do not correspond to QN92 warm events. We suspect that the discrepancies result from a combination of the uncertain nature of our reconstructions (which, after all, resolve only about 40% of the instrumental Niño-3 variance for the calibration period; Table 10.3) and possible weaknesses in the QN92 chronology itself (note, for example, the imperfect correspondence between the QN92 warm-event chronology and the negative SOI peaks during the verification period; Fig. 10.13). However, we note that the correspondence between our reconstruction and the QN92 warm-event chronology is significant well above the 99% confidence level based on Monte Carlo permutation tests (Table 10.3).

The pattern for 1652 (shown, as in all events, relative to a zero reference corresponding to the 1902–80 climatology) indicates warming in the Niño-3 region but an unusual lack of warm anomalies along Pacific coastal South and Central America. This is similarly true for 1804 and 1828, which show cooling or neutral anomalies in most areas other than the Niño-3 region, largely because these events are superimposed on the considerably cold global baseline conditions of the early nineteenth century discussed earlier. The 1828 pattern, which also falls in the center of the 1801–50 period earlier noted (Fig. 10.19) for its unusual global ENSO correlation pattern, is particularly atypical. The 1877 and 1884 events exhibit particularly strong tropical eastern Pacific warm tongue anomalies, which are more typical of the twentieth century (e.g., Fig. 10.11) pattern. In summary, the patterns of different reconstructed ENSO years are quite variable. This variability is probably due in part to the inherent event-to-event noise, in part to the obscuring influence of other processes influencing global temperatures, and in part to real changes in ENSO teleconnections over time (e.g., Fig. 10.19).

Figure 10.26 shows global temperature anomaly patterns for two selected cold events during the eighteenth century. Note that the patterns depart significantly from the mirror images of warm events such as are shown in Figure 10.25, in large part because of the

(a) 1652 TEMPERATURE ANOMALY PATTERN

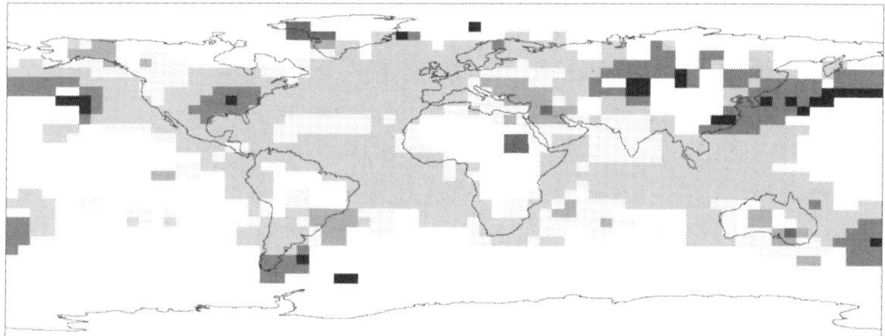

(b) 1720 TEMPERATURE ANOMALY PATTERN

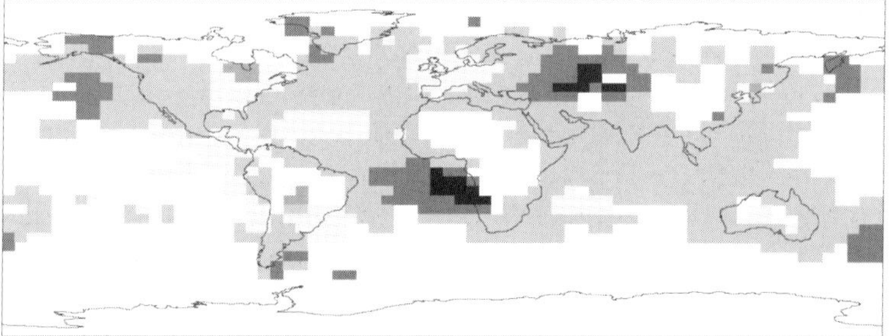

(c) 1747 TEMPERATURE ANOMALY PATTERN

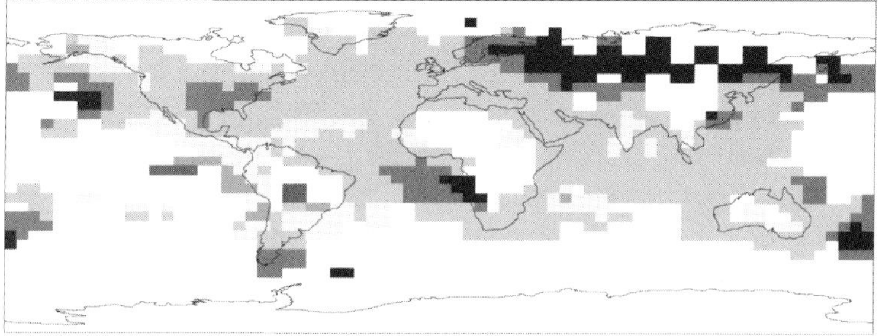

(d) 1791 TEMPERATURE ANOMALY PATTERN

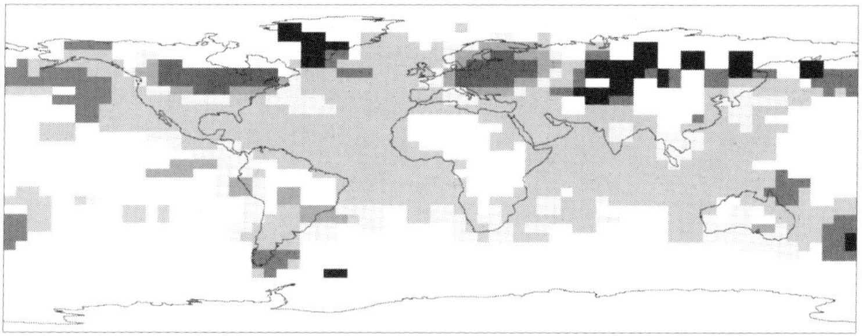

Fig. 10.25 Global temperature anomaly patterns for selected QN92 documented strong (S) and very strong (VS) El Niño years. (a) 1652, (b) 1720, (c) 1747, (d) 1791,

(e) 1804 TEMPERATURE ANOMALY PATTERN

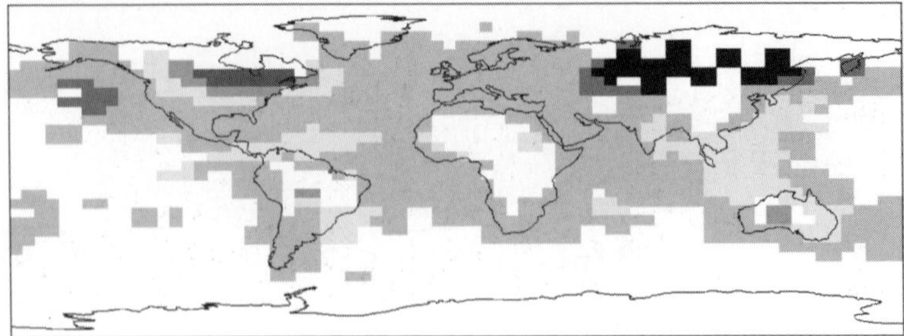

(f) 1828 TEMPERATURE ANOMALY PATTERN

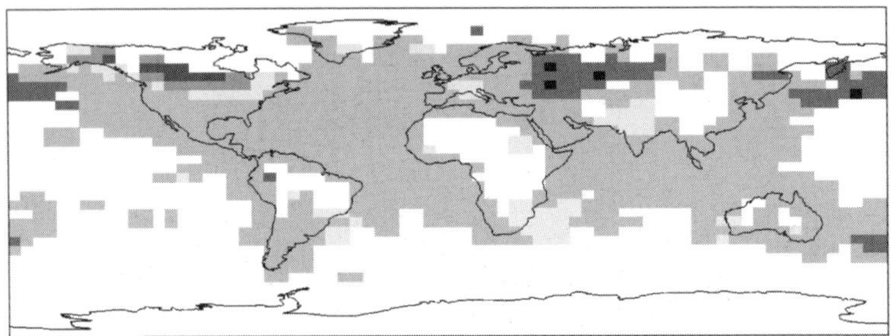

(g) 1877 TEMPERATURE ANOMALY PATTERN

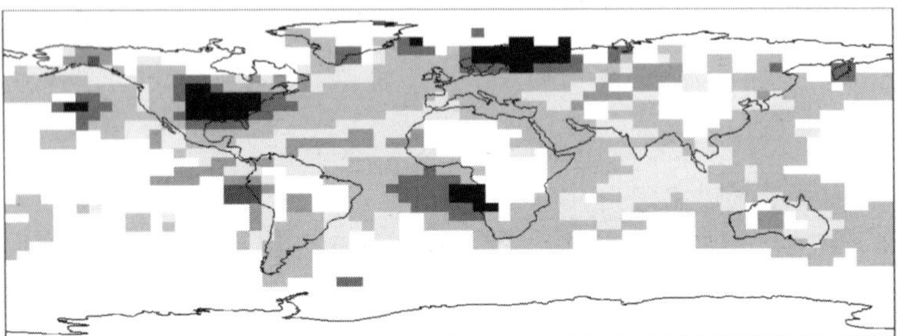

(h) 1884 TEMPERATURE ANOMALY PATTERN

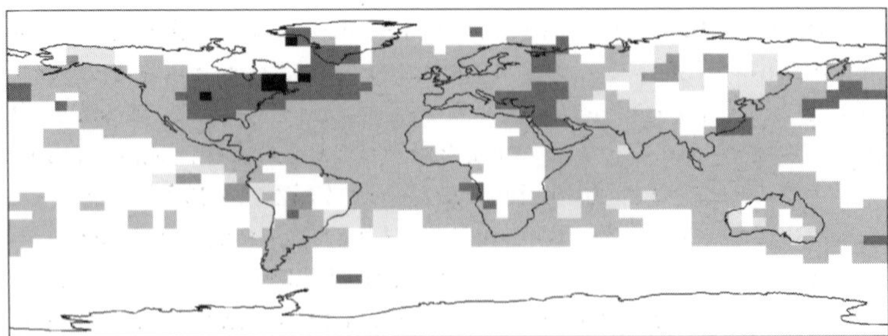

Fig. 10.25 (*cont.*) (e) 1804, (f) 1828, (g) 1877, (h) 1884. Color scale (same as in Fig. 10.19) indicates anomalies in degrees Celsius relative to 1902–80 climatological mean.

1732 TEMPERATURE ANOMALY PATTERN

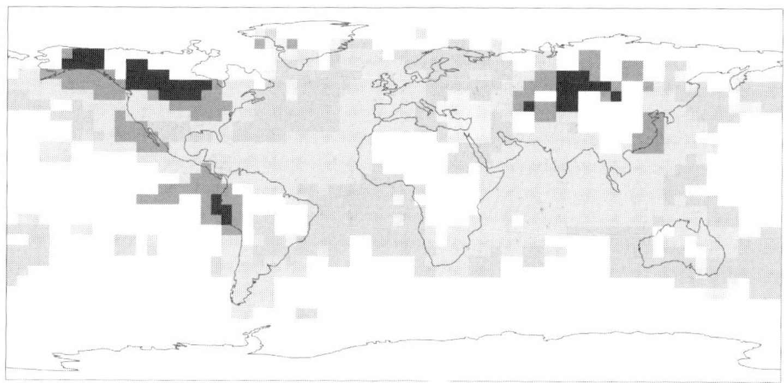

1777 TEMPERATURE ANOMALY PATTERN

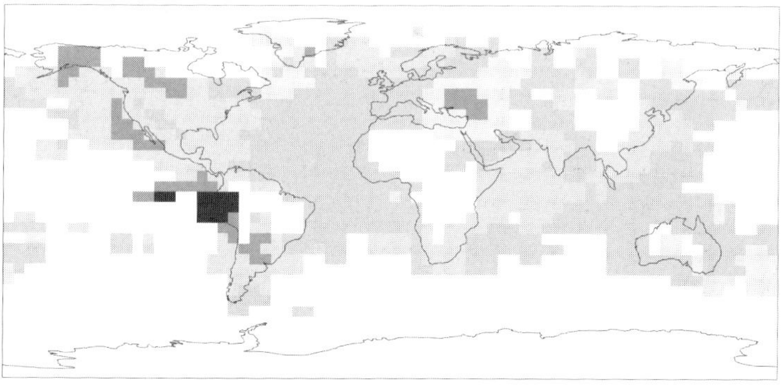

Fig. 10.26 Global temperature anomaly patterns for two reconstructed cold ENSO (La Niña) years:1732 and 1777. Conventions are as in Figure 10.25.

colder climatology of the period. For example, neutral conditions in the central North Pacific relative to the twentieth-century climatology would actually correspond to more familiar warm conditions in the region if plotted relative to the colder eighteenth-century climatology. There is no independent historical corroboration of cold events, as the QN92 index documents only El Niño, and not La Niña, episodes. The cold events typically, as we would expect, fall between QN92 warm events, and frequently right after warm events. For example, the notable 1732 and 1777 reconstructed cold events shortly follow the 1728 VS and 1775 S warm events, respectively. The eighteenth century was a time of relatively pronounced interannual variability in Niño-3 (see Fig. 10.20), so we might tend to believe that the delayed oscillator mechanism was particularly active at this time. Hence, it is reasonable to conclude that strong warm events should tend to be followed by particularly strong cold events during this period.

Conclusions

The multiproxy-based reconstructions of El Niño and its large-scale patterns of influence described in this chapter provide insight into a number of issues related to ENSO, including its natural variability and the varying large-scale teleconnections on multidecadal and century timescales. Relationships with global or hemispheric mean temperatures are also evident, as are possible relationships with natural and external forcing of climate, including anthropogenic factors. Certain ENSO-related patterns and features appear to have exhibited significant trends during the twentieth century, although classic ENSO indices show only modest warming trends in comparison with the dramatic warming trend in hemispheric and global temperature during the past century. There is some evidence that a pattern associated with negative tropical feedbacks may be abating an El Niño–like warming trend in the tropical Pacific.

Equally interesting in the climate reconstructions is evidence of changes in the amplitude of interannual ENSO variability, global teleconnections of ENSO, and the amplitude and frequency of extremes. For example, the incidence of large warm and cold events appears to have increased during the past century, while the frequency of long warm ENSO spells has decreased. There is evidence, however, that some of these trends may be reversing during the most recent decades. The breakdown of interannual ENSO variability during the nineteenth century appears to have had particularly significant impacts on the incidence of extremes and on global teleconnection patterns of ENSO during that period. We find some support for the proposition that this breakdown may have been associated with the same external forcings that led to generally cold global temperatures at the time. This period may thus provide an analog for the behavior of ENSO and the possible breakdown of typical mechanisms of ENSO variability under the impacts of external – including anthropogenic – forcing of climate.

As increasingly rich networks of high-quality, seasonally resolved proxy data become available, both global temperature pattern and ENSO-scale reconstructions should be possible with considerably reduced uncertainties. Of particular utility for ENSO-scale reconstruction will be the availability of a greater number of well-dated coral isotopic indicators in the tropical Pacific to supplement the sparse existing network. With reduced uncertainties, indices of past ENSO variability should allow the unusual 1982/1983 and 1997/1998 warm events and the protracted 1990–95 "warm spell" to be placed in a proper longer term perspective. With such a perspective, we should be better able to assess whether or not these seemingly anomalous recent events represent significant and potentially anthropogenic-related climate changes.

Acknowledgments We are indebted to Rosanne D'Arrigo, David Fisher, Gordon Jacoby, Judith Lean, Alan Robock, David Stahle, Charles Stockton, Eugene Vaganov, Ricardo Villalba, and the numerous contributors to the International Tree-Ring Data Bank for many of the proxy data series used in this study. We thank Mike Dettinger for a thoughtful and thorough review of the manuscript. We are also grateful to Frank Keimig, Martin Munro, and Richard Holmes for their technical assistance. This work was supported by the National Science Foundation and the Department of Energy. M.E.M. acknowledges support through the Alexander Hollaender Distinguished Postdoctoral Research

Fellowship program of the U.S. Department of Energy. This work is a contribution to the Analysis of Rapid and Recent Climatic Change (ARRCC) project, sponsored by the National Science Foundation and the U.S. National Oceanic and Atmospheric Administration.

References

ALLAN, R. and D'ARRIGO, R. D., 1999: "Persistent" ENSO sequences: How unusual was the 1990–1995 El Niño? *Holocene*. **9**, 101–118.

ALLAN, R., LINDESAY, J., and PARKER, D., 1996: *El Niño Southern Oscillation and Climate Variability*. Melbourne: CSIRO Publishing.

ANGELL, J. K., 1990: Variation in global tropospheric temperature after adjustment for the El Niño influence, 1958–1989. *Geophysical Research Letters*, **17**, 1093–1096.

BARNETT, T. P., SANTER, B., JONES, P. D., and BRADLEY, R. S., 1996: Estimates of low frequency natural variability in near-surface air temperature.*Holocene*, **6**, 255–263.

BARNSTON, A. G. and LIVEZEY, R. E., 1987: Classification, seasonality and persistence of low-frequency atmospheric circulation patterns. *Monthly Weather Review*, **115**, 1083–1126.

BONINSEGNA, J. A., 1992: South American dendroclimatological records. In Bradley, R. S. and Jones, P. D. (eds.), *Climate Since A.D. 1500*. London: Routledge, 246–268.

BRADLEY, R. S., 1996: Are there optimum sites for global paleotemperature reconstruction? *In* Jones, P. D., Bradley, R. S., and Jouzel, J. (eds.), *Climatic Variations and Forcing Mechanisms of the Last 2000 Years*. Heidelberg: Springer-Verlag, 603–624.

BRADLEY, R. S. and JONES, P. D., 1993: 'Little Ice Age' summer temperature variations: Their nature and relevance to recent global warming trends. *Holocene*, **3**, 367–376.

BRADLEY, R. S., MANN, M. E., and PARK, J., 1994: A spatiotemporal analysis of ENSO variability based on globally distributed instrumental and proxy temperature data. *EOS Supplement*, **75**, 383.

BRIFFA, K. R. et al., 1992a: Fennoscandian summers from AD 500: Temperature changes on short and long timescales. *Climate Dynamics*, **7**, 111–119.

BRIFFA, K. R., JONES, P. D., and SCHWEINGRUBER, F. H., 1992b: Tree-ring density reconstructions of summer temperature patterns across western North America since 1600. *Journal of Climate*, **5**, 735–753.

BRIFFA, K. R., JONES, P. D., SCHWEINGRUBER, F. H., SHIYATOV, S. G., and COOK, E. R., 1995: Unusual twentieth-century summer warmth in a 1000-year temperature record from Siberia. *Nature*, **376**, 156–159.

CANE, M. et al., 1997: Twentieth-century sea surface temperature trends. *Science*, **275**, 957–960.

CHANG, P., JI, L., LI, H., and FLUGEL, M., 1996: Chaotic dynamics versus stochastic processes in El Niño–Southern Oscillation in coupled ocean–atmosphere models. *Physica D*, **98**, 301–320.

CHANG, P., JI, L., and LI, H., 1997: A decadal climate variation in the tropical Atlantic Ocean from thermodynamic air–sea interactions. *Nature*, **385**, 516–518.

CLEAVELAND, M. K., COOK, E. R., and STAHLE, D. W., 1992: Secular variability of the Southern Oscillation detected in tree-ring data from Mexico and the southern United States. *In* Diaz, H. F., and Markgraf, V. (eds.), *El Niño: Historical and Paleoclimatic Aspects of the Southern Oscillation*. Cambridge: Cambridge University Press, 271–291.

COLE, J. and COOK, E., 1997: The coupling between ENSO and U.S. drought: How stable is it over the past century? *EOS Supplement*, **78**.

COLE, J. E., FAIRBANKS, R. G., and SHEN, G. T., 1993: Recent variability in the Southern Oscillation: Isotopic results from a Tarawa atoll coral. *Science*, **260**, 1790–1793.

COOK, E. R., BRIFFA, K. R., and JONES, P. D., 1994: Spatial regression methods in dendroclimatology: A review and comparison of two techniques. *International Journal of Climatology*, **14**, 379–402.

COOK, E. R., BRIFFA, K. R., MEKO, D. M., GRAYBILL, D. A., and FUNKHOUSER, G., 1995: The 'segment length' curse in long tree-ring chronology development for paleoclimatic studies. *Holocene*, **5**, 229–237.

COOK, E. R. et al., 1991: Climatic change in Tasmania inferred from a 1089-year tree-ring chronology of Huon pine. *Science*, **253**, 1266–1268.

D'ARRIGO, R. D., COOK, E. R., JACOBY, G. C., and BRIFFA, K. R., 1993: NAO and sea surface temperature signatures in tree-ring records from the North Atlantic sector. *Quaternary Science Review*, **12**, 431–440.

DELWORTH, T. D. and MANN, M. E., 1999: Observed and simulated multidecadal variability in the North Atlantic. In press.

DELWORTH, T. D., MANABE, S., and STOUFFER, R. J., 1993: Interdecadal variations of the thermohaline circulation in a coupled ocean–atmosphere model. *Journal of Climate*, **6**, 1993–2011.

DELWORTH, T. D., MANABE, S., and STOUFFER, R. J., 1997: Multidecadal climate variability in the Greenland Sea and surrounding regions: A coupled model simulation. *Geophysical Research Letters*, **24**, 257–260.

DIAZ, H. F. and MARKGRAF, V., 1993: *El Niño: Historical and Paleoclimatic Aspects of the Southern Oscillation*. Cambridge: Cambridge University Press, 476 pp.

DIAZ, H. F. and PULWARTY, R. S., 1994: An analysis of the time scales of variability in centuries-long ENSO-sensitive records in the last 1000 years. *Climatic Change*, **26**, 317–342.

DUNBAR, R. B., WELLINGTON, G. M., COLGAN, M. W., and GLYNN, P. W., 1994: Eastern Pacific sea surface temperature since 1600 A.D. The δ^{18}O record of climate variability in the Galapagos corals. *Paleoceanography*, **9**, 291–315.

FISHER, D. A. et al., 1996: Inter-comparison of ice core delta O-18 and precipitation records from sites in Canada and Greenland over the last 3500 years and over the last few centuries in detail using EOF techniques. *In* Jones, P. D., Bradley, R. S., and Jouzel, J. (eds.), *Climatic Variations and Forcing Mechanisms of the Last 2000 Years*. Heidelberg: Springer-Verlag, 297–328.

FISHER, D. A. et al., 1998: Penny ice cap cores, Baffin Island, Canada, and the Wisconsin Foxe Dome connection: Two states of Hudson Bay ice cover. *Science*, **279**, 692–965.

FOLLAND, C. K., PARKER, D. E., and KATES, F. E., 1984: Worldwide marine temperature fluctuations 1856–1981. *Nature*, **310**, 670–673.

FRITTS, H., 1991: *Reconstructing Large-Scale Climatic Patterns from Tree Ring Data*. Tucson: University of Arizona Press.

FRITTS, H. C., BLASING, T. J., HAYDEN, B. P., and KUTZBACH, J. E., 1971: Multivariate techniques for specifying tree-growth and climate relationships and for reconstructing anomalies in paleoclimate. *Journal of Applied Meteorology*, **10**, 845–864.

FRITTS, H. C. and SHAO, X.-M., 1992: *In* Bradley, R. S., and Jones, P. D. (eds.), *Climate since A.D. 1500*. London: Routledge, 269–294.

GODDARD, L. and GRAHAM, N. E., 1998: El Niño in the 1990s. *Journal of Geophysical Research*, **102**, 10423–10436.

GRAHAM, N. E., 1994: Decadal-scale variability in the tropical and North Pacific during the 1970s and 1980s: Observations and model results. *Climate Dynamics*, **10**, 135–162.

GU, D. and PHILANDER, S. G. H., 1997: Interdecadal climate fluctuations that depend on exchanges between the tropics and extratropics. *Science*, **275**, 805–807.

GUIOT., J., 1988: The combination of historical documents and biological data in the reconstruction of climate variations in space and time. *Palaeoklimaforschung*, **7**, 93–104.

HALPERT, M. S. and ROPELEWSKI, C. F., 1992: Surface temperature patterns associated with the Southern Oscillation. *Journal of Climate*, **5**, 577–593.

HEISS, G. A., 1994: Coral reefs in the Red Sea: Growth, production and stable isotopes. GEOMAR Report 32, 1–141.

HUGHES, M. K. and DIAZ, H. F., 1994: Was there a "Medieval Warm Period" and if so, where and when? *Climatic Change*, **26**, 109–142.

HUGHES, M. K., VAGANOV, E. A., SHIYATOV, S. A., TOUCHAN, R., and FUNKHOUSER, G., 1999: Twentieth-century summer warmth in northern Yakutia in a 600-year context. *The Holocene*, **9**, 603–608.

HURRELL, J. W., 1995: Decadal trends in the North Atlantic Oscillation, regional temperatures and precipitation. *Science*, **269**, 676–679.

Intergovernmental Panel on Climate Change (IPCC), 1995: *Climate Change 1995: The Science of Climate Change*, Houghton, J. T. et al. (eds.). Cambridge: Cambridge University Press.

JACOBY, G. C. and D'ARRIGO, R., 1989: Reconstructed Northern Hemisphere annual temperature since 1671 based on high-latitude tree-ring data from North America. *Climatic Change*, **14**, 39–59.

JACOBY, G. C. and D'ARRIGO, R. D., 1990: Teak: A tropical species of large-scale dendroclimatic potential. *Dendrochronologia*, **8**, 83–98.

JACOBY, G. C., D'ARRIGO, R. D., and TSEVEGYN, D., 1996: Mongolian tree rings and 20th-century warming. *Science*, **9**, 771–773.

JIN, F. F., NEELIN, J. D., and GHIL, M., 1994: El Niño on the devil's staircase: Annual subharmonic steps to chaos. *Science*, **264**, 70–72.

JONES, P. D., 1989: The influence of ENSO on global temperatures. *Climate Monitor*, **17**, 80–89.

JONES, P. D., 1998: It was the best of times, it was the worst of times. *Science*, **280**, 544–545.

JONES, P. D. and BRADLEY, R. S., 1992: *In* Bradley, R. S. and Jones, P. D. (eds.), *Climate since* A.D. *1500*, London: Routledge, 246–268.

JONES, P. D. and BRIFFA, K. R., 1992: Global surface air temperature variations during the 20th century: Part I – Spatial, temporal and seasonal details. *Holocene*, **1**, 165–179.

KAMEDA, T., NARITA, H., SHOJI, J., NISHIO, F., and WATANABE, O., 1992: *Proc. International Symposium on the Little Ice Age Climate*, Tokyo Metropolitan University, Tokyo, Japan.

KAPLAN, A., CANE, M. A., KUSHNIR, Y., BLUMENTHAL, M. B., and RAJAGALOPAN, B., 1998: Analyses of global sea surface temperatures 1856–1991. *Journal of Geophysical Research*, **103**, 18567–18589.

KNUTSON, T., MANABE, S., and GU, D., 1997: Simulated ENSO in a global coupled ocean–atmosphere model: Multidecadal amplitude modulation and CO_2 sensitivity. *Journal of Climate*, **10**, 138–161.

KUSHNIR, Y., 1994: Interdecadal variations in North Atlantic sea surface temperature and associated atmospheric conditions. *Journal of Climate*, **7**, 141–157.

LATIF, M. and BARNETT, T. P., 1994: Causes of decadal climate variability over the North Pacific and North America. *Science*, **266**, 634–637.

LAU, N. C. and NATH, M. J., 1994: A modeling study of the relative roles of tropical and extratropical SST anomalies in the variability of the global atmosphere–ocean system. *Journal of Climate*, **7**, 1184–1207.

LAU, N. C., PHILANDER, S. G. H., and NATH, M. J., 1992: Simulation of ENSO-like phenomena with a low-resolution coupled GCM of the global ocean and atmosphere. *Journal of Climate*, **5**, 284–307.

LINSLEY, B. K., DUNBAR, R. B., and MUCCIASONE, D. A., 1994: A coral-based reconstruction of Inter-Tropical Convergence Zone variability over Central America since 1707. *Journal of Geophysical Research*, **99**, 9977–9994.

LOUGH, J. M., 1991: Rainfall variations in Queensland, Australia: 1891–1986. *International Journal of Climatology*, **11**, 745–768.

MANLEY, G., 1959: Mean temperature of Central England 1698–1952. *Quarterly Journal of the Royal Meteorological Society*, **79**, 242–261.

MANN, M. E., BRADLEY, R. S., and HUGHES, M. K., 1998a: Global-scale temperature patterns and climate forcing over the past six centuries. *Nature*, **392**, 779–787.

MANN, M. E., BRADLEY, R. S., and HUGHES, M. K., 1999: Northern Hemisphere temperatures during the past millennium: Inferences, uncertainties, and limitations. *Geophysical Research Letters*, **26**, 759–762.

MANN, M. E., BRADLEY, R. S., HUGHES, M. K., and JONES, P. D., 1998b: Global temperature patterns. *Science*, **280**, 2029–2030.

MANN, M. E. and LEES, J., 1996: Robust estimation of background noise and signal detection in climatic time series. *Climatic Change*, **33**, 409–445.

MANN, M. E. and PARK, J., 1994: Global scale modes of surface temperature variability on interannual to century time scales. *Journal of Geophysical Research*, **99**, 25819–25833.

MANN, M. E. and PARK, J., 1996: Joint spatio-temporal modes of surface temperature and sea level pressure variability in the Northern Hemisphere during the last century. *Journal of Climate*, **9**, 2137–2162.

MANN, M. E. and PARK, J., 1999: Oscillatory spatiotemporal signal detection in climate studies: A multiple-taper spectral domain approach. *Advances in Geophysics*, **41**, 1–131.

MANN, M. E., PARK, J., and BRADLEY, R. S., 1995: Global interdecadal and century-scale oscillations during the past five centuries. *Nature*, **378**, 266–270.

MEEHL, G. A. and BRANSTATOR, G. W., 1992: Coupled climate model simulation of El Niño/Southern Oscillation: Implications for paleoclimate. *In* Diaz, H. F. and Markgraf, V. (eds.), *El Niño: Historical and Paleoclimatic Aspects of the Southern Oscillation*. Cambridge: Cambridge University Press, 69–91.

MEEHL, G. A. and WASHINGTON, W. M., 1996: El Niño–like climate change in a model with increased atmospheric CO_2 concentrations. *Nature*, **382**, 56–60.

MEKO, D., 1992: Spectral properties of tree-ring data in the United States Southwest as related to El Niño/Southern Oscillation. *In* Diaz, H. F. and Markgraf, V. (eds.), *El Niño: Historical and Paleoclimatic Aspects of the Southern Oscillation*. Cambridge: Cambridge University Press, 349–375.

MICHAELSEN, J., 1989: Long-period fluctuations in El Niño amplitude and frequency reconstructed from tree-rings. *In* Peterson, D. H. (ed.), *Aspects of Climate Variability in the*

Pacific and the Western Americas, Geophysical Monograph 55. Washington, DC: American Geophysical Union, 69–74.

MICHAELSEN, J. and THOMPSON, L. G., 1992: A comparison of proxy records of El Niño/Southern Oscillation. *In* Diaz, H. F. and Markgraf, V. (eds.), *El Niño: Historical and Paleoclimatic Aspects of the Southern Oscillation*. Cambridge: Cambridge University Press, 7–28.

NORTON, D. A. and PALMER, J. G., 1992: Dendroclimatic evidence from Australasia. *In* Bradley, R. S. and Jones, P. D. (eds.), *Climate since* A.D. *1500*. London: Routledge, 463–482.

ORTLIEB, L., 2000: The documented historical record of El Niño events in Peru: An update of the Quinn record (sixteenth through nineteenth centuries), Chapter 7 of this volume.

PARK, J., LINDBERG, C. R., and VERNON, F. L., III, 1987: Multitaper spectral analysis of high-frequency seismograms. *Journal of Geophysical Research*, **92**, 12675–12684.

PARK, J. and MANN, M. E., 1999: Interannual temperature events and shifts in global temperature: A multiple wavelet correlation approach. *Earth Interactions*. In Press.

PFISTER, C., 1992: Monthly temperature and precipitation in central Europe from 1525–1979: Quantifying documentary evidence on weather and its effects. *In* Bradley, R. S. and Jones, P. D. (eds.), *Climate since* A.D. *1500*. London: Routledge, 118–142.

PHILANDER, S. G. H., 1990: *El Niño, La Niña, and the Southern Oscillation*. New York: Academic Press.

PREISENDORFER, R. W., 1988: *Principal Component Analysis in Meteorology and Oceanography*. Amsterdam: Elsevier.

QUINN, W. H. and NEAL, V. T., 1992: The historical record of El Niño events. *In* Bradley, R. S., and Jones, P. D. (eds.), *Climate since* A.D. *1500*. London: Routledge, 623–648.

QUINN, T. M., TAYLOR, F. W., and CROWLEY, T. J., 1993: A 173 year stable isotope record from a tropical South Pacific coral. *Quaternay Science Reviews*, **12**, 407–418.

QUINN, T. M., TAYLOR, F. W., CROWLEY, T. J., and LINK, S. M., 1996: Evaluation of sampling resolution in coral stable isotope records: A case study using monthly stable isotope records from New Caledonia and Tarawa. *Paleoceanography*, **11**, 529–542.

RAJAGOPALAN, B., LALL, U., and CANE, M. A., 1997: Anomalous ENSO occurrences: An alternative view. *Journal of Climate*, **10**, 2351–2357.

RAJAGOPALAN, B., MANN, M. E., and LALL, U., 1998: A multivariate frequency-domain approach to long-lead climatic forecasting. *Weather and Forecasting*, **13**, 58–74.

SCHLESINGER, M. E. and RAMANKUTTY, N., 1994: An oscillation in the global climate system of period 65–70 years. *Nature*, **367**, 723–726.

SCHWEINGRUBER, F. H., BRIFFA, K. R., and JONES, P. D., 1991: Yearly maps of summer temperatures in western Europe from A.D. 1750 to 1975 and western North America from 1600 to 1982. *Vegetatio*, **92**, 5–71.

SMITH, T. M., REYNOLDS, R. W., LIVEZEY, R. E., and STOKES, D. C., 1996: Reconstruction of historical sea surface temperatures using empirical orthogonal functions. *Journal of Climate*, **9**, 1403–1420.

STAHLE, D. W. and CLEAVELAND, M. K., 1993: Southern Oscillation extremes reconstructed from tree rings of the Sierra Madre Occidental and southern Great Plains. *Journal of Climate*, **6**, 129–140.

STAHLE, D. W., CLEAVELAND, M. K., and HEHER, J. G., 1988: North Carolina climate changes reconstructed from tree rings A.D. 372 to 1985. *Science*, **240**, 1517–1519.

STAHLE, D. W., D'ARRIGO, R. D., KRUSIC, P. J., CLEAVELAND, M. K., COOK, E. R., ALLAN, R. J., COLE, J. E., DUNBAR, R. B., THERELL, M. D., GAY,

D. A., MOORE, M. D., STOKERS, M.A., BURNS, B.T., and THOMPSON, L. G., 1998: Experimental dendroclimatic reconstruction of the Southern Oscillation. *Bulletin of the American Meteorological Society*, **79**, 2137–2152.

TARUSSOV, A., 1992: The Arctic and Svalbard to Severnaya Zemlya: Climatic reconstructions from ice cores. *In* Bradley, R. S. and Jones, P. D. (eds.), *Climate since A.D. 1500*. London: Routledge, 505–516.

THOMSON, D. J., 1982: Spectrum estimation and harmonic analysis. *IEEE Proc.,* **70**, 1055–1096.

THOMPSON, L. G., 1992: Ice core evidence from Peru and China. *In* Bradley, R. S. and Jones, P. D. (eds.), *Climate Since A.D. 1500*. London: Routledge, 517–548.

TOURRE, Y., RAJAGOPALAN, B., and KUSHNIR, Y., 1999: Dominant patterns of climate variability in the Atlantic over the last 136 years. *Journal of Climate*, **12**, 2285–2299.

TRENBERTH, K. E., 1990: Recent observed interdecadal climate changes in the Northern Hemisphere. *Bulletin of the American Meteorological Society*, **71**, 988–993.

TRENBERTH, K. E. and HOAR, T. J., 1996: The 1990–1995 El Niño–Southern Oscillation event: Longest on record. *Geophysical Research Letters*, **23**, 57–60.

TRENBERTH, K. E. and HOAR, T. J., 1997: El Niño and climate change. *Geophysical Research Letters*, **24**, 3057–3060.

TRENBERTH, K. E. and HURRELL, J. W., 1994: Decadal atmosphere–ocean variations in the Pacific. *Climate Dynamics*, **9**, 303–319.

TZIPERMAN, E., STONE, L., CANE, M. A., and JARSOH, H., 1994: El Niño chaos: Overlapping of resonances between the seasonal cycle and the Pacific ocean–atmosphere oscillator. *Science*, **264**, 72–74.

ZEBIAK, S. E. and CANE, M. A., 1987: A model El Niño–Southern Oscillation. *Monthly Weather Review*, **115**, 2262–2278.

ZEBIAK, S. E. and CANE, M. A., 1991: Natural climate variability in a coupled model. *In* Schlesinger, M. E. (ed.), *Greenhouse Gas–Induced Climatic Change: A Critical Appraisal of Simulations and Observations.* Amsterdam: Elsevier, 457–469.

11

Modulation of ENSO Variability on Decadal and Longer Timescales

RICHARD KLEEMAN* AND SCOTT B. POWER

Bureau of Meteorology Research Centre,
GPO Box 1289K, Melbourne 3000, AUSTRALIA

Abstract

The mechanisms for decadal and longer timescale variability in the El Niño/Southern Oscillation (ENSO) phenomenon are examined. Decadal variability may be caused by factors internal to the tropical climate system, such as synoptic variability, or by those external to it, such as the so-called North Pacific gyre mode. Model evidence for these mechanisms is carefully considered.

Longer term variability of ENSO caused by such factors as global warming is also examined. Modeling evidence to date suggests little future change in ENSO due to this change in forcing. The limitations of models making such null projections are critically analyzed.

Introduction

The past fifteen years has seen significant advances in our understanding of the interannual aspects of El Niño/Southern Oscillation (ENSO). This has been achieved mainly as a result of the immensely successful Tropical Ocean–Global Atmosphere (TOGA) program. These advances have been accompanied by great progress in recent years in the development of realistic coupled ocean–atmosphere models, which are fundamental to increasing our understanding of the ENSO-coupled climate phenomenon.

While the dominant timescale usually associated with ENSO is interannual (roughly 2 to 7 years), it is quite clear from observational studies that the spectral peak is broad

* Present affiliation: Courant Institute for Mathematical Sciences, New York University, New York, NY 10012, U.S.A.

and has power at decadal and longer time scales. Although an understanding of this low-frequency signal is more limited, significant progress has been made in the past five years. This effort has been built upon the foundations of our knowledge regarding interannual timescales of climatic variability.

Understanding the longer term variability of ENSO is of great importance to advancing the depth of our knowledge of global climate fluctuations. This is because ENSO is such a large climate signal and has powerful global teleconnections. A case in point is illustrated in Graham (1995), who provides evidence that much of the total global warming observed since 1970 may be explained by warming in the tropical Pacific where ENSO has its center of action.

Our ability to increase understanding in this area is unfortunately limited by the current length of quality observations of the tropical Pacific of sufficient spatial density. A consequence of this limitation is that researchers have relied heavily on model experiments. Given that such models often have problems in simulating the detail of climate processes such as ENSO accurately, this situation is clearly less than desirable. Consequently, efforts currently under way to recover further historical data (such as the Global Oceanographic Data Archeology and Rescue [GODAR (1994)] project) are widely viewed as a high priority for the climate research community.

This chapter is separated into two main sections: (1) decadal variability of ENSO and (2) variability of ENSO on century timescales. It seems likely that the former kind of variability is mainly naturally induced, whereas the latter is very likely influenced by variations in anthropogenic forcing (principally greenhouse gases and industrial aerosols). This supposition underlies the material presented. In a third smaller section, we consider whether the recent unusual behavior of ENSO is a consequence of decadal or much longer timescale variability. Finally, we summarize the key points of this chapter in the last section.

Decadal Variability of ENSO

Decadal variability in ENSO, illustrated in Figure 11.1 (adapted from Kestin et al. 1998), is obtained by calculating the power spectrum of various measures of ENSO (Southern Oscillation Index [SOI], sea surface temperature [SST] in the NINO-3 region [NINO3], and central Pacific rainfall), which have time series typically of the order of a century. What is immediately apparent is that although the spectrum peaks at periods of about 4 years, it is quite broad and extends significantly to lower frequencies.

In the literature, two specific aspects of this decadal variability have received significant attention:

(1) *Change in the amplitude of the interannual signal.* The 1920–40 period was generally quiescent, while the more recent period 1970 to present has witnessed quite large excursions such as the 1975 cold event and the 1982–83 warm event. Further details on this may be found in Chapter 1 of this volume and in Trenberth and Hurrell (1994).

(2) *Direct decadal variation in the ENSO signal.* Such variability has been particularly apparent in the record since 1945 and takes the form of certain decades

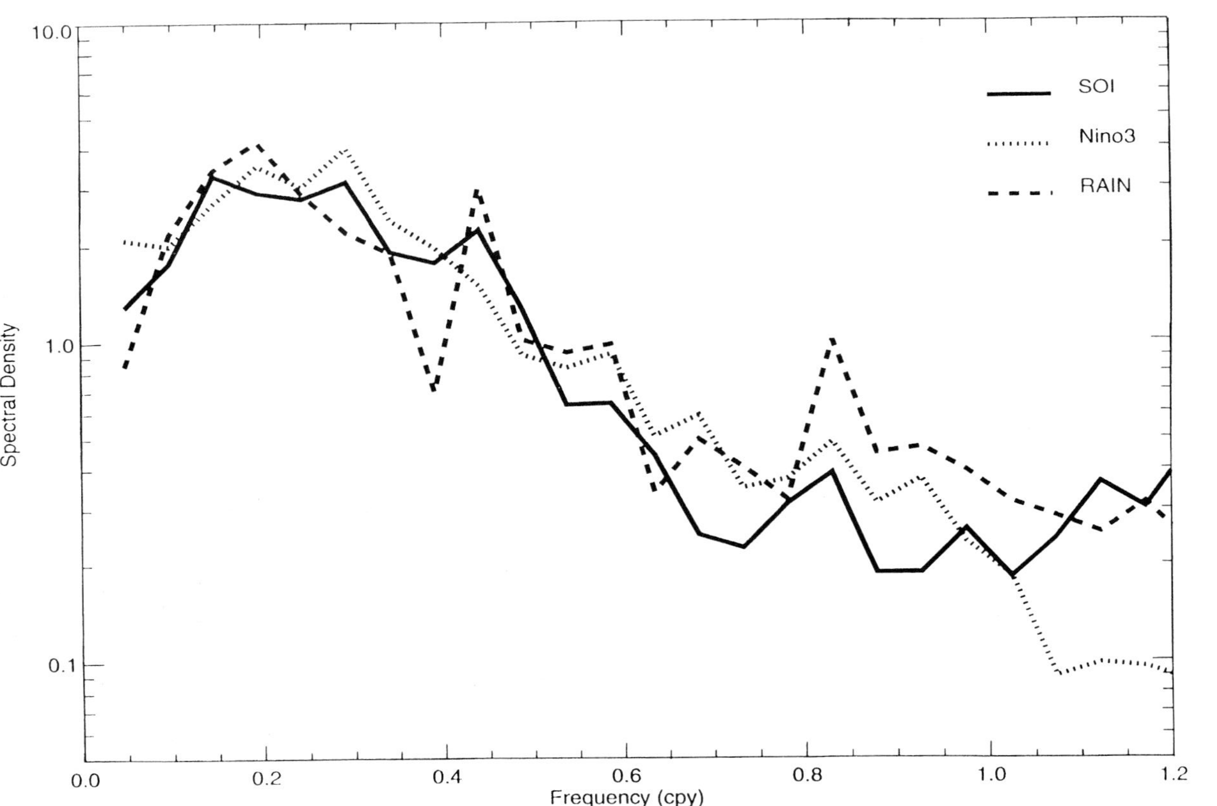

Fig. 11.1 Three spectral estimates for ENSO derived from century-long time series for the Southern Oscillation Index (SOI) (solid line), NINO3 (dotted line), and central equatorial Pacific rainfall (dashed line). Note that the trend has been removed. (Derived from Kestin et al. 1998.)

being mainly warm in the eastern tropical/subtropical Pacific (1960s, 1980s, and 1990s), while others are mainly cold (1950s and 1970s). In addition, the spatial pattern associated with this variation tends to be much broader meridionally and less associated with eastern equatorial fluctuations (see next section below).

In this section we shall review physical mechanisms that have been suggested to explain this variability. In discussing these mechanisms, we shall find it convenient to separate them into two separate conceptual categories: (1) those generated internally to the deep tropical Pacific by stochastic influences or by chaotic dynamics and (2) those generated externally to the deep tropics by such things as variations in the midlatitude gyres and thermohaline circulations. A central question here is how such external variability communicates its influence to the deep tropics. A third subsection deals with the possible practical implications of decadal variability for ENSO predictability.

Internal Influences

In the past fifteen years significant progress has been made in the simulation of ENSO (see National Research Council 1996); however, a question that has received less attention concerns the irregularity of the ENSO signal. In terms of Figure 11.1, much attention has understandably focused on the evident spectral peak but less on why it is so broad.

Three studies in the late 1980s and early 1990s were perhaps the first to address this issue. First, Schopf and Suarez (1988) conducted a sensitivity study with their intermediate[1] coupled model and showed that the irregular ENSO cycle depicted could be made perfectly regular by replacing the dynamically active atmospheric component with a steady-state Gill (1980) type atmosphere. Such a result suggested that the irregularity was due to the stochastic-like forcing of the coupled system by atmospheric transients. A similar result was also obtained by Battisti (1988), who found that his intermediate coupled model, which produced very regular ENSO-like oscillations, could be made to exhibit irregular behavior by the addition of external stochastic forcing. A little later, Zebiak and Cane (1991) conducted a very long integration of their intermediate coupled model and demonstrated that irregular behavior could be produced in a coupled model without atmospheric transients (the Zebiak atmospheric model produces a steady-state response to SST anomalies). This result suggested that chaotic influences in the low-frequency climate system may be significant in producing irregularity.

Recently, both possible causes of irregularity have received considerable attention in the literature. The stochastic scenario has been investigated by Kleeman and Power (1994), Blanke et al. (1997), Eckert and Latif (1997), and Kleeman and Moore (1997). In general, these studies concluded that stochastic input could produce realistic looking irregularity. The chaotic scenario has been investigated by Münich et al. (1991), Jin

[1] Such models include only the physical process thought essential for ENSO. This is in contrast to general circulation models, which are relatively complete in their physical treatment.

et al. (1994), and Tzipermann et al. (1994). These authors demonstrated that such behavior was possible in a number of different intermediate coupled models and investigated the possible dynamical causes (see next section below).

Kleeman and Moore (1999) have compared results from a stochastically forced coupled model with the observations and showed that the observed variation in the amplitude of ENSO can be well explained by such a model. Figure 11.2 depicts a time series of NINO3 from their model together with its spectrum. It is evident that there are periods of a decadal timescale when the ENSO oscillation is quite large (and regular) and also times when it is quite small. In addition, the spectrum depicted in Figure 11.2b shows strong similarities with the observed one (Fig. 11.1). A perhaps significant difference is that the observed power for NINO3 at periods longer than the interannual one is somewhat enhanced relative to the model results. This suggests that external influences may still be significant (see next subsection).

The chaotic mechanism has been examined in quite some depth by Jin et al. (1996), who provide evidence that the interaction between the model annual cycle and its interannual oscillation may be the main mechanism by which chaos is produced. Two pieces of evidence from their study suggest, however, that the stochastic mechanism may be the more likely scenario for ENSO irregularity. First, spectra from chaotic runs of their model show distinct (but broad) peaks at multiples of the annual period. Such behavior is not apparent in Figure 11.1. In addition, Jin et al. demonstrate that chaos tends to occupy quite small regions of parameter space for their intermediate model. Most of this space is filled by perfectly regular oscillations with periods being a multiple of the annual period. The chaos occurs on the boundary between regular oscillations of differing periods. It should be stressed that such conclusions remain preliminary, as they apply only to a particular coupled model. It is noteworthy, however, that most models with steady-state atmospheric components exhibit regular ENSO oscillations.

In summary, it appears quite plausible that the observed variation in the amplitude of ENSO on decadal timescales may be due to factors internal to the deep tropics, such as stochastic forcing by atmospheric transients or chaotic dynamics of the climate system.

External Influences

There is now reasonable evidence that modes of variability with longer timescales than that of ENSO have a significant influence in the midlatitude and tropical Pacific. This may be illustrated by a singular value decomposition (SVD) analysis of joint modes of SST and wind stress. Such an analysis finds modes of variability in both the atmosphere and ocean. The first three modes in terms of variance explained are depicted in Figure 11.3. The data sets used for this extend between 1949 and 1991 and are amongst the better such generally available records (the Parker et al. [1995] and da Silva et al. [1994] data sets). The associated time series (for SST) are depicted in Figure 11.4.

The first mode is clearly associated with the classical ENSO interannual signal, but interestingly, the time series shows a definite trend (which will be discussed in the next section). The second mode shows a trend superimposed on high-frequency variability and seems to correspond to some extent with a mode of long-term variability discussed

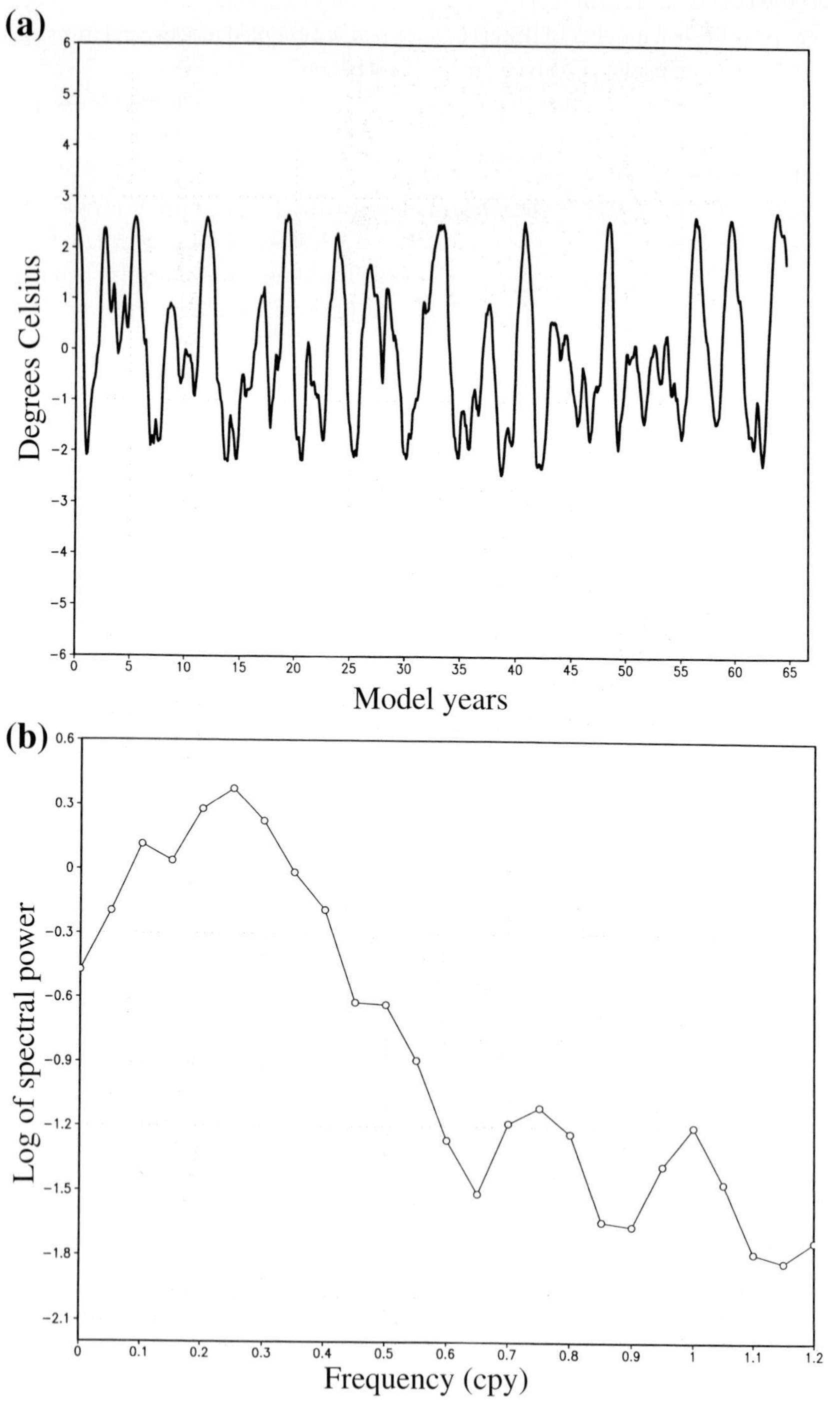

Fig. 11.2 (a) A typical time series of NINO3 derived from a stochastically forced ENSO in a coupled ocean–atmosphere model. (b) The spectrum derived from the time series of (a).

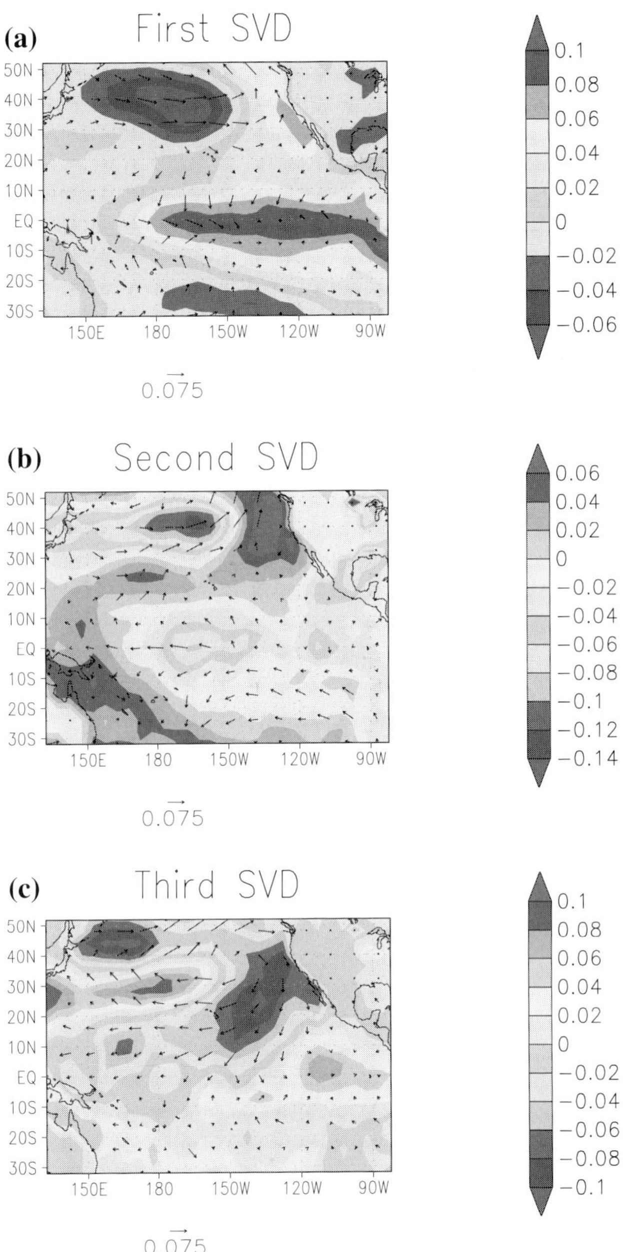

Fig. 11.3 (a) The spatial patterns of sea surface temperature (SST) and wind stress for the first singular value decomposition (SVD) mode of the Pacific (see text). (b) Same as (a), but for the second SVD mode. (c) Same as (a), but for the third SVD mode.

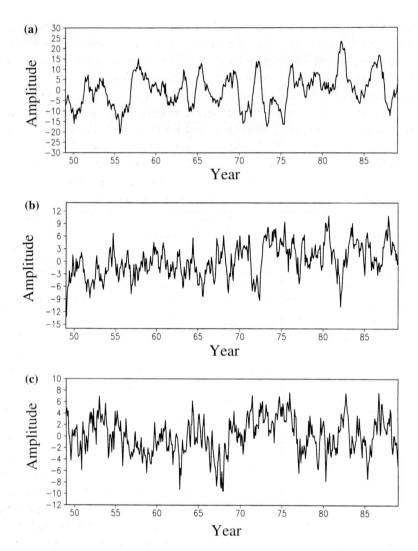

Fig. 11.4 The associated time series for (a) the first singular value decomposition (SVD) mode (the SST series is displayed), (b) the second SVD mode, and (c) the third SVD mode.

recently by Cane et al. (1997). This will be discussed further in the next section. The third mode has a clear decadal time series and is concentrated in the North Pacific. We shall focus on this variability and its possible ENSO implications in this section.

Modes of variability with a similar time and space scale to the third mode above have been noted in quite a number of observational studies (see, for example, Trenberth and Hurrell [1994], Mann and Park [1996], and Zhang and Levitus [1997]). This mode has sometimes been referred to as the North Pacific gyre mode.

A coupled ocean–atmosphere mode of variability with similar space-time structure has been noted and partially analyzed by Latif and Barnett (1994) and Latif and Barnett

(1996) within a global coupled general circulation model (GCM). These authors suggest a reasonably complex mechanism for their mode, but principally it involves the spin-up and spin-down of the North Pacific gyre. The resemblance between this model result and the observations has given rise to a number of interesting simplified coupled model results and analyses that have explored the coupled feedback processes involved in the creation of such modes within models. Using an atmospheric GCM (AGCM), Latif and Barnett (1994) claim that the midlatitude atmosphere can react quite strongly to midlatitude SST anomalies, producing significant wind stress anomalies which then feed back to the ocean and produce a significant positive loop; this process appears crucial to their coupled mode. Some dispute remains about the strength of this atmospheric feedback (other atmospheric models are unable to produce as strong a response); however, the SVD analysis discussed previously suggests that something similar may occur in the real world. This is because the observed wind stress pattern associated with the observed SST pattern shows significant similarities with the pattern obtained from the atmospheric GCM forced with an SST pattern similar to that of the SVD.

Recently, Münich et al. (1998) used a particularly simple model to claim that this feedback loop is all that is required to produce the observed variability. The model uses the gyre adjustment time (determined by Rossby wave propagation) to set a decadal time-scale for model oscillations. Some differences between the spatial patterns obtained by these authors and the observations together with an highly idealized atmospheric model make their conclusions preliminary, however. Similar results have been obtained by Xu et al. (1998), who used a more realistic hybrid coupled model to show that this feedback loop tends to dominate a heat flux mechanism positive feedback originally proposed by Latif and Barnett (1994). This latter feedback is made possible because the wind speed variations implied by the wind stress patterns discussed above tend to be collocated with SST anomalies. Thus turbulent heat flux changes (particularly evaporation) tend to reinforce SST anomalies. Such an effect is illustrated in Figure 11.5, which shows the analogous third SVD mode obtained when wind speed rather than wind stress is used as the atmospheric variable. The similarity with decadal gyre mode in Figure 11.3c is obvious, and the correspondence between SST and wind speed is remarkable. While one interpretation of this is that the mode is simply the result of wind speed changes forcing SST, the fact that an AGCM forced by similar SST patterns produces similar wind speed patterns strongly supports the notion of a genuine positive feedback loop between ocean and atmosphere. These modeling and observational results suggest that although the heat flux feedback may not be central to the noted coupled modes, it would appear to have a considerable destabilizing influence. A further significant result obtained by Xu et al. (1997) was that the tropics are not required for the existence of their coupled mode. Such a result supports the idea that the gyre mode is physically distinct from the interannual ENSO mode.

The relationship between this decadal variability and ENSO remains unclear, although there is some evidence to suggest a connection. To demonstrate this, we first discuss recent work on the spatial patterns of ENSO decadal variability. Graham (1994)

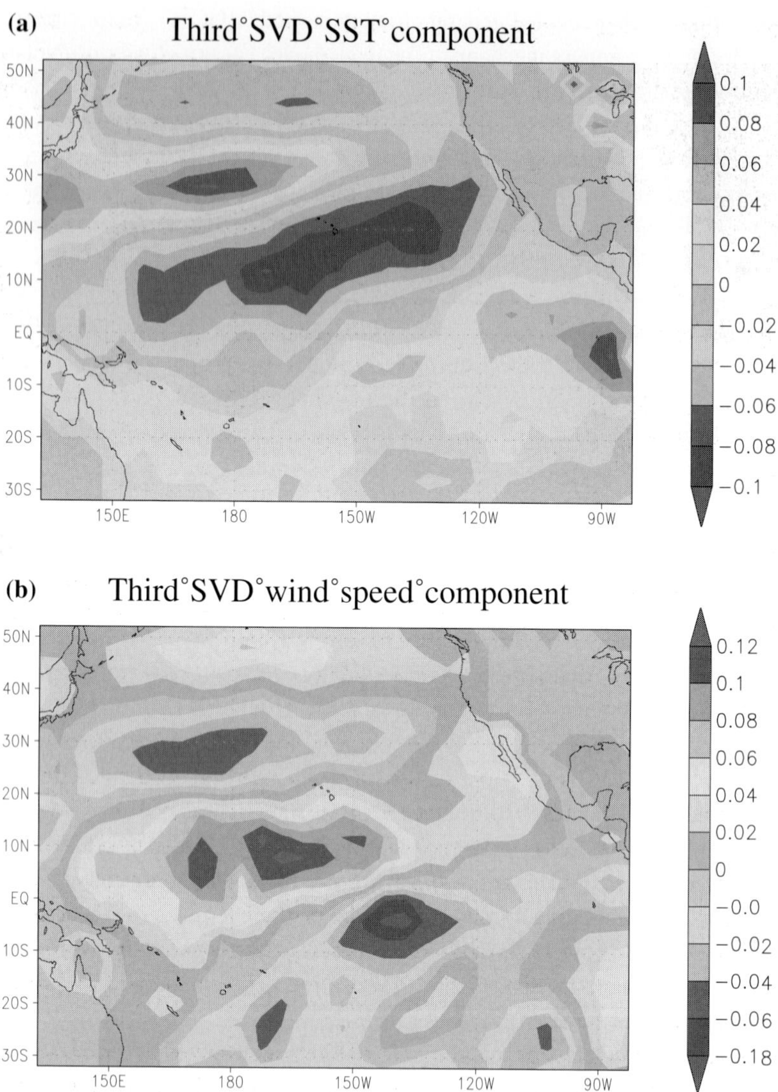

Fig. 11.5 The spatial patterns of (a) sea surface temperature (SST) and (b) wind speed derived from the third singular value decomposition (SVD) mode of SST and wind speed.

(see also Trenberth and Hurrell 1994) studied the apparent change in tropical Pacific climate variables that occurred around 1976. More recently, Kleeman et al. (1997) conducted a similar study for the early 1990s and showed that the anomalies of this recent five-year period were similar to those noted by Graham for the earlier period. This similarity suggested that these epoch-like changes were really part of a consistent longer time variability. The nature of this variability for SST can be clearly seen in the trend between 1970 and 1994, which is displayed in Figure 11.6. A characteristic

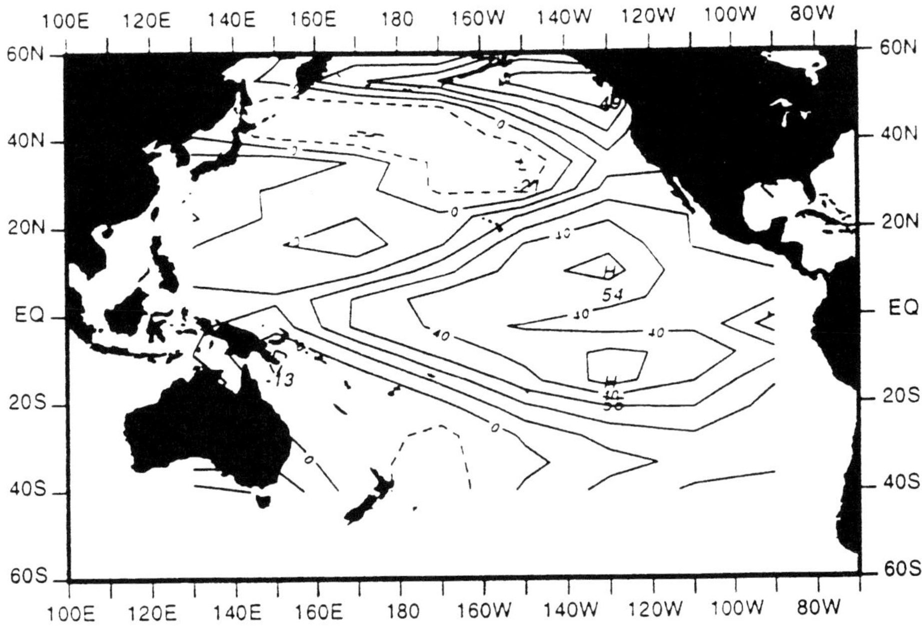

Fig. 11.6 The trend in sea surface temperature (SST) from 1970 to 1994. Units are degrees Celsius per 1,000 years.

"horseshoe" shape pattern of very broad meridional extent may be noted. In the subtropical eastern Pacific, this variability explains a large portion (50%) of the total SST variance in this region, and so it cannot be ignored in the context of ENSO. The pattern is likely to cause a significant atmospheric response, because large anomalies are also located in the sensitive high climatological SST western Pacific. In this context, it is worth noting that Graham (1995) has attributed a large amount of global warming since 1970 to warming in the tropical Pacific. This warming is posited to occur directly and via global teleconnections.

The longer term context of this recent variability has been analyzed by Latif et al. (1997), who argue that it is primarily decadal in nature. Such a hypothesis contrasts with that proposed by Trenberth and Hoar (1996), who contend that the anomalies of the 1990s were exceptional in the 150-year observational record. These issues are discussed further in the next subsection. Latif et al. (1997) also showed that the decadal mode of variability in SST had strong links with other climate variables of great social and economic consequence, such as northeast Australian rainfall. Such a relationship has of course been previously noted for interannual variability by, for example, McBride and Nicholls (1983) and Ropelewski and Halpert (1987). This link was apparently evident in the first half of the current decade, when a particularly severe drought occurred in Queensland.

Latif et al. conducted a Principal Oscillation Pattern (POP) analysis for the global region of 30°S to 60°N. The decadal mode of variability that was identified showed strong similarities in the tropical Pacific with Figure 11.6. Curiously, in the North Pacific the variability pattern resembles that of the third SVD mode discussed previously. In addition, the time series associated with the modes are also very similar (cf. Fig. 11.4c) and are very similar to the decadal aspect of northeast Australian rainfall. Note also the correspondence with the decadal preponderance of warm and cold conditions discussed at the beginning of this section. These correspondences suggest an interaction between the gyre mode and ENSO, but this remains a very open area of research.

In this context, considerable recent basic oceanographic research has taken place into how the midlatitudes and subtropics may communicate with the equatorial Pacific. McCreary and Lu (1994) and Liu et al. (1994) have demonstrated using modeling studies that a large fraction of the water associated with the equatorial undercurrent originates in the subduction zones of the subtropics (at roughly 25°N and 25°S). All of this water mass reaches the surface in the eastern equatorial Pacific and consequently affects SST there. The water is returned to higher latitudes mainly by Ekman drift in the upper layers, and the total circulation is referred to as the subtropical cell (STC). Such a circulation also has been inferred indirectly from observations by a number of authors, including Fine et al. (1981, 1983) and Deser et al. (1996). Clearly, variations in either the strength of this cell or in the temperature of the subducted water can influence ENSO. There are, therefore, quite plausible oceanographic mechanisms by which midlatitude decadal variability may interact with ENSO.

A further possibility recently discussed by Gu and Philander (1997) is that a delayed negative feedback of equatorial SST anomalies can occur via ENSO atmospheric teleconnections. This feedback causes SST anomalies of the opposite sign to the equatorial anomalies in the midlatitudes. This water is then hypothesized to enter the STC via subduction, and the long time scale associated with water mass transport in this cell causes a decadal oscillation to develop. The mechanism is thus very similar conceptually to that proposed for ENSO itself (see Suarez and Schopf 1988); that is, it is a delayed action oscillator.

As should be clear from the discussion above, the nature of external interactions with ENSO remains an open and exciting area of research. There are, however, grounds for believing that these interactions will prove to be important.

Decadal Variation in ENSO Predictability

A practical implication of the decadal variability of ENSO discussed previously concerns variations in the predictability of ENSO. A number of studies (Balmaseda et al. 1995, Chen et al. 1995, and Ji et al. 1996) have clearly demonstrated a large decadal variation in the skill of dynamical ENSO predictions. Thus the 1970s and 1990s were considerably less predictable than the 1980s, particularly on timescales of 10 months or longer.

The reason for this variation is as yet unclear, but two possible hypotheses have been very recently advanced. Goddard and Graham (1997) suggest that the warming of the 1990s discussed in the previous subsection (cf. Fig. 11.6) has interfered with the presumed dominant mode of interannual variability (the delayed action oscillator discussed above) and that this interference has degraded forecast skill. Kleeman and Moore (1999) provide evidence that the amplitude of the dominant ENSO normal mode is a significant parameter in determining predictability (skill is high when this amplitude is high). They further show that this amplitude exhibits significant decadal variation, which is directly related to the skill variation discussed earlier. The authors also show that such variation can be simply due to the kind of stochastic irregularity of ENSO discussed in the first subsection previously.

Longer Term Variability in ENSO

Climate models show a marked global warming in response to increased carbon dioxide and other greenhouse gases, even if the offsetting effects of sulfate aerosols are taken into account. But what do they tell us about what might happen to ENSO and about the way ENSO impacts upon regions removed from the tropical Pacific?

To help answer these and related questions, we will summarize the changes evident in coupled general circulation models (CGCMs) in which carbon dioxide levels are elevated. We will see that most studies suggest that changes to ENSO will be modest or even nonexistent and that remote associations to ENSO variability might change, but this too may be a small effect. In the next section we will discuss four areas of concern that lower confidence in these scenarios: (1) the widely held view that ENSO dynamics are distorted in the CGCMs used to date; (2) the apparent contrast between the model results and what has actually happened in the tropical Pacific this century; (3) the potential importance of natural multidecadal variability in signal detection; and (4) the use of flux adjustments to help prevent climate drift.

A brief summary of the rapid progress being made both within institutions and through new international programs aimed at improving models, and thereby increasing confidence in model scenarios, is presented. The unusual climatic conditions that occurred during the early 1990s as well as the possibility of an interaction between ENSO and the oceans' thermohaline circulation will also be discussed.

What Do Climate Models Tell Us about What Might Happen to ENSO under Greenhouse Warming?

Perhaps the best known of all ENSO models (Zebiak and Cane 1987) depicts ENSO as an internal oscillation of the tropical Pacific atmosphere–ocean system. It is an anomaly model in the sense that the variability occurs about specified mean background conditions. Zebiak and Cane (1991) mimicked global warming in this model by modifying the background conditions through the use of highly idealized representations of the global warming evident in climate models. They found that the nature of the resulting variability changed substantially. In one case, warm events were larger in magnitude, more regular, and lasted longer, while in another case the variability was reduced and was far less regular than in the control integration. These experiments were of an exploratory nature primarily because the imposed changes were only crude approximations to those apparent in the climate models of that time but also because the CZ (Cane–Zebiak) model itself represents a simplification of the tropical atmosphere–ocean system. This insightful truncation of the physics was done with the objective of producing a model in which the interannual variability could be examined. In the wider context of global warming, however, some of the truncated physics can be expected to play a significant role that might not necessarily be adequately represented by straightforward changes to the background state. Presumably, there are parameters within the model that need to be tuned to obtain variability that matches those observed, and there is no guarantee that the parameter settings used for today's climate will be the most appropriate for a warmer climate.

Nevertheless, the study highlighted the possibility that ENSO might change significantly due to global warming and the need for further investigation of this fascinating and important problem. Since that time there have been numerous studies using more sophisticated CGCMs. Meehl et al. (1993) used a CGCM in conjunction with an AGCM coupled to a mixed-layer ocean model and an AGCM forced by using prescribed SST anomalies. They addressed a number of basic but very important questions, showing that the ENSO-like variability evident in the model persisted, and the associated climatic fluctuations evident in the tropics and subtropics were qualitatively similar when carbon dioxide levels were doubled. Neither Meehl et al. (1993) nor Tett (1995), using a different CGCM, was able to detect any significant change in the magnitude of NINO3. Tokioka et al. (1995) used a model with much greater resolution in the tropical Pacific, concluding that "no definite indication is found that ENSO is affected by the CO_2-induced warming." More recently, Knutson et al. (1997) examined the impact of a CO_2 quadrupling on ENSO, finding that the modeled interannual SST variability is slightly reduced in the warm climate compared to the control. The length of this particular integration (1,000 years) allowed them to show that global warming did not appear to detune ENSO (i.e., did not shift its preferred frequency band away from that evident in the control climate).

So far, we have only discussed what might happen to ENSO itself. What about the changes associated with ENSO, both in the tropics and elsewhere? Even if the dynamics and statistical properties of ENSO remain unchanged, the fact that the "mean climate" has changed may modify how these fluctuations in the tropical Pacific are communicated to the rest of the world.

Two studies have addressed this important question. In the first, Meehl et al. (1993) found that there was an intensification of associated events in the tropics and subtropics. For example, areas that were dry during a warm event in the control climate became even drier during a warm event in the integration with increased carbon dioxide. In the second study (Smith et al. 1997), the intensification was restricted to the Australian–Indonesian region.

On the basis of these results, we might conclude that ENSO will persist and if the magnitude of the SST variability is changed at all it will be slightly reduced, with no significant change in its preferred frequency. It is unclear if events associated with ENSO will be intensified outside the western Pacific.

How Confident Are We in These Results?

Distortion of Equatorial Wave Properties

To begin to answer this question it is worth considering theoretical work that helps to underpin our understanding of ENSO dynamics. A paradigm often proposed to explain ENSO variability is the so-called delay action oscillator (Suarez and Schopf 1988), or DAO, mentioned earlier. In this view, warm and cold events are brought to an end upon the arrival of internal ocean Kelvin waves from the western Pacific. These waves are thought to occur after a westward propagating (ocean) Rossby wave strikes the western boundary. This mechanism has been shown to operate in the Zebiak and Cane (1987) model as well as in other simplified coupled systems (Schopf and Suarez 1988) and is thought to underlie the predictive skill that these models exhibit on seasonal timescales (see Kleeman 1993 and Latif and Graham 1992).

To model this kind of ENSO, you would therefore need to be able to support both Kelvin and Rossby waves in the ocean component of the CGCM. How much resolution is actually required? To answer this, we need to consider some of the background theory for the equatorial ocean "waveguide" (e.g., Ng and Hsieh 1994; Gill 1982; Philander 1990). All of the CGCMs described above (apart from the Japanese model) use variants of the NOAA Geophysical Fluid Dynamics Laboratory (GFDL) ocean model code (Bryan 1969), in which a finite difference approximation to the governing equations in the horizontal (and vertical) is employed. Finite difference models distort equatorial waves for two main reasons: First, there may be insufficient meridional resolution to do justice to their structure, and second, numerical constraints require that a larger than desirable horizontal viscosity coefficient (and diffusion coefficient to a lesser extent) be used.

The impact of frictional dissipation in the zonal momentum equation can be approximated by

$$\begin{aligned} o(dU/dt)_{\text{fric}} &\approx U/\tau_{\text{fric}} \\ &\approx o(\kappa^* d^2U/dy^2) \\ &\approx \kappa^* U/(L_y)^2, \end{aligned}$$

where L_y is a typical meridional scale, τ_{fric} is the frictional timescale, and κ is the horizontal viscosity of the model. This then gives an equation for the frictional spin-down time:

$$\tau_{\text{fric}} = (L_y)^2/\kappa.$$

The smaller this parameter is, the less time it takes for waves to be damped out of the system.

If we suppose that the Kelvin wave has a meridional scale of 200–600 km, then we can estimate the frictional spin-down time for viscosities typically used in models in order to examine ENSO in the climate change context. Coarse resolution models typically use a horizontal viscosity of about 5×10^5 m^2 s^{-1}, which gives times ranging from about 1 to 10 days. It is therefore not surprising that most of the coarse resolution models do not appear to exhibit a DAO, because the typical basin crossing timescale for Kelvin modes is of the order of a month or longer. Thus in models with such viscosities one would expect to see Kelvin-like waves that were very rapidly damped and thus unable to sustain a DAO. Instead, most of the variability in such models may be ascribed to the so-called SST-mode (Neelin 1991), in which internal ocean waves are not required. There are exceptions to this construct. For example, Tett (1995) provided evidence suggesting that Kelvin waves were playing a role, but no evidence was presented that any significant role was played by Rossby waves. This makes sense because such waves have scales that are much smaller than the scale of a Kelvin wave, and so one might therefore expect frictional decay to be even more severe. If we suppose that the first internal Rossby mode has a meridional scale given by 100–300 km, then the spin-down time (for the same viscosity) is one quarter of the values given for the Kelvin mode. Also, the gravest Rossby mode propagates at one-third the speed of the Kelvin wave, so this exacerbates the decline in energy further. This argument therefore implies that very little Rossby wave energy will remain after the wave traverses any reasonable fraction of the Pacific basin. In summary, unless τ_{fric} for Kelvin and low-order Rossby modes is greater than the order of the basin transit time for such waves, the existence of a DAO in such models is problematic.

Evidence apparently contradictory to this was given by Knutson et al. (1997), who claim that some sort of DAO, complete with eastern propagation of thermal heat content centered on the equator and with westward propagation off the equator (consistent with the expected structures of Kelvin and Rossby waves), explains part of the ENSO variability seen in their model. It will be interesting to see how the westward propagation is achieved in this coarse model, as it is unclear how Rossby waves at least can play a significant role in the DAO evident. It may be that their off-equatorial response is purely a forced phenomenon and not wavelike at all as is required for the DAO.

Nevertheless, the apparent existence of DAO-like variability in this model (and perhaps even in the model employed by Moore [1995] after the coupling strength between the wind stress and SST was artificially strengthened by a factor of 4) and the general consistency with results obtained in which the interannual variability has a different mechanism help to increase confidence that the results obtained thus far may indeed provide reasonable guidance as to what might occur in future studies using improved ocean components. Certainly this remains a point of concern, but many groups have recently increased the resolution of their ocean components, which has allowed them to reduce viscosities and diffusivities, and so we can look forward to potentially more reliable assessments in the near future. An example is the model of Meehl and Washington (1996), in which the much smaller viscosity of 4×10^3 m^2 s^{-1} is used. The corresponding τ_{fric} for a Kelvin wave in this model is 4–35 months, which is vastly superior to the coarse-resolution models discussed above.

Sampling Error

A second area of concern arises from the prominent interdecadal variability that appears in very long integrations of simple coupled models (Zebiak and Cane 1991) and in

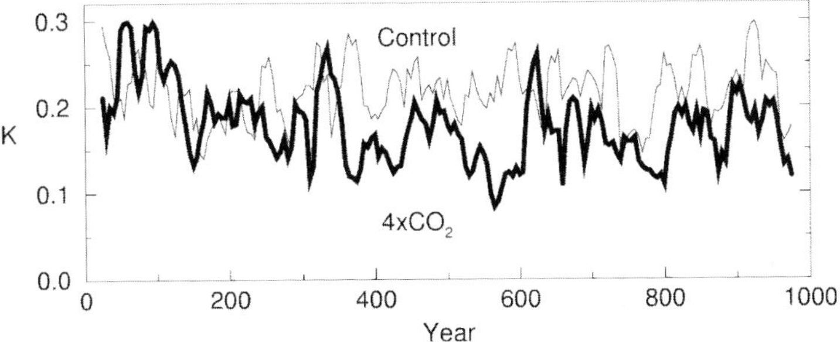

Fig. 11.7 Time series of the standard deviation of 2- to 7-year filtered SST (7°N–7°S; 173°E–120°E) obtained by using a moving window of width 20 years. The light line corresponds to the control experiment with unchanged atmospheric CO_2 concentration; the dark line corresponds to an experiment with four times the atmospheric concentration of CO_2. (Redrawn from Knutson et al. 1997.)

coarse CGCMs (see Fig. 11.7, adapted from Knutson et al. [1997]; see also Tokioka et al. [1995]). As was pointed out by Knutson et al. (1997), interdecadal variability is of great concern in the current context, first, because it means that even a century-long integration of a model perturbed with increased CO_2 may not be long enough to get a clear indication of the long-term response; conversely, it would mean that 100 years of observations may not be long enough to get a good estimate of the global warming signal. These conclusions are based on the assumption that the impact of global warming will differ depending on the phase or state of the multidecadal, naturally occurring variability.

Knutson et al. (1997) were more interested in the very long term response of ENSO after the model had been given time to reach a new quasi-equilibrium. This approach contrasts with most other studies, which have focused more on what might occur over the coming century, as CO_2 levels increase, before stabilization (i.e., they have been interested in the transient rather than the equilibrium response). If the phase of the multidecadal variability in the climate system is determined to some extent by the initial conditions in these transient experiments, then only an ensemble of a few integrations will be required to determine the change in ENSO due to global warming over the coming century. There is some evidence that this scenario may be the case, with Tokioka et al. (1995) showing that the phase of their dominant mode of decadal variability in a century-long integration was largely unaffected by an imposed CO_2 increase. The point to note is that for a 100-year period, not all phases of the interdecadal variability may be possible, and a smaller ensemble than might be expected on the basis of the integration described by Knutson et al. (1997) may be enough to determine the ENSO response.

If the naturally occurring variability in ENSO is actually unpredictable beyond a few decades or less, then many simulations will be required to ensure that all the phases of the multidecadal variability are properly sampled. This would be necessary if the ENSO response to warming depends on the phase of the longer term variability. This possibility helps to boost the value of international projects that combine results from several models, since this collection can be viewed as providing an ensemble base from which the actual signal might be detected above this background "noise."

Apparent Discrepancies between Observations and Model Responses

The third area of concern arises from a comparison of CGCM responses to what has actually occurred in the tropical Pacific this century. Virtually all CGCMs to date exhibit a reduction in the east–west SST gradient across the Pacific (let's call it dT/dx) in response to global warming (see, e.g., Fig. 11.8, which is adapted from Meehl and Washington [1996]). This result is in stark contrast to what appears to have actually happened in the observational record since the Second World War (Bottomley et al. 1990; Latif et al. 1997), which shows a modest increase in the same quantity. Indeed, some evidence has been presented (Kaplan et al. 1998; see also Cane et al. 1997) that the signal that remains after the ENSO signal is removed actually shows a cooling in the eastern tropical Pacific during the twentieth century. Of course, this is subject to a good deal of uncertainty, as there is very little data coverage for these regions during the early part of this century. This lack of data can be very clearly seen in Figure 11.9 (adapted from Bottomley et al. 1990).

Two main reasons can be given for this apparent discrepancy. First, as is indicated above, 50 years may not be long enough to obtain the "greenhouse fingerprint" in the tropical Pacific because of inherent interdecadal variability. In fact, Knutson et al. (1997) show that although there is a long-term reduction in dT/dx, it takes approximately 100 years for this reduction to become evident in their single greenhouse simulation. The second possibility, favored by researchers affiliated with Lamont-Doherty in the United States (Cane et al. 1997; Seager and Murtugudde 1997; Clement et al. 1996; Kaplan et al. 1998), is that the CGCMs incorporating coarse-resolution OGCMs are wrong in this respect. This school of thought emphasizes that the observed modest increase in dT/dx is entirely consistent with the response evident in simplified coupled systems used in highly idealized scenarios of global warming. In these models the response in the central eastern Pacific is dominated by a positive feedback mechanism that operates between the zonal wind stress and dT/dx. In the central eastern Pacific, oceanic upwelling of cold water displaces the surface mixed layer water warmed by the increase in surface heating, so that its response is lower than in the west where upwelling does not occur. This upwelling boosts dT/dx, driving an atmospheric response that includes stronger trades, which drive increased upwelling. As a result, the model's response to uniform warming is an increase in dT/dx with the temperature in the east actually reduced!

Seager and Murtugudde (1997) examined the response of a different simplified "coupled" model and again found that dT/dx increased even if allowance was made for water subducted in the subtropics to warm up before being upwelled along the equator. The reason for this was that the surface heat flux formulation employed damped sea surface temperature anomalies in the subtropics more rapidly than in the tropics, and so the near-equilibrium temperature of the subducted water was substantially smaller than the waters that they displaced in the tropics. It is interesting to note that on

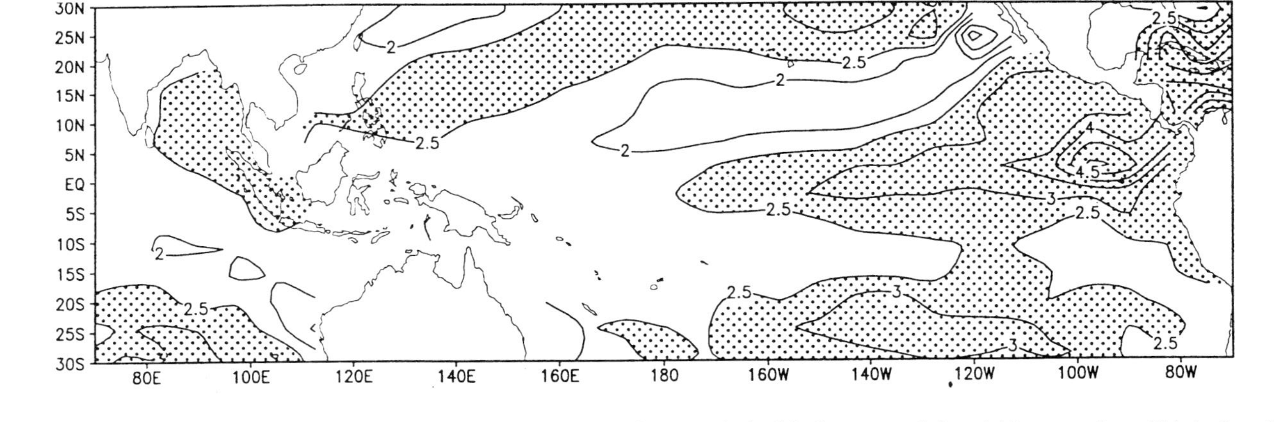

Fig. 11.8 The change in tropical Pacific SSTs in response to a transient increase in atmospheric CO_2 from a coupled model incorporating a 1° latitude and longitude resolution ocean mode. Contour interval is 0.5°C. (From Meehl and Washington 1996.)

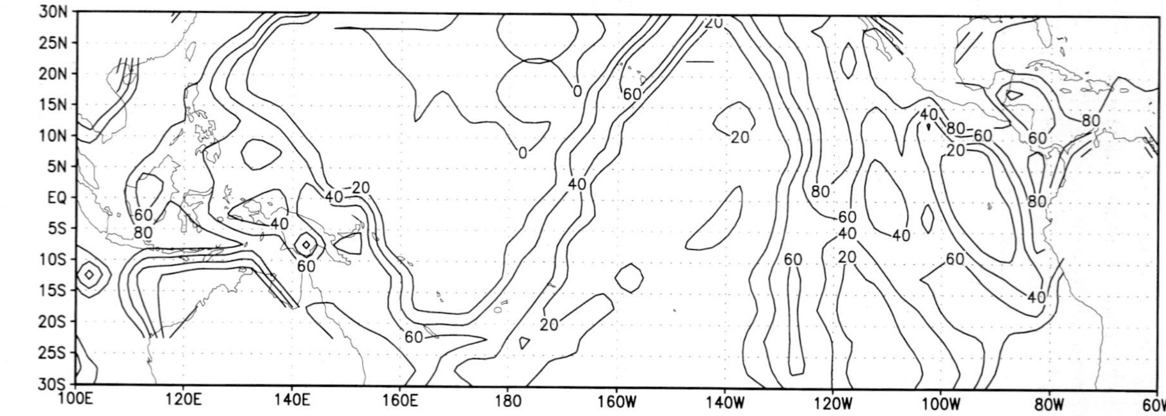

Fig. 11.9 Percentage of seasons from 1911 to 1920 with SST data (on a $5^\circ \times 5^\circ$ spatial scale). Data set is the comprehensive GOSTA. (From Bottomley et al. 1990.)

50-year timescales a latitude-dependent surface heat flux is not required to enhance dT/dx (Power and Hirst 1997). In this study a spatially uniform heating anomaly of 2 W/m^2 was added at the ocean surface for a period of 50 years to a coarse-resolution version of the GFDL modular ocean model; it too showed an increased dT/dx. This increase was offset by a dynamical feedback in the subtropical meridional cell in which the upwelling rate is reduced and the cooling effect dampened. Furthermore, a coupled model incorporating a very similar ocean model with precisely the same horizontal resolution (Power et al. 1995) was used in a transient greenhouse experiment (Colman et al. 1995), and dT/dx was reduced. The contrasting difference in behavior probably lies in wind stress changes in the tropics or in the fact that a uniform heating rate is too simplistic, and not in meridional contrasts in the heating rate.

Cane et al. (1997) and Seager and Murtugudde (1997) argued that the positive feedback mechanism between wind stress and dT/dx discussed above is weaker than it should be in the CGCMs primarily because of the coarse resolution of their oceanic components. The existence of this feedback mechanism in reality seems very plausible. In fact, we saw in the previous section that an analysis of the joint variability exhibited by the SST and wind speed in the observations since the 1940s reveals just such a mode. What is less clear is if this mode is not adequately represented in the CGCM with coarse OGCMs. It would be very surprising if the mechanism were not actually evident, given that all that is required is equatorial upwelling (which coarse-grid models have) and a sensitivity of the Walker Circulation to dT/dx. An alternative explanation is that the mechanism is modeled but that it is swamped by other effects, chief of which may be that the surface heating is not uniform as is assumed in these studies. The reconstruction by Cane et al. (1997) of the non-ENSO trend in SST during the past century may indeed represent a realistic mode of variability, but it may be overestimated in their reconstruction.

The models may be correct, but the greenhouse fingerprint in dT/dx may not yet be clearly evident. This suggestion is supported by the fact that in the period since 1945 virtually all of the Northern Hemisphere oceans have actually cooled (Latif et al. 1997; Wallace et al. 1996). But if one then looks at what has happened over a longer period, one sees that there has been a warming over the North Pacific (Bottomley et al. 1990; Nicholls et al. 1996). In other words, the 1945–1992 trend is not representative of the trend over a longer period.

Ongoing Advancement

The emphasis so far in this section has been on peer-reviewed work up until early 1997. It is important to realize that a great deal of effort is being spent at many institutions aimed at improving the performance of CGCMs. A number of new international initiatives have also arisen with the aim of improving the performance of CGCMs (the Coupled Model Intercomparison Project [CMIP] and ENSIP [see acronym list at the end of this chapter] for example), which add to existing programs that will continue to aid CGCM development (e.g., AMIP, FANGIO, WOCE, CLIVAR, IPCC; see list of acronyms). Indeed, it is probably fair to say that CGCMs and their various subcomponents have never come under closer scrutiny, and there has never been as much effort specifically focused on their improvement than at the present time. Some of this development specifically addresses the concerns mentioned here – ocean resolution and realism of the modeled ENSO. The need for longer integrations and for pooling results from several models is now widely recognized (e.g., IPCC 1990, 1995; Whetton et al. 1996; Knutson et al. 1997).

Another area of concern has been the use of flux adjustment strategies (e.g., Manabe and Stouffer 1988; Sausen et al. 1988) aimed at reducing spurious climate drift, which primarily arises from systematic biases within the model components and perhaps in the formulas used to couple them. It is reassuring to note that models have improved to the point where century-long integrations exhibit only small surface climate drift without any form of adjustment (Bryan 1998).[2] This trend will undoubtedly continue. Consequently, our ability to answer "what if" questions about ENSO and global warming will continue to improve at a rapid rate.

Can the Thermohaline Circulation Affect ENSO?

Whether the thermohaline circulation affects ENSO is a very interesting question that has not received much attention primarily because models have, until recently, been developed to look at ENSO specifically or to look at global circulation issues, but not both. Furthermore, the time required to equilibrate the ocean's thermohaline circulation in an OGCM takes on the order of 1,000 (model) years, so unless carefully designed, the experiments would be very costly.

It is, nevertheless, not unreasonable to suppose that the thermohaline circulation can affect ENSO (e.g., Gray and Sheaffer 1997). This circulation, after all, helps to determine the temperature and salinity structure of the global ocean, thereby contributing to the background state that ENSO must operate in. It is known, for example, that the rate at which North Atlantic Deep Water (NADW) is formed – an important element in the thermohaline circulation – can affect the vertical temperature structure in the tropical Pacific of global OGCMs (Washington et al. 1994; see also Semtner and Chervin 1992). What is less clear is how large this influence actually is. Observational estimates of radioactive tracer distributions in the ocean suggest that equatorial upwelling penetrates far too deeply in coarse OGCMs (Toggweiler and Samuels 1993). The model used by Washington et al. (1994) has a higher resolution than those discussed in Toggweiler and Samuels (1993), but they consider changes in the rate of production of about 10 Sv, which is much larger than is normally associated with naturally occurring variability in the thermohaline circulation (e.g., Delworth et al. 1993). Even if changes of this magnitude occur (as they might under extreme CO_2 scenarios; Manabe and Stouffer 1994), or in different climatic epochs (e.g., Manabe and Stouffer 1988), it is unclear whether they will have a significant impact on ENSO.

It is also interesting to note that the perturbations considered by Washington et al. (1994) communicated rather rapidly (<10 years) to the equatorial Pacific. This result

[2] Eastern equatorial Pacific SST drifts by only 1.9 degrees, while western values drift by less than 1 degree. Thus a reasonable east–west gradient is maintained.

suggests that the modulation of ENSO via oceanic teleconnections to fluctuations in NADW formation should not necessarily proceed on much longer advective timescales.

Even if no deep upwelling occurs in the equatorial Pacific, it is widely believed that some NADW enters the equatorial Pacific (e.g., Tomczak and Godfrey 1994; Schmitz 1995) and thereby helps to partially determine the vertical temperature profile. Furthermore, we know that fluctuations in the production of NADW are associated with significant alterations to the atmospheric conditions over the Atlantic (Rind et al. 1986; Manabe and Stouffer 1988; Power et al. 1995) and over the Southern Ocean (e.g., Marotzke and Willebrand 1991; Power and Kleeman 1993). So if these changes produce significant remote teleconnections in the equatorial Pacific, the thermohaline circulation might affect ENSO indirectly via the atmosphere. At present there does not seem to be a study that has specifically focused on this topic, but it would be surprising if the same could be said in a few years.

ENSO in the 1990s

The first half of the current decade witnessed a particularly prolonged period in which the SOI was negative (1991–95). The duration of this event under some definitions was unprecedented in the historical record. These atmospheric conditions were accompanied by persistent positive SST anomalies in large areas of the off-equatorial eastern and equatorial central-western Pacific (see the section "Decadal Variability in ENSO"). A persistent El Niño under its classical definition was, however, not present: Far-eastern Pacific equatorial temperature anomalies fluctuated during the period between positive *and* negative values. There were, however, during this period three very short-lived El Niño events of weak to moderate amplitude (see Goddard and Graham 1997).

There is considerable controversy in the literature regarding the nature of the 1990s event(s). Trenberth and Hoar (1996) constructed an autoregressive moving average (ARMA) statistical model of Darwin pressure based on about 120 years of data. They then showed that events like the one in the 1990s occurred once every 1,000 years or so in the model simulation. On the basis of this result they ruled out decadal variability as the causative mechanism. Latif et al. (1997) conducted a POP analysis of SST data since 1949 (the most reliable long-term period) and obtained a decadal mode with a spatial structure very similar to the 1990s pattern of SST anomalies. This mode is distinct from the normal interannual El Niño mode, which showed up in their analysis as the dominant POP. These authors also calculated the trend in SST for the period 1949–91 and found some similarity with the 1990s pattern, at least in the eastern Pacific. They showed, however, that in terms of explained variance the decadal mode is more important in this region. The conclusion of this study therefore is that decadal variability played a large role in the 1990s event (larger in fact than the dominant El Niño mode). A point to be borne in mind here is (as was pointed out in the previous section) that 40 years may be too short a period to adequately define the trend.

Resolution of these apparently contradictory viewpoints awaits further research. Perhaps part of the answer may be that the signal of global warming resembles that of decadal variability and ENSO, as has been suggested by Meehl and Washington (1996) (and partially supported by Latif et al. 1997). Over a 150-year period, therefore, the extent of eastern Pacific warming may be comparable with the decadal swing that has recently been observed.

Summary and Outlook

Decadal variability of ENSO has been explained in the literature by two major mechanisms: (1) internal equatorial Pacific influence such as stochastic forcing by atmospheric transients or chaotic climate dynamics and (2) external midlatitude variability forcing. A potentially important example of this is decadal variations in the North Pacific Ocean gyre.

Longer term variability of ENSO due to changes in anthropogenic forcing has been addressed in a number of GCM studies. These studies find little change in either the magnitude or spectrum of ENSO. There remain significant caveats to this null result, including model deficiencies, sampling problems due to short model runs, and the fact that there may be discrepancies between the observed trend in the tropical Pacific and model projections.

In conclusion, it should be stressed that the field reviewed in this chapter is currently the subject of intensive international research efforts. An example of these efforts is the recently formulated long-term Climate Variability and Predictability Program (CLIVAR), which has as one of its major foci the subject matter discussed in this chapter. In view of this, we expect that significant new results in this area will be available in the next five to ten years.

Glossary of Acronyms

AGCM: atmospheric general circulation model. A reasonably complete physical model of the global atmosphere.

AMIP: Atmospheric Modeling Intercomparison Project. A collection of complex atmospheric models from most major climate research laboratories.

ARMA: autoregressive moving average. A stochastic statistical model usually used to fit observational data.

CGCM: coupled general circulation model. A comprehensive model of the global climate systems involving at least the ocean and atmosphere and often sea ice.

CLIVAR: Climate Variability and Predictability Program. An international umbrella organization aimed at fostering research into the global climate system.

CMIP: Coupled Model Intercomparison Project. Similar to AMIP but involving coupled models.

DAO: delayed action oscillator. A model oscillator involving a direct positive feedback and a delayed negative feedback. Often used as a paradigm for ENSO.

ENSIP: Similar to CMIP but aimed at examining the depiction of ENSO in coupled models.

FANGIO: Feedback Analysis of GCMs for Intercomparison with Observations. A series of reasonably simple feedback experiments conducted on a large number of AGCMs in order to elucidate physical processes within the climate system.

GODAR: Global Oceanographic Data Archeology and Rescue Project. A research project aimed at recovering marine data from national archives such as those of the U.S. Navy.

IPCC: International Panel on Climate Change. A large collection of international climate experts who aim to provide a balanced assessment of the science of global climate change.

NADW: North Atlantic Deep Water. An ocean water mass formed in the North Atlantic by deep convection processes.

OGCM: ocean general circulation model. Same as AGCM but for the oceans.

POP: principal oscillation pattern. A statistical technique used to extract the dynamical or oscillating patterns of spatial variability from observational data.

SVD: singular value decomposition. A statistical technique used in the present context to extract joint patterns of variability from the observational data.

TOGA: Tropical Ocean–Global Atmosphere program. A large-scale modeling and observation program carried out between 1985 and 1994 that aimed to gain a better physical understanding of ENSO.

WOCE: World Ocean Circulation Experiment. An extensive global oceanographic observational program.

Acknowledgments The authors wish to thank Tahl Kestin, The Cooperative Research Centre for Southern Hemisphere Meteorology in Australia, for producing Figure 11.1, which is derived from Kestin et al. (1998), and Briony Horton at the UK Meteorological Office for the data used in Figure 11.9.

References

BALMASEDA, M. A., DAVEY, M. K., and ANDERSON, D. L. T., 1995: Decadal and seasonal dependence of ENSO prediction skill. *Journal of Climate*, **8**, 2705–2715.

BATTISTI, D. S., 1988: Dynamics and thermodynamics of a warming event in a coupled tropical atmosphere ocean model. *Journal of Atmospheric Science*, **45**, 2889–2929.

BLANKE, B., NEELIN, J. D., and GUTZLER, D., 1997: Estimating the effect of stochastic windstress forcing on ENSO irregularity. *Journal of Climate*, **10**, 1473–1486.

BOTTOMLEY, M., FOLLAND, C. K., HSUING, J., NEWELL, R. E., and PARKER, D. E., 1990: *Global Ocean Surface Temperature Atlas "GOSTA."* A Joint Project of the Meteorological Office and MIT.

BRYAN, F., 1998: Climate drift in a multi-century integration of the NCAR Climate System Model. *Journal of Climate*, **11**, 1455–1471.

BRYAN, K., 1969: Climate and the ocean circulation. III: The ocean model. *Monthly Weather Review*, **97**, 806–827.

CANE, M. A., CLEMENT, A. C., KAPLAN, A., KUSHNIR, Y., POZNYAKOV, D., SEAGER, R., ZEBIAK, S. E., and MURTUGUDDE, R., 1997: Twentieth-century sea surface temperature trends. *Science*, **275**, 957–960.

CHEN, D., ZEBIAK, S. E., BUSALACCHI, A. J., and CANE, M. A., 1995: An improved procedure for El Niño forecasting. *Science*, **269**, 1699–1702.

CLEMENT, A. C., SEAGER, R., CANE, M. A., and ZEBIAK, S. E., 1996: An ocean dynamical thermostat. *Journal of Climate*, **9**, 2190–2196.

COLMAN, R. A., POWER, S. B., McAVANEY, B. J., and DAHNI, R. R., 1995: A non-flux corrected transient CO_2 experiment using the BMRC coupled atmosphere/ocean GCM. *Geophysical Research Letters*, **22**, 3047–3050.

COX, M. D., 1984: A primitive equation, three-dimensional model of the ocean. *GFDL Ocean Group Technical Report No.1*, GFDL, Princeton, NJ: Princeton University.

da SILVA, A. M., YOUNG, C. C., and LEVITUS, S., 1994: *Atlas of Surface Marine Data 1994*, Volumes 1 and 3. NOAA Atlas NESDIS 6 and 8. Available from NOCD, NOAA/NESDIS E/OC21, Washington, DC 20235.

DELWORTH, T., MANABE, S., and STOUFFER, R. J., 1993: Interdecadal variations of the thermohaline circulation in a coupled ocean–atmosphere model. *Journal of Climate*, **6**, 1993–2011.

DESER, C., ALEXANDER, M. A., and TIMLIN, M. S., 1996: Upper ocean thermal variations in the North Pacific during 1970–1991. *Journal of Climate*, **9**, 1840–1855.

ECKERT, C. and LATIF, M., 1997: Predictability of a stochastically forced hybrid coupled model of El Niño. *Journal of Climate*, **10**, 1488–1504.

FINE, R. A., PETERSON, W. H., ROOTH, C. G. H., and OSTLUND, H. G., 1983: Cross-equatorial tracer transport in the upper waters of the Pacific Ocean. *Journal of Geophysical Research*, **88**, 763–769.

FINE, R. A., REID, J. L., and OSTLUND, H. G., 1981: Circulation of tritium in the Pacific Ocean. *Journal of Physical Oceanography*, **11**, 3–14.

GILL, A. E., 1980: Some simple solutions for heat-induced tropical circulation. *Quarterly Journal of the Royal Meteorological Society*, **106**, 447–462.

GILL, A. E., 1982: *Atmosphere–Ocean Dynamics*. Orlando: Academic Press, 662 pp.

GODDARD, L. and GRAHAM, N. E., 1997: El Niño in the 1990s. *Journal of Geophysical Research*, **102**, 10423–10436.

GRAHAM, N. E., 1994: Decadal-scale climate variability in the tropical Pacific during the 1970s and 1980s: Observations and model results. *Climate Dynamics*, **10**, 135–162.

GRAHAM, N. E., 1995: Simulation of recent global temperature trends. *Science*, **267**, 666–671.

GRAY, W. M. and SHEAFFER, J. D., 1997: The Atlantic Ocean thermohaline circulation as a flywheel for multi-decadal global and regional climate changes. *In Proceeding of the 7th American Meteorological Society Conference on Climate Variations*, Long Beach, CA.

GU, D. and PHILANDER, S. G. H., 1997: Interdecadal climate fluctuations that depend on exchanges between the tropics and extratropics. *Science*, **275**, 805–807.

I. P. C. C., 1990: *Climate Change: The IPCC Scientific Assessment.* Houghton, J. T., Jenkins, G. J., and Ephraums, J. J., (eds.), Cambridge: Cambridge University Press.

I. P. C. C., 1995: *Climate Change 1995: The Science of Climate Change.* Houghton, J. T., Meira Filho, L. G., Callander, B. A., Harris, N., Kattenberg, A., and Maskell, K. (eds.). Cambridge: Cambridge University Press, 572 pp.

JI, M., LEETMAA, A., and KOUSKY, V. E., 1996: Coupled model forecasts of ENSO during the 1980s and 1990s at the National Meteorological Center. *Journal of Climate*, **9**, 3105–3120.

JIN, F. F., NEELIN, J. D., and GHIL, M., 1994: El Niño on the Devil's staircase: Annual subharmonic steps to chaos. *Science*, **264**, 70–72.

JIN, F. F., NEELIN, J. D., and GHIL, M., 1996: El Niño/Southern Oscillation and the annual cycle: Subharmonic frequency locking and aperiodicity. *Physica D*, **98**, 442–465.

KAPLAN, A., CANE, M. A., KUSHNIR, Y., CLEMENT, A. C., BLUMENTHAL, M. B., and RAJAGOPALAN, B., 1998: Analyses of global sea surface temperature 1856–1991. *Journal of Geophysical Research*, **103**, 18567–18589.

KESTIN, T. S., KAROLY, D. J., YANO, J. I., and RAYNER, N. A., 1998: Time frequency variability of ENSO and stochastic simulations. *Journal of Climate*, **11**, 2258–2272.

KLEEMAN, R., 1993: On the dependence of hindcast skill on ocean thermodynamics in a coupled ocean–atmosphere model. *Journal of Climate*, **6**, 2012–2033.

KLEEMAN, R., COLMAN, R., SMITH, N. R., and POWER, S. B., 1997: A recent change in the mean state of the Pacific Ocean: Observational evidence and atmospheric and oceanic responses. *Journal of Geophysical Research (Oceans)*, **101**, 20483–20499.

KLEEMAN, R. and MOORE, A. M., 1997: A theory for the limitation of ENSO predictability due to stochastic atmospheric transients. *Journal of Atmospheric Science*, **54**, 753–767.

KLEEMAN, R. and MOORE, A. M., 1999: A new method of determining the reliability of dynamical ENSO predictions. *Monthly Weather Review*, **127**, 694–705.

KLEEMAN, R. and POWER, S. B., 1994: Limits to predictability in a coupled ocean–atmosphere model due to atmospheric noise. *Tellus*, **46A**, 529–540.

KNUTSON, T. R., MANABE, S., and GU, D., 1997: Simulated ENSO in a global coupled ocean–atmosphere model: Multidecadal amplitude modulation and CO_2 sensitivity. *Journal of Climate*, **10**, 131–161.

LATIF, M. and BARNETT, T. P., 1994: Causes of decadal climate variability over the North Pacific and North America. *Science*, **266**, 634–637.

LATIF, M. and BARNETT, T. P., 1996: Decadal variability over the North Pacific and North America: Dynamics and predictability. *Journal of Climate*, **9**, 2407–2423.

LATIF, M. and GRAHAM, N. E., 1992: How much predictive skill is contained in the thermal structure of an OGCM? *Journal of Physical Oceanography*, **22**, 951–962.

LATIF, M., KLEEMAN, R., and ECKERT, C., 1997: Greenhouse warming, decadal variability or El Niño? An attempt to understand the anomalous 1990s. *Journal of Climate*, **10**, 2221–2239.

LAVERY, B. M., KARIKO, A. P., and NICHOLLS, N., 1992: A high-quality historical rainfall dataset for Australia. *Australian Meteorological Magazine*, **40**, 33–39.

LEVITUS, S., GELFELD, R., BOYER, T. P., and JOHNSON, D., 1994: *Results of the NODC and IOC Oceanographic Data Archeology and Rescue Project: Report I*, NODC, Silver Springs, MD, U.S.A., 73 pp.

LIU, Z., PHILANDER, S. G. H., and PACANOWSKI, R. C., 1994: A GCM study of the tropical-subtropical upper ocean water exchange. *Journal of Physical Oceanography*, **24**, 2606–2623.

MANABE, S. and STOUFFER, R. J., 1988: Two stable equilibria of a coupled atmosphere–ocean model. *Journal of Climate*, **1**, 841–866.

MANABE, S. and STOUFFER, R. J., 1994: Multiple century response of a coupled ocean–atmosphere model to an increase of atmospheric carbon dioxide. *Journal of Climate*, **7**, 5–23.

MANN, M. E. and PARK, J., 1996: Joint spatiotemporal modes of surface temperature and sea level pressure variability in the Northern Hemisphere during the last century. *Journal of Climate*, **9**, 2137–2162.

MAROTZKE, J. and WILLEBRAND, J., 1991: Multiple equilibria of the global thermohaline circulation. *Journal of Physical Oceanography*, **21**, 1372–1385.

McBRIDE, J. L. and NICHOLLS, N. N., 1983: Seasonal relationships between Australian rainfall and the Southern Oscillation. *Monthly Weather Review*, **111**, 1998–2004.

McCREARY, J. P. and LU, P., 1994: Interaction between the subtropical and equatorial ocean circulations: The subtropical cell. *Journal of Physical Oceanography*, **24**, 466–497.

MEEHL, G. A. and WASHINGTON, W. M., 1996: El Niño–like climate change in a model with increased atmospheric CO_2 concentrations. *Nature*, **382**, 56–60.

MEEHL, G. A., BRANSTATOR, G. W., and WASHINGTON, W. M., 1993: Tropical Pacific interannual variability and CO_2 climate change. *Journal of Climate*, **6**, 42–63.

MOORE, A. M., 1995: Tropical interannual variability in a global coupled GCM: Sensitivity to mean climate state. *Journal of Climate*, **8**, 807–828.

MÜNICH, M., CANE, M. A., and ZEBIAK, S. E., 1991: A study of self excited oscillations in a tropical ocean–atmosphere system. Part II: Nonlinear cases. *Journal of Atmospheric Science*, **48**, 1238–1248.

MÜNICH, M., LATIF, M., VENZKE, S., and MAIER-REIMER, E., 1998: Decadal oscillations in a simple coupled model. *Journal of Climate*, **11**, 3309–3319.

NATIONAL RESEARCH COUNCIL, 1996: *Learning to Predict Climate Variability Associated with El Niño and the Southern Oscillation*. Washington, DC: National Academy Press, 171 pp.

NEELIN, J. D., 1991: The slow sea surface temperature mode and the fast-wave limit: Analytic theory for tropical interannual oscillations and experiments in a hybrid coupled model. *Journal of Atmospheric Science*, **48**, 584–606.

NG, M. K. F. and HSIEH, W. W., 1994: The equatorial Kelvin wave in finite difference models. *Journal of Geophysical Research*, **99**, 14173–14186.

NICHOLLS, N., GRUZA, G. V., JOUZEL, J., KARL, T. R., OGALLO, L. A., and PARKER, D. E., 1996: Observed climate variability and change. *In* Houghton, J. T., Meira Filho, L. G., Callander, B. A., Harris, N., Kattenberg, A., and Maskell, K. (eds.), *Climate Change 1995: The Science of Climate Change 1995*. Cambridge: Cambridge University Press, 572 pp.

PARKER, D. E., FOLLAND, C. K., BEVAN, A., WARD, M. N., JACKSON, M., and MASKELL, K., 1995: Marine surface data for analysis of climatic fluctuations on interannual to century time scales. *In* Martinson, D. G., Bryan, K., Ghil, M., Hall, M. M., Karl, T. R., Sarachik, E. S., Sorooshian, S., and Talley, L. D. (eds.), *Natural Climate Variability on Decade-to-Century Time Scales*. Washington DC: National Academic Press, 241–252.

PHILANDER, S. G., 1990: *El Niño, La Niña, and the Southern Oscillation*. San Diego: Academic Press, 293 pp.

POWER, S. B. and HIRST, A. C., 1997: Eddy parameterization and the oceanic response to idealized global warming. *Climate Dynamics*, **13**, 417–428.

POWER, S. B. and KLEEMAN, R., 1993: Multiple equilibria in a global OGCM. *Journal of Physical Oceanography*, **23**, 1670–1681.

POWER, S. B., KLEEMAN, R., COLMAN, R. A., and McAVANEY, B. J., 1995: Modeling the surface heat flux response to long-lived SST anomalies in the North Atlantic. *Journal of Climate*, **8**, 2161–2180.

RIND, D., PETEET, D., BROECKER, W., McINTYRE, A., and RUDDIMAN, W., 1986: The impact of cold North Atlantic sea surface temperatures on climate: Implications for the Younger Dryas cooling (11–10k). *Climate Dynamics*, **1**, 3–33.

ROPELEWSKI, R. W. and HALPERT, M. S., 1987: Global and regional scale precipitation patterns associated with the El Niño/Southern Oscillation. *Monthly Weather Review*, **115**, 1606–1626.

SAUSEN, R., BARTHELS, R. K., and HASSELMANN, K., 1988: Coupled ocean–atmosphere models with flux correction. *Climate Dynamics*, **2**, 154–163.

SCHMITZ, W. J., Jr., 1995: On the interbasin-scale thermohaline circulation. *Reviews of Geophysics*, **33**, 151–174.

SCHOPF, P. S. and SUAREZ, M. J., 1988: Vacillations in a coupled ocean–atmosphere model. *Journal of Atmospheric Science*, **45**, 549–566.

SEAGER, R. and MURTUGUDDE, R., 1997: Ocean dynamics, thermocline adjustment, and regulation of tropical SST. *Journal of Climate*, **10**, 521–534.

SEMTNER, A. J., Jr., and CHERVIN, R. M., 1992: Ocean general circulation from a global eddy-resolving model. *Journal of Geophysical Research*, **97**, 5493–5550.

SMITH, I. N., DIX, M., and ALLAN, R. J., 1997: The effect of greenhouse SSTs on ENSO simulations with an AGCM. *Journal of Climate*, **10**, 342–352.

SUAREZ, M. J. and SCHOPF, P. S., 1988: A delayed action oscillator for ENSO. *Journal of Atmospheric Science*, **45**, 3283–3287.

TETT, S., 1995: Simulations of El Niño–Southern Oscillation–like variability in a global AOGCM and its response to CO_2 increase. *Journal of Climate*, **8**, 1473–1502.

TOGGWEILER, J. R. and SAMUELS, B., 1993: New radiocarbon constraints on the upwelling of abyssal water to the ocean's surface. *In* Heimann, M. (ed.), *The Global Carbon Cycle, NATO ASI Series, Volume 115*. Berlin: Springer-Verlag.

TOKIOKA, T., NODA, A., KITOJ, A., NIKAIDOU, A., NAKAGAWA, S., MOTOI, T., and YUKIMOTO, S., 1995: A transient CO_2 experiment with the MRI CGCM. *Journal of the Meteorological Society of Japan*, **73**, 817–826.

TOMCZAK, M. and GODFREY, J. S., 1994: *Regional Oceanography: An Introduction*. New York: Pergamon Press, 422 pp.

TRENBERTH, K. E. and HOAR, T. J., 1996: The 1990–1995 El Niño–Southern Oscillation event: Longest on record. *Geophysical Research Letters*, **23**, 57–60.

TRENBERTH, K. E. and HURRELL, J. W., 1994: Decadal atmosphere–ocean variations in the Pacific. *Climate Dynamics*, **9**, 303–319.

TZIPERMANN E., STONE, L., CANE, M. A., and JAROSH, H., 1994: El Niño chaos: Overlapping of resonances between the annual cycle and the Pacific ocean–atmosphere oscillator. *Science*, **264**, 72–74.

WALLACE, J. M., ZHANG, Y., and BAJUK, L., 1996: Interpretation of interdecadal trends in Northern Hemisphere surface air temperature. *Journal of Climate*, **9**, 249–259.

WASHINGTON, W. M., MEEHL, G. A., VERPLANK, L., and BETTGE, T. W., 1994: A world ocean model for greenhouse sensitivity studies: Resolution intercomparison and the role of diagnostic forcing. *Climate Dynamics*, **9**, 321–344.

WHETTON, P., ENGLAND, M., O'FARRELL, S., WATTERSON, I., and PITTOCK, A. B., 1996: Global comparison of the regional rainfall results of enhanced greenhouse coupled and mixed layer ocean experiments: Implications for climate change scenario development. *Climatic Change*, **33**, 497–519.

XU, W., BARNETT, T. P., and LATIF, M., 1998: Decadal variability in the North Pacific as simulated in a hybrid coupled model. *Journal of Climate*, **11**, 297–312.

ZEBIAK, S. E. and CANE, M. A., 1987: A model El Niño/Southern Oscillation. *Monthly Weather Review*, **115**, 2262–2278.

ZEBIAK, S. E. and CANE, M. A., 1991: Natural climate variability in a coupled model. *In* Schlesinger, M. E. (ed.), *Greenhouse Gas Induced Climatic Change: Critical Appraisal of Simulations and Observations*. Amsterdam: Elsevier, 457–470.

ZHANG, R. H. and LEVITUS, S., 1997: Structure and cycle of decadal variability of upper ocean temperature in the North Pacific. *Journal of Climate*, **10**, 710–727.

12

Global Climate Change and El Niño: A Theoretical Framework

DE-ZHENG SUN

Cooperative Institute for Research in Environmental Sciences, Climate Diagnostics Center, University of Colorado, Boulder, Colorado 80309, U.S.A.

Abstract

To better understand what drives El Niño, an analytical model of the coupled ocean–atmosphere system over the equatorial Pacific is constructed. The equatorial atmosphere is approximated as a linear feedback system whose surface winds are driven by sea surface temperature (SST) gradients and whose thermal effect is to restore the entire equatorial SST to its maximum value – the SST of the warm pool. The upper ocean is represented by a shallow-water model capped by a mixed layer with a constant depth. The zonal mean stratification of the thermocline is maintained by upwelling from the deep ocean. The model captures the oscillatory behavior of the present tropical Pacific climate – the El Niño Southern Oscillation (ENSO). The main features of the oscillation in the model agree well with the observed El Niño, including the period of the oscillation and the phase relationship between the variations of SST and the variations in the depth of the thermocline. Moreover, the model predicts that the climate of the eastern tropical Pacific has two regimes: One is warm and steady, and the other is cold and oscillating, consistent with the inference from geoarcheological data that El Niño did not exist during the early to mid-Holocene, when the global and regional climates were warmer than today. The transition from the steady climate to the oscillating climate takes place when the temperature contrast between the surface warm pool and the deep ocean exceeds a critical value. A stability analysis reveals that the zonal SST contrast and the accompanying wind-driven currents have to be sufficiently strong to become oscillatory, and that requires a sufficiently large difference between the SST of the warm pool and the temperature of the deep ocean. On the timescale of millennia, a sufficiently cold equatorial deep ocean implies sufficiently cold high latitudes, a condition that is met by the present climate, but possibly not by the climate of the early to mid-Holocene.

In the oscillating regime, the magnitude of El Niño is found to increase monotonically with increases in the difference between the SST of the warm pool and the temperature of the deep ocean. The increase in the magnitude of El Niño is accompanied by an increase in the zonal SST contrast in the equatorial region. The implication of these results for the response of El Niño to an increase in the greenhouse effect is discussed.

Introduction

The global ramifications of El Niño Southern Oscillation (ENSO) have been well recognized (Glantz et al. 1991; Rasmusson and Wallace 1983). What is much less studied is the dependence of this oscillatory behavior of the tropical Pacific climate on the basic parameters that characterize the global climate. What will happen to ENSO after significant global warming takes place? Was ENSO during the last ice age more energetic or weakened? These are outstanding questions in the study of climate change, and yet we have just began to study them. Pivotal to addressing these questions is a clear understanding of what drives ENSO. The purpose of this article is to put forward a simple hypothesis on what drives ENSO.

The existence of ENSO has been attributed to the dynamic coupling between the atmosphere and ocean in the equatorial Pacific region (Neelin et al. 1998; Jin 1996; Neelin 1991; Battisti 1988; Suarez and Schopf 1988; Zebiak and Cane 1987; Bjerknes 1969). The dynamic coupling refers to the positive feedback loop among surface wind stress, SST gradients, and oceanic upwelling. Over the tropical Pacific Ocean, the surface winds are driven by SST gradients (Lindzen and Nigam 1987). As a consequence, changes in SST gradients affect the strength of surface winds. Because upwelling is driven by the surface winds, changes in the strength of the surface winds affect the strength of the upwelling, which in return affects the SST gradients. The prevailing theory of El Niño, the delayed oscillator hypothesis (Battisti 1988; Suarez and Schopf 1988), states that the positive feedback loop among surface wind stress, SST gradients, and oceanic upwelling is responsible for the growth of an SST anomaly, while the phase lag between upper ocean heat content and surface winds provides the mechanism for the phase transition. The theory also links the period of the oscillation to the memory of the subsurface ocean.

The dynamic coupling between the atmosphere and ocean in the equatorial Pacific, however, may not be sufficient for the existence of El Niño. Recent geoarcheological and geomorphic evidence suggests that El Niño may not have been present prior to about 5,000 years ago, when the global and regional climates were warmer than today (Sandweiss et al. 1996, 1997; Keefer et al. 1998). The picture that emerges from this suggestion is that the eastern tropical Pacific may have two distinct regimes: One is warm and steady; the other is colder and oscillating. Moreover, the former regime is accompanied by a warmer global climate, and the latter is associated with a colder global climate. The early to mid-Holocene was only slightly warmer than today, and as usual the warming was most noticeable at higher latitudes (the polar region was about 2–3°C warmer than today [Klimanov 1982; Velichko 1984; Borzenkova and Zubakov 1984]). The suggested connection to the higher latitude temperature is particularly interesting because it assigns a potential role to the temperature of the equatorial deep ocean in determining the state of the surface ocean. The property of deep water of the tropical Pacific Ocean is connected to the surface condition in the high-latitude oceans through the conveyor belt–like thermohaline circulation (Broecker 1987). Paleotemperature records indicate that the deep ocean temperature of the tropical oceans responds sensitively to coolings in the high latitudes (Hoffert 1990; Berger 1981).

The question is then whether a significantly warmer equatorial deep ocean temperature can result in a warmer eastern Pacific that has no El Niño. The lack of a ready answer from existing paradigms of El Niño to this very specific question highlights a deficiency in the reigning paradigms, which is a clear exposition of what fundamentally drives El Niño. Though the existing paradigms of El Niño suggest that El Niño depends on the climatological state, which aspect of the climatological state serves as the main thermal forcing of El Niño is not clear. We will attempt to isolate the main thermal forcing that drives El Niño and examine the behavior of the tropical SST as a function of the strength of the thermal forcing. We will suggest that El Niño is a thermally forced oscillation: It is fundamentally driven by the contrast between the tropical maximum SST and the temperature of the equatorial deep ocean.

A Heuristic Model

We begin with some heuristic arguments on what determines the magnitude of El Niño. It has long been recognized that El Niño is associated with a zonal redistribution of warm water in the equatorial Pacific (Wyrtki 1985). To help fix this picture in mind, we may divide the surface equatorial Pacific into two boxes: the western Pacific box, with temperature T_w, and the eastern Pacific, with temperature T_E (Fig. 12.1a). Temperatures T_w and T_E may be regarded respectively as the characteristic climatological value of the warm pool and cold tongue. To the extent that the El Niño warming is due to an eastward displacement of the warm pool (Fig. 12.1b), we may equate the magnitude of El Niño, $\tilde{T}$, to the difference between T_w and T_E:

$$\tilde{T} \sim T_w - T_E. \tag{1}$$

The normal or climatological condition of the equatorial Pacific may be safely regarded as the average between the El Niño and the La Niña conditions. It follows that the climatological SST in the eastern Pacific may be equated to the average of the SST in the eastern Pacific during El Niño (T_E (Niño)) and La Niña (T_E(Niña)). During La Niña, the thermocline is almost exposed to the surface in the eastern Pacific. For our purpose here, we will make an approximation and assume that the thermocline does get exposed to the surface during La Niña (Fig. 12.1c). This gives T_E(Niña) $\sim T_{\text{sub}}$, where T_{sub} is the subsurface temperature. Further nothing that T_E (Niño) $\sim T_w$, we have

$$T_E \sim \frac{1}{2}(T_w + T_{\text{sub}}). \tag{2}$$

Combining Eq. (1) with Eq. (2), we have

$$\tilde{T} \sim \frac{1}{2}(T_w - T_{\text{sub}}). \tag{3}$$

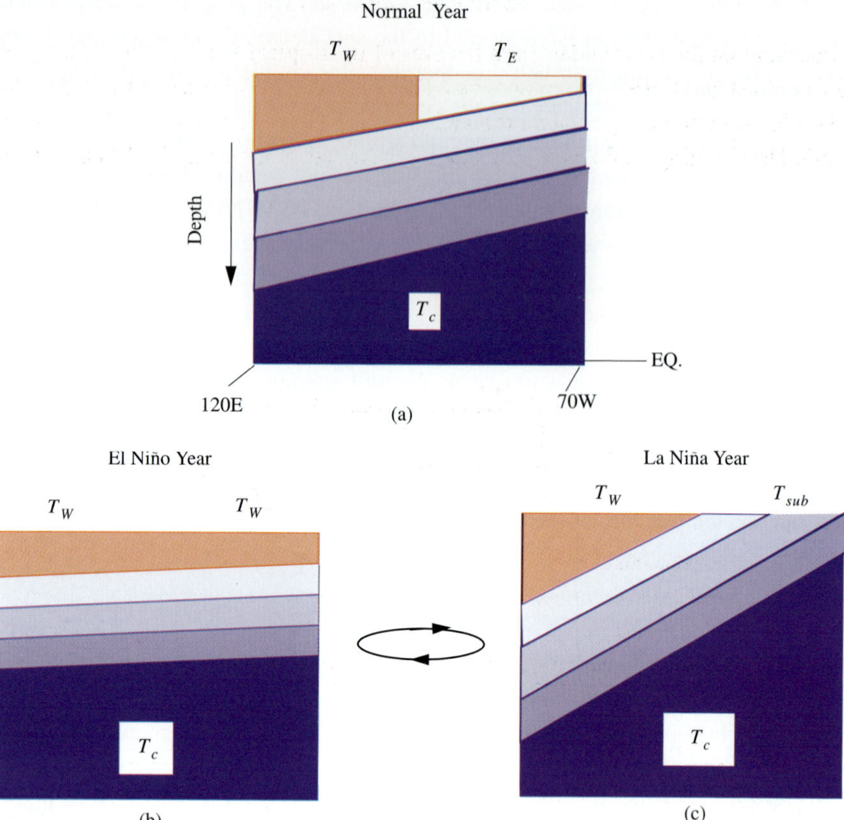

Fig. 12.1 A schematic illustration of the thermal structure of the equatorial Pacific during (a) a normal year, (b) an El Niño year, and (c) a La Niña year.

The value of T_{sub} depends on the east–west slope of the thermocline as well as the vertical stratification of the thermocline (Fig. 12.1c). The slope of the thermocline is balanced gravitationally by the zonal wind stress and thus is largely determined by the zonal SST contrast, or $T_w - T_{\text{sub}}$. Now we see that the value of T_{sub} depends on a factor that in turn depends on the value of T_{sub} – a nonlinearity that is responsible for the self-organization that we will introduce later. To the leading order, the vertical stratification is given by the difference between T_w and the deep ocean temperature T_c. This raises the role of temperature of the deep ocean T_c in determining the value of T_{sub} and thereby the magnitude of El Niño. The next section describes these heuristic arguments using differential equations so that the analysis can be carried out quantitatively. A quantitative analysis will allow us to see that the difference between T_w and T_c needs to be sufficiently large to enable the coupled equatorial ocean–atmosphere to find itself in an oscillatory state, a point that can be brought out only when the nonlinearity and the effects of thermal and mechanical damping are considered quantitatively.

A More Rigorous Model

Focusing on the two fundamental features of the tropical Pacific climate, the zonal SST contrast and ENSO, we may reduce the coupled tropical ocean–atmosphere to a low-order system through spatial truncation (Sun 1997a,b; Sun and Liu 1996; Neelin 1991). The resulting model is schematically illustrated in Figure 12.2. The upper ocean

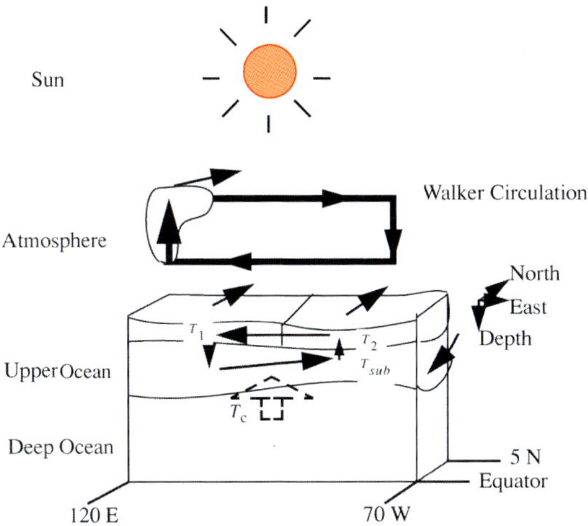

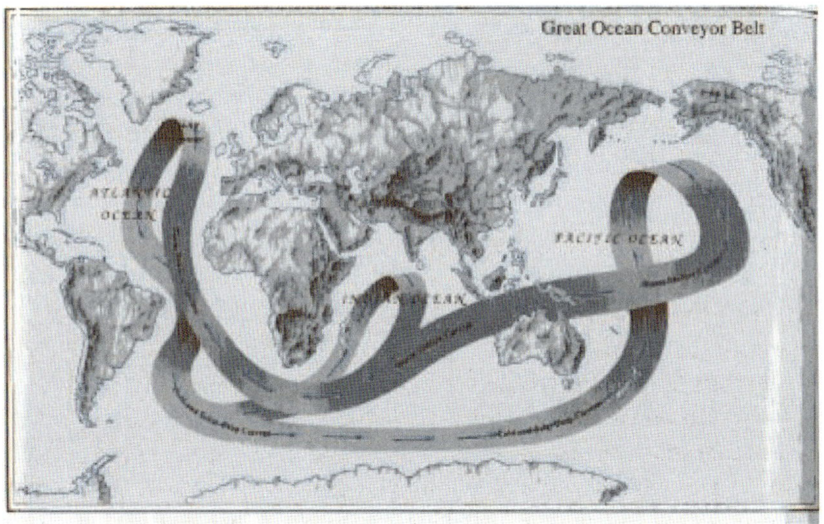

Fig. 12.2 A schematic diagram of the coupled model. Solid arrows represent the Walker circulation in the atmosphere and the equatorial wind-driven currents in the upper ocean. The latter are constituted by the upwelling in the east Pacific, the westward surface drift, the poleward surface drift, the returning equatorward subsurface flow, and the equatorial undercurrent. The large dashed arrow represents the background thermohaline circulation, which may be regarded as the Pacific branch of the global conveyor belt (shown at the bottom of the figure).

model is a box model version of the model of Cane (1979), a shallow-water model for the equatorial upper ocean embedding a mixed layer with a fixed depth. The zonal mean vertical stratification is maintained by a balance between surface heating from above and upwelling of cold water from the equatorial deep ocean. The equatorial deep ocean is connected to the high-latitude oceans through the global thermohaline circulation (Fig. 12.2b). Following Sun (1997a,b), the surface ocean is divided into two regions with equal areas: the western surface ocean (120°E–155°W) and the eastern surface ocean (155°W–70°W). These two regions are represented respectively by two boxes with temperatures T_1 and T_2. The atmosphere is approximated by a linear feedback system whose surface winds are driven by SST gradients and whose thermal effect is to force the entire equatorial SST toward its maximum value – the SST of the warm pool (Sun 1997b; Sun and Liu 1996; Neelin 1991).

The heat budget of the surface ocean may be formulated as follows:

$$\frac{dT_1}{dt} = \frac{Q_1 A_1}{C_p \rho V_1} + sq(T_2 - T_1), \tag{4}$$

$$\frac{dT_2}{dt} = \frac{Q_2 A_2}{C_p \rho V_1} + q(T_{\text{sub}} - T_2), \tag{5}$$

where $Q_1 A_1$ and $Q_2 A_2$ are respectively the net heat flux into the western surface ocean and into the eastern surface ocean, with A_1 and A_2 being respectively the surface areas of the two regions; V_1 and V_2 are the volumes of the two boxes for the surface ocean; and C_p and ρ are the specific heat and density of water respectively. We assume that $V_1 = V_2$ and $A_1 = A_2$. The second term on the right side of Eqs. (4) and (5) represents the heat transport by the ocean currents; $q = w/H_1$, where w is the upwelling velocity and H_1 is the depth of the surface ocean; $sq = u/L_x$, with u and L_x being respectively the zonal velocity and the half zonal width of the tropical Pacific Ocean. Thus $s = uH_1/wL_x$, the ratio between the east–west flow and the total upwelling. It assumes a constant value here. There is observational evidence suggesting that zonal advection may play an important role in establishing the timescale of El Niño (Picaut et al. 1996). Note that the inflow of box 1 comes from box 2 and thus carries temperature T_2. Correspondingly, the heat moved from box 2 to box 1 is $u(V_1/L_x)\rho C_p T_2$, where $u(V_1/L_x)$ is the flow rate. At the same time, through the outflow of the same rate, the heat moved out of box 1 is $u(V_1/L_x)\rho C_p T_1$. Consequently, the net heat gained by box 1 from the ocean currents is $u(V_1/L_x)\rho C_p(T_2 - T_1)$. Dividing this term by $V_1 \rho C_p$, one obtains the second term on the right-hand side of Eq. (4). The same consideration was employed to obtain the corresponding terms in Eq. (5). Such a formulation of the oceanic transport may also result from an upstream differencing of the advection terms in the temperature equation of the surface ocean. The term T_{sub} in Eq. (5) is the temperature of the water upwelled into the mixed layer.

The surface heat flux into the ocean is assumed to be proportional to the difference between the local SST and the SST of the warm pool:

$$Q_1 = C_p \rho H_1 c(T_w - T_1), \tag{6}$$

$$Q_2 = C_p \rho H_1 c(T_w - T_2), \tag{7}$$

where T_w is the SST of the warm pool, H_1 is the depth of the mixed layer, C_p is the specific heat, and c is a constant whose reciprocal has the unit of time ($1/c$ is the characteristic time for the surface flux to remove an SST anomaly). Such a formulation is consistent with the observation that the net heat flux over the warm pool is nearly zero (Ramanathan et al. 1995). The formulation produces a zonal distribution of heat flux that is consistent with that of Esbensen and Kushnir (1981). The formulation also implies that the surface flux is negatively correlated with the SST variations in the equatorial region, which is also consistent with the observed relationship between the interannual variations of surface flux and the SST (Sun and Trenberth 1998).

As has been demonstrated by previous studies, the positive feedback loop among surface winds, ocean currents, and zonal SST gradients is crucial for the occurrence of El Niño. We capture the essence of this feedback loop by assuming that the strength of the upwelling is proportional to the strength of the surface wind and that the strength of the surface wind is proportional to the SST gradients. This results in the following equation for q:

$$q = \frac{\alpha}{a}(T_1 - T_2), \tag{8}$$

where α measures the sensitivity of wind stress to changes in the SST gradients, and a defines the adjustment timescale of the ocean currents to changes in the surface winds.

Using $\Phi(z)$ to represent the zonal mean temperature profile of the subsurface upper ocean and assuming that the effect of the zonal variation of the depth of the upper ocean is simply to displace the profile $\Phi(z)$ vertically (Zebiak and Cane 1987), we have

$$T_{\text{sub}} = \Phi(-H_1 + h_2'), \tag{9}$$

where h_2' is the deviation of the depth of the upper eastern Pacific Ocean from its reference value H. We may regard H as the zonal mean depth of the upper ocean. The profile $\Phi(z)$ may be parameterized as follows:

$$\Phi(z) = T_w - \frac{T_w - T_c}{2}\left(1 - \tanh\left(\frac{z + z_0}{H^*}\right)\right), \tag{10}$$

where T_c is temperature below the upper ocean. We will loosely call T_c the temperature of the equatorial deep ocean. Here H^* and z_0 are constants that have the units of depth. The basic physics embodied in Eq. (10) is that the vertical temperature gradients in the equatorial ocean result from atmospheric heating from above and upwelling of cold water from below. The exponential form is based on observations (Zebiak and Cane 1987) and is also consistent with results from theoretical and numerical models of the thermocline (Munk 1966; Verdiere 1988). The variation of the depth of the upper ocean (or the thermocline) are governed by the following equations:

$$h_2' - h_1' = -\frac{H_1}{H_2} H \frac{\alpha}{b^2}(T_1 - T_2), \tag{11}$$

$$\frac{1}{r}\frac{dh_1'}{dt} = -h_1' + \frac{H_1}{2H_2} H \frac{\alpha}{b^2}(T_1 - T_2). \tag{12}$$

Equation (11) represents the balance between the zonal pressure gradients and the zonal wind stress. Here h_1' is the deviation of the depth of the upper ocean in the western Pacific from its reference value H; $H_2 = H - H_1$; and $b = c_k/L_x$, where c_k is the speed of the first baroclinic Kelvin wave and L_x is the half width of the basin. Equation (12) is an approximate way to represent the slow adjustment of the depth of the thermocline to its equilibrium value determined by the surface wind stress and mass conservation; r defines the timescale of this slow adjustment (Jin 1996).

With Eqs. (6) and (7), we may rewrite Eqs. (4) and (5) as

$$\frac{dT_1}{dt} = c(T_w - T_1) + sq(T_2 - T_1), \tag{13}$$

$$\frac{dT_2}{dt} = c(T_w - T_2) + q(T_{\text{sub}} - T_2). \tag{14}$$

Equilibrium SST as a Function of T_c

Figure 12.3 shows the equilibrium SST of the coupled system as a function of T_c. The value of T_w is fixed at 29.5°C. The use of a constant T_w is based on paleotemperature records indicating that the value of T_w was strongly regulated (Crowley and North 1991). Starting from a very warm value for T_c, the equatorial Pacific is characterized by a very warm SST without zonal SST contrast. Zonal SST gradients are developed when T_c

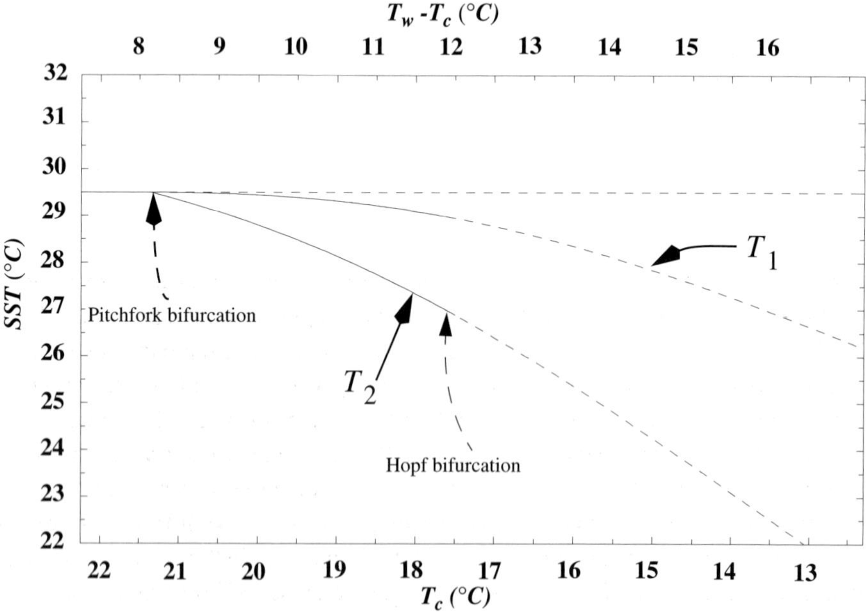

Fig. 12.3 Equilibrium SST of the coupled system as a function of deep ocean temperature T_c. Warm-pool SST T_w is fixed at 29.5°C. Dashed lines indicate that the equilibrium is unstable. The system has a Pitchfork bifurcation at $T_c \sim 21.3$°C and a Hopf bifurcation at $T_c \sim 17.5$°C. After the Hopf bifurcation, the system starts to oscillate. Parameters used are: $1/c = 150$ days, $1/r = 300$ days, $\alpha/a = 3.0 \times 10^8\ \text{K}^{-1}\text{s}^{-1}$, $\alpha/b^2\ H_1/2H_2(1 + H_1/H_2) = 11.5\ \text{mK}^{-1}$, $H^* = 65$ m, $H_1 = 50$ m, and $z_0 = 75$ m.

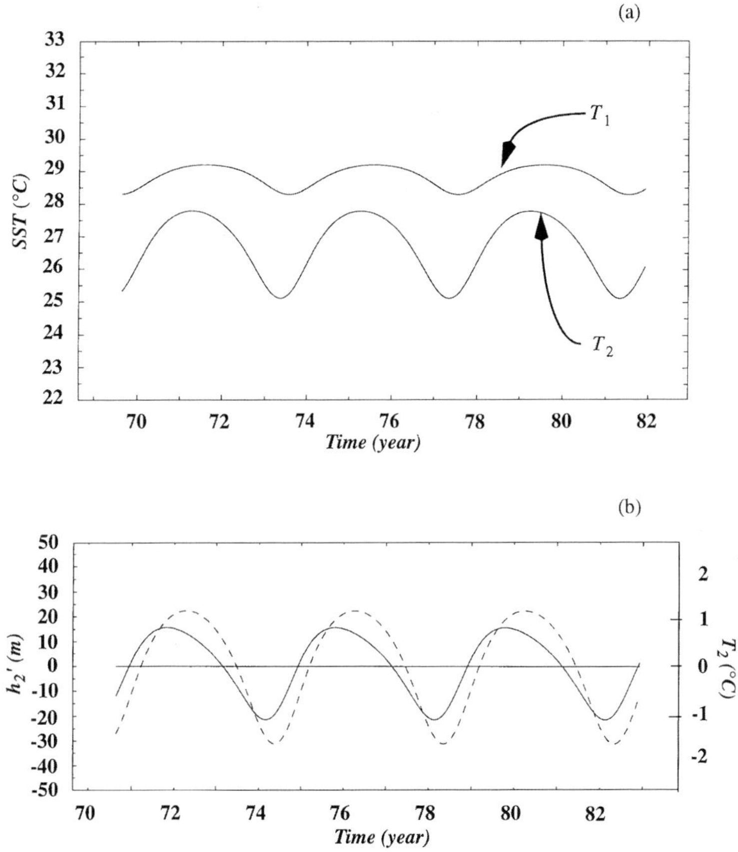

Fig. 12.4 Oscillations with $T_c = 17.3$°C. (a) Variations of T_1 and T_2. (b) Variations of h_2' (anomalies, solid line). Dashed line denotes anomalies of T_2.

is below 21.3°C. The zonal SST gradients increase quickly with further decreases in T_c. When T_c is below approximately 17.5°C, the coupled system starts to oscillate. Oscillations at $T_c = 17.3$°C are plotted in Figure 12.4. The oscillations have a period of about 4 years with a slight westward phase propagation (Fig. 12.4a). The variations of the depth of the thermocline in the eastern half of the ocean also lead slightly the variations of SST in that region (Fig. 12.4). All these features agree well with those of observed El Niño (Wang and Fang 1996; Rasmusson and Carpenter 1982). In the oscillating regime, the magnitude of the oscillation increases with further decreases in T_c (Fig. 12.5). The typical amplitude of ENSO anomaly in the present climate is about 2°C. Figure 12.5 raises the possibility that the present climate may be close to the critical point. The early to mid-Holocene was about 2–3°C warmer in the polar latitudes (Klimanov 1982; Velichko 1984; Borzenkova and Zubakov 1984; R. Webb, personal communication). Since the temperature of the equatorial deep ocean has to equilibrate with the temperature of the polar oceans through the global circulation, it is plausible that in the early to mid-Holocene, the value of T_c may have been significantly warmer than the critical value needed to sustain an oscillatory surface climate.

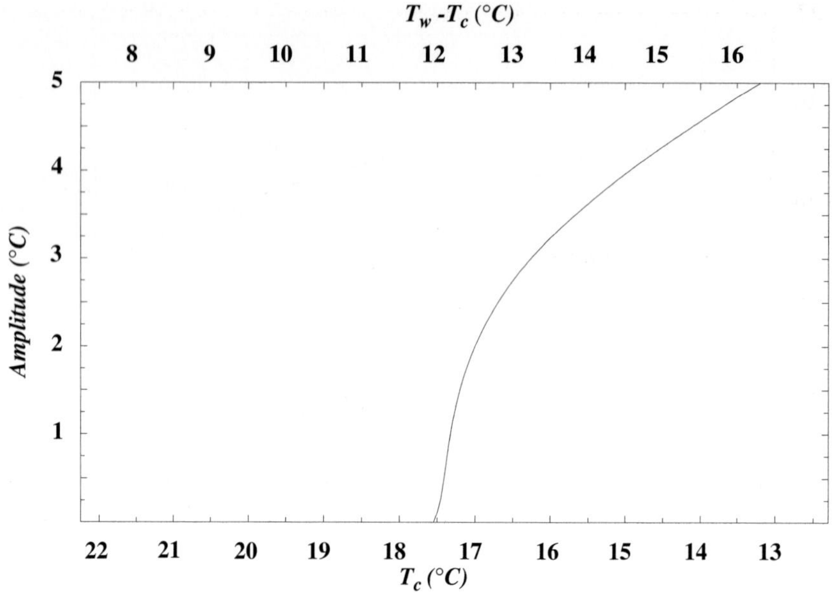

Fig. 12.5 Amplitude of the oscillation as a function of T_c. The amplitude is defined here as the half value of the difference between the maximum and minimum values of T_2. The value of T_w is fixed at 29.5°C. Other parameters are the same as in Figure 12.3.

The Physics of El Niño

To understand the physics of the two bifurcations in Figure 12.3, we replace Eq. (10) by a linear profile, so that Eq. (9) can be written as

$$T_{\text{sub}} = T_{s0} + \gamma h_2', \tag{15}$$

where $T_{s0} = \lambda T_w + (1 - \lambda)T_c$ and $\gamma = \gamma^*(1 - \lambda)(T_w - T_c)/H_2$. Here λ and γ^* are numerical constants that are related to z_0 and H^* in Eq. (10). Equation (15) may be regarded as a first-order approximation of Eq. (10) with T_{s0} and γ being respectively the characteristic temperature and the lapse rate of the upper subsurface ocean. With Eq. (15) for T_{sub}, the coupled system has qualitatively the same behavior as the original system (Fig. 12.6).

A schematic illustration of the reorganization of the system as the deep ocean temperature becomes systematically colder is given in Figure 12.7. In state 1, the heat transfer is a balance between downward background diffusion and upwelling of the background circulation. In this stage, the deep ocean has minimum control on the surface temperature and therefore minimum control on the surface climate. Because of the positive feedback loop among SST, surface winds, and oceanic upwelling, this state is potentially unstable. When the deep ocean temperature becomes sufficiently cold and the vertical stratification becomes sufficiently large, the system possesses more potential energy and the positive feedback becomes strong enough to overcome the effects of thermal and mechanical damping. This enables the system to create zonal

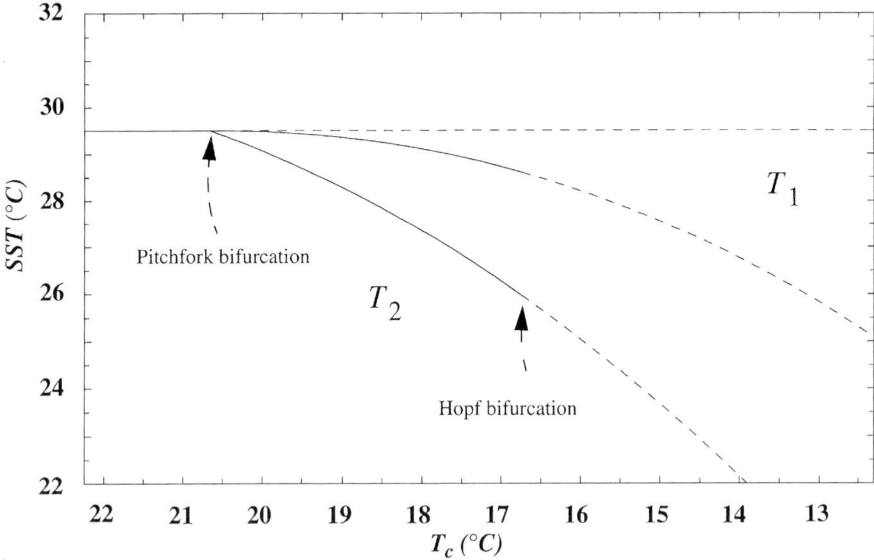

Fig. 12.6 Equilibrium SST of the coupled system as a function of T_c. Dashed lines indicate that the equilibrium is unstable. This plot differs from Figure 12.3 in that Eq. (15) is used for calculating the value of T_{sub}. T_w is fixed at 29.5°C. Other parameters are: $1/c = 150$ days, $1/r = 300$ days, $\alpha/a = 3.0 \times 10^8\ \text{K}^{-1}\text{s}^{-1}$, $\alpha/b^2\ H_1/2H_2(1 + H_1/H_2) = 12.5\ \text{mK}^{-1}$, and $\gamma^*/H_2 = 4/175\ \text{m}^{-1}$.

SST contrast, winds, and currents. The system enters a state with steady zonal SST contrast and steady currents. In this state, the deep ocean has more control on the surface climate, and the warm pool retreats to the west. The zonal SST contrast and the strength of the currents both increase with further decreases in the deep ocean temperature. When the strength of the currents becomes so strong that "overshooting" occurs, the system enters an oscillatory state. The situation is analogous to a waterwheel (Strogatz 1994).

A more quantitative analysis may be obtained by nondimensionalizing Eqs. (8), (11), (12), (13), (14), and (15) (Appendix 1). After nondimensionalization, we find that the dynamic behavior of the system is determined by four nondimensional parameters: $R = \alpha(T_w - T_{s0})/ac$, $\Lambda = p\kappa\gamma^*$ (where $p = H_1/2H_2(1 + H_1/H_2)$ and $\kappa = ac/b^2$), $\sigma = r/c$, and $s = uH_1/wLx$.

The nondimensional parameter R is analogous to the Rayleigh number in the classical problem of Rayleigh–Bénard convection (Rayleigh 1916). It measures how hard the system is driven, relative to the dissipation. In the present case, the driving force for the circulation is from the dynamic tension between the warmth that the greenhouse effect and the solar radiation tend to create (T_w) and the coolness that the subsurface ocean tries to retain (T_{s0}). The dissipation is from the mechanical and thermal damping. The thermal damping depends on the radiative feedbacks (Sun and Liu 1996; Sun 1997a). It is readily shown that SST gradients start to develop when $R = 1$ (the first bifurcation in Fig. 12.3). This is easy to understand because the driving force has to

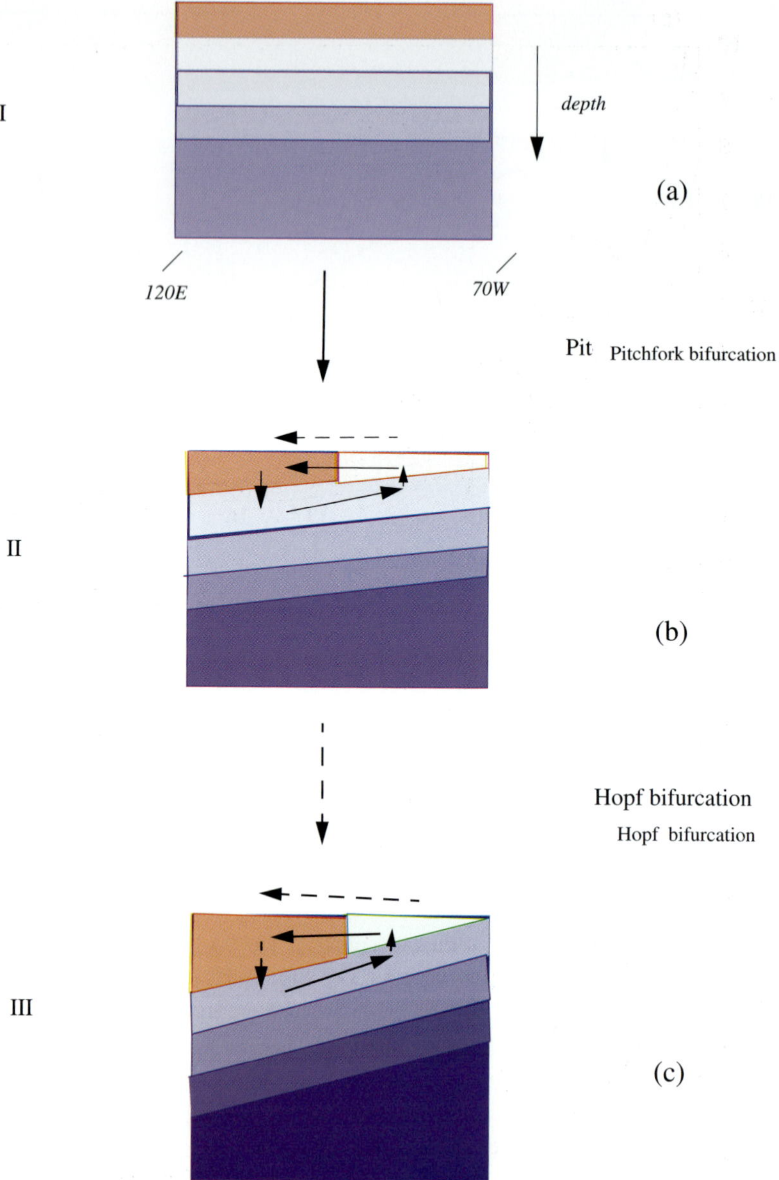

Fig. 12.7 A schematic illustration of the reorganization of the system as the deep ocean temperature becomes systematically colder. The dashed arrow represents the zonal wind. The solid arrows stand for the zonal branch of the wind-driven circulation.

be strong enough to overcome dissipation in order to create and maintain a steady circulation.

The oscillation regime is switched on when $R = R_c \cong C/\Lambda$, where C is a parameter whose value depends only on the values of s and σ (Fig. 12.8). The parameter s measures how strong the eastern and the western Pacific are coupled through the surface

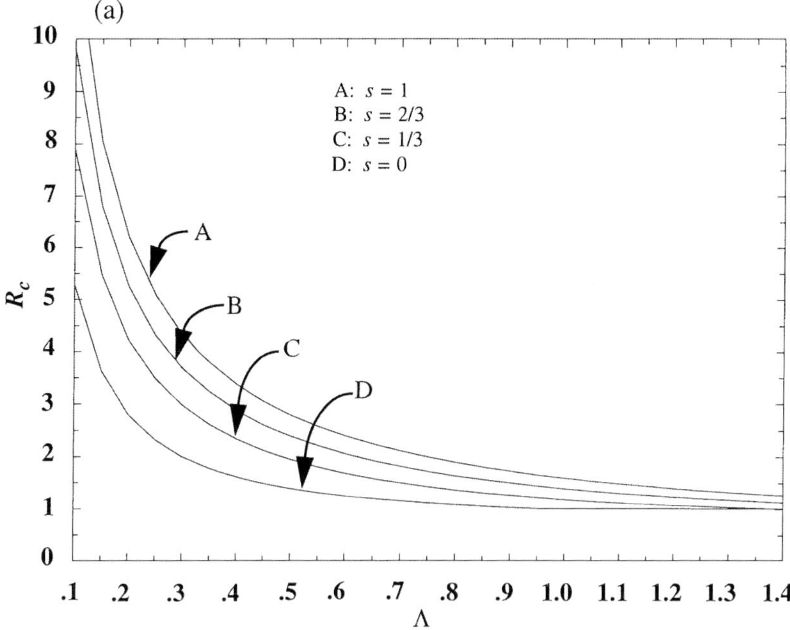

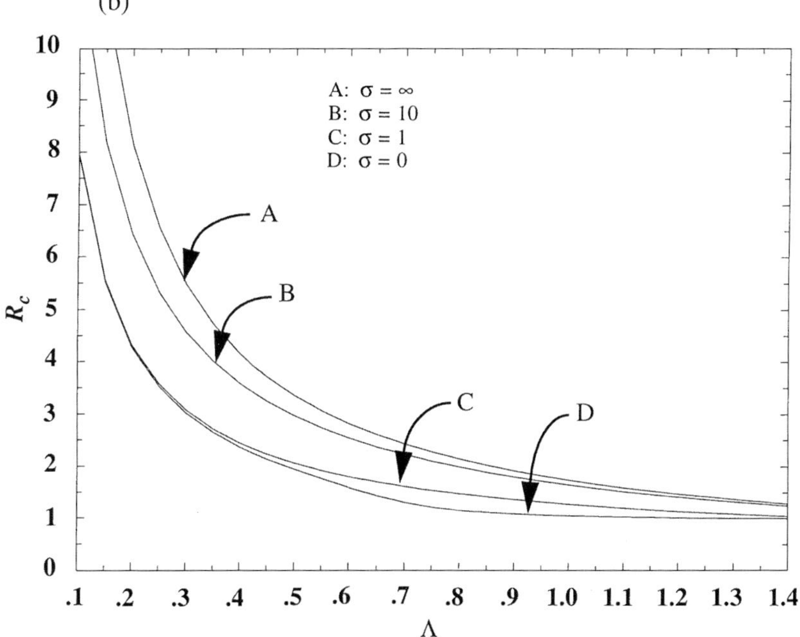

Fig. 12.8 (a) Dependence of the critical value of R (at which the Hopf bifurcation takes place) on the values of Λ and s. The value of σ is fixed at 1/2. (b) Dependence of the critical value of R (at which the Hopf bifurcation takes place) on the values of Λ and σ. The value of s is fixed at 1/3.

current, while σ measures the memory of the subsurface ocean. For $s = 1/3, \sigma = 1/2$ (the values used in Fig. 12.4), C is about 1. The parameter C increases with increases in either s or σ. As long as $s \neq 0$, C remains a finite value even when the ocean has no memory (i.e., $r = \infty$ or $\sigma = \infty$). For example, with $s = 1/3, C = 1.5$ for $\sigma = \infty$. It is easy to show that $\partial T_{\text{sub}}/\partial T_2 = \Lambda R$. Therefore, when the circulation in equilibrium is perturbed by a cooling in T_2^* in the amount of $-\delta T_2^*$, the corresponding change in T_{sub}^* will be $-\Lambda R \delta T_2^*$. A decrease in T_{sub} enhances the cooling effect of the upwelling upon T_2. Opposing the cooling effect is the accompanying increase in the surface heat flux. When ΛR is sufficiently large, the enhanced cooling from the upwelling is able to overcome the opposing effect from the accompanying increase in the surface heat flux and consequently the initial cooling in T_2 will be further amplified. The instability is an oscillatory one because, after some time, the decrease in T_{sub} is slowed down and eventually stopped by the adjustment of T_1 to changes in the cooling from the zonal advection and by the adjustment of h_1' to changes in the zonal SST contrast. The corresponding saturation and decrease in the cooling from the upwelling allows the surface heating to catch up to, stop, and eventually reverse the cooling in T_2. In short, it is critical for the instability of the steady circulation that the temperature of the upwelled water depend strongly on the strength of the flow rate (i.e., a large Λ) and that the flow rate (or the thermal forcing that drives the flow) be sufficiently large relative to the thermal and mechanical damping (i.e., a large R). A similar mechanism is responsible for the onset of oscillation in the Lorenz system (Lorenz 1963), which is a low-order approximation of Rayleigh–Bénard convection. Thus, fundamentally, El Niño arises from an intensified competition between the warming effects from the atmosphere and the cooling effects from the subsurface ocean. The regime transitions discussed above are further illustrated schematically in Figure 12.7.

Note also that $R_c \sim b^2 = (C_k/L_x)^2$. Therefore, the minimum value of $T_w - T_c$ needed to sustain the oscillating regime is inversely proportional to the width of the basin ($T_w - T_c \sim 1/\alpha(C_k/L_x)^2$). This explains the absence of self-sustaining El Niño–like phenomena in the tropical Atlantic Ocean whose width is only one-third that of the tropical Pacific Ocean.

At the critical point $R = R_c = C/\Lambda$, where the oscillation takes place, the period of the oscillation decreases with increases in σ (Fig. 12.9). This is consistent with early analysis of El Niño (Neelin 1991; Jin and Neelin 1993). The period also depends significantly on the parameter s (Fig. 12.9a). For $\Lambda = 0.8, \sigma = 1/2$, the period corresponding to $s = 1/3$ is about $11/c$, while the period corresponding to $s = 1$ is reduced to $6/c$. This provides a theoretical basis for a suggestion by a previous study (Picaut et al. 1996) that zonal advection may play an important role in determining the timescale of ENSO. For a fixed s or σ, the period also increases quickly with increases in Λ (Fig. 12.9b). Doubling Λ can result in a doubling of the period. This reveals the importance of the width of the basin and the speed of the Kelvin wave in setting the timescale of the oscillation (recall $\Lambda \sim (L_x/C_k)^2$). With basin width, s, and σ fixed, changes in T_c or R result in no significant changes in the period of oscillation.

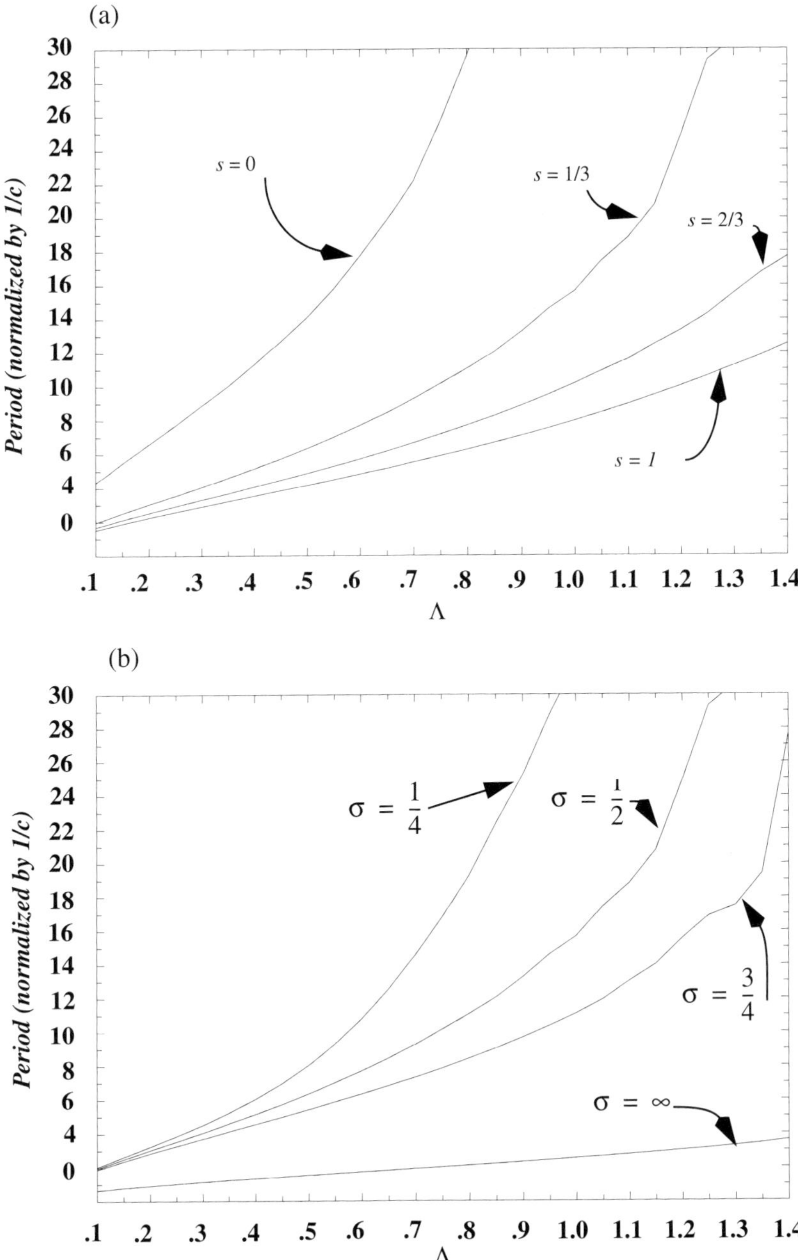

Fig. 12.9 (a) Dependence of the period of the oscillation on Λ and s. The value of σ is fixed at 1/2. (b) Dependence of the period of the oscillation on Λ and σ. The value of s is fixed at 1/3.

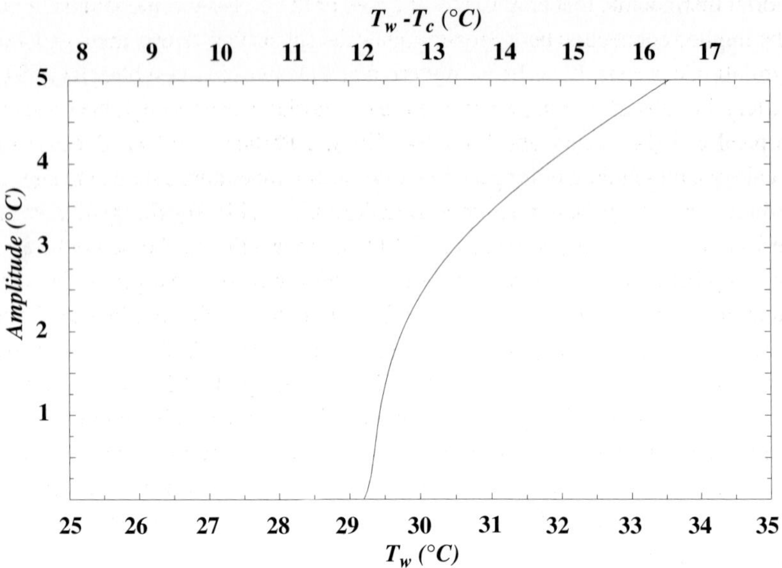

Fig. 12.10 Amplitude of the oscillation as a function of T_w. The amplitude is defined here as the half value of the difference between the maximum and minimum values of T_2. The value of T_c is fixed at 17.3°C. Other parameters are the same as in Figure 12.3.

The nondimensional analysis reveals that El Niño is thermally driven by the difference between T_w and T_c, implying that increasing the value of T_w has the same effect on the magnitude of El Niño as decreasing the value of T_c. Figure 12.10 shows the dependence of the magnitude of El Niño on the value of T_w with the value of T_c being fixed. Comparing Figure 12.10 with Figure 12.5 confirms the results from the nondimensional analysis. To the extent that T_c can be related to the temperature of high latitudes, this result suggests that El Niño is fundamentally driven by the equator-to-pole temperature difference. This in turn raises the question of whether El Niño plays an important role in determining the equator-to-pole temperature difference and the global heat balance. Recent analyses of the heat budget of the tropical Pacific Ocean have suggested a positive answer to this question. Sun and Trenberth (1998) found that during the 1986–87 El Niño, there was a large increase in the poleward heat transport in the equatorial region in both the atmosphere and the ocean. Sun (1999) further showed that most of the heat removed by El Niño from the equatorial region to the subtropical region does not return to the equatorial region, nor is it lost to space in the subtropics: Most of the heat eventually gets transported to the extratropics.

Discussion

The focus of this chapter has been to provide a simple theoretical framework for understanding ENSO variability on millennium and longer timescales. For conceptual clarity, we adopted a simple parametrization for the zonal mean stratification of the equatorial thermocline that emphasizes the role of the global thermohaline circulation and the implied connection between the equatorial subsurface temperature and the SST of high-latitude oceans. It has been suggested that changes in the subtropical SST may also affect the subsurface temperature of the equatorial ocean through the wind-driven subtropical cell (McCreary and Lu 1994; Gu and Philander 1997). Because on the timescale of a millennium or longer, the subtropical temperature usually changes in the same direction as the polar temperature (Budyko and Izrael 1991), the qualitative results we derived here for the dependence of ENSO on global climate should not be affected by the simplification we adopted. On the decadal timescale, changes in the value of the equatorial subsurface temperature may be dominated by changes in the subtropical SST. Accordingly, when applying the present theory (i.e., ENSO is driven by the difference between T_w and T_c) to the decadal variability of El Niño, the value of T_c should be linked to the subtropical SST. Independent of the relative importance of the subtropical cell and the global thermohaline circulation in determining the temperature of the equatorial subsurface ocean, the present theory predicts that an increase in the magnitude of El Niño should be accompanied by an increase in the zonal SST contrast in the equatorial Pacific.

The positive dependence of the magnitude of El Niño on the difference between T_w and T_c is qualitatively consistent with the result of Zebiak and Cane (1987). Using an intermediate coupled model, they found that reducing the mean vertical stratification of the upper ocean thermocline results in a less energetic oscillation. The positive correlation between changes in the zonal SST contrast in the equatorial region, and the magnitude of El Niño is qualitatively consistent with the finding of Knutson et al. (1997). Using a coupled GCM, they found in their CO_2 sensitivity experiments that a reduction in the magnitude of El Niño is accompanied by a reduction in the equatorial zonal SST contrast in the time mean state. (Knutson et al. [1997] did not show the corresponding changes in the equatorial subsurface temperature. Judging from the spatial distribution of the SST changes in the CO_2 sensitivity experiments, it is likely that the equatorial subsurface temperature in their model increased more than the SST in the warm pool.) To the extent that global warming eventually reduces the equator-to-pole temperature difference and thereby the difference between the warm pool SST and the temperature of the equatorial subsurface ocean, the present analysis suggests that global warming will eventually reduce the magnitude of El Niño. The analysis also suggests, however, that the initial response of the magnitude of El Niño to global warming can be the opposite. Because it is the surface that first feels the enhanced radiative heating, it is possible that the response of the equatorial subsurface temperature to an increase in the greenhouse effect may lag the response of the warm pool SST. It follows that the magnitude of El Niño may increase in the initial stage of global warming (until the increase in the subsurface temperature catches up with the increase in the warm pool SST).

Though it appears that the difference between T_w and T_c provides the major thermal forcing, the nondimensional analysis reveals that the magnitude of the El Niño also depends on other parameters. Specifically, the analysis shows that an increase in the coupling strength between the atmosphere and ocean has the same effect on the magnitude of El Niño as an increase in the difference between T_w and T_c. The other potentially important parameter is γ^*, which measures the lapse rate of the uppermost part of the thermocline scaled by the mean vertical stratification of the entire upper ocean (recall that $\Lambda \sim \gamma^*$). Though there is no obvious reason to suspect that these two parameters will change significantly in response to global warming, recognizing their role may be important in understanding differences in the simulated El Niño by different climate models. Note that in anomaly models of El Niño such as the model of Zebiak and Cane (1987), the mean thermocline is fixed. Correspondingly, γ^* decreases as T_w increases. The difference between the present results and those from Clement et al. (1996) on the response of El Niño to an increase in T_w are likely due to the differences in treating the response of the thermocline structure. In our model, the value of γ^*, not the value of γ, remains constant as the value of T_w increases.

Appendix 1: Nondimensional Analysis

Introducing $\tau = ct$, $q^* = q/c$, $T_1^* = (T_1 - T_w)/(T_w - T_{s0})$, $T_2^* = (T_2 - T_w)/(T_w - T_{s0})$, $T_s^* = (T_{\text{sub}} - T_w)/(T_w - T_{s0})$, $h_2^* = h_2'/H_2$, and $h_1^* = h_1'/H_2$, we have

$$\frac{dT_1^*}{d\tau} = -T_1^* + sq^*(T_2^* - T_1^*), \tag{16}$$

$$\frac{dT_2^*}{d\tau} = -T_2^* + q^*(T_s^* - T_2^*), \tag{17}$$

$$\frac{dh_1^*}{d\tau} = -\sigma\left[h_1^* - p\kappa R(T_1^* - T_2^*)\right], \tag{18}$$

$$q^* = R(T_1^* - T_2^*), \tag{19}$$

$$T_s^* = -1 + \gamma^* h_2^*, \tag{20}$$

$$h_2^* = h_1^* - 2p\kappa R(T_1^* - T_2^*), \tag{21}$$

where $R = \alpha(T_w - T_{s0})/ac$, $\kappa = ac/b^2$, $\sigma = r/c$, and $p = (\frac{H_1}{2H_2})(1 + \frac{H_1}{H_2})$. Plugging Eqs. (19), (20), and (21) into Eqs. (16), (17), and (18), and replacing $\gamma^* h_1^*$ by η^*, we have

$$\frac{dT_1^*}{d\tau} = -T_1^* - sR^*(T_2^* - T_1^*)^2, \tag{22}$$

$$\frac{dT_2^*}{d\tau} = -T_2^* + R(T_1^* - T_2^*)\left[-1 + \eta^* - 2\Lambda R(T_1^* - T_2^*) - T_2^*\right], \tag{23}$$

$$\frac{d\eta^*}{d\tau} = -\sigma\left[\eta^* - \Lambda R(T_1^* - T_2^*)\right], \tag{24}$$

where $\Lambda = p\kappa\gamma^*$. After such a manipulation, we see that the dynamic behavior of the coupled system is determined by four nondimensional parameters: R, Λ, σ, and s.

Appendix 2: Definition of Symbols

T_w: warm pool SST (tropical maximum SST)
T_c: deep ocean temperature
T_1: equatorial western Pacific SST (120°E–155°W)
T_2: equatorial eastern Pacific SST (155°W–70°W)
T_{sub}: subsurface temperature
H_1: depth of the mixed layer
H: zonal mean depth of the upper ocean
$H_2 = H - H_1$
L_x: half width of the basin
u: zonal velocity
w: upwelling velocity
q: w/H_1
s: uH_1/wL_x
$1/a$: adjustment timescale of the surface ocean currents to changes in the surface winds
$1/c$: timescale of removing an SST anomaly by surface fluxes
$1/r$: timescale of the slow adjustment in the ocean
α: sensitivity of the surface wind stress to changes in the SST gradients
h_1': deviation of the depth of the upper ocean in the western Pacific from its reference value H
h_2': deviation of the depth of the upper ocean in the eastern Pacific from its reference value H
c_k: phase speed of the first baroclinic Kelvin wave
b: c_k/L_x
p: $(H_1/2H_2)(1 + H_1/H_2)$
k: ac/b^2
σ: r/c
T_{s0}: characteristic value of the subsurface temperature
λ: $(T_{s0} - T_c)/(T_w - T_c)$
γ: lapse rate of the subsurface ocean
γ_0: $(T_w - T_{s0})/H_2$
γ^*: γ/γ_0
R: $\alpha(T_w - T_{s0})/ac$
Λ: $p\kappa\gamma^*$

Acknowledgments The author would like to thank Dr. H. Diaz and Dr. D. Enfield for their very helpful comments. This research was supported by NOAA.

References

BATTISTI, D. S., 1988: The dynamics and thermodynamics of a warm event in a coupled ocean–atmosphere model. *Journal of Atmospheric Sciences*, **45**, 2889–2919.

BERGER, W. H., 1981: Paleoceanography: The deep-sea record. *In* Emiliani, C. (ed.), *The Oceanic Lithosphere: The Sea*, Vol. 7. New York: Wiley-Interscience, pp. 1437–1519.

BJERKNES, J., 1969: Atmospheric teleconnections from the equatorial Pacific. *Monthly Weather Review*, **97**, 163–172.

BORZENKOVA, I. I. and ZUBAKOV, V. A., 1984: Climatic optimum of the Holocene as a model of the global climate at the beginning of the 21st century. *Meteorologlya i Gidrologiya*, **8**, 12–35 (in Russian).

BROECKER, W. S., 1987: The biggest chill. *Natural History Magazine*, October, 74–82.

BUDYKO, M. I. and IZRAEL, Y. A., 1991: *Anthropogenic Climate Change*. Tucson: University of Arizona Press, 277–318.

CANE, M., 1979: The response of an equatorial ocean to simple wind stress patterns. Part 1: Model formulation and analytical results. *Journal of Marine Research*, **37**, 233–252.

CLEMENT, A. C., SEAGER, R., CANE, M. A., and ZEBIAK, S. E., 1996: An ocean dynamical thermostat. *Journal of Climate*, **9**, 2190–2196.

CROWLEY, T. J. and NORTH, G. R., 1991: *Paleoclimatology*. New York: Oxford University Press, 339 pp.

ESBENSEN, S. K. and KUSHNIR, Y., 1981: *The Heat Budget of the Global Ocean: An Atlas Based on Estimates from Surface Marine Observations*. Climate Research Institute report, No. 29, Oregon State University.

GLANTZ, M. H., KATZ, R. W., and NICHOLLS, N. (eds.), 1991: *Teleconnections Linking Worldwide Climate Anomalies*. Cambridge: Cambridge University Press, 533 pp.

GU, D. and PHILANDER, S. G. H., 1997: Interdecadal climate fluctuations that depend on exchanges between the Tropics and the extratropics. *Science*, **275**, 805–807.

HOFFERT, M. I., 1990: Climate change and ocean bottom water formation: Are we missing something? *In* Schlesinger, M. E. (ed.), *Climate–Ocean Interaction*. Dordrecht: Kluwer Academic Publishers, 385 pp.

JIN, F. F., 1996: Tropical ocean–atmosphere interaction, the Pacific cold tongue, and the El Niño–Southern Oscillation. *Science*, **274**, 76–78.

JIN, F. F. and NEELIN, J. D., 1993: Modes of interannual tropical ocean–atmosphere interaction: A unified view. Part III: Analytical results in fully coupled cases. *Journal of the Atmospheric Sciences*, **50**, 3523–3540.

KEEFER, D. K., de FRANCE, S. D, MOSELEY, M. E., RICHARSON III, J. B., SCATTERLEE, D. R., and DAY-LEWIS, A., 1998: Early maritime economy and El Niño events at Quebrada Tacahuay, Peru. *Science*, **281**, 1833–1835.

KLIMANOV, V. A., 1982: Climate of eastern Europe during the climatic optimum of the Holocene (from palynological data). *In Development of the Environment of the USSR in the Late Pleistocene and Holocene*. Moscow: Nauka, pp. 251–258 (in Russian).

KNUTSON, T. R., MANABE, S., and GU, D., 1997: Simulated ENSO in a global coupled ocean–atmosphere model: Multidecadal amplitude modulation and CO_2 sensitivity. *Journal of Climate*, **10**, 131–161.

LINDZEN, R. S. and NIGAM, S., 1987: On the role of sea surface temperature gradients in forcing low-level winds and convergence in the tropics. *Journal of the Atmospheric Sciences*, **44**, 2418–2436.

LORENZ, E., 1963: Deterministic nonperiodic flow. *Journal of the Atmospheric Sciences*, **20**, 131–141.

McCREARY, J. P. and LU, P., 1994: Interaction between the subtropical and equatorial circulations: The subtropical cell. *Journal of Physical Oceanography*, **24**, 466–497.

MUNK, W. H., 1966: Abyssal recipes. *Deep Sea Research*, **13**, 707–730.

NEELIN, J. D., 1991: The slow sea surface temperature mode and the fast wave limit. *Journal of the Atmospheric Sciences*, **48**, 584–606.

NEELIN, J. D., BATTISTI, D. S., HIRST, A. C., JIN, F. F., WAKATA, Y., YAMAGATA, T., and ZEBIAK, S., 1998: ENSO theory. *Journal of Geophysical Research*, **103**, 14261–14290.

PICAUT, J., LOUSLALEN, M., MENKES, C., DELCROIX, T., and McPHADEN, M. J., 1996: Mechanism of the zonal displacements of the Pacific warm pool: Implications for ENSO. *Science*, **274**, 1486–1489.

RASMUSSON, E. M. and CARPENTER, T. H., 1982: Variations in tropical sea surface temperature and surface winds associated with the Southern Oscillation/El Niño. *Monthly Weather Review*, **110**, 354–383.

RAMANATHAN, V., SUBASILAR, B., ZHANG, G. J., CONANT, W., CESS, R. D., KIEHL, J. T., GRASSL, H., and SHI, L., 1995: Warm pool heat budget and shortwave cloud forcing: A missing physics. *Science*, **267**, 499–503.

RASMUSSON, E. M. and WALLACE, J. M., 1983: Meteorological aspects of the El Niño/Southern Oscillation. *Science*, **222**, 1195–1202.

RAYLEIGH, LORD, 1916: On convection currents in a horizontal layer of fluid when the higher temperature is on the under side. *Philosophical Magazine*, **32**, 529–546.

SANDWEISS, D. H., RICHARDSON, J. B., III, REITZ, E. J., ROLLINS, H. B., MAASCH, K. A., 1996: Geoarchaeological evidence from Peru for a 5000 years B. P. onset of El Niño. *Science*, **273**, 1531–1533.

SANDWEISS, D. H., RICHARDSON, J. B., III, REITZ, E. J., ROLLINS, H. B., MAASCH, K. A., 1997: Determining the early history of El Niño: Response. *Science*, **276**, 966–967.

STROGATZ, S. H., 1994: *Nonlinear Dynamics and Chaos*. Reading: Addison-Wesley, 498 pp.

SUAREZ, M. J. and SCHOPF, P., 1988: A delayed action oscillator for ENSO. *Journal of the Atmospheric Sciences*, **45**, 3283–3287.

SUN, D.-Z., 1997a: Tropical zonal SST gradients: A coupled response to radiative feedbacks and the meridional gradients in insolation. *Preprints of the Seventh Conference on Climate Variations*. American Meteorological Society, Long Beach, California, pp., 282–284.

SUN, D.-Z., 1997b: El Niño: A coupled response to radiative heating? *Geophysical Research Letters*, **24**, 2031–2034.

SUN, D.-Z. and LIU, Z., 1996: Dynamic ocean–atmosphere coupling, a thermostat for the tropics. *Science*, **272**, 1148–1150.

SUN, D.-Z. and TRENBERTH, K. E., 1998: Coordinated heat removal from the equatorial Pacific during the 1986–87 El Niño. *Geophysical Research Letters*, **25**, 2659–3062.

SUN, D.-Z., 1999: The heat sources and sinks of the 1986–87 El Niño. *Journal of Climate*. (accepted)

VELICHKO, A. A. (ed.), 1984: *Later Quaternary Environments of the Soviet Union*. Minneapolis: University of Minnesota, 327 pp.

VERDIERE, A. C., 1988: Buoyancy driven planetary flow. *Deep Sea Research*, **46**, 215–265.

WANG, B. and FANG, Z., 1996: Chaotic oscillations of tropical climate: A dynamic system theory for ENSO. *Journal of the Atmospheric Sciences*, **53**, 2786–2802.

WYRTKI, K., 1985: Water displacements in the Pacific and the genesis of El Niño cycles. *Journal of Geophysical Research*, **90**, 7129–7132.

ZEBIAK, S. E. and CANE, M. A., 1987: A model El Niño–Southern Oscillation. *Monthly Weather Review*, **115**, 2262–2278.

13

The Past ENSO Record: A Synthesis

VERA MARKGRAF

Institute of Arctic and Alpine Research (INSTAAR), University of Colorado, Boulder, CO 80309, U.S.A.

HENRY F. DIAZ

Climate Diagnostics Center, Oceanic and Atmospheric Research, National Oceanic and Atmospheric Administration, 325 Broadway, Boulder, CO 80303, U.S.A.

Abstract

A broad range of paleoclimate indicators is available to study the El Niño/Southern Oscillation (ENSO) phenomenon. These climate proxy records, each of which is sensitive to somewhat different aspects of climatic variations, can be used for studying ENSO – to answer specific questions, such as, How long has ENSO been operating in its present form? In this chapter, we evaluate the information provided by the various proxy records, highlighting their possible strengths and weaknesses, and pointing out areas where outstanding research questions into the nature of ENSO variability still remain.

At present, the preponderance of evidence suggests that during glacial times, the ENSO phenomenon (if it was operating as an alternating east–west source of heat and moisture to the atmosphere) did not leave the same spatial or temporal expression in the paleoclimate record that is evident more recently. Conditions in the early Holocene, prior to ~6,000 years before present (BP), are indicative of changed atmospheric and oceanic patterns, substantially different from those of today. Only after about 6,000 years ago did the climatic associations related to the changes in sea surface temperature and atmospheric circulation patterns related to ENSO that are seen today begin to be systematically recorded in the paleoclimate record.

Introduction

One of the major expressions of global interannual climate variability is related to the coupled ocean–atmosphere phenomenon of El Niño/Southern Oscillation (ENSO). Because of the far-reaching effects of ENSO on societies, concerns about its future behavior have promoted research into the ENSO phenomenon's long-term history, especially of those intervals when climate boundary conditions (orbitally determined seasonal insolation, global ice cover, sea level, aerosols, etc.) were markedly different from those of today (Diaz and Markgraf 1992).

In discussing past aspects of ENSO, as reconstructed from paleoclimate proxy records, several concerns need to be addressed. First, how robust is the ENSO phenomenon in relation to major changes in boundary conditions? Is ENSO an unstable oscillation between cold or warm tropical ocean states that can alternately become permanently "El Niño–like," or permanently "La Niña–like," or an intermittent oscillation, as it is now (Enfield 1992; Sun, this volume)? Second, how can the ENSO phenomenon, as expressed in paleoclimate proxy records, be isolated from a range of other climatic forcing parameters, when boundary conditions are different? Third, how robust are ENSO climate teleconnections under changed boundary conditions (Markgraf et al. 1992; Meehl and Branstator 1992; Allan, this volume; Kleeman and Power, this volume; Mann et al., this volume; Ortlieb, this volume)? Finally, if there is a connection between ENSO and solar activity (Anderson 1992; Enfield 1992; Mann et al. 1998), can we assume comparable effects related to Milankovitch insolation changes?

The paleoclimate record can be used to understand both high- and low-resolution changes in the ENSO phenomenon. High-resolution records (for instance, tree rings, fire scars, varved sediments, corals, ice cores) register manifestations (droughts, fires, floods, temperature and precipitation fluctuations, altered sea surface temperatures (SSTs), changes in ocean salinity, etc.) that can be attributed to single ENSO events. However, high-resolution records for long time periods are rare and probably will not be generated for all regions. Low-resolution records can also be used, but these need to be interpreted with more caution. For instance, older riverine deposits indicative of flooding episodes are less likely to be preserved, and thus there is a bias in the record in favor of less frequent but larger events. Pollen records tend to be derived mainly from long-lived plants, such as trees, and, even when yearly samples can be obtained, yearly fluctuations are not apparent in the pollen record. However, the lack of annual resolution does not mean that some broad interpretation of the presence or absence of high-frequency ENSO events cannot be made. Some plant species and plant communities are indicative of variable climatic conditions. Either they are more tolerant to environmental changes and environmental stresses or they recover quickly after they are disturbed. When such plant species or communities spread, it is a good indication that overall variability has increased. For example, charcoal records that can be interpreted to represent regional fire histories can be interpreted to reflect changes in ENSO modes (e.g., McGlone et al. 1992).

A combination of approaches is needed for a comprehensive assessment of ENSO behavior. Multiproxy analysis of single records, which includes analysis of a variety of indicators that respond to different aspects of ENSO (e.g., marine records; Baumgartner et al. 1985), is particularly useful. Multiproxy analyses dramatically reduce the possibility that some other (non-ENSO) environmental factors are responsible. Detailed comparison of spatially distributed high-resolution paleoclimate proxy records, such as tree rings from the Southwest United States with ice core records from Peru or with marine records from southern California (Baumgartner et al. 1989; Michaelsen and Thompson 1992), also can be used effectively to study past ENSO behavior. Low-resolution studies of ENSO demand the use of networks of sites, especially those from regions known for their sensitivity to ENSO teleconnections, such as Australia/New Zealand with South America (e.g., McGlone et al. 1992). Whereas an ensemble of low-resolution records, even from high-response regions, may fall short of the certainty that can be obtained from multiproxy high-resolution records, when subjected to intercomparison, they play a useful part in developing the larger picture.

Paleo-ENSO Records

Glacial Intervals

Seasonal latitudinal distributions of solar radiation during the last full glacial interval (ca. 20,000 years ago, 20 ka) were similar to those of today. However, surface boundary conditions were very different: Continental ice sheets in eastern North America, Greenland, and western Europe were at their maximum extent; sea level was substantially lower than it is today; SSTs were lower overall, including those at tropical latitudes (Anderson and Webb 1994); and polar sea ice was extended equatorwards. The glacial period's pole-to-equator temperature gradients were likely much steeper than they are today, and circulation features, including the polar fronts and the westerlies, were also shifted equatorwards. In the equatorial Pacific Ocean specifically, glacial intervals (marine oxygen stages 2, 4, 6, 8, 10) were characterized by 3 to 4°C lower SSTs and markedly stronger upwelling, leading to increased productivity and suggesting enhanced trade winds (Rea et al. 1991; Patrick and Thunell 1997; Ohkouchi et al. 1997; Perks and Keeling 1998; see also Guilderson et al.'s (1994) study of Atlantic Ocean SST gradients). Despite lower SSTs, Patrick and Thunell (1997) suggest that the east–west temperature gradient was similar to what it is today and that, as a consequence, the zonal Walker circulation must have continued to operate.

One of the first discussions of the past expression of ENSO was presented by Quinn (1971). Quinn suggested that the paleoenvironmental evidence for enhanced upwelling and enhanced ocean productivity during full glacial times in the tropical Pacific Ocean, indicating permanently enhanced trade winds, would imply that ENSO was "arrested" in its high-index (La Niña) phase. Anderson et al. (1990, 1992) came to a similar conclusion based on high-resolution analysis of marine sediments off the California coast. Enhanced upwelling along the coast of California is today related to cold-event (La Niña) intervals. Apparently, enhanced upwelling is also documented for the last glacial maximum between 14 and 28 ka, although the higher resolution analysis shows substantial decadal variability as well (Anderson et al. 1990).

Terrestrial evidence for the full glacial interval from Australasia and South America, however, does not show paleoclimate patterns that resemble today's ENSO teleconnection patterns (Markgraf et al. 1992; McGlone et al. 1992). Paleovegetation records for Java (van der Kaars and Dam 1997), Papua New Guinea (Hope 1996; Haberle

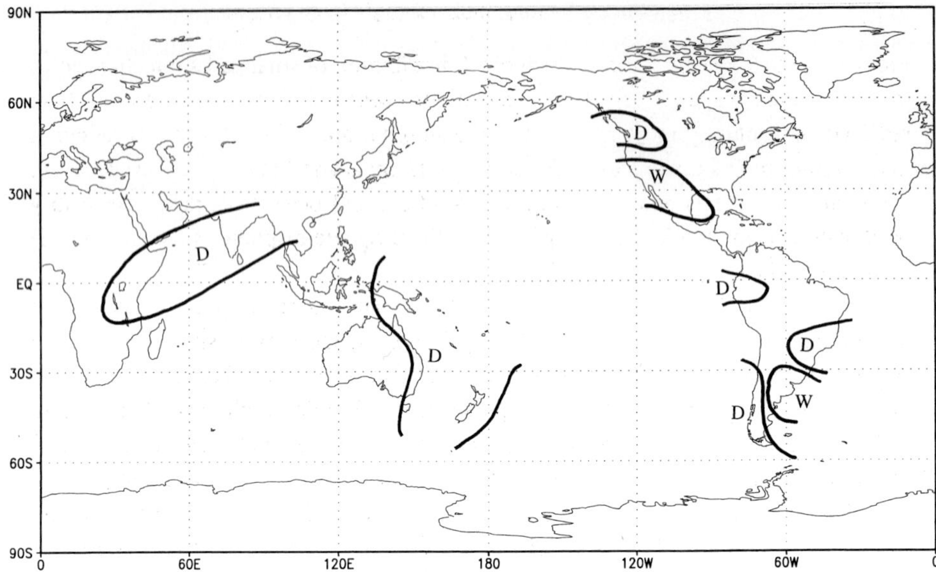

Fig. 13.1 Map depicting the prevailing climate, relative to modern, in areas around the globe with a significant modulation of precipitation during ENSO events. Areas inferred to have been drier and wetter than they are today during full glacial times are denoted by Ds and Ws, respectively.

1998), northeastern Australia (Kershaw 1994), and South America (Ledru et al. 1998; Markgraf 1991), and the Huascarán ice core record from Peru (Thompson et al. 1995), all show invariably colder and drier conditions during the last glacial (see Fig. 13.1). If, as is suggested by the marine records, La Niña conditions had become a permanent mode, only parts of South America should have become arid, while northern Australia and Melanesia should have experienced high levels of precipitation. An explanation for this discrepancy between marine and terrestrial records could relate to the fact that a reduction of 3 to 4°C in SST in the western Pacific warm pool would result in a much weaker coupling between SST and convection (Kraus 1973; Graham 1995; Sun, this volume), with an overall reduction of precipitation over both the western tropical Pacific and the adjacent land areas.

The argument that ENSO must have continuously persisted throughout the Cenozoic (Nicholls 1992), because, for example, the fauna and flora of Australia are highly adapted to precipitation variability, probably underestimates the potential range of climatic conditions plants and animals can either adapt to or respond to with strategies such as migration into suitable habitats (see Markgraf et al. 1995). During times when glacial climates have existed, the expression of ENSO variability might have been quite different from what it is now. Although extreme full glacial intervals occurred only about every 100,000 years and had a duration of only 10,000 to 15,000 years, the remaining times, with exception of the short interglacial periods, were also rather glacial in character compared to interglacial periods.

Late Glacial Transitional Period (16 to 10 ka)

From 15 ka to 9 ka, the seasonal latitudinal distribution of solar radiation changed as the Earth–Sun distance lessened and the axial tilt steepened. The Northern Hemisphere received substantially more radiation in the summer and substantially less radiation in the winter by 9 ka, whereas in the Southern Hemisphere the opposite occurred. As continental ice sheets and sea ice began to retreat and oceans warmed, seasonality of climate became markedly stronger in the Northern Hemisphere but weaker in the Southern Hemisphere.

Late glacial terrestrial and marine records are characterized by a high degree of environmental variability; today, especially in the Pacific region, such variability is commonly attributed to ENSO forcing. Flood deposits in southern Peru dated to before 12 ka were interpreted to indicate the presence of El Niño conditions (Keefer et al. 1998). Lin et al. (1998), in comparing the trans-U.S. wet event between 14 and 13.5 ka with paleoclimatic patterns in a few ENSO-sensitive regions, suggested that perhaps a mechanism "akin to ENSO" was operating at that time. Pollen analysis of a marine core from the Santa Barbara basin off southern California documents repeated *Pinus* and *Juniperus* peaks between 17.1 ka (20,400 calendar years before present, 20.4 kcal) and 8.8 ka (10.1 kcal), recurring with a periodicity of 1,070 to 1,550 years (Heusser and Sirocko 1997). The highest amplitude and frequency of peaks are recorded between 14.6 ka (17 kcal) and 12 ka (14.5 kcal). Although today's oak- and pine-dominated vegetation assemblage, and hence the modern precipitation regime, does not appear until after 13 ka (15 kcal), the authors interpret these peaks to reflect intervals of high precipitation (pine) alternating with intervals of low precipitation (juniper). Because the timing of these pine peaks is apparently comparable with peaks of δ^{18}O in a marine core from the western Arabian Sea, Heusser and Sirocko (1997) have suggested that this pan-Pacific covariability reflects ENSO-type forcing. Stable isotope, marine faunal, and sedimentologic analyses of the same core from the Santa Barbara basin, however, show a different pattern, and in particular lack the variability (Kennett and Ingram 1995; Behl and Kennett 1996). Prior to 13.4 ka (15.8 kcal), the last glacial maximum (LGM), and between 11 and 10 ka (12.5 and 10.5 kcal), the Younger Dryas interval, SSTs were colder than they are today, and sediments throughout are bioturbated. Between 13 and 11 ka (15 and 12.5 kcal), the Bølling/Allerød interval, and after 10 ka (10 kcal), the onset of the Holocene, SSTs were warm and sediments were varved. Because the changes in this record are similar to changes in the Greenland ice core records, Kennett and Ingram (1995) suggest that the Santa Barbara basin record has been influenced by changes in the North Atlantic affecting the North Pacific. If ENSO forcing was at work, causing millennial fluctuations, the sedimentologic record would reflect the same peaks as the pollen record.

Terrestrial records exist that can shed light on the ENSO versus the North Atlantic mechanism of climate variability during times of deglaciation. In southern California, a high-resolution paleolimnological record for Owens Lake (Bradbury and Forester, 1999) shows major changes during the late glacial interval. However, these changes do not compare to the high variability of the marine pollen record but rather resemble the sedimentologic changes in the marine core. If rapid precipitation changes in fact were responsible for the pollen peaks in the marine record, Owens Lake levels, which are very sensitive to changes in runoff, should have responded in a similar fashion. The lack of comparable fluctuations in the levels of Owens Lake suggests that the pollen peaks in the Santa Barbara basin record were not caused by regional precipitation changes but perhaps by glacial meltwater events (Bradbury and Forester, 1999). Another high-resolution pollen record quoted for its late Quaternary occurrence of high environmental variability is that for Lake Tulane in Florida (Grimm et al. 1993). In this record, the frequency and duration of pine peaks, reflecting short-term precipitation increases, are markedly different from those for the Santa Barbara basin. The similarity of the reconstructed regional climate variability of the Lake Tulane record to changes in North Atlantic marine records, specifically the Heinrich events, is interpreted as suggesting a climate link to the North Atlantic rather than to ENSO forcing.

A high-resolution pollen record for a fossil glacial meltwater channel in southern Chile (Moreno 1997) shows several peaks of Myrtaceae pollen between 19.5 and 20 ka, at 15.5 ka, and at 14.5 ka. Myrtaceae pollen represents a group of small trees that grows in temperate rain forests but also riparian trees along lakes and rivers in the steppe region. These records are interpreted to reflect short-term regional changes in precipitation. Again these peaks are not comparable with those for the Santa Barbara basin, although southern Patagonia is also an ENSO-sensitive region. An additional complication with this record arises from this group's ecology, since these peaks could be explained as responses to lake level changes related to the melting piedmont glaciers, rather than to regional climate variability. Another high-resolution pollen and stable isotope record for a high-latitude peat bog in southernmost South America (Markgraf and Kenny 1997) shows high environmental and temperature variability between 13 and 10 ka. Resemblance of the stable isotope record from this peat core with stable isotope records for coastal Antarctica and Greenland (Steig et al. 1998) was interpreted as suggesting interhemispheric linkages related to the thermohaline circulation. Although late glacial records are too few to allow a decision on the issue of forcing, it is less likely that ENSO is involved. Variability alone does not provide sufficient proof for ENSO forcing, especially as neither the frequency, nor the timing, nor the climatic patterns are comparable in the records.

The Holocene Record

The Early Holocene (10 to 8 ka)

The primary difference between early Holocene and present-day boundary conditions is the seasonal distribution of solar radiation. Between 12 and 9 ka, the Earth–Sun distance was less than it is today during the Northern Hemisphere summer and the axial tilt was steeper. Hence, seasonality of climate was markedly stronger in the Northern Hemisphere but weaker in the Southern Hemisphere. At about 9 ka, insolation in the temperate latitudes of the Northern Hemisphere was about 8% higher in July (summer) and 10% lower in January (winter) than it is today. At the same time, insolation in the Southern Hemisphere temperate latitudes was about 6% lower in January (summer) and 4% higher in July (winter). Paleoclimate model simulations suggest that this insolation forcing produced enhanced summer precipitation in the Northern Hemisphere lower temperate latitudes, but it reduced summer precipitation in the Southern Hemisphere (Kutzbach et al. 1993).

Early Holocene surface boundary conditions were characterized by near present-day global land and ocean temperatures, extent of mountain glaciers, and circumpolar sea ice. However, continental ice sheets in the Northern Hemisphere had not yet completely disintegrated, and global sea level was still below modern values. Consequently, the area of shallow seas to the north of Australia was still more extensive than it is today, and the land bridge between Australia and New Guinea persisted until after 9 ka. Both these factors greatly influenced ocean circulation and temperatures in the western Pacific warm pool region, with early Holocene SSTs in the western tropical Pacific still about 1°C lower than they are today (Aharon 1983; Markgraf et al. 1992). These lower SSTs greatly contrast with those at the higher latitudes of the southern oceans, which already were markedly warmer than they are today from 13 ka onward (Labracherie et al. 1989).

The temporal resolutions of most early Holocene paleoclimate records are low, ranging from decades to centuries. Inferences on the expression of ENSO during that time, therefore, are based primarily on interregional comparisons of paleoclimate modes, especially in terms of precipitation (McGlone et al. 1992). Mode changes of ENSO to either dampened (or enhanced) behavior would be expected to result in a decline (or an increase) of floods or droughts, depending on the specific region. Atmospheric circulation can be inferred, as a consequence, to be more zonal (or more meridional), with reduced (or enhanced) climate variability. In addition to comparing past patterns of changes in effective moisture from regions known today to be sensitive to ENSO climate teleconnections, changes in past climate variability can also be documented in environmental records by using charcoal analysis to reconstruct past fire frequency. Tree-ring and fire scar analyses have shown that changes in fire frequency, which reflect changes in precipitation variability, can be related to ENSO variability (Swetnam and Betancourt 1992; Kitzberger et al. 1998; Veblen et al. 1999).

By comparing early Holocene vegetation, lake level, and fire history records for New Zealand, Australia, and South America, McGlone et al. (1992) inferred the possible character of ENSO in some detail for this interval. Their study concluded that between 10 and 8 ka, circum-Pacific precipitation patterns, lack of environmental variability, and absence of fires all suggest an absence of ENSO as we know it today. Lowlands in northeastern and southeastern Australia were drier and perhaps also cooler than they are today at the onset of the Holocene (Kershaw 1994; Shulmeister and Lees 1995). In New Zealand, leeward eastern districts remained dry until 8,000–7,500 BP. Mainland windward wetter areas appear to have been cloudy, moist, and mild, with low variability, as were the far southern islands south of 50 S, although annual rainfall seems to have been no higher than it is now (McGlone and Neall 1994; McGlone et al. 1997b; McGlone and Moar 1998).

During the early Holocene, the climates in the midlatitudes of southwestern South America were drier and warmer than they are today, with drought-adapted taxa expanding southwards (Markgraf 1991; Villagran 1991). Central and southern Brazil were also markedly drier overall between 10 and 7 ka than they are today. In contrast, records for Salitre, the southernmost site, indicate cool and moist conditions thought to relate to a high frequency of Antarctic polar air incursions (Ledru et al. 1998). The paleotemperature record from the Huascarán ice core record in Peru shows, from 10 ka onward, a trend to warmer temperatures, which reach a maximum between 8.6 and 5.2 ka (Thompson et al. 1995; Thompson et al., this volume). A major dust peak in the ice core at 9 ka suggests a major period of drought. Pollen records for northern Peru also suggest warmer, but also moister, conditions than exist today (Hansen and Rodbell 1995), and lake level records for northern Chile also suggest humid conditions for the early Holocene (Grosjean 1998).

In the early Holocene, regions of the southwestern United States where modern maxima of precipitation occur during the summer were wetter than they are today because of enhanced monsoon rainfall. In contrast, western U.S. regions with dry summer conditions were drier than they are today, likely because of an expanded North Pacific high-pressure system (Barnosky et al. 1987; Whitlock et al. 1995). Also, southern Alaska was warmer and drier than today (Barnosky et al. 1987).

Figure 13.2 illustrates the precipitation patterns that are thought to have prevailed in the early Holocene, relative to the modern climate. It is tempting to ascribe the early Holocene regional moisture patterns in North America (dry in the U.S. Northwest and

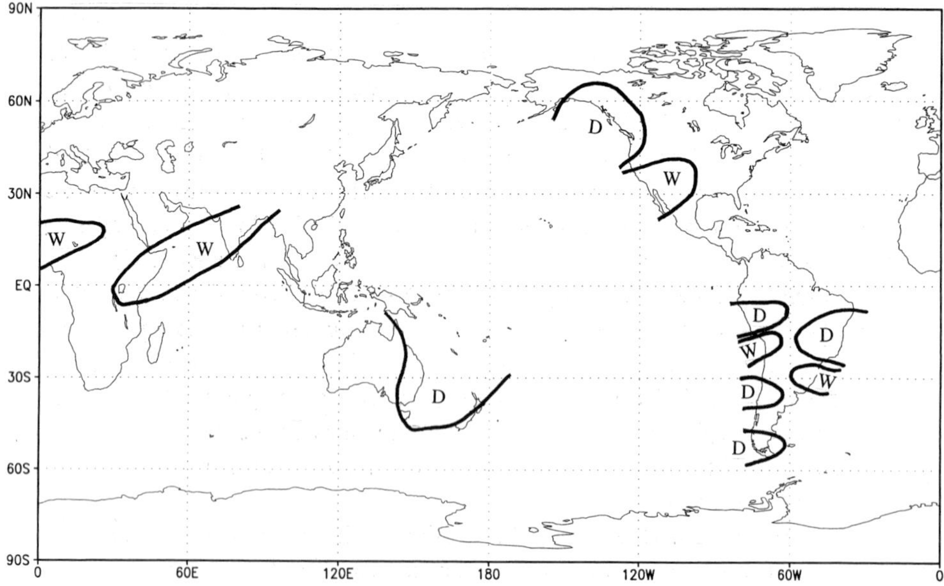

Fig. 13.2 Map depicting climatic conditions during the early Holocene. Areas inferred to have been drier and wetter than they are today during full glacial times are denoted by Ds and Ws, respectively.

wet in the U.S. Southwest), and in Australia/New Zealand (generally drier and cooler), to the predominance of El Niño–like conditions. However, regional paleoclimate patterns in South America (dry in the midlatitudes of Chile and in central and southern Brazil, and lowered temperatures in Peru) do not corroborate this hypothesis. Instead, these patterns most likely relate to characteristic features of the prevailing atmospheric circulation, and the ENSO system might not have operated at that time, at least in its present form. Cooler temperatures in the western Pacific warm pool, low seasonality in the Southern Hemisphere, and delayed onset of monsoonal activity in India until about 9 ka, thought to relate to the delayed warming of the Tibetan Plateau (Overpeck et al. 1996), would also suggest an absence of ENSO and its climate teleconnections (see Sun, this volume). Probably the strongest argument for an absence of ENSO oscillations in the early Holocene is reduced monsoon activity, given the close link between the two systems (Charles et al. 1997; Sirocko 1996; Webster et al. 1998).

Mid-Holocene (8 to 5 ka)

For the period between 8 and 5 ka, a controversy has developed about the interpretation of geomorphologic and geoarchaeological evidence from shell middens along the coast of Peru and its relevance to ENSO. These shell middens contain thermally anomalous molluscan assemblages (TAMAs), which have been interpreted to indicate that between 8 and 5 ka ocean temperatures were continuously much warmer than they are today; these warm ocean temperatures suggest a poleward extension of tropical water masses to at least 10°C (Sandweiss et al. 1996, 1997). Recent geomorphic evidence from other archaeological sites in southern Peru shows greatly reduced frequency of flood deposits between 8 and 5 ka, suggesting also that ENSO in its present expression did not exist (Keefer et al. 1998). In contrast, a different interpretation of the TAMA evidence suggests that they are related to the specific coastal morphology of an emerging coast with narrow and shallow, and hence warm embayments, and therefore yield no information on SSTs in the open ocean (de Vries and Wells 1990; de Vries et al. 1997; Wells and Noller 1997). Furthermore, a debris flow in northern Peru has been dated at 7.5 ka (Wells 1990), which is interpreted to reflect a major El Niño event, suggesting that ENSO operated also at that time and even earlier (Wells and Noller 1997).

Data for mid-Holocene corals suggest that areas near the present-day location of the western Pacific warm pool were warmer (by over 1°C) and saltier (because of greater evaporation) than at present (Gagan et al. 1998). The authors also inferred lowered interannual SST variability before about 5 ka compared to the present. Other studies consistent with this picture of the lowered ENSO variability before about 5 ka are those of Shulmeister and Lees (1995) and McGlone et al. (1992), which suggest stable wet conditions in northern Australia.

Other types of terrestrial records could shed light on this controversy. One line of evidence comes from the Huascarán ice core from northern Peru (Thompson et al. 1995). Although the ice core evidence reflects primarily climate conditions in the Amazon basin and tropical Atlantic Ocean temperatures, analysis of data for the past 100 years documents a convincing link to ENSO variability (Thompson et al., this volume).

During this mid-Holocene interval, temperatures were substantially and uniformly warmer than they are today, and high dust accumulation occurred during several periods between 7.5 and 7 ka, 6.2 and 5.7 ka, 5 and 4.6 ka, and at 4.2 ka, suggesting repeated extremely dry events in the tropical lowlands. Evidence for dry and warm conditions also comes from pollen records for northern Peru and Colombia (Hansen and Rodbell 1995; Markgraf 1991), and from paleolimnological records for Lake Titicaca and lakes in northern Chile; these records indicate low lake levels between 8 and 4 ka (Mourguiart et al. 1995; Valero-Garcés et al. 1996; Grosjean et al. 1997). Geomorphological evidence in river sections in the Chilean Atacama Desert, however, document repeated storm events, recurring every 500 to 1,200 years from at least 6 ka onwards (Grosjean et al. 1997).

Conditions that were wetter than they were during the early Holocene were recorded for the Galápagos and northern South America (Lake Valencia, Venezuela; Leyden 1985; Bradbury et al. 1981), for southern Brazil (Ledru et al. 1998), and for southern South America (Markgraf 1991). Records for the U.S. Northwest suggest increases in effective moisture during the early Holocene, judging from the replacement of steppe by ponderosa pine woodland (Barnosky 1985), and California was also wetter and warmer overall than it is today (Thompson et al. 1993). The deserts of the Southwest, in contrast, were becoming drier toward 6 ka, although summer precipitation apparently was still somewhat higher than it is today (Thompson et al. 1993).

The terrestrial records for the mid-Holocene in Australia and New Zealand show generally higher moisture levels than existed during the early Holocene (McGlone et al. 1992; Shulmeister 1999). The tropical lowlands in northern Australia were interpreted to reflect an increase in effective precipitation after 8 ka, which reached a maximum between 5 and 4 ka (Kershaw 1994; Shulmeister and Lees 1995). Southeastern Australia also shows an increase in moisture after 8 ka, but the moisture maximum was reached earlier (i.e., between 7 and 5 ka; Shulmeister 1999). Leeward eastern districts of New Zealand became progressively wetter between 7.5 and 4 ka (McGlone et al. 1995; McGlone et al. 1997a). Windward western and northern districts show little change, except that there is some indication of increased seasonality and drier summers (Newnham et al. 1989; McGlone and Neall 1994).

Vegetation and lake level records for Colombia to Chile all indicate drier conditions with millennial-scale recurrence of droughts in the Amazon lowlands. These data would suggest an absence of ENSO variability, or quasi-permanent El Niño conditions (i.e., more zonally extensive warm water than exists today in the central and eastern equatorial Pacific). Comparison of moisture patterns in North America and Australia/New Zealand, however, does not provide a consistent pattern comparable to the present patterns. A possible explanation for the contradicting evidence may relate to the frequency of ENSO events. If the recurrence intervals of ENSO events were at the scale of centuries to millennia, lower resolution records may not have responded by showing an overall increase or decrease in moisture or change in environmental variability.

Finally, a recently published study of El Niño–driven alluviation in an alpine lake in Ecuador (Rodbell et al. 1999) argues persuasively that the modern El Niño period was established about 5,000 years ago. The evidence that supports this conclusion comes from an analysis of laminae deposited in the lake by debris flows over approximately the past 15,000 years. The main findings of this study are that the frequency of storm-induced deposition events on this lake, which is today located in a region that is highly sensitive to the recurrent heavy rains from El Niño episodes, underwent a major change about 5,000 ago, from variability on decadal and longer timescales, to variability with a characteristic timescale of 2–8.5 years.

Late Holocene (5 ka to Present)

Beginning at about 5 ka, and especially after 3 ka, consistent evidence indicates that ENSO had begun to operate much as it does now. Figures 13.3 and 13.4 illustrate the principal areas throughout each hemisphere that show significant El Niño–year precipitation anomalies for the summer and winter seasons. After about 5 ka, terrestrial paleoenvironmental records for Australasia and the Americas generally show that enhanced paleoenvironmental variability becomes a major characteristic of these records (McGlone et al. 1992; Shulmeister 1999). Increasing fire frequency and disturbance in Australasian and South American pollen records, which cannot be ascribed to anthropogenic factors alone, were likely the result of increasing climatic variability (McGlone et al. 1992). In general, pollen records show increases in vegetation that is tolerant of edaphic and climatic extremes. Especially in temperate southern South America, anomalous combinations of plant assemblages began to occur, indicating that

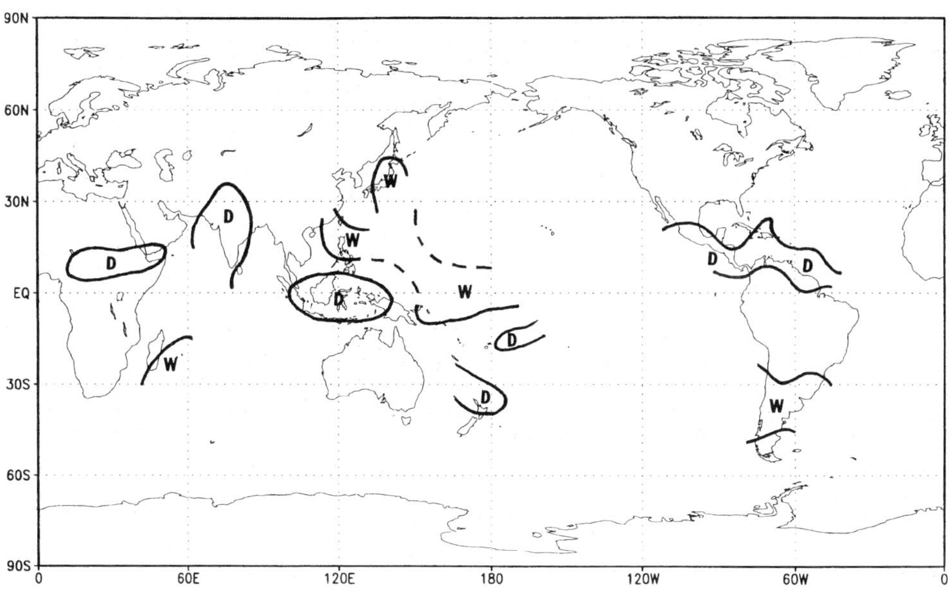

Fig. 13.3 Map depicting areas of the world with significant ENSO modulation of precipitation. Shown is the pattern for the northern summer (southern winter) season of June–August (JJA-0). Drier than normal and wetter than normal areas are denoted by Ds and Ws, respectively.

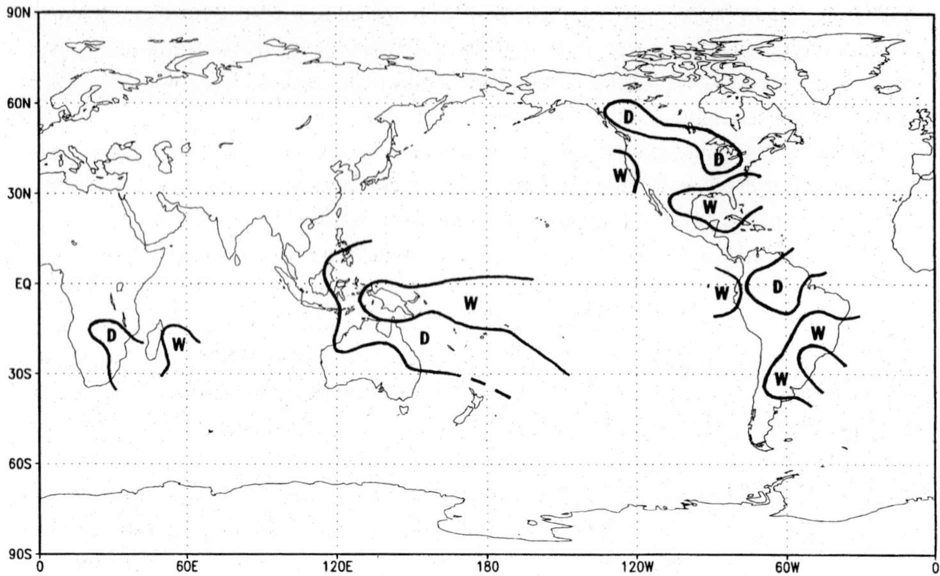

Fig. 13.4 Same as Figure 13.3, but for the northern winter (southern summer) season of December–February (DJF+1).

increased moisture and increased drought and fire frequency occurred simultaneously in different parts of this region (McGlone et al. 1992). These changes suggest the presence of variations in climate patterns on interdecadal, century, and even millennial timescales. The inferred variability in climate patterns can be related to ENSO teleconnection patterns, as well as to the type of lower frequency changes recently identified as being associated with slow changes in North Pacific SST patterns (the Pacific Decadal Oscillation, or PDO, described by Mantua et al. 1997) All these scenarios could be explained by the presence of ENSO and decadal ENSO mode fluctuations (see Allan, this volume).

Numerous sequences of flood deposits in Peru (Wells 1990), and strandline sequences in Peru (Macharé and Ortlieb 1990) and in Brazil (Martin et al. 1993), have been interpreted as documenting repeated extreme ENSO events (or ENSO modes). Along the coast of northern Peru, eleven strandlines have been dated to between 5 ka and the present, indicating a recurrence of extreme El Niño events of between 100 and 600 years apart (Ortlieb et al. 1995). Flood deposits in northern Peru indicate extreme El Niño events recurring every 1,300 years, intermediate events every 600 years, and less strong events every 75 years (Wells 1990). Late Holocene beach ridge terraces in the Rio Doce coastal plain of central Brazil also indicate ENSO occurrences during intervals when the coast is emerging (between 5.1 and 3.9 ka, between 3.6 and 2.7 ka, and since 2.5 ka). Between 5.1 and 3.9 ka, seven episodes were recorded of alongshore transport reversal; between 3.6 and 2.7 ka, the direction of alongshore transport was apparently permanently northwards, suggesting permanent El Niño–like conditions; and since 2.5 ka, four alongshore transport reversals were recorded (Martin et al.

1993). Sedimentologic evidence of landslides in lake sediments from New Zealand was interpreted to reflect episodes of high storm frequency during the past 2,250 years, related to La Niña events (Eden and Page 1998). The highest frequency of events occurred between 2 and 1.8 ka, followed by events at 2.1, 1.4, 1, and 0.3 ka.

The most convincing evidence for the presence of ENSO comes from proxy records with annual or even seasonal resolution. These records include ice cores, corals, tree rings, and annually layered (varved) sediments from lakes and ocean basins.

Ice core records for tropical South America include two sites: Huascarán in northern Peru (Thompson et al. 1995) and Quelccaya in southern Peru (Thompson et al. 1992). The Quelccaya cores have been analyzed for ENSO variability back to about A.D. 1450, and the Huascarán core back to 1925 (Thompson et al., this volume). El Niño events are characterized by reduced snow accumulation, increased dust concentrations, and less negative $\delta^{18}O$ ratios, all of which reflect climate conditions over the Amazon lowlands. Correlation between ENSO parameters such as Southern Oscillation Index (SOI) and SST (Pacific and Atlantic) is excellent, suggesting that these ice core proxies can be used to reconstruct paleo-ENSO behavior. Comparison of dust concentrations, $\delta^{18}O$, and accumulation indicates certain periods when the relation between these parameters changes. During periods that were cooler overall, such as between A.D. 1550 and 1880, accumulation was low overall, but dust concentrations were low as well. Together, the data suggest that El Niño events must have been quite rare; a La Niña–type mode was the dominant feature for this interval. Archival data from travel logs and for Nile River floods (Quinn 1992) support the evidence of few El Niño events at that time, as do tree-ring records from southern Patagonia (Villalba 1994). Differences in recurrence intervals of El Niño events are also seen in the coral data. A 400-year-long record from the Galápagos Islands shows that the dominant oscillatory ENSO mode shifted from 6 years to 4.6 years by A.D. 1700, and to 3.4 years after 1850 (Diaz and Pulwarty 1992; Dunbar et al. 1994). Changes in recurrence apparently also influence interhemispheric correlation patterns of ENSO events. Comparison of temperature-sensitive tree-ring records for Alaska and southern Patagonia shows excellent correlation at different periodicities, except for the interval between 1860 and 1930 (Villalba et al. 2000). During that time, the interannual oscillatory mode of ENSO apparently increased (Stahle et al. 1998b), suggesting that teleconnections depend on the recurrence interval (see Allan, this volume; Mann et al., this volume; and Ortlieb, this volume). However, despite the weaker correlations between the tree-ring time series at some times, correlation of these tree-ring time series with SST time series for the Pacific Ocean indicates considerable coherence (Diaz and Pulwarty 1994; Villalba et al. 2000), suggesting that perhaps the large-scale teleconnection fields are more stable than single time series responses (Mann et al. 1998).

The Past 1,000 Years

The past 1,000 years offer many opportunities for utilizing the paleoclimate record to learn more about the long-term behavior of ENSO. A number of high-resolution proxy records are available (annual and seasonal resolution) that can be put to good use (e.g., Bradley 1996; Mann et al. 1998; Mann et al., this volume) to develop a more spatially and temporally complete view of climatic variations. The multiproxy approach provides a far more robust signal of ENSO through time and may make it possible to identify how much background climate changes affect ENSO teleconnections (Mann et al. 1998; Stahle et al. 1998b).

The past 1,000 years of climatic history encompass three (and perhaps four) potentially large-scale climatic episodes, which have been documented in the literature to varying degrees. At about the start of the second millennium A.D., the climate of the Northern Hemisphere may have been experiencing a warm epoch known as the Medieval Warm Period (MWP). A review of the subject published by Hughes and Diaz (1994) concluded that there was insufficient evidence at the time to conclude, with a high degree of certainty, that the Earth's climate during this time interval was warmer than it is today. However, although the available evidence makes it difficult to show that on hemispheric scales the climate was warmer than it is today, it does indicate that there were warmer conditions than exist now for select regions around the world (e.g., western Europe).

Regional evidence gathered from regions such as Egypt, where records of the level of the Nile River during its annual high water stage were kept (or survived) from as early as A.D. 622, indicates that there was a minimum in the frequency of very low or very high flows during the MWP. Quinn (1992) used this Nile record to infer possible large-scale ENSO episodes, based on the association of ENSO with rainfall in the highlands of Ethiopia, the major source region of the Nile River. Given the apparent low variability in the Nile flows during the time of the MWP, one might infer normal conditions prevailing in the equatorial Pacific. Such a state would mean typical upwelling conditions prevailing along the equator and along the Peruvian coast, with prolonged periods of aridity possible in those areas (see Ortlieb, this volume).

If the North Atlantic Ocean was warmer than it is now, as evidence gathered from western Europe and Greenland suggests, this SST pattern, in combination with normal to cooler SST patterns in the equatorial Pacific, could have promoted the development of prolonged drought episodes in the southwestern United States (Diaz and Gutzler 2000). At the same time, wetter than normal conditions could have prevailed in the U.S. Northwest, see for example, Dettinger et al. (1998). Evidence exists for severe drought episodes during the MWP in the U.S. Southwest from the southern Rockies to California (Hughes and Brown 1992; Graumlich 1993; Hughes and Diaz 1994; Stine 1994; Hughes and Graumlich 1996).

The fourteenth and fifteenth centuries appear to comprise a transitional period leading into the Little Ice Age (LIA) of the sixteenth to nineteenth centuries. The actual climatic character of these two centuries is not well documented at present, except for certain areas in Europe and Asia. Accounts of low flood stages in the Nile River show a relative increase from previous centuries, which may indicate more frequent occurrences of El Niño events at this time. In the southwest United States, the period of protracted low-moisture years ends with tree-ring records and fluvial indicators both showing generally wetter conditions than during the MWP.

The LIA centuries may have been characterized by increased climatic variability, with extremely dry and wet periods (and particularly cold decades) recorded in many parts of the world (Bradley and Jones 1992; Jones et al. 1996). During this time, it begins to be possible to reconstruct specific elements of ENSO patterns from historical documentary evidence (Quinn et al. 1987; Quinn and Neal 1992; Ortlieb, this volume). In addition, recent work that relies on a combined multiproxy record approach using sophisticated statistical techniques also provides glimpses of possible spatial variability in large-scale ENSO climatic patterns (see Mann et al., this volume). Severe droughts may also have been common over larger portions of the United States during the early phase of the LIA, as is suggested by tree-ring analysis for the eastern United States (Stahle et al. 1998a), and by changes in diatom sequences in lake sediments in the north-central United States (Laird et al. 1996). Recent precipitation patterns in the contiguous United States (since the 1920s) show a predominant ENSO signal of wet El Niño and dry La Niña years in the U.S. Southwest, together with a tendency for dryness (wetness) to develop in the middle-eastern portions of the United States during El Niño (La Niña) years (see Kiladis and Diaz 1989). It appears that this pattern may have been less prevalent in the nineteenth century, based on analysis of tree-ring data (Cole and Cook 1998).

Some work has been done to evaluate the expression of ENSO effects on both sides of the Pacific basin prior to the availability of instrumental records (Quinn 1992; Whetton and Rutherfurd 1994). There is evidence for rather abrupt changes in the climatic regime in southern California toward the end of the sixteenth century, as recorded in the varved sediment record of the Santa Barbara basin (Baumgartner et al. 1989; Biondi et al. 1997), and coinciding with severe and sustained drought conditions in the U.S. Southwest (Meko et al. 1995; Hughes and Funkhouser 1998). Nevertheless, ENSO teleconnection patterns in these earlier centuries are ambiguous, due to the still incomplete nature of the available proxy and early instrumental records (Whetton et al. 1996; Ortlieb, this volume).

While knowledge of the detailed expression on long timescales of climatic variations related to the ENSO phenomenon remains incomplete, the available data suggest that changes may have occurred in the spatial and temporal patterns of the equatorial Pacific SST variability. Evidence for associated atmospheric teleconnections in different parts of the globe, due to climatic changes associated with the LIA episode and the MWP, are less well understood. The veracity of these changes in ENSO teleconnection patterns remains to be firmly established (see Mann et al. 1998; Mann et al., this volume). Nevertheless, if changes in ENSO teleconnection patterns are found, this would have important implications for predicting the impacts of ENSO in the future in connection with both natural climatic variability (Hoerling and Kumar, this volume) and anthropogenic changes in climate (Knutson and Manabe 1998).

Conclusions

Given the differences in surface boundary conditions, with greatly altered land/ocean distribution especially in the western Pacific, it is very likely that during full and late glacial times, when climates were colder overall, the expression of ENSO variability may have been quite different from its expression in the modern climate. At no time do the terrestrial paleoclimate records show precipitation teleconnection patterns that are comparable to the modern patterns, and high paleoclimatic variability might be due to other causes.

Some theoretical support exists for the absence of ENSO – at least in its modern expression – during cold tropical SST epochs (Kraus 1973; Neelin et al. 1998; Sun, this volume). Also, some evidence indicates that, although the large-scale circulation patterns associated with ENSO are quite robust over time (Dettinger et al. 1998), some of the associations may weaken, or perhaps disappear altogether at times (Mann et al., this volume; Ortlieb, this volume). The question of the robustness of the ENSO teleconnections was a subject considered in our earlier book (Meehl and Branstator 1992). The subject remains a current topic of research, as evidence of low-frequency modes in North Pacific SST variability (Graham 1994; Latif and Barnett 1996; Gu and Philander 1997; Latif et al. 1997; Kleeman and Power, this volume) has been uncovered in the past several years.

The recent behavior of the ENSO system has focused increased attention on the subject of low-frequency variations of the phenomenon. The occurrence of two major El Niño episodes in the past twenty years (1982–83 and 1997–98), the infrequent occurrence of the cold SST phase (La Niña) during this time, and the impetus provided by anticipated changes in climate due to increasing atmospheric greenhouse gas concentrations has spurred research into past variations in the functioning of ENSO. This chapter represents an effort to once again examine varied aspects of the ENSO phenomenon on different timescales and to consider the paleoclimate evidence in terms of whether or not it displays sensitivity to ENSO-scale climate variability. Our main goal has been to examine possible changes in ENSO behavior, in the context of long-term changes in climate, not only to better understand the connections between ENSO and the state of the climate system but also to provide a historical context against which to judge future changes in both these areas.

Acknowledgments We thank Robin Webb and Matt McGlone for their helpful reviews of our manuscript. V. M. acknowledges support from NSF-ATM 9526139.

References

AHARON, P., 1983: Surface ocean temperature variations during the late Wisconsinan stages. $^{16}O/^{18}O$ isotopic evidence from coral reef terraces in eastern Papua New Guinea. *In* Chappell, J. M. A. and Grindrod, A. (eds.), *CLIMANZ I*. Australian National University, Canberra: Research School of Pacific Studies, 3–4.

ALLAN, R., 2000: ENSO and climatic variability in the past 150 years. *In* Diaz, H. F. and Markgraf, V. (eds.), *El Niño and the Southern Oscillation: Multiscale Variability and Global and Regional Impacts*. Cambridge: Cambridge University Press, Chapter 1, this volume.

ANDERSON, D. M. and WEBB, R. S., 1994: Ice-age tropics revisited. *Nature*, **367**, 23–24.

ANDERSON, R. Y., 1992: Long-term changes in the frequency of occurrence of El Niño events. *In* Diaz, H. F. and Markgraf, V. (eds.), *El Niño: Historical and Paleoclimatic Aspects of the Southern Oscillation*. Cambridge: Cambridge University Press, 193–200.

ANDERSON, R. Y., LINSLEY, B. K., and GARDNER, J. V., 1990: Expression of seasonal and ENSO forcing in climatic variability at lower than ENSO frequencies: Evidence from marine varves off California. *In* Meyers, P. A. and Benson, L. V. (eds.), Paleoclimates: The record from lakes, ocean, and land. *Palaeogeography, Palaeoclimatology, Palaeoecology*, **78**, 287–300.

ANDERSON, R. Y., SOUTAR, A., and JOHNSON, T. C., 1992: Long-term changes in El Niño/Southern Oscillation: Evidence from marine and lacustrine sediments. *In* Diaz, H. F. and Markgraf, V. (eds.), *El Niño: Historical and Paleoclimatic Aspects of the Southern Oscillation*. Cambridge: Cambridge University Press, 419–434.

BARNOSKY, C. W., 1985: Late Quaternary vegetation in the southwestern Columbia basin, Washington. *Quaternary Research*, **23**, 109–122.

BARNOSKY, C. W., ANDERSON, P. M., and BARTLEIN, P. J., 1987: The northwestern U.S. during deglaciation: Vegetational history and paleoclimatic implications. *In* Ruddiman, W. F. and Wright, H. E., Jr. (eds.), *North America and Adjacent Oceans During the Last Deglaciation*. Geological Society of America, The Geology of North America K-3, 289–321.

BAUMGARTNER, T. R., FERREIRA-BARTINA, V., SCHRADER, H., and SOUTAR, A., 1985: A 20-year varve record of siliceous phytoplankton variability in the central Gulf of California. *Marine Geology*, **64**, 113–129.

BAUMGARTNER, T. R., MICHAELSEN, J., THOMPSON, L. G., SHEN, G. T., SOUTAR, A., and CASEY, R. E., 1989: The recording of interannual climatic change by high resolution natural systems: Treerings, coral bands, glacial ice layers, and marine varves. *In* Peterson, D. (ed.), *Aspects of Climate Variability in the Pacific and Western Americas*. Geophysical Monograph 55, Washington, DC: American Geophysical Union, 1–15.

BEHL, R. J. and KENNETT, J. P., 1996: Brief interstadial events in the Santa Barbara basin, NE Pacific, during the past 60k. *Nature*, **379**, 243–246.

BIONDI, F., LANGE, C. B., HUGHES, M. K., and BERGER, W. H., 1997: Inter-decadal signals during the last millennium (A.D. 1117–1192) in the varve record of Santa Barbara basin, California. *Geophysical Research Letters*, **24**, 193–196.

BRADBURY, J. P. and FORESTER, R. M., 1999: Paleolimnological record of climate change from Owens Lake, California, for the last 50,000 years. Washington, DC: Smithsonian Institution Series, Smithsonian Institution.

BRADBURY, J. P., LEYDEN, B., SALGADO-LABOURIAU, M. L., LEWIS, W. M., Jr., SCHUBERT, C., BINFORD, M. W., FREY, D. G., WHITEHEAD, D. R., and WEIBEZAHN, F. H., 1981: Late Quaternary environmental history of Lake Valencia, Venezuela. *Science*, **214**, 1299–1305.

BRADLEY, R. S., 1996: Are there optimum sites for global paleotemperature reconstructions? *In* Jones, P. D., Bradley, R. S., and Jouzel, J. (eds.), *Climatic Variations and Forcing Mechanisms of the Last 2000 Years*. Berlin: Springer Verlag, 603–624.

BRADLEY, R. S. and JONES, P. D. (eds.), 1992: *Climate Since A.D. 1500*. Routledge, London, 706 pp.

BUSH, A. B. G. and PHILANDER, S. G. H., 1998: The role of ocean–atmosphere interactions in tropical cooling during the last glacial maximum. *Science*, **279**, 1341–1344.

CHARLES, C. D., HUNTER, D. E., and FAIRBANKS, R. G., 1997: Interaction between the ENSO and the Asian monsoon in a coral record of tropical climate. *Science*, **277**, 925–928.

COLE, J. E. and COOK, E. R., 1998: The changing relationship between ENSO variability and moisture balance in the continental United States. *Geophysical Research Letters*, **25**, 4529–4532.

DETTINGER, M. D., CAYAN, D. R., DIAZ, H. F., and MEKO, D. M., 1998: North–south precipitation patterns in western North America on interannual-to-decadal timescales. *Journal of Climate*, **11**, 3095–3111.

de VRIES, T. J., ORTLIEB, L., DIAZ, A., WELLS, L., and HILLAIRE-MARCEL, C., 1997: Determining the early history of El Niño. *Science*, **276**, 965–966.

de VRIES, T. J. and WELLS, L., 1990: Thermally-anomalous Holocene molluscan assemblages from coastal Peru: Evidence for paleogeographic, not climatic change. *Palaeogeography, Palaeoclimatology, Palaeoecology*, **81**, 11–32.

DIAZ, H. F. and GUTZLER, D. S., 2000: Temperature and precipitation regimes of the 1950s drought in the U.S. Southwest. *In* Betancourt, J. L. (ed.), *The 1950s Drought in the American Southwest: Hydrological, Ecological, and Socioeconomic Impacts*. Tucson: University of Arizona Press.

DIAZ, H. F. and MARKGRAF, V. (eds.), 1992: *El Niño: Historical and Paleoclimatic Aspects of the Southern Oscillation*. Cambridge: Cambridge University Press, 476 pp.

DIAZ, H. F. and PULWARTY, R. S., 1992: A comparison of Southern Oscillation and El Niño signals in the tropics. *In* Diaz, H. F. and Markgraf, V. (eds.), *El Niño: Historical and Paleoclimatic Aspects of the Southern Oscillation*. Cambridge: Cambridge University Press, pp. 175–192.

DIAZ, H. F. and PULWARTY, R. S., 1994: An analysis of the time scales of variability in centuries-long ENSO-sensitive records. *Climatic Change*, **26**, 317–342.

DIAZ, H. F., MARKGRAF, V., and HUGHES, M. K., 1992: Synthesis and future prospects. *In* Diaz, H. F. and Markgraf, V. (eds.), *El Niño: Historical and Paleoclimatic Aspects of the Southern Oscillation*. Cambridge: Cambridge University Press, 463–471.

DUNBAR, R. B., WELLINGTON, G. M., COLGAN, M. W., and GLYNN, P. W., 1994: Eastern Pacific sea surface temperature since 1600 A.D.: The δ^{18}O record of climate variability of Galápagos corals. *Paleoceanography*, **9**, 291–316.

EDEN, D. N. and PAGE, M. H., 1998: Palaeoclimatic implications of a storm erosion record from late Holocene lake sediments, North Island, New Zealand. *Palaeo3*, **139**, 37–58.

ENFIELD, D. B., 1992: Historical and prehistorical overview of El Niño/Southern Oscillation. *In* Diaz, H. F. and Markgraf, V. (eds.), *El Niño: Historical and Paleoclimatic Aspects of the Southern Oscillation*. Cambridge: Cambridge University Press, 95–118.

GAGAN, M. K., AYLIFFE, L. K., HOPLEY, D., CALI, J. A., MORTIMER, G. E., CHAPPELL, J., McCULLOCH, M. T., and HEAD, M. J., 1998: Temperature and surface–ocean water balance of the mid-Holocene tropical western Pacific. *Science*, **279**, 1014–1018.

GRAHAM, N. E., 1994: Decadal-scale climate variability in the tropical and North Pacific during the 1970s and 1980s: Observations and model results. *Climate Dynamics*, **10**, 135–162.

GRAHAM, N. E., 1995: Simulation of recent global temperature trends. *Science*, **267**, 666–671.

GRAUMLICH, L. J., 1993: A 1000-year record of temperature and precipitation in the Sierra Nevada. *Quaternary Research*, **39**, 249–255.

GRIMM, E. C., JACOBSON, G. L., Jr., WATTS, W. A., HANSEN, B. C. S., and MAASCH, K. A., 1993: A 50,000-year record of climate oscillations from Florida and its temporal correlation with the Heinrich events. *Science*, **261**, 198–200.

GROSJEAN, M., 1998: Late-glacial and Holocene climate changes on the southern Altiplano (20°–28°S). Abstracts: PEP-1 Meeting, Mérida, Venezuela, 1998, 4 pp.

GROSJEAN, M., NÚÑEZ, L., CARTAJENA, I., and MESSERLI, B., 1997: Mid-Holocene climate and culture change in the Atacama Desert, northern Chile. *Quaternary Research*, **48**, 239–246.

GU, D. and PHILANDER, S. G. H., 1997: Interdecadal climate fluctuations that depend on exchanges between the tropics and extratropics. *Science*, **275**, 805–807.

GUILDERSON, T. R., FAIRBANKS, R. G., and RUBENSTONE, J. L., 1994: Reconciling tropical sea surface temperature estimates for the last glacial maximum. *Science*, **263**, 663–665.

HABERLE, S., 1998: Late Quaternary vegetation change in the Tari Basin, Papua New Guinea. *Palaeogeography, Palaeoclimatology, Palaeoecology*, **137**, 1–24.

HANSEN, B. C. S. and RODBELL, D. T., 1995: A late-glacial/Holocene pollen record from the eastern Andes of northern Peru. *Quaternary Research*, **44**, 216–227.

HEUSSER, L. E. and SIROCKO, F., 1997: Millennial pulsing of environmental change in southern California from the past 24 k.y.: A record of Indo-Pacific ENSO events? *Geology*, **25**, 243–246.

HOERLING, M. P. and KUMAR, A., 2000: Understanding and predicting extratropical teleconnections related to ENSO. *In* Diaz, H. F. and Markgraf, V. (eds.), *El Niño and the Southern Oscillation: Multiscale Variability and Global and Regional Impacts*. Cambridge: Cambridge University Press, Chapter 2, this volume.

HOPE, G. S., 1996: History of Nothofagus in New Guinea and New Caledonia. *In* Veblen, T. T., Hill, R. S., and Read, J. (eds.), *The Ecology and Biogeography of Nothofagus Forests*. New Haven: Yale University Press, 257–270.

HUGHES, M. K. and BROWN, P. M., 1992: Drought frequency in central California since 101 B.C. recorded in giant sequoia tree rings. *Climate Dynamics*, **6**, 161–167.

HUGHES, M. K. and DIAZ, H. F., 1994: Was there a Medieval Warm Period, and if so, where and when? *Climatic Change*, **26**, 109–142.

HUGHES, M. K. and FUNKHOUSER, G., 1998: Extremes of moisture availability reconstructed from tree rings for recent millennia in the Great Basin of western North America. *In* Beniston, M. and Innes, J. (eds.), *The Impacts of Climate Variability on Forests*. Berlin: Springer-Verlag, 99–107.

HUGHES, M. K. and GRAUMLICH, L. J., 1996: Multimillennial dendroclimatic studies from the western United States. *In* Jones, P. D., Bradley, R. S., and Jouzel, J. (eds.), *Climatic Variations and Forcing Mechanisms of the Last 2000 Years*. Berlin: Springer-Verlag, 109–124.

JONES, P. D., BRADLEY, R. S., and JOUZEL, J. (eds.), 1996: *Climatic Variations and Forcing Mechanisms of the Last 2000 Years*. Berlin: Springer-Verlag, 649 pp.

KEEFER, D. K., de FRANCE, S. D., MOSELEY, M. E., RICHARDSON, J. B., III, SATTERLEE, D. R., and DAY-LEWIS, A., 1998: Early maritime economy and El Niño events at Quebrada Tacahuay, Peru. *Science*, **281**, 1833–1835.

KENNETT, J. P. and INGRAM, G. L., 1995: A 20,000 yr record of ocean circulation and climate change from the Santa Barbara Basin. *Nature*, **377**, 510–514.

KERSHAW, A. P., 1994: Pleistocene vegetation of the humid tropics of northeastern Queensland, Australia. *Palaeogeography, Palaeoclimatology, Palaeoecology*, **109**, 399–412.

KILADIS, G. N. and DIAZ, H. F., 1989: Global climatic anomalies associated with extremes in the Southern Oscillation. *Journal of Climate*, **2**, 1069–1090.

KITZBERGER, T., VEBLEN, T. T., and VILLALBA, R., 1998: Climatic influences on fire regimes along a rainforest-to-xeric woodland gradient in northern Patagonia. *Journal of Biogeography*, **24**, 35–47.

KLEEMAN, R. and POWER, S., 2000: Modulation of ENSO on decadal and longer timescales. *In* Diaz, H. F. and Markgraf, V. (eds.), *El Niño and the Southern Oscillation: Multiscale Variability and Global and Regional Impacts*. Cambridge: Cambridge University Press, Chapter 11, this volume.

KNUTSON, T. R. and MANABE, S., 1998: Model assessment of decadal variability and trends in the tropical Pacific Ocean. *Journal of Climate*, **11**, 2273–2296.

KRAUS, E. B., 1973: Comparison between ice age and present general circulations. *Nature*, **245**, 129–133.

KUTZBACH, J. E., GUETTER, P. J., BEHLING, P. J., and SELIN, R., 1993: Simulated climatic changes: Results of the COHMAP climate model experiments. *In* Wright, H. E., Jr., Kutzbach, J. E., Webb, T., III, Ruddiman, W. F., Street-Perrott, F. A., and Bartlein, P. J. (eds.), *Global Climates Since the Last Glacial Maximum*. Minneapolis: University of Minnesota Press, 5–11.

LABRACHERIE, M., LABEYRIE, L. D., DURPAT, J., BARD, E., ARNOLD, M., PICHON, J.-J., and DUPLESSY, J.-C., 1989: The last deglaciation in the Southern Ocean. *Paleoceanography*, **4**, 629–638.

LAIRD, K. R., FRITZ, S. C., MAASCH, K. A., and CUMMING, B. F., 1996: Greater drought intensity and frequency before A.D. 1200 in the northern Great Plains. *Nature*, **384**, 552–554.

LATIF, M. and BARNETT, T. P., 1996: Decadal climate variability over North Pacific and North America: Dynamics and predictability. *Journal of Climate*, **9**, 2407–2423.

LATIF, M., KLEEMAN, R., and ECKERT, C., 1997: Greenhouse warming, decadal variability, or El Niño? An attempt to understand the anomalous 1990s. *Journal of Climate*, **10**, 2221–2239.

LEDRU, M. P., SALGADO-LABOURIAU, M.-L., and LORSCHEITTER, M. L., 1998: Vegetation dynamics in southern and central Brazil during the last 10,000 yr B. P. *Review of Palaeobotany and Palynology*, **99**, 131–142.

LEYDEN, B. W., 1985: Late Quaternary aridity and Holocene moisture fluctuations in the Lake Valencia basin, Venezuela. *Ecology*, **66**, 1279–1295.

LIN, JO C., BROECKER, W. S., HEMMING, S. R., HAJDAS, I., ANDERSON, R. F., KELLEY, M., and BONANI, G., 1998: A reassessment of U-Th and ^{14}C ages for late-glacial high-frequency hydrological events at Searles Lake, California. *Quaternary Research*, **49**, 11–23.

MACHARÉ, J. and ORTLIEB, L., 1990: Global change studies in northwestern Peru: A high potential for records of former El Niño events. *Revista Geofísica* (Mexico), **32**, 153–171.

MANN, M. E., BRADLEY, R. S., and HUGHES, M. K., 1998: Global-scale temperature patterns and climate forcing over the past six centuries. *Nature*, **392**, 779–787.

MANN, M. E., BRADLEY, R. S., and HUGHES, M. K., 2000: Long-term variability in the El Niño/Southern Oscillation and associated teleconnections. *In* Diaz, H. F. and Markgraf, V. (eds.), *El Niño and the Southern Oscillation: Multiscale Variability and Global and Regional Impacts*. Cambridge: Cambridge University Press, Chapter 10, this volume.

MANTUA, N. J., HARE, S. R., ZHANG, Y., WALLACE, J. M., and FRANCIS, R. C., 1997: A Pacific interdecadal climate oscillation with impacts on salmon production. *Bulletin of the American Meteorological Society*, **78**, 1069–1079.

MARKGRAF, V., 1991: Late Pleistocene environmental and climatic evolution in southern South America. *Bamberger Geographische Schriften*, **11**, 271–281.

MARKGRAF, V. and KENNY, R., 1997: Character of rapid vegetation and climate change during the late-glacial in southernmost South America. *In* Huntley, B., Cramer, W., Morgan, A. V., Prentice, H. C., and Allen, J. R. M. (eds.), *Past and Future Rapid Environmental Changes*. NATO ASI Series I 47, 81–90.

MARKGRAF, V., DODSON, J. R., KERSHAW, A. P., McGLONE, M. S., and NICHOLLS, N., 1992: Evolution of late Pleistocene and Holocene climates in the circum–South Pacific land areas. *Climate Dynamics*, **6**, 193–211.

MARKGRAF, V., McGLONE, M. S., and HOPE, G. S., 1995: Neogene paleoenvironmental and paleoclimatic change in southern temperate ecosystems – A southern perspective. *Trends in Ecology and Evolution*, **10**, 143–147.

MARTIN, L., FOURNIER, M., MOURGUIART, P., SIFEDDINE, A., and TURCQ, B., 1993: Southern Oscillation signal in South American palaeoclimatic data of the last 7000 years. *Quaternary Research*, **39**, 338–346.

McGLONE, M. S., KERSHAW, A. P., and MARKGRAF, V., 1992: El Niño/Southern Oscillation climatic variability in Australasian and South American paleoenvironmental records. *In* Diaz, H. F. and Markgraf, V. (eds.), *El Niño: Historical and Paleoclimatic Aspects of the Southern Oscillation*. Cambridge: Cambridge University Press, 435–462.

McGLONE, M. S., MARK, A. F., and BELL, D., 1995: Late Pleistocene and Holocene vegetation history, Central Otago, South Island, New Zealand. *Journal of the Royal Society of New Zealand*, **25**, 1–22.

McGLONE, M. S. and MOAR, N. T., 1998: Dry land Holocene vegetation history, Central Otago and the Mackenzie Basin, South Island, New Zealand. *New Zealand Journal of Botany*, **36**, 91–111.

McGLONE, M. S., MOAR, N. T., and MEURK, C. D., 1997a: Growth and vegetation history of Alpine Mires on the Old Man Range, Central Otago, New Zealand. *Arctic and Alpine Research*, **29**, 32–44.

McGLONE, M. S., MOAR, N. T, WARDLE, P., and MEURK, C. D., 1997b: The late-glacial and Holocene vegetation and environmental history of Campbell Island, far southern New Zealand. *The Holocene*, **7**, 1–12.

McGLONE, M. S. and NEALL, V. E., 1994: The late Pleistocene and Holocene vegetation of Taranaki, North Island, New Zealand. *New Zealand Journal of Botany*, **32**, 251–269.

MEEHL, G. A. and BRANSTATOR, G. W., 1992: Coupled climate model simulation of El Niño/Southern Oscillation: Implications for paleoclimate. *In* Diaz, H. F. and Markgraf, V. (eds.), *El Niño: Historical and Paleoclimatic Aspects of the Southern Oscillation*. Cambridge: Cambridge University Press, 69–91.

MEKO, D., STOCKTON, C. W., and BOGGESS, W. R., 1995: The tree-ring record of severe and sustained drought. *Water Resources Bulletin*, **31**, 789–801.

MICHAELSEN, J. and THOMPSON, L. G., 1992: A comparison of proxy records of El Niño/Southern Oscillation. *In* Diaz, H. F. and Markgraf, V. (eds.), *El Niño: Historical and Paleoclimatic Aspects of the Southern Oscillation*. Cambridge: Cambridge University Press, 323–348.

MORENO, P. I., 1997: Vegetation and climate near Lago Llanquihue in the Chilean Lake District between 20,200 and 9500 ^{14}C yr B. P. *Journal of Quaternary Science*, **12**, 485–500.

MOURGUIART, P., ARGOLLO, J., CARBONEL, P., CORRÈGE, T., and WIRRMANN, D., 1995: El Lago Titicaca durante el Holoceno: Una historia compleja. *In* Argollo, J. and Mourguiart, P. (eds.), *Cambios Cuaternarios en America del* Sur. ORSTOM, La Paz, 173–188.

NEELIN, J. D., BATTISTI, D. S., HIRST, A. C., JIN, F.-F., WAKATA, Y., YAMAGATA, T., and ZEBIAK, S. E., 1998: ENSO theory. *Journal of Geophysical Research*, **103**, 14262–14290.

NEWNHAM, R. M., LOWE, D. J., and GREEN, J. D., 1989: Palynology vegetation and climate of the Waikato lowlands, North Island, New Zealand, since c. 18,000 years ago. *Journal of the Royal Society of New Zealand*, **19**, 127–150.

NICHOLLS, N., 1992: Historical El Niño/Southern Oscillation variability in the Australasian region. *In* Diaz, H. F. and Markgraf, V. (eds.), *El Niño: Historical and Paleoclimatic Aspects of the Southern Oscillation*. Cambridge: Cambridge University Press, 151–174.

OHKOUCHI, N., KAWAMURA, K., and TAIRA, A., 1997: Fluctuations of terrestrial and marine biomarkers in the western tropical Pacific during the last 23,300 years. *Paleoceanography*, **12**, 623–630.

ORTLIEB, L., 2000: The documented historical record of El Niño events in Peru: An update of the Quinn record (sixteenth through nineteenth centuries). *In* Diaz, H. F. and Markgraf, V. (eds.), *El Niño and the Southern Oscillation: Multiscale Variability and Global and Regional Impacts*. Cambridge: Cambridge University Press, Chapter 7, this volume.

ORTLIEB, L., HOCQUENGHEM, A. M., and MINAYA, A., 1995: Toward a revised historical chronology of El Niño events registered in western South America. XIV INQUA Congress, Berlin, 1995. Abstract vol., *Terra Nostra*, 2/95, p. 113.

OVERPECK, J., ANDERSON, D., TRUMBORE, S., and PRELL, W., 1996: The southwest Indian monsoon over the last 18,000 years. *Climate Dynamics*, **12**, 213–225.

PATRICK, A. and THUNELL, R. C., 1997: Tropical Pacific sea surface temperatures and upper water column thermal structure during the last glacial maximum. *Paleoceanography*, **12**, 649–657.

PERKS, H. M. and KEELING, R. F., 1998: A 400 kyr record of combustion oxygen demand in the western equatorial Pacific: Evidence for a precessionally forced climate response. *Paleoceanography*, **13**, 63–69.

QUINN, W. H., 1971: Late Quaternary meteorological and oceanographic developments in the equatorial Pacific. *Nature*, **229**, 330–332.

QUINN, W. H., 1992: A study of Southern Oscillation–related climatic activity for A.D. 622–1900 incorporating Nile River flood data. *In* Diaz, H. F. and Markgraf, V. (eds.), *El Niño: Historical and Paleoclimatic Aspects of the Southern Oscillation*. Cambridge: Cambridge University Press, 119–149.

QUINN, W. H. and NEAL, V. T., 1992: The historical record of El Niño events. *In* Bradley, R. S. and Jones, P. D. (eds.), *Climate Since A.D. 1500*. London: Routledge, 623–648.

QUINN, W. H., NEAL, V. T., and ANTÚNEZ de MAYOLO, S., 1987: El Niño occurrences over the past four and a half centuries. *Journal of Geophysical Research*, **92**, 14449–14461.

REA, D. K., PISIAS, N. G., and NEWBERRY, T., 1991: Late Pleistocene paleoclimatology of the central equatorial Pacific: Flux patterns of biogenic sediments. *Paleoceanography*, **6**, 227–244.

RODBELL, D. T, SELTZER, G. O., ANDERSON, D. M., ABBOTT, M. B., ENFIELD, D. B., and NEWMAN, J. H., 1999: An ~15,000-year record of El Niño–driven alluviation in southwestern Ecuador. *Science*, **283**, 516–520.

SANDWEISS, D. H., RICHARDSON, J. B., III, REITZ, E. J., ROLLINS, H. B., and MAASCH, K. A., 1996: Geoarchaeological evidence from Peru for a 5000 years B.P. onset of El Niño. *Science*, **237**, 1531–1533.

SANDWEISS, D. H., RICHARDSON, J. B., III, REITZ, E. J., ROLLINS, H. B., and MAASCH, K. A., 1997: Determining the early history of El Niño: Response. *Science*, **276**, 966–967.

SIROCKO, F., 1996: Past and present subtropical summer monsoons. *Science*, **274**, 937–938.

SHULMEISTER, J., 1999: Australasian evidence for mid-Holocene climate change implies precessional control of Walker circulation in the Pacific. *Quaternary International*, **57/58**, 81–91.

SHULMEISTER, J. and LEES, B. G., 1995: Pollen evidence from tropical Australia for the onset of an ENSO-dominated climate at c. 4000 B.P. *The Holocene*, **5**, 10–18.

STAHLE, D. W., CLEAVELAND, M. K., BLANTON, D. B., THERRELL, M. D., and GAY, D. A., 1998a: The Lost Colony and Jamestown droughts. *Science*, **280**, 564–567.

STAHLE, D. W., D'ARRIGO, R. D., KRUSIC, P. J., CLEAVELAND, M. K., COOK, E. R., ALLAN, R. J., COLE, J. E., DUNBAR, R. B., THERRELL, M. D., GAY, D. A., MOORE, M. D., STOKES, M. A., BURNS, B. T., VILLANUEVA-DIAZ, J., and THOMPSON, L. G., 1998b: Experimental dendroclimatic reconstruction of the Southern Oscillation. *Bulletin of the American Meteorological Society*, **79**, 2137–2152.

STEIG, E. J., BROOK, E. J., WHITE, J. W. C., SUCHER, C. M., BENDER, M. L., LEHMAN, S. J., MORSE, D. L., WADDINGTON, E. D., and CLOW, G. D., 1998: Synchronous climate changes in Antarctica and the North Atlantic. *Science*, **282**, 92–95.

STINE, S., 1994: Extreme and persistent drought in California and Patagonia during medieval time. *Nature*, **369**, 546–549.

SUN, D.-Z., 2000: Global climate change and El Niño: A theoretical framework. *In* Diaz, H. F. and Markgraf, V. (eds.), *El Niño and the Southern Oscillation: Multiscale Variability, and Global and Regional Impacts*. Cambridge: Cambridge University Press, Chapter 12, this volume.

SWETNAM, T. W. and BETANCOURT, J. L., 1992: Temporal patterns of El Niño/Southern Oscillation detected in tree-ring data from Mexico and the southern United States. *In* Diaz, H. F. and Markgraf, V. (eds.), *El Niño: Historical and Paleoclimatic Aspects of the Southern Oscillation*. Cambridge: Cambridge University Press, 259–270.

THOMPSON, L. G., HENDERSON, K. A., MOSLEY-THOMPSON, E., and LIN, P.-N., 2000: The tropical ice core record of ENSO. *In* Diaz, H. F. and Markgraf, V. (eds.), *El Niño and the Southern Oscillation: Multiscale Variability and Global and Regional Impacts*. Cambridge: Cambridge University Press, Chapter 9, this volume.

THOMPSON, L. G., MOSELY-THOMPSON, E., DAVIS, M. E., LIN, P.-N., HENDERSON, K. A., COLE-DAI, J., BOLZAN, J. F., and LIU, K.-B., 1995: Late glacial stage and Holocene tropical ice core records from Huascarán, Peru. *Science*, **269**, 46–50.

THOMPSON, L. G., MOSLEY-THOMPSON, E., and THOMPSON, P. A., 1992: Reconstructing interannual climate variability from tropical and subtropical ice-core records. *In* Diaz, H. F. and Markgraf, V. (eds.), *El Niño: Historical and Paleoclimatic Aspects of the Southern Oscillation*. Cambridge: Cambridge University Press, 295–322.

THOMPSON, R. S., WHITLOCK, C., BARTLEIN P. J., HARRISON, S. P., and SPAULDING, W. G., 1993: Climate changes in the western United States since 18,000 yr B.P. *In* Wright, H. E., Jr., Kutzbach, J. E., Webb, T., III, Ruddiman, W. F., Street-Perrott, F. A., and Bartlein, P. J. (eds.), *Global Climates Since the Last Glacial Maximum*. Minneapolis: University of Minnesota Press, 468–513.

VALERO-GARCÉS, B. L., GROSJEAN, M., SCHWALB, A., GEYH, M., MESSERLI, B., and KELTS, K., 1996: Limnogeology of Laguna Miscanti: Evidence for mid- to late Holocene moisture changes in the Atacama Altiplano (northern Chile). *Journal of Paleolimnology*, **16**, 1–21.

van der KAARS, S. and DAM, R., 1997: Vegetation and climate change in West Java, Indonesia during the last 135,000 years. *Quaternary International*, **37**, 67–71.

VEBLEN T. T., KITZBERDER, T., VILLALBA, R., and DONNEGAN, J., 1999: Fire history in northern Patagonia: The roles of humans and climatic variation. *Ecological Monographs*, **69**, 47–67.

VILLAGRAN, C., 1991: Historia de los bosques templados del sur de Chile durante el Tardiglacial y Postglacial. *Revista Chilena de Historia Natural*, **64**, 447–460.

VILLALBA, R., 1994: Tree ring and glacial evidence for the Medieval Epoch and the Little Ice Age in southern South America. *Climate Change*, **26**, 183–198.

VILLALBA, R., D'ARRIGO, R., COOK, E. R., WILES, G., and JACOBY, G. C., 2000: Decadal-scale climatic variability along the extra-tropical western coast of the Americas over past centuries inferred from tree-ring records. *In* Markgraf, V. (ed.), *Present and Past Inter-Hemispheric Climate Linkages in the Americas and Their Societal Effects*. San Diego: Academic Press. In press.

WEBSTER, P. J., MAGAÑA, V. O., PALMER, T. N, SHUKLA, J., TOMAS, R. A., YANAI, M., and YASUNARI, T., 1998: Monsoons: Processes, predictability, and the prospects for prediction. *Journal of Geophysical Research*, **103**, 14451–14510.

WELLS, L., 1990: Holocene history of the El Niño phenomenon as recorded in flood sediments of northern coastal Peru. *Geology*, **18**, 1134–1137.

WELLS, L. E. and NOLLER, J. S., 1997: Determining the early history of El Niño. *Science*, **276**, 966.

WHETTON, P. H. and RUTHERFORD, I., 1994: Historical ENSO teleconnections in the Eastern Hemisphere. *Climatic Change*, **28**, 221–253.

WHETTON, P. H., ALLAN, T. J., and RUTHERFORD, I., 1996: Historical ENSO teleconnections in the Eastern Hemisphere: Comparison with latest El Niño series of Quinn. *Climatic Change*, **32**, 103–109.

WHITLOCK, C., BARLEIN, P. J., and van NORMAN, K. H., 1995: Stability of Holocene climate regimes in the Yellowstone region. *Quaternary Research*, **43**, 433–436.

Index